LIE THEORY AND ITS APPLICATIONS IN PHYSICS V

Proceedings of the Fifth International Workshop

LIE THEORY AND ITS APPLICATIONS IN PHYSICS V

Varna, Bulgaria 16 – 22 June 2003

Organized by

Institute of Nuclear Research and Nuclear Energy
Bulgarian Academy of Sciences

Edited by

H.-D. Doebner
Technical University of Clausthal, Germany

V. K. Dobrev
University of Northumbria, UK
and
Bulgarian Academy of Sciences, Bulgaria

World Scientific

NEW JERSEY • LONDON • SINGAPORE • BEIJING • SHANGHAI • HONG KONG • TAIPEI • CHENNAI

Published by

World Scientific Publishing Co. Pte. Ltd.
5 Toh Tuck Link, Singapore 596224
USA office: Suite 202, 1060 Main Street, River Edge, NJ 07661
UK office: 57 Shelton Street, Covent Garden, London WC2H 9HE

British Library Cataloguing-in-Publication Data
A catalogue record for this book is available from the British Library.

LIE THEORY AND ITS APPLICATIONS IN PHYSICS V

ISBN 981-238-936-9

Printed by MultiPrint Services

Preface

The Workshop series 'Lie Theory and Its Application in Physics' should serve the community of theoretical physicists, mathematical physicists and mathematicians working on mathematical models for physical systems based on geometrical methods and in the field of Lie theory.

The series reflects the trend towards a geometrisation of the mathematical description of physical systems and objects. A geometric approach to a system yields in general some notion of symmetry which is very helpful in understanding its structure. Geometrisation and symmetries are meant in their widest sense, i.e., classical geometry, differential geometry, groups and quantum groups, infinite-dimensional (super-)algebras, and their representations. Furthermore we include the necessary tools from functional analysis and number theory. This is a big interdisciplinary and interrelated field.

The present volume reports on the latest developments in this field. The material is collected from the International Workshop organized by the Institute of Nuclear Research and Nuclear Energy of the Bulgarian Academy of Sciences (BAS) in June 2003, at the Guest House of BAS near Varna on the Bulgarian Black Sea Coast. The number of participants was about 60 and they came from 19 countries. The opening lecture was given by Professor V.G. Kac.

The material is arranged as follows. The first Chapter is devoted to *Lie theory*, in particular, vertex algebras, Lie superalgebras, symmetric spaces. The second Chapter is devoted to applications to *field theory*, in particular, conformal field theory, noncommutative field theory, Feynman graphs, the AdS/CFT correspondence. The third Chapter is devoted to applications to *string theory*, in particular, their relation to large N gauge theories, the string/gauge duality, the boundary ground ring, integrable systems, branes, etc. The fourth Chapter is devoted to applications to *integrable systems*, in particular, to the XXZ model, the KP-hierarchy, integrable systems with impurity, the Hall effect. The fifth Chapter is devoted to applications to *quantum mechanical and probabilistic systems*, in particular, to generalizations of quantum mechanics (discrete and/or noncommutative), the Wigner quantum oscillator, measures for infinite-dimensional configurations, Brownian motion. The sixth Chapter is devoted to *quantum groups and related objects*, in particular, q-deformations of algebras used in matrix chain models, q-boson-fermion realizations, representations of liftings of quantum planes, exotic bialgebras which are not deformations of classical (super-)structures though are related to such.

The material in the preceding volumes 'Lie Theory and Its Applications in Physics I,II,III' (World Scientific 1996,1998,2000) was organized in a similar way. (The fourth Workshop was integral part of the 2nd International Symposium "Quantum Theory and Symmetries", also published by World Scientific in 2002.) A comparison between these volumes shows how certain topics evolve and that new structures were found and used. For the present volume we mention an emphasis on *vertex algebras*, *conformal superalgebras*, *conformal field theory*, *string theory*.

For possible future workshops one may think to extend and to diversify the field of applications, e.g., to biology and to economy, in order to show how mathematical tools taken from geometry and from symmetry considerations, which were successful in physics, lead to very fruitful models and results also there.

Acknowledgments:

We express our gratitude to the

- Institute of Nuclear Research and Nuclear Energy
- Abdus Salam International Centre for Theoretical Physics
- International Association of Mathematical Physics
- European Mathematical Society
- Kozloduy NPP Plc

for financial help. We thank the Bulgarian Academy of Sciences for providing its Guest House which contributed very much to the stimulating and pleasant atmosphere during the Workshop. We thank the publisher, World Scientific, for assistance and of speedy publication. Last, but not least, we thank the members of the Local Organizing Committee who, through their efforts, made the workshop run smoothly and efficiently.

H.-D. Doebner and **V.K. Dobrev**
Clausthal and Newcastle, March 2004

Contents

III. String Theory

IV. Integrable Systems

V. Quantum Mechanical and Probabilistic Systems

VI. Quantum Groups and Related Objects

I. Lie Theory

TWISTED MODULES OVER LATTICE VERTEX ALGEBRAS

BOJKO BAKALOV
Department of Mathematics, NCSU, Raleigh, NC 27695, USA
bojko_bakalov@ncsu.edu

VICTOR G. KAC
Department of Mathematics, MIT, Cambridge, MA 02139, USA
kac@math.mit.edu

Dedicated to Professor Ivan Todorov on the occasion of his 70th birthday.

1. Introduction

A vertex algebra is essentially the same as a chiral algebra in conformal field theory [2, 10]. Vertex algebras arose naturally in the representation theory of infinite-dimensional Lie algebras and in the construction of the "moonshine module" for the Monster finite simple group [3, 9].

If V is a vertex algebra and Γ is a finite group of automorphisms of V, the subalgebra V^Γ of Γ-invariant elements in V is called an *orbifold vertex algebra*. Orbifolds play an important role in string theory; in the physics literature they were introduced in one of the earliest papers on conformal field theory [5, 6]. Recently, there have been numerous mathematical papers on orbifolds. All these papers are concerned in some way or another with the problem of describing the representations of V^Γ in terms of the vertex algebra V and the group Γ. However, the solution is known only in very special cases and it is highly nontrivial.

Let Q be an integral lattice. Then one can construct a vertex algebra V_Q called a *lattice vertex algebra* [3, 9, 12]. If $\bar{\Gamma}$ is a finite group of isometries of Q, its elements can be lifted to automorphisms of V_Q. One obtains a group $\Gamma \subset \operatorname{Aut} V$, which is a central extension of $\bar{\Gamma}$. In [1], we construct a collection of V_Q^Γ-modules, and we compute their characters and modular transformations. Some of these modules play an important role in the attempts of a conformal field theory understanding of the fractional quantum

Hall effect (see [4] and the references therein).

The present paper is the first step in the construction of the V_Q^Γ-modules [1]. Here we classify the so-called *twisted* V_Q*-modules*; our main result is Theorem 4.2. In the case when Q is a root lattice of a simple finite-dimensional Lie algebra and σ is an element of its Weyl group, our results agree with those of [13, 14]. Some of our results were obtained independently in [15].

2. Preliminaries on Vertex Algebras

In this section we recall the definition of a vertex algebra and some of its properties, following the book [12]. Below $z, w, \ldots$ denote formal commuting variables. All vector spaces are over the field $\mathbb{C}$ of complex numbers.

2.1. *Local Fields*

Let V be a vector space. A *field* on V is a formal power series in z, z^{-1} of the form

$$\phi(z) = \sum_{m\in\mathbb{Z}} \phi_{(m)} z^{-m-1}, \qquad \phi_{(m)} \in \operatorname{End}(V), \tag{2.1}$$

such that

$$\phi_{(m)} a = 0 \quad \text{for all } a \in V \text{ and } m \gg 0. \tag{2.2}$$

Note that all fields form a vector space invariant under ∂_z.

Two fields $\phi(z)$ and $\psi(z)$ are called *local* with respect to each other if

$$(z-w)^N \big[\phi(z), \psi(w)\big] = 0 \quad \text{for } N \gg 0 \tag{2.3}$$

as a formal power series in z, z^{-1}, w, w^{-1}. We will also say that the pair (ϕ, ψ) is local. Obviously, if (ϕ, ψ) is local, then so are (ψ, ϕ) and $(\partial\phi, \psi)$.

The locality (2.3) is equivalent to the commutator formula

$$\big[\phi(z), \psi(w)\big] = \sum_{n\geq 0} \big(\phi(w)_{(n)}\psi(w)\big)\partial_w^{(n)}\delta(z-w), \tag{2.4}$$

where $\partial_w^{(n)} = \partial_w^n / n!$,

$$\delta(z-w) = \sum_{k\in\mathbb{Z}} z^k w^{-k-1} \tag{2.5}$$

is the formal δ-function, and

$$\phi(w)_{(n)}\psi(w) = \operatorname{Res}_z (z-w)^n [\phi(z), \psi(w)], \qquad n \geq 0, \tag{2.6}$$

is the nth *product* of the fields ϕ, ψ. Here, as usual, Res_z denotes the coefficient at z^{-1}. Note that the sum in (2.4) is finite.

One can define nth product of fields for any $n \in \mathbb{Z}$:

$$\begin{aligned}\phi(w)_{(n)}\psi(w) &= \mathrm{Res}_z\, \phi(z)\psi(w)\, i_{z,w}(z-w)^n \\ &\quad - \mathrm{Res}_z\, \psi(w)\phi(z)\, i_{w,z}(z-w)^n.\end{aligned} \tag{2.7}$$

Here $i_{z,w}$ means that we expand in the domain $|z| > |w|$, i.e.,

$$i_{z,w}(z-w)^n = \sum_{k=0}^{\infty} \binom{n}{k} z^{n-k}(-w)^k, \tag{2.8}$$

while

$$i_{w,z}(z-w)^n = \sum_{k=0}^{\infty} \binom{n}{k} z^{k}(-w)^{n-k}. \tag{2.9}$$

In particular, $\delta(z-w) = i_{z,w}(z-w)^{-1} - i_{w,z}(z-w)^{-1}$.

The (-1)st product is called the *normally ordered product* and is denoted by $:\phi\psi:$. One has

$$:\phi(w)\psi(w): = \phi(w)^+\psi(w) + \psi(w)\phi(w)^-, \tag{2.10}$$

where

$$\phi(w)^+ = \sum_{m<0} \phi_{(m)} w^{-m-1}, \quad \phi(w)^- = \sum_{m\geq 0} \phi_{(m)} w^{-m-1}. \tag{2.11}$$

One also has

$$\phi(w)_{(-n-1)}\psi(w) = :(\partial_w^{(n)}\phi(w))\psi(w):, \qquad n \geq 0. \tag{2.12}$$

2.2. *Vertex Algebras*

Definition 2.1. A *vertex algebra* is a vector space V (space of states) endowed with a vector $|0\rangle \in V$ (vacuum vector), an endomorphism T (infinitesimal translation operator) and a linear map from V to the space of fields on V (state-field correspondence)

$$a \mapsto Y(a,z) = \sum_{m\in\mathbb{Z}} a_{(m)} z^{-m-1}, \tag{2.13}$$

such that the following properties hold:

locality axiom: all fields $Y(a,z)$ are local with respect to each other,

translation covariance:

$$[T, Y(a,z)] = \partial_z Y(a,z), \tag{2.14}$$

vacuum axioms:

$$Y(|0\rangle, z) = \mathrm{id}_V, \quad T|0\rangle = 0, \tag{2.15}$$

$$Y(a,z)|0\rangle - a \in zV[[z]]. \tag{2.16}$$

Here are some corollaries of the definition:

$$Ta = a_{(-2)}|0\rangle, \quad Y(Ta, z) = \partial_z Y(a,z). \tag{2.17}$$

We also have the skew-symmetry

$$Y(a,z)b = e^{zT} Y(b, -z)a. \tag{2.18}$$

The most important property of a vertex algebra V is the following *Borcherds identity* (which along with the vacuum axioms provides an equivalent definition of a vertex algebra):

$$\begin{aligned} &\mathrm{Res}_{z-w}\, Y(Y(a, z-w)b, w)\, i_{w,z-w} F(z,w) \\ &\quad = \mathrm{Res}_z\, Y(a,z) Y(b,w)\, i_{z,w} F(z,w) \\ &\quad - \mathrm{Res}_z\, Y(b,w) Y(a,z)\, i_{w,z} F(z,w) \end{aligned} \tag{2.19}$$

for any $a, b \in V$ and any rational function $F(z,w)$ with poles only at $z=0$, $w=0$ or $z=w$.

Let us give some consequences of the Borcherds identity. Taking $F(z,w) = (z-w)^n \delta(u-z)$, viewed as a series in u, u^{-1}, we obtain

$$\begin{aligned} &Y(a,u)Y(b,w)\, i_{u,w}(u-w)^n - Y(b,w)Y(a,u)\, i_{w,u}(u-w)^n \\ &\quad = \sum_{m=0}^{\infty} Y(a_{(m+n)}b, w)\, \partial_w^{(m)} \delta(u-w). \end{aligned} \tag{2.20}$$

For $n=0$ this gives the commutator formula

$$\big[Y(a,u), Y(b,w)\big] = \sum_{m=0}^{\infty} Y(a_{(m)}b, w)\, \partial_w^{(m)} \delta(u-w), \tag{2.21}$$

(which implies locality). Taking Res_u of (2.20), we get

$$Y(a,w)_{(n)} Y(b,w) = Y(a_{(n)}b, w), \qquad n \in \mathbb{Z}. \tag{2.22}$$

2.3. *Conformal Vertex Algebras*

Definition 2.2. A vertex algebra V is *conformal* of central charge (or rank) $c \in \mathbb{C}$ if there exists a vector $\nu \in V$ (the conformal vector) with the following properties:

(i) The modes L_n, $n \in \mathbb{Z}$, of the field $L(z) \equiv Y(\nu, z) = \sum_{n\in\mathbb{Z}} L_n z^{-n-2}$ give a representation of the Virasoro algebra with central charge c. (The field $L(z)$ is called the energy-momentum field in CFT.)

(ii) L_{-1} is the infinitesimal translation operator T.

(iii) The operator L_0 is diagonalizable with a non-negative spectrum. (L_0 is called the energy operator or Hamiltonian.)

In a conformal vertex algebra V, (2.21) implies

$$[L_0, Y(a,z)] = zY(L_{-1}a, z) + Y(L_0 a, z). \tag{2.23}$$

The operator L_0 defines a gradation of V, $V = \bigoplus_{\Delta\geq 0} V(\Delta)$, such that

$$L_0|_{V(\Delta)} = \Delta\, \mathrm{id}_{V(\Delta)}\,. \tag{2.24}$$

For $a \in V(\Delta)$, the field $Y(a,z)$ is of *conformal weight* Δ, i.e., $\deg a_{(m)} = \Delta - m - 1$ for all $m \in \mathbb{Z}$. The field $L(z)$ is of conformal weight 2 and $\nu = L_{-2}|0\rangle$.

3. Twisted Modules over Vertex Algebras

In this section we study twisted modules over vertex algebras (cf. [7, 9]). It seems that some of our results are new even in the untwisted case.

3.1. *Definition in Terms of Borcherds Identity*

Let V be a vertex algebra and σ be an automorphism of V of finite order N. We let $\epsilon = e^{2\pi i/N}$ and $V_j = \{a \in V \mid \sigma a = \epsilon^{-j} a\}$, $0 \leq j \leq N-1$.

An *N-twisted field* $\phi(z)$ on a vector space M is a formal power series in $z^{1/N}, z^{-1/N}$ of the form

$$\phi(z) = \sum_{m\in\frac{1}{N}\mathbb{Z}} \phi_{(m)} z^{-m-1}, \qquad \phi_{(m)} \in \mathrm{End}(M), \tag{3.1}$$

such that

$$\phi_{(m)} v = 0 \quad \text{for all } v \in M \text{ and } m \gg 0. \tag{3.2}$$

For any integer k, we will denote by $\phi(e^{2\pi i k} z)$ the field obtained from $\phi(z)$ by substituting $z^{1/N}$ with $\epsilon^k z^{1/N}$, i.e., $\phi(e^{2\pi i k} z) = \sum_{m\in\frac{1}{N}\mathbb{Z}} \phi_{(m)} e^{-2\pi i k m} z^{-m-1}$.

Definition 3.1. A *σ-twisted V-module* is a vector space M endowed with a linear map from V to the space of N-twisted fields on M,

$$a \mapsto Y^M(a,z) = \sum_{m\in\frac{1}{N}\mathbb{Z}} a^M_{(m)} z^{-m-1}, \tag{3.3}$$

such that for all $a \in V$

$$Y^M(\sigma a, z) = Y^M(a, e^{2\pi i} z), \tag{3.4}$$
$$Y^M(|0\rangle, z) = \mathrm{id}_M, \tag{3.5}$$

and the following twisted Borcherds identity holds for any $a \in V_j$, $b \in V$, $0 \le j \le N-1$, and any rational function $F(z,w)$ with poles only at $z=0$, $w=0$ or $z=w$:

$$\begin{aligned} &\mathrm{Res}_{z-w}\, Y^M(Y(a,z-w)b,w)\, i_{w,z-w}\, z^{j/N} F(z,w) \\ &= \mathrm{Res}_z\, Y^M(a,z) Y^M(b,w)\, i_{z,w}\, z^{j/N} F(z,w) \\ &- \mathrm{Res}_z\, Y^M(b,w) Y^M(a,z)\, i_{w,z}\, z^{j/N} F(z,w). \end{aligned} \tag{3.6}$$

Of course, for $\sigma = 1$, a 1-twisted module is called just a *module*.

Remark 3.1. The Borcherds identity (3.6) is equivalent to the following collection of identities ($a \in V_j$, $b \in V$, $c \in M$, $m \in \frac{j}{N} + \mathbb{Z}$, $n \in \mathbb{Z}$, $k \in \frac{1}{N}\mathbb{Z}$):

$$\begin{aligned} \sum_{i=0}^{\infty} \binom{m}{i} (a_{(n+i)}b)^M_{(m+k-i)} c &= \sum_{i=0}^{\infty} (-1)^i \binom{n}{i} a^M_{(m+n-i)} (b^M_{(k+i)} c) \\ &- \sum_{i=0}^{\infty} (-1)^{i+n} \binom{n}{i} b^M_{(k+n-i)} (a^M_{(m+i)} c). \end{aligned} \tag{3.7}$$

When V is a conformal vertex algebra we will assume that the automorphism σ preserves the conformal vector: $\sigma(\nu) = \nu$. Then, for any σ-twisted V-module M, the modes L^M_n of $Y^M(\nu, z) = \sum_{n\in\mathbb{Z}} L^M_n z^{-n-2}$ give a representation in M of the Virasoro algebra with central charge $c = \operatorname{rank} V$.

Definition 3.2. A *σ-twisted module over a conformal vertex algebra V* is a σ-twisted V-module M in the sense of Definition 3.1 satisfying the additional requirement that the operator L^M_0 be diagonalizable with finite-dimensional eigenspaces.

3.2. *Consequences of the Definition*

Let M be a σ-twisted V-module. In this subsection we will derive some consequences of Definition 3.1.

First of all, note that, by (3.4), $Y^M(a,z) z^{j/N}$ contains only integer powers of z for $a \in V_j$. For $0 \le j \le N-1$, we introduce the N-twisted δ-functions

$$\delta_j(z-w) = z^{j/N} w^{-j/N} \delta(z-w) = \sum_{k \in \frac{j}{N} + \mathbb{Z}} z^k w^{-k-1}. \tag{3.8}$$

Then

$$\mathrm{Res}_z\, Y^M(a,z)\delta_j(w-z) = Y^M(\pi_j a, w), \tag{3.9}$$

where

$$\pi_j = \frac{1}{N}\sum_{k=0}^{N-1} \epsilon^{kj}\sigma^k \tag{3.10}$$

is the projection onto V_j.

Putting $F(z,w) = z^{-j/N}(z-w)^n\delta_j(u-z)$ in the twisted Borcherds identity (3.6), we obtain a twisted version of (2.20):

$$\begin{aligned} &Y^M(a,u)Y^M(b,w)\, i_{u,w}(u-w)^n - Y^M(b,w)Y^M(a,u)\, i_{w,u}(u-w)^n \\ &\quad = \sum_{m=0}^{\infty} Y^M(a_{(m+n)}b, w)\, \partial_w^{(m)}\delta_j(u-w), \qquad a \in V_j, n \in \mathbb{Z}. \end{aligned} \tag{3.11}$$

The collection of identities (3.11) for all $n \in \mathbb{Z}$ is equivalent to (3.6). Note that, as in the untwisted case, all the fields $Y^M(a,z)$ are local with respect to each other, since, letting $n = 0$ in (3.11), we get the twisted commutator formula ($a \in V_j$):

$$[Y^M(a,u), Y^M(b,w)] = \sum_{m=0}^{\infty} Y^M(a_{(m)}b, w)\, \partial_w^{(m)}\delta_j(u-w). \tag{3.12}$$

Let us define the nth product ($n \in \mathbb{Z}$) of two twisted fields $Y^M(a,w)$, $Y^M(b,w)$ for $a \in V_j$ as $\mathrm{Res}_u\, u^{j/N}w^{-j/N}$ of the left hand side of (3.11). For $n = -1$, this definition coincides with (2.10), (2.11), but for $n < -1$ it differs from (2.12). By (3.11), we have for $a \in V_j$

$$Y^M(a,w)_{(n)}Y^M(b,w) = \sum_{m=0}^{\infty} \binom{j/N}{m} w^{-m}\, Y^M(a_{(m+n)}b, w). \tag{3.13}$$

In particular, letting $b = |0\rangle$ and $n = -2$ in (3.13), we get

$$Y^M(Ta,z) = \partial_z Y^M(a,z). \tag{3.14}$$

When V is a conformal vertex algebra, the commutator formula (3.12) implies:

$$[T^M, Y^M(a,z)] = \partial_z Y^M(a,z), \tag{3.15}$$

$$[L_0^M, Y^M(a,z)] = z\partial_z Y^M(a,z) + Y^M(L_0^M a, z), \tag{3.16}$$

where $T^M := L_{-1}^M$.

3.3. *Associativity of Twisted Fields*

We will try to invert formula (3.13), i.e., to express $Y^M(a_{(n)}b, w)$ in terms of $Y^M(a,w)_{(n)}Y^M(b,w)$. To this end, we multiply both sides of (3.13) by $z^{-n-1}w^{j/N}$ and sum over $n \in \mathbb{Z}$. Using the properties of the δ-function and Taylor's formula, we get

$$i_{w,z}(z+w)^{j/N}\, Y^M(Y(a,z)b,w) = i_{z,w}(z+w)^{j/N}\, Y^M(a,z+w)Y^M(b,w) - \sum_{k=0}^{\infty} Y^M(b,w)a^M_{(k+j/N)}\,(-\partial_w)^{(k)}\delta(z-(-w)), \qquad a \in V_j. \tag{3.17}$$

Of course, we cannot divide (3.17) by $(z+w)^{j/N}$. Note, however, that the sum in (3.17) becomes finite when applied to any $v \in M$. Using that $(z+w)^n(-\partial_w)^{(k)}\delta(z-(-w)) = 0$ for $n > k$, we obtain the *associativity* of twisted fields

$$\begin{aligned} i_{w,z}&(z+w)^{n+j/N}\, Y^M(Y(a,z)b,w)v \\ &= i_{z,w}(z+w)^{n+j/N}\, Y^M(a,z+w)Y^M(b,w)v, \\ &\qquad a \in V_j, b \in V, v \in M, n \gg 0. \end{aligned} \tag{3.18}$$

Here we can take the minimal $n \geq 0$ such that $a^M_{(k+j/N)}v = 0$ for $k \geq n$.

Note that the powers of z and w in both sides of (3.18) are bounded from below. Therefore, we can multiply both sides by $i_{w,z}(z+w)^{-n-j/N}$ to get $(a \in V_j, b \in V, v \in M)$:

$$\begin{aligned} &Y^M(Y(a,z)b,w)v \\ &= \left(i_{w,z}(z+w)^{-n-j/N}\right) i_{z,w}(z+w)^{n+j/N}\, Y^M(a,z+w)Y^M(b,w)v. \end{aligned} \tag{3.19}$$

3.4. *Definition in Terms of Associativity*

We will show that the associativity (3.18), together with (3.4), (3.5), implies the twisted Borcherds identity (3.6).

First let $b = |0\rangle$ in (3.18). Using (3.5) and the identity $Y(a,z)|0\rangle = e^{zT}a$ (cf. (2.18)), we get:

$$i_{w,z}(z+w)^{n+j/N}\, Y^M(e^{zT}a,w)v = i_{z,w}(z+w)^{n+j/N}\, Y^M(a,z+w)v.$$

Notice that the right-hand side contains only non-negative integer powers of $z+w$. This implies $Y^M(e^{zT}a,w) = i_{w,z}Y^M(a,z+w)$ and, in particular, Eq. (3.14).

Now let $a \in V_j, b \in V_k, v \in M$, and let $n,p \in \mathbb{Z}$ be such that $a^M_{(m+j/N)}v = 0$ for $m \geq n$, $b^M_{(m+k/N)}v = 0$ for $m \geq p$. Replacing in (3.18) b

with $e^{uT}b$, we obtain:

$$\begin{aligned} & i_{z,w}i_{w,u}(z+w)^{n+j/N}\,Y^M(a,z+w)Y^M(b,u+w)v \\ & \qquad = i_{w,z}i_{w,u}i_{z,u}(z+w)^{n+j/N}\,Y^M(Y(a,z-u)b,u+w)v. \end{aligned} \tag{3.20}$$

Note that, if we multiply the left-hand side by $i_{w,u}(u+w)^{p+k/N}$, it will contain only non-negative integer powers of $u+w$ and hence of w. Therefore it makes sense to put $w=0$:

$$\begin{aligned} & z^{n+j/N}u^{p+k/N}\,Y^M(a,z)Y^M(b,u)v \\ & \qquad = i_{w,z}i_{w,u}i_{z,u}(z+w)^{n+j/N}(u+w)^{p+k/N} \\ & \qquad\quad \times Y^M(Y(a,z-u)b,u+w)v\big|_{w=0}. \end{aligned} \tag{3.21}$$

Interchanging the roles of a and b and using (2.18), we get:

$$\begin{aligned} & z^{n+j/N}u^{p+k/N}\,Y^M(b,u)Y^M(a,z)v \\ & \qquad = i_{w,z}i_{w,u}i_{u,z}(z+w)^{n+j/N}(u+w)^{p+k/N} \\ & \qquad\quad \times Y^M(Y(a,z-u)b,u+w)v\big|_{w=0}. \end{aligned} \tag{3.22}$$

Notice that

$$i_{w,z}(z+w)^{n+j/N} = i_{w,u}\sum_{i\in\mathbb{Z}_+}\binom{n+j/N}{i}(u+w)^{n-i+j/N}(z-u)^i$$

and

$$i_{z,u}(z-u)^{-m-1} - i_{u,z}(z-u)^{-m-1} = \partial_u^{(m)}\delta(z-u), \qquad m\geq 0.$$

Using this and (3.21), (3.22), we get for $l\in\mathbb{Z}$:

$$\begin{aligned} & i_{z,u}(z-u)^l z^{n+j/N}u^{p+k/N}\,Y^M(a,z)Y^M(b,u)v \\ & \qquad - i_{u,z}(z-u)^l z^{n+j/N}u^{p+k/N}\,Y^M(b,u)Y^M(a,z)v \\ & = i_{w,u}\sum_{\substack{i,m\in\mathbb{Z}\\ i\geq 0,\, m\geq i+l}}\binom{n+j/N}{i}(u+w)^{n+p-i+(j+k)/N} \\ & \qquad \times Y^M(a_{(m)}b,u+w)v\,\partial_u^{(m-i-l)}\delta(z-u)\big|_{w=0}. \end{aligned}$$

From here it is easy to deduce (3.11), which implies (3.6).

Therefore, we have the following equivalent definition of a σ-twisted V-module.

Proposition 3.1. *A σ-twisted V-module is the same as a vector space M endowed with a linear map* (3.3) *from V to the space of N-twisted fields on M, satisfying* (3.4), (3.5) *and* (3.18).

4. Twisted Modules over a Lattice Vertex Algebra

In the first subsection we introduce the main object of our study: the lattice vertex algebra V_Q. The remainder of the section is devoted to the classification of all σ-twisted V_Q-modules, where σ is an automorphism of the lattice Q.

4.1. *Lattice Vertex Algebras*

The purpose of this subsection is to fix the notation and review some properties of lattice vertex algebras. For more details, see [12].

Let Q be an integral lattice of rank l. We denote the bilinear form on Q by $(\cdot|\cdot)$, and write $|\alpha|^2 = (\alpha|\alpha)$ for $\alpha \in Q$. We extend the bilinear form to $\mathfrak{h} = \mathbb{C} \otimes_{\mathbb{Z}} Q$ by $\mathbb{C}$-bilinearity. There exists a bimultiplicative function $\varepsilon\colon Q \times Q \to \{\pm 1\}$ satisfying

$$\varepsilon(\alpha,\alpha) = (-1)^{|\alpha|^2(|\alpha|^2+1)/2}, \qquad \alpha \in Q. \tag{4.1}$$

Then by bimultiplicativity

$$\varepsilon(\alpha,\beta)\varepsilon(\beta,\alpha) = (-1)^{(\alpha|\beta)+|\alpha|^2|\beta|^2}, \qquad \alpha,\beta \in Q. \tag{4.2}$$

Introduce the twisted group algebra $\mathbb{C}_\varepsilon[Q]$: it has a basis $\{e^\alpha\}_{\alpha\in Q}$ over $\mathbb{C}$ and multiplication

$$e^\alpha e^\beta = \varepsilon(\alpha,\beta)e^{\alpha+\beta}, \qquad \alpha,\beta \in Q. \tag{4.3}$$

Let $\hat{\mathfrak{h}} = \mathfrak{h}[t,t^{-1}] \oplus \mathbb{C}K$ be the *Heisenberg current algebra*; this is a Lie algebra with the bracket

$$[ht^m, h't^n] = m\delta_{m,-n}(h|h')K, \quad [ht^m, K] = 0, \qquad h,h' \in \mathfrak{h}. \tag{4.4}$$

It has a unique irreducible representation of level 1 (i.e., with $K = 1$) on the Fock space $S = S(\mathfrak{h}[t^{-1}]t^{-1})$ such that $\mathfrak{h}[t^{-1}]t^{-1}$ acts by multiplication and $\mathfrak{h}[t]1 = 0$. This representation extends to the space $V_Q = S \otimes \mathbb{C}_\varepsilon[Q]$ by

$$(ht^m)(s \otimes e^\alpha) = \big(ht^m + \delta_{m,0}(h|\alpha)\big)s \otimes e^\alpha \qquad \text{for } m \geq 0. \tag{4.5}$$

We define a representation of the algebra $\mathbb{C}_\varepsilon[Q]$ in V_Q by left multiplication:

$$e^\gamma(s \otimes e^\alpha) = \varepsilon(\gamma,\alpha)s \otimes e^{\alpha+\gamma}. \tag{4.6}$$

This gives rise to a representation in V_Q of the associative algebra $\mathcal{A}_Q = U(\hat{\mathfrak{h}})\otimes\mathbb{C}_\varepsilon[Q]$, which is a "twisted" tensor product of the universal enveloping algebra $U(\hat{\mathfrak{h}})$ of $\hat{\mathfrak{h}}$ and the algebra $\mathbb{C}_\varepsilon[Q]$ by the relation

$$e^\alpha(ht^m) = (ht^m - \delta_{m,0}(\alpha|h))e^\alpha. \tag{4.7}$$

Here and further e^α (respectively ht^m) stands for $1 \otimes e^\alpha$ (respectively $ht^m \otimes 1$). The algebra $\mathcal{A}_Q$ has a $\mathbb{Z}_2$-gradation (i.e., it is an associative superalgebra) defined by

$$p(u \otimes e^\alpha) = |\alpha|^2 \mod 2\mathbb{Z}. \tag{4.8}$$

This induces a $\mathbb{Z}_2$-gradation on V_Q:

$$p(s \otimes e^\alpha) = |\alpha|^2 \mod 2\mathbb{Z}. \tag{4.9}$$

Introduce the following fields on V_Q (called *currents*):

$$h(z) = \sum_{m\in\mathbb{Z}} (ht^m) z^{-m-1}, \qquad h \in \mathfrak{h}. \tag{4.10}$$

Then the commutation relations (4.4) for $K = 1$ can be rewritten as

$$[h(z), h'(w)] = (h|h')\, \partial_w \delta(z-w), \qquad h, h' \in \mathfrak{h}. \tag{4.11}$$

Hence all the fields $h(z)$ are local with respect to each other.

For $\alpha \in Q$, introduce the *vertex operator*

$$\begin{aligned} Y_\alpha(z) &= e^\alpha :\exp \textstyle\int \alpha(z): \\ &\equiv e^\alpha z^\alpha \exp\Big(\sum_{n<0} (\alpha t^n) \frac{z^{-n}}{-n}\Big) \exp\Big(\sum_{n>0} (\alpha t^n) \frac{z^{-n}}{-n}\Big) \end{aligned} \tag{4.12}$$

of parity $|\alpha|^2 \mod 2\mathbb{Z}$. The vertex operator $Y_\alpha(z)$ is a field on V_Q, and it is local with respect to $h(z)$ because

$$[h(z), Y_\alpha(w)] = (h|\alpha) Y_\alpha(w) \delta(z-w), \qquad h \in \mathfrak{h}, \alpha \in Q. \tag{4.13}$$

Moreover, the fields $Y_\alpha(z)$ are local among themselves. This follows from (4.2) and the formula

$$Y_\alpha(z) Y_\beta(w) = \varepsilon(\alpha,\beta)(z-w)^{(\alpha|\beta)}\, e^{\alpha+\beta} :\exp\big(\textstyle\int \alpha(z) + \int \beta(w)\big):. \tag{4.14}$$

Theorem 4.1. *The fields $Y(ht^{-1}, z) = h(z)$ $(h \in \mathfrak{h})$, of parity 0, and $Y(e^\alpha, z) = Y_\alpha(z)$ $(\alpha \in Q)$, of parity $p(e^\alpha) = |\alpha|^2 \mod 2\mathbb{Z}$, generate a vertex algebra structure on $V_Q = S \otimes \mathbb{C}_\varepsilon[Q]$ with the vacuum vector $|0\rangle = 1 \otimes 1$ and the operator T defined by*

$$[T, ht^m] = -m\, ht^{m-1}, \quad Te^\alpha = (\alpha t^{-1}) e^\alpha, \qquad h \in \mathfrak{h}, \alpha \in Q. \tag{4.15}$$

This vertex algebra is conformal of rank $l = \operatorname{rank} Q$ with the conformal vector

$$\nu = \frac{1}{2} \sum_{i=1}^{l} (a^i t^{-1})(b^i t^{-1}), \tag{4.16}$$

where $\{a^i\}$, $\{b^i\}$ are dual bases of $\mathfrak{h}$.

Let σ be an automorphism of the lattice Q of finite order N. For $0 \le j \le N-1$, let

$$\mathfrak{h}_j = \{h \in \mathfrak{h} \mid \sigma h = \epsilon^{-j} h\}, \qquad \epsilon = e^{2\pi \mathrm{i}/N}. \tag{4.17}$$

Then $\mathfrak{h}_j$ and $\mathfrak{h}_k$ are orthogonal unless $j + k \equiv 0 \mod N\mathbb{Z}$. Since there exists a unique, up to equivalence, ± 1-valued 2-cocycle ε on Q satisfying (4.1), the cocycles $\varepsilon(\alpha, \beta)$ and $\varepsilon(\sigma\alpha, \sigma\beta)$ are equivalent. Hence there exists a function $\eta\colon Q \to \{\pm 1\}$ such that

$$\eta(\alpha)\eta(\beta)\varepsilon(\alpha,\beta) = \eta(\alpha+\beta)\varepsilon(\sigma\alpha,\sigma\beta), \qquad \alpha,\beta \in Q. \tag{4.18}$$

Moreover, η can be chosen in such a way that

$$\eta(\alpha) = 1 \quad \text{for } \alpha \in Q \cap \mathfrak{h}_0. \tag{4.19}$$

Proposition 4.1. *Any automorphism σ of Q can be lifted to an automorphism of the vertex algebra V_Q so that*

$$\sigma(ht^m) = \sigma(h)t^m, \quad \sigma(e^\alpha) = \eta(\alpha)^{-1} e^{\sigma\alpha}, \qquad h \in \mathfrak{h}, \alpha \in Q. \tag{4.20}$$

It fixes the conformal vector ν: $\sigma(\nu) = \nu$.

Proof. Follows from the observation that σ defines an automorphism of the associative superalgebra $\mathcal{A}_Q$. □

Remark 4.1. If $\sigma\colon Q \to Q$ is an automorphism of order N, then its lifting $\sigma\colon V_Q \to V_Q$ defined by (4.20) has order N or $2N$.

4.2. *Twisted Vertex Operators*

Let σ be an automorphism of Q lifted to an automorphism of V_Q, as in Proposition 4.1. We use the notation from Section 3 for $V = V_Q$. Let M be a σ-twisted V_Q-module (see Definition 3.2). We will study the fields $Y^M(ht^{-1}, z)$ and $Y^M(e^\alpha, z)$ for $h \in \mathfrak{h}, \alpha \in Q$.

Comparing the commutator formulas (2.21) and (3.12), we see that Eqs. (4.11), (4.13) immediately imply

$$[Y^M(ht^{-1}, z), Y^M(h't^{-1}, w)] = (h|h')\, \partial_w \delta_j(z-w), \tag{4.21}$$

$$[Y^M(ht^{-1}, z), Y^M(e^\alpha, w)] = (h|\alpha)\, Y^M(e^\alpha, w)\delta_j(z-w), \tag{4.22}$$

$$h \in \mathfrak{h}_j, h' \in \mathfrak{h}, \ \alpha \in Q.$$

These formulas can be restated as:

$$[h^M_{(m)}, (h')^M_{(n)}] = (\pi_{Nm}h|h')\, m\delta_{m,-n}, \tag{4.23}$$

$$[h^M_{(m)}, Y^M(e^\alpha, w)] = (\pi_{Nm}h|\alpha)\, w^m Y^M(e^\alpha, w), \tag{4.24}$$

$$h, h' \in \mathfrak{h},\ m, n \in \tfrac{1}{N}\mathbb{Z},\ \alpha \in Q,$$

where π_j is the projection of $\mathfrak{h}$ onto $\mathfrak{h}_{j \bmod N\mathbb{Z}}$. (Note that this π_j is the restriction to $(\mathfrak{h}t^{-1})|0\rangle$ of the π_j defined by (3.10)). In the sequel we will use the notation $h_0 = \pi_0 h$ for $h \in \mathfrak{h}$.

Also note that, for $a = ht^{-1}$, (3.4) is equivalent to

$$(\sigma h)^M_{(m)} = h^M_{(m)}\, e^{-2\pi \mathrm{i} m}, \qquad h \in \mathfrak{h},\ m \in \tfrac{1}{N}\mathbb{Z}. \tag{4.25}$$

Lemma 4.1. *There exist operators U^M_α $(\alpha \in Q)$ on M such that*

$$Y^M(e^\alpha, z) = z^{b_\alpha} U^M_\alpha E^M_\alpha(z) \tag{4.26}$$

where

$$b_\alpha = (|\alpha_0|^2 - |\alpha|^2)/2 \tag{4.27}$$

and

$$\begin{aligned} E^M_\alpha(z) &= :\exp \textstyle\int Y^M(\alpha t^{-1}, z): \\ &\equiv z^{\alpha^M_{(0)}} \exp\Big(\sum_{n \in \frac{1}{N}\mathbb{Z}_{<0}} \alpha^M_{(n)} \frac{z^{-n}}{-n} \Big) \exp\Big(\sum_{n \in \frac{1}{N}\mathbb{Z}_{>0}} \alpha^M_{(n)} \frac{z^{-n}}{-n} \Big). \end{aligned} \tag{4.28}$$

The operators U^M_α satisfy

$$[h^M_{(m)}, U^M_\alpha] = \delta_{m,0}(h_0|\alpha)\, U^M_\alpha, \qquad h \in \mathfrak{h},\ m \in \tfrac{1}{N}\mathbb{Z}, \tag{4.29}$$

$$U^M_{\sigma\alpha} = \eta(\alpha) U^M_\alpha e^{2\pi \mathrm{i}(b_\alpha + \alpha^M_{(0)})}. \tag{4.30}$$

Proof. Define the operators

$$U^M_\alpha(z) = \exp\Big(\sum_{n \in \frac{1}{N}\mathbb{Z}_{<0}} \alpha^M_{(n)} \frac{z^{-n}}{n} \Big) Y^M(e^\alpha, z)\, z^{-\alpha^M_{(0)}} \exp\Big(\sum_{n \in \frac{1}{N}\mathbb{Z}_{>0}} \alpha^M_{(n)} \frac{z^{-n}}{n} \Big).$$

Then, by (4.23), (4.24),

$$[h^M_{(m)}, U^M_\alpha(z)] = \delta_{m,0}(h_0|\alpha)\, U^M_\alpha(z), \qquad h \in \mathfrak{h}.$$

In particular, $U^M_\alpha(z)$ commutes with $\alpha^M_{(n)}$ for $n < 0$; hence, we have

$$Y^M(e^\alpha, z) = U^M_\alpha(z) E^M_\alpha(z). \tag{4.31}$$

We will deduce Eqs. (4.26) and (4.29) once we show that $U_\alpha^M := z^{-b_\alpha} U_\alpha^M(z)$ is independent of z. To this end we use the translation invariance (3.14). By (4.15), we have $Te^\alpha = (\alpha t^{-1})e^\alpha = \alpha_{(-1)}e^\alpha$. Using (3.14), (4.15) and (3.13), we find

$$\begin{aligned}\partial_z Y^M(e^\alpha, z) &= Y^M(Te^\alpha, z) = Y^M(\alpha_{(-1)}e^\alpha, z) \\ &= {:}Y^M(\alpha t^{-1}, z)Y^M(e^\alpha, z){:} - \sum_{j=0}^{N-1} \frac{j}{N}(\pi_j\alpha|\alpha) z^{-1} Y^M(e^\alpha, z).\end{aligned}$$

Therefore

$$\partial_z U_\alpha^M(z) = -\sum_{j=0}^{N-1} \frac{j}{N}(\pi_j\alpha|\alpha) z^{-1} U_\alpha^M(z).$$

Using that $(\pi_j\alpha|\alpha) = (\pi_{N-j}\alpha|\alpha)$ for $1 \le j \le N-1$, it is easy to see that

$$\sum_{j=0}^{N-1} \frac{j}{N}(\pi_j\alpha|\alpha) = -b_\alpha. \tag{4.32}$$

This proves that $U_\alpha^M = z^{-b_\alpha} U_\alpha^M(z)$ is independent of z.

Finally, to prove (4.30), we apply (3.4) for $a = e^\alpha$, using that $\sigma(e^\alpha) = \eta(\alpha)^{-1}e^{\sigma\alpha}$ and that by (4.25) we have $E_{\sigma\alpha}^M(z) = e^{-2\pi i \alpha_{(0)}^M} E_\alpha^M(e^{2\pi i}z)$. □

Lemma 4.2. *We have*

$$Y^M(e^\alpha, z)Y^M(e^\beta, w) = i_{z,w} f_{\alpha,\beta}(z,w)\, z^{b_\alpha} w^{b_\beta}\, U_\alpha^M U_\beta^M\, E_{\alpha,\beta}^M(z,w), \tag{4.33}$$

where

$$f_{\alpha,\beta}(z,w) = \prod_{k=0}^{N-1} \left(z^{1/N} - \epsilon^k w^{1/N}\right)^{(\sigma^k\alpha|\beta)}, \tag{4.34}$$

$$E_{\alpha,\beta}^M(z,w) = {:}\exp\left(\textstyle\int Y^M(\alpha t^{-1}, z) + \int Y^M(\beta t^{-1}, w)\right){:}. \tag{4.35}$$

Proof. Standard exercise, using the fact that $e^A e^B e^{-A} = e^{\operatorname{ad} A} e^B = e^{[A,B]} e^B$ for any two operators A, B commuting with $[A,B]$. □

Lemma 4.3. *We have*

$$\begin{aligned}&Y^M(Y(e^\alpha, z)e^\beta, w) \qquad (4.36)\\ &= \varepsilon(\alpha,\beta) B_{\alpha,\beta}^{-1}\, i_{w,z} f_{\alpha,\beta}(z+w, w)\, (z+w)^{b_\alpha} w^{b_\beta}\, U_{\alpha+\beta}^M\, E_{\alpha,\beta}^M(z+w, w),\end{aligned}$$

where

$$B_{\alpha,\beta} = \frac{f_{\alpha,\beta}(z,w)}{(z-w)^{(\alpha|\beta)}}\bigg|_{z^{1/N}=w^{1/N}=1} = N^{-(\alpha|\beta)} \prod_{k=1}^{N-1} \left(1-\epsilon^k\right)^{(\sigma^k\alpha|\beta)}. \tag{4.37}$$

Proof. We use the same argument as in the proof of Lemma 4.1. First, using (3.12), (4.24) and (4.5), we compute the commutator ($h \in \mathfrak{h}_j$, $n \in \frac{j}{N} + \mathbb{Z}$):

$$\begin{aligned}
&[h^M_{(n)}, Y^M(Y(e^\alpha, z)e^\beta, w)] \\
&\quad = \sum_{m=0}^{\infty} \binom{n}{m} w^{n-m}\, Y^M\big(h_{(m)}(Y(e^\alpha, z)e^\beta), w\big) \\
&\quad = \sum_{m=0}^{\infty} \binom{n}{m} w^{n-m}\big((h|\alpha)z^m + \delta_{m,0}(h|\beta)\big) Y^M(Y(e^\alpha, z)e^\beta, w) \\
&\quad = \big(i_{w,z}(z+w)^n(h|\alpha) + w^n(h|\beta)\big) Y^M(Y(e^\alpha, z)e^\beta, w).
\end{aligned}$$

It follows that

$$Y^M(Y(e^\alpha, z)e^\beta, w) = U^M_{\alpha,\beta}(z, w) E^M_{\alpha,\beta}(z+w, w),$$

where the operator $U^M_{\alpha,\beta}(z, w)$ satisfies

$$[h^M_{(n)}, U^M_{\alpha,\beta}(z, w)] = \delta_{n,0}(h_0|\alpha+\beta)\, U^M_{\alpha,\beta}(z, w).$$

Next, we note that (by (4.15), (3.14), (2.14))

$$\begin{aligned}
Y^M\big(Y(e^\alpha, z)(\beta_{(-1)}e^\beta), w\big) &= Y^M\big(Y(e^\alpha, z)(Te^\beta), w\big) \\
&= (\partial_w - \partial_z) Y^M(Y(e^\alpha, z)e^\beta, w).
\end{aligned}$$

A similar computation as above, using (3.13) for $n = -1$, shows that for $h \in \mathfrak{h}_j$ one has

$$\begin{aligned}
Y^M\big(Y(e^\alpha, z)(h_{(-1)}e^\beta), w\big) &= {:}Y^M(ht^{-1}, w) Y^M(Y(e^\alpha, z)e^\beta, w){:} \\
&- \left(z^{-1} i_{w,z}\left(1 + \frac{z}{w}\right)^{j/N}(h|\alpha) + \frac{j}{N} w^{-1}(h|\beta)\right) Y^M(Y(e^\alpha, z)e^\beta, w).
\end{aligned}$$

Therefore

$$\begin{aligned}
&(\partial_w - \partial_z) U^M_{\alpha,\beta}(z, w) \\
&\quad = -\sum_{j=0}^{N-1} \left(z^{-1} i_{w,z}\left(1 + \frac{z}{w}\right)^{j/N}(\pi_j\beta|\alpha) + \frac{j}{N} w^{-1}(\pi_j\beta|\beta)\right) U^M_{\alpha,\beta}(z, w).
\end{aligned}$$

With some more computation, we see that

$$\begin{aligned}
-\sum_{j=0}^{N-1} z^{-1} i_{w,z}\left(1 + \frac{z}{w}\right)^{j/N}(\pi_j\beta|\alpha) &= i_{w,z} i_{w,z+w} \sum_{k=0}^{N-1} \frac{(1/N)\epsilon^k w^{1/N-1}(\sigma^k\alpha|\beta)}{(z+w)^{1/N} - \epsilon^k w^{1/N}} \\
&= i_{w,z} \frac{(\partial_w - \partial_z) f_{\alpha,\beta}(z+w, w)}{f_{\alpha,\beta}(z+w, w)}
\end{aligned}$$

On the other hand, by (4.32), we have

$$-\sum_{j=0}^{N-1} \frac{j}{N} w^{-1}(\pi_j\beta|\beta) = (\partial_w - \partial_z)w^{b_\beta}/w^{b_\beta}.$$

It follows that

$$U^M_{\alpha,\beta}(z,w)/i_{w,z}f_{\alpha,\beta}(z+w,w)w^{b_\beta}$$

depends only on $z+w$.

Finally, note that

$$Y^M(Y(e^\alpha,z)e^\beta,w) = z^{(\alpha|\beta)}\varepsilon(\alpha,\beta)Y^M(e^{\alpha+\beta},w) + \text{higher powers of } z,$$

while

$$i_{w,z}f_{\alpha,\beta}(z+w,w) = z^{(\alpha|\beta)}B_{\alpha,\beta}\, w^{(\alpha_0-\alpha|\beta)} + \text{higher powers of } z.$$

Since $(\alpha_0-\alpha|\beta) = b_{\alpha+\beta} - b_\alpha - b_\beta$, this completes the proof. □

Corollary 4.1. *In any σ-twisted V_Q-module M, one has*

$$U^M_\alpha U^M_\beta = \varepsilon(\alpha,\beta)B^{-1}_{\alpha,\beta}\,U^M_{\alpha+\beta}, \qquad \alpha,\beta\in Q. \tag{4.38}$$

Proof. Follows immediately from (4.33), (4.36) and the associativity (3.18). □

Remark 4.2. In the proofs of Lemmas 4.2 and 4.3, we used only the commutator formulas (4.23), (4.24), the translation invariance (3.14), and formula (3.13) for $n=-1$, $a=ht^{-1}$, $b=e^\beta$.

4.3. *The Heisenberg Pair* $(\hat{\mathfrak{h}}_\sigma, G_\sigma)$

The results of the previous subsection motivate the following definitions.

The *σ-twisted current algebra* $\hat{\mathfrak{h}}_\sigma$ consists of all σ-invariant elements from $\mathbb{C}K \oplus \mathfrak{h}[t^{1/N}, t^{-1/N}]$, where σ acts as

$$\sigma(ht^m) = \sigma(h)e^{2\pi i m}t^m, \quad \sigma(K)=K, \qquad h\in\mathfrak{h},\ m\in\tfrac{1}{N}\mathbb{Z}. \tag{4.39}$$

In other words, $\hat{\mathfrak{h}}_\sigma$ is spanned over $\mathbb{C}$ by K and the elements ht^m such that $h\in\mathfrak{h}_j$, $m\in\frac{j}{N}+\mathbb{Z}$. This is a Lie algebra with bracket (4.4).

Let $G = \mathbb{C}^\times \times \exp\mathfrak{h}_0 \times Q$ be the set consisting of elements $c e^h U_\alpha$ ($c \in \mathbb{C}^\times$, $h \in \mathfrak{h}_0$, $\alpha \in Q$). We define a multiplication in G by the formulas:

$$e^h e^{h'} = e^{h+h'}, \tag{4.40}$$

$$e^h U_\alpha e^{-h} = e^{(h|\alpha)} U_\alpha, \tag{4.41}$$

$$U_\alpha U_\beta = \varepsilon(\alpha,\beta) B_{\alpha,\beta}^{-1} U_{\alpha+\beta}. \tag{4.42}$$

Then G is a group. From (4.42) we get the commutator

$$C_{\alpha,\beta} := U_\alpha U_\beta U_\alpha^{-1} U_\beta^{-1} = (-1)^{|\alpha|^2|\beta|^2} \prod_{k=0}^{N-1} \left(-\epsilon^k\right)^{-(\sigma^k\alpha|\beta)}. \tag{4.43}$$

We give another expression for $C_{\alpha,\beta}$ which will be useful in the sequel:

$$C_{\alpha,\beta} = (-1)^{|\alpha|^2|\beta|^2} e^{\pi \mathrm{i}(\alpha_0|\beta)} e^{2\pi \mathrm{i}(\alpha_*|\beta)} \quad \text{for } \alpha = \alpha_0 + (1-\sigma)\alpha_*,\ \alpha_0 \in \mathfrak{h}_0, \alpha_* \in \mathfrak{h}_0^\perp. \tag{4.44}$$

Denote by Q_{ev} the sublattice of Q consisting of all even elements, i.e., α such that $|\alpha|^2 \in 2\mathbb{Z}$.

Lemma 4.4. *The center $Z(G)$ of G consists of all elements of the form $c e^{2\pi \mathrm{i}\lambda_0} U_{(1-\sigma)\lambda}$, where $c \in \mathbb{C}^\times$ and $\lambda \in (Q_{\mathrm{ev}})^*$ is such that $\alpha := (1-\sigma)\lambda \in Q$ and $\lambda \in Q^*$ if $\alpha \in Q_{\mathrm{ev}}$.*

Proof. If $e^{2\pi \mathrm{i}h} U_\alpha$ is in the center, then (4.41) implies $\alpha_0 = 0$. Then $\alpha = (1-\sigma)\alpha_*$ for some uniquely defined $\alpha_* \in \mathfrak{h}_0^\perp$. Letting $\lambda = h + \alpha_*$, we get $e^{2\pi \mathrm{i}h} U_\alpha = e^{2\pi \mathrm{i}\lambda_0} U_{(1-\sigma)\lambda}$. Using (4.41) and (4.44), we see that $e^{2\pi \mathrm{i}\lambda_0} U_{(1-\sigma)\lambda}$ commutes with U_β iff

$$(\lambda|\beta) + |\alpha|^2|\beta|^2/2 \in \mathbb{Z} \quad \text{for } \beta \in Q. \tag{4.45}$$

Since

$$|\alpha|^2 = ((1-\sigma)\alpha_*|(1-\sigma)\alpha_*) = 2(\alpha_*|\alpha_*) - 2(\alpha_*|\sigma\alpha_*) = 2(\alpha_*|\alpha) = 2(\lambda|\alpha),$$

equation (4.45) is equivalent to $\lambda \in Q^*$ if $\alpha \in Q_{\mathrm{ev}}$ and to $\lambda \in (Q_{\mathrm{ev}})^*$ if $\alpha \in Q \setminus Q_{\mathrm{ev}}$. □

In particular, by Lemma 4.4, all elements of the form $e^{2\pi \mathrm{i}\alpha_0} U_{(1-\sigma)\alpha}$ ($\alpha \in Q$) are central in G. We let G_σ be the factor of G over the central subgroup

$$N_\sigma := \{\eta(\alpha) U_{\sigma\alpha}^{-1} U_\alpha e^{2\pi \mathrm{i}(b_\alpha + \alpha_0)} \mid \alpha \in Q\}. \tag{4.46}$$

Note that $N_\sigma \cap \mathbb{C}^\times = \{1\}$.

We endow Q with the discrete topology so that G and G_σ are Lie groups with a Lie algebra $\mathbb{C} \oplus \mathfrak{h}_0$. By (4.46), (4.27) and (4.19), we have

$$e^{2\pi i\alpha} = 1 \quad \text{in } G_\sigma \text{ for } \alpha \in Q \cap \mathfrak{h}_0. \tag{4.47}$$

It is easy to see that the connected component of the unit in G_σ is equal to $\mathbb{C}^\times$ times the torus

$$T_\sigma := \exp 2\pi i(\mathfrak{h}_0 / Q \cap \mathfrak{h}_0). \tag{4.48}$$

The group G acts on $\hat{\mathfrak{h}}_\sigma$ by conjugation:

$$(ce^h U_\alpha)(h't^m + c'K)(ce^h U_\alpha)^{-1} = h't^m + \delta_{m,0}(h'_0|\alpha)K + c'K. \tag{4.49}$$

This action is compatible with the adjoint action of $\mathbb{C} \oplus \mathfrak{h}_0$ on $\hat{\mathfrak{h}}_\sigma$ (which is trivial), hence, $(\hat{\mathfrak{h}}_\sigma, G)$ is a *Heisenberg pair* in the sense of [8]. The same is true for $(\hat{\mathfrak{h}}_\sigma, G_\sigma)$ because N_σ acts trivially on $\hat{\mathfrak{h}}_\sigma$. A module M over $\hat{\mathfrak{h}}_\sigma$ or over $(\hat{\mathfrak{h}}_\sigma, G_\sigma)$ will be called *restricted* if the action of $\mathfrak{h}_0$ is diagonalizable and for any $v \in M$, $(ht^m)v = 0$ for $h \in \mathfrak{h}$ and sufficiently large $m \in \frac{1}{N}\mathbb{Z}$.

Now we can summarize the results of Section 4.2 as follows.

Proposition 4.2. *Any σ-twisted V_Q-module M is naturally a restricted module over the Heisenberg pair $(\hat{\mathfrak{h}}_\sigma, G_\sigma)$ of level 1 (i.e., both $K \in \hat{\mathfrak{h}}_\sigma$ and $1 \in G_\sigma$ act as 1). Conversely, any restricted $(\hat{\mathfrak{h}}_\sigma, G_\sigma)$-module of level 1 can be endowed with the structure of a σ-twisted V_Q-module. This establishes an equivalence of the corresponding abelian categories.*

Proof. 1. Let M be a σ-twisted V_Q-module. By definition, the action of L_0^M is diagonalizable with finite-dimensional eigenspaces. Since $\mathfrak{h}_0$ commutes with L_0^M, its action is diagonalizable too. By (4.23), the modes $h_{(m)}^M$ ($h \in \mathfrak{h}$, $m \in \frac{1}{N}\mathbb{Z}$) provide a restricted representation of $\hat{\mathfrak{h}}_\sigma$ of level 1. The action of $U_\alpha \in G_\sigma$ is given by the operator U_α^M, see (4.29), (4.30), (4.38).

2. Conversely, let M be a restricted $(\hat{\mathfrak{h}}_\sigma, G_\sigma)$-module of level 1. Denote the image of ht^m in End M by $h_{(m)}^M$, and that of U_α by U_α^M. This allows us to define the fields $Y^M(ht^{-1}, z)$ and then $Y^M(e^\alpha, z)$ by (4.26) ($h \in \mathfrak{h}, \alpha \in Q$), They satisfy (3.4), (3.12); in particular, they are local. The map Y^M can be extended uniquely to the whole V_Q by applying (3.19) repeatedly for $a \in \mathfrak{h}t^{-1}$. Note that (3.19) implies the translation invariance (3.14) for $a \in \mathfrak{h}t^{-1}$. Then the proof of Lemma 4.1 shows that (3.14) holds for $a = e^\alpha$, and hence for any $a \in V_Q$. It follows from Remark 4.2 that Lemmas 4.2 and 4.3 hold. This implies the associativity (3.18). By Proposition 3.1, M is a σ-twisted V_Q-module. □

4.4. *The Groups $G^\perp$ and $G_\sigma^\perp$*

Before we proceed to the classification of all restricted $(\hat{\mathfrak{h}}_\sigma, G_\sigma)$-modules of level 1, we need to study the groups G and G_σ in more detail.

Let $G^\perp \subset G$ be the subgroup of G consisting of all cU_α with $c \in \mathbb{C}^\times$, $\alpha \in Q \cap \mathfrak{h}_0^\perp$. Clearly, the centralizer of $\hat{\mathfrak{h}}_\sigma$ in G equals $\exp \mathfrak{h}_0 \times G^\perp$ (cf. (4.49)). In other words, $G^\perp$ is the outer centralizer of the torus $\exp \mathfrak{h}_0$ in G.

Denote by $G_\sigma^\perp$ the image of $G^\perp$ in G_σ. It can be described as the factor of $G^\perp$ over the central subgroup (cf. (4.46), (4.27))

$$N_\sigma^\perp := N_\sigma \cap G^\perp = \{\eta(\alpha)(-1)^{|\alpha|^2} U_{\sigma\alpha}^{-1} U_\alpha \mid \alpha \in Q \cap \mathfrak{h}_0^\perp\}. \tag{4.50}$$

The centralizer of $\hat{\mathfrak{h}}_\sigma$ (and of T_σ) in G_σ is equal to $T_\sigma \times G_\sigma^\perp$. Notice that $G_\sigma^\perp$ is a central extension (by $\mathbb{C}^\times$) of the finite abelian group $(Q \cap \mathfrak{h}_0^\perp)/(1-\sigma)(Q \cap \mathfrak{h}_0^\perp)$.

Definition 4.1. (i) Let P_σ be the set of all λ that appear in Lemma 4.4, i.e., the set of all $\lambda \in (Q_{\text{ev}})^*$ such that $(1-\sigma)\lambda \in Q$ and $\lambda \in Q^*$ if $(1-\sigma)\lambda \in Q_{\text{ev}}$. Note that P_σ is a sublattice of $(Q_{\text{ev}})^*$ containing Q.

(ii) Let $Q_\sigma = (1-\sigma)P_\sigma \subset Q$.

(iii) Let $Z_\sigma = P_\sigma/Q$ be the subgroup of $((Q_{\text{ev}})^*/Q)^\sigma$ consisting of classes $\lambda + Q$ such that $\lambda \in Q^*$ if $(1-\sigma)\lambda \in Q_{\text{ev}}$. In particular, when the lattice Q is even, $Z_\sigma = (Q^*/Q)^\sigma$ is the group of σ-invariant elements in Q^*/Q.

Similarly to Lemma 4.4, we can describe the centers of G_σ, $G^\perp$ and $G_\sigma^\perp$.

Lemma 4.5. (i) $Z(G_\sigma) \simeq Z(G)/N_\sigma \simeq \mathbb{C}^\times \times Z_\sigma$.

(ii) $Z(G^\perp) = \{cU_\alpha \mid c \in \mathbb{C}^\times, \alpha \in Q_\sigma\} \simeq \mathbb{C}^\times \times Q_\sigma$.

(iii) $Z(G_\sigma^\perp) \simeq Z(G^\perp)/N_\sigma^\perp \simeq \mathbb{C}^\times \times Q_\sigma/(1-\sigma)(Q \cap \mathfrak{h}_0^\perp)$.

Proof. (i) follows from Lemma 4.4, (4.44) and the fact that $N_\sigma \cap \mathbb{C}^\times = \{1\}$.

(ii) A similar argument as in the proof of Lemma 4.4 shows that the center of $G^\perp$ consists of all elements of the form $cU_{(1-\sigma)\lambda}$, where $c \in \mathbb{C}^\times$ and $\lambda \in (Q_{\text{ev}} \cap \mathfrak{h}_0^\perp)^*$ is such that $\alpha := (1-\sigma)\lambda \in Q$ and $\lambda \in (Q \cap \mathfrak{h}_0^\perp)^*$ if $\alpha \in Q_{\text{ev}}$. Next, we use the following lemma.

Lemma 4.6. *Let $\mathfrak{h}'$ be a subspace of $\mathfrak{h}$ such that the restriction of the bilinear form on it is nondegenerate. Denote by π' the orthogonal projection of $\mathfrak{h}$ onto $\mathfrak{h}'$. Then for any lattice $L \subset \mathfrak{h}$, one has $L^* \cap \mathfrak{h}' = (\pi' L)^{*'}$, where $*'$ means that the dual is taken in $\mathfrak{h}'$.*

Proof. Follows from the fact that $(h|\alpha) = (h|\pi'\alpha)$ for $h \in \mathfrak{h}'$, $\alpha \in L$. $\square$

Now, by Lemma 4.6, $(Q \cap \mathfrak{h}_0^\perp)^* = \pi_\perp(Q^*)$, and similarly for $Q_{\rm ev}$, where $\pi_\perp$ is the orthogonal projection from $\mathfrak{h}$ to $\mathfrak{h}_0^\perp$. Noting that $(1-\sigma)\pi_\perp = 1-\sigma$ completes the proof of part (ii).

Part (iii) follows from part (ii), (4.44) and the fact that $N_\sigma^\perp \cap \mathbb{C}^\times = \{1\}$.

Corollary 4.2. *There is a natural exact sequence*

$$1 \to \exp 2\pi \mathrm{i}(Q^* \cap \mathfrak{h}_0) \to Z(G) \xrightarrow{p} Z(G^\perp) \to 1\,, \tag{4.51}$$

where the homomorphism p is given by

$$c\, e^{2\pi \mathrm{i}\lambda_0} U_{(1-\sigma)\lambda} \overset{p}{\mapsto} c\, U_{(1-\sigma)\lambda}\,, \qquad c \in \mathbb{C}^\times\,,\ \lambda \in P_\sigma\,. \tag{4.52}$$

The sequence (4.51) *splits, so we have a non-canonical isomorphism* $Z(G) \simeq \exp 2\pi \mathrm{i}(Q^* \cap \mathfrak{h}_0) \times Z(G^\perp)$.

Proof. Clearly, the kernel of p consists of $e^{2\pi \mathrm{i}\lambda_0}$ with $\lambda \in Q^*$ such that $(1-\sigma)\lambda = 0$. Let $s\colon \pi_\perp(P_\sigma) \to P_\sigma$ be a linear section of the projection $\pi_\perp$. For $\lambda' \in \pi_\perp(P_\sigma)$, let $\lambda = s(\lambda') \in P_\sigma$. Then $\lambda' = \pi_\perp(\lambda)$ and $(1-\sigma)\lambda = (1-\sigma)\lambda'$. The map $c\,U_{(1-\sigma)\lambda'} \mapsto c\, e^{2\pi \mathrm{i}\lambda_0} U_{(1-\sigma)\lambda}$ is a splitting of (4.51). $\square$

Since $G_\sigma^\perp$ is a central extension of a finite abelian group, its representations are completely reducible and the irreducible ones are classified by the characters of $Z(G_\sigma^\perp)$. We will consider only representations on which $1 \in G_\sigma^\perp$ acts as the identity operator. The irreducible ones are classified by the finite abelian group $Q_\sigma/(1-\sigma)(Q \cap \mathfrak{h}_0^\perp)$. All of them have the same dimension $d(\sigma)$, which satisfies

$$d(\sigma)^2 = |G_\sigma^\perp / Z(G_\sigma^\perp)| = |(Q \cap \mathfrak{h}_0^\perp)/Q_\sigma|. \tag{4.53}$$

Definition 4.2. The non-negative integer $d(\sigma)$ is called the *defect* of σ (cf. [13]).

4.5. *Representations of* $(\hat{\mathfrak{h}}_\sigma, G_\sigma)$

In this subsection we show that the category of all restricted $(\hat{\mathfrak{h}}_\sigma, G_\sigma)$-modules of level 1 is semisimple, and we classify the irreducible ones.

Let $\hat{\mathfrak{h}}_\sigma^-$ (resp. $\hat{\mathfrak{h}}_\sigma^+$) be the subalgebra of $\hat{\mathfrak{h}}_\sigma$ consisting of all elements ht^m with $m > 0$ (resp. $m < 0$). It is well known (see, e.g., [11]) that any restricted $\hat{\mathfrak{h}}_\sigma$-module M of level 1 is induced from its vacuum subspace

$$\Omega_M = \{v \in M \mid \hat{\mathfrak{h}}_\sigma^- v = 0\}. \tag{4.54}$$

More precisely,

$$M \simeq \operatorname{Ind}_{\mathfrak{h}_0 \oplus \hat{\mathfrak{h}}_\sigma^-}^{\hat{\mathfrak{h}}_\sigma} \Omega_M \simeq S(\hat{\mathfrak{h}}_\sigma^+) \otimes \Omega_M. \tag{4.55}$$

The subalgebra $\mathfrak{h}_0$ acts on Ω_M diagonally, and the $\hat{\mathfrak{h}}_\sigma$-module M is completely reducible (it is irreducible iff $\dim \Omega_M = 1$).

Now assume that M is a restricted $(\hat{\mathfrak{h}}_\sigma, G_\sigma)$-module of level 1. It follows from (4.49) that Ω_M is a G_σ-module. By definition, $1 \in G_\sigma$ acts as 1 and the torus T_σ acts diagonally (cf. (4.48)). We will call such G_σ-modules *restricted*.

If Ω is a restricted G_σ-module, it has a compatible $\mathfrak{h}_0$-action, because $\mathfrak{h}_0$ is the Lie algebra of the torus T_σ. We let $\hat{\mathfrak{h}}_\sigma^-$ act trivially on Ω and form the induced $\hat{\mathfrak{h}}_\sigma$-module $M(\Omega) = \operatorname{Ind}_{\mathfrak{h}_0 \oplus \hat{\mathfrak{h}}_\sigma^-}^{\hat{\mathfrak{h}}_\sigma} \Omega$. Using (4.49), we can extend the action of G_σ from Ω to $M(\Omega)$. Then $M(\Omega)$ becomes a restricted $(\hat{\mathfrak{h}}_\sigma, G_\sigma)$-module of level 1.

Proposition 4.3. *The functors $M \mapsto \Omega_M$ and $\Omega \mapsto M(\Omega)$ establish an equivalence of abelian categories between the category of restricted $(\hat{\mathfrak{h}}_\sigma, G_\sigma)$-modules of level* 1 *and the category of restricted G_σ-modules.*

Therefore we are left with describing restricted G_σ-modules.

Proposition 4.4. *Any restricted G_σ-module is completely reducible and is determined by the action of the center of G_σ. Isomorphism classes of restricted irreducible G_σ-modules are parameterized by the (finite) set Z_σ.*

Let Ω be a restricted G_σ-module. For $\mu \in \mathfrak{h}_0$, we denote by Ω_μ the weight μ subspace of Ω :

$$\Omega_\mu := \{v \in \Omega \mid e^h v = e^{(h|\mu)} v \text{ for } h \in \mathfrak{h}_0\}. \tag{4.56}$$

Then $(\mu|\alpha) \in \mathbb{Z}$ for $\alpha \in Q \cap \mathfrak{h}_0$, i.e., $\mu \in (Q \cap \mathfrak{h}_0)^* = \pi_0(Q^*)$ by Lemma 4.6.

Lemma 4.7. (i) *$U_\alpha \Omega_\mu = \Omega_{\mu + \pi_0 \alpha}$ for all $\alpha \in Q$, $\mu \in \pi_0(Q^*)$. In particular, the subgroup $G_\sigma^\perp \subset G_\sigma$ preserves each Ω_μ.*

(ii) *Let $\Omega_{\mu+\pi_0(Q)} = \sum_{\alpha \in Q} \Omega_{\mu+\pi_0\alpha} \subset \Omega$. Then $\Omega_{\mu+\pi_0(Q)}$ is a G_σ-submodule of Ω.*

(iii) *The G_σ-module $\Omega_{\mu+\pi_0(Q)}$ is irreducible if and only if the $G_\sigma^\perp$-module Ω_μ is irreducible.*

Proof. (i) It follows from (4.41) that $U_\alpha \Omega_\mu \subset \Omega_{\mu+\pi_0\alpha}$ for any $\alpha \in Q$. Since by (4.42) U_α^{-1} is proportional to $U_{-\alpha}$, we get $U_\alpha \Omega_\mu = \Omega_{\mu+\pi_0\alpha}$.

(ii) follows from (i) and the definition of G_σ (see (4.40)–(4.42)).

(iii) First, let Ω_μ be an irreducible $G_\sigma^\perp$-module. Assume that Λ is a nontrivial G_σ-submodule of $\Omega_{\mu+\pi_0(Q)}$. Using the action of T_σ, we can write $\Lambda = \sum_{\alpha\in Q} \Lambda_{\mu+\pi_0\alpha}$ where each $\Lambda_{\mu+\pi_0\alpha} \subset \Omega_{\mu+\pi_0\alpha}$. Moreover, $\Lambda_{\mu+\pi_0\alpha} = U_\alpha \Lambda_\mu$. In particular, Λ_μ is a $G_\sigma^\perp$-submodule of Ω_μ. But Ω_μ is irreducible; hence, $\Lambda_\mu = \Omega_\mu$ and $\Lambda = \Omega_{\mu+\pi_0(Q)}$.

Conversely, assume that the $G_\sigma^\perp$-module Ω_μ is not irreducible. Since $G_\sigma^\perp$ is a central extension of a finite abelian group, its representations are completely reducible. If $\Omega_\mu = \bigoplus_i L_i$ as a $G_\sigma^\perp$-module, let $L^i = \sum_{\alpha\in Q} U_\alpha L_i$. Then each L^i is a G_σ-submodule of $\Omega_{\mu+\pi_0(Q)}$ and $\Omega_{\mu+\pi_0(Q)} = \bigoplus_i L^i$ as a G_σ-module. □

Remark 4.3. $\Omega_{\mu+\pi_0(Q)}$ is isomorphic to the induced module $\operatorname{Ind}_{T_\sigma\times G_\sigma^\perp}^{G_\sigma} \Omega_\mu$.

From Lemma 4.7(iii) and its proof, we see that the G_σ-module $\Omega_{\mu+\pi_0(Q)}$ is completely reducible. Since $\Omega = \bigoplus_{[\mu]\in\pi_0(Q^*)/\pi_0(Q)} \Omega_{[\mu]}$, it is also completely reducible.

Now let Ω be an irreducible G_σ-module. Then $\Omega = \Omega_{\mu+\pi_0(Q)}$ for some $\mu \in \pi_0(Q^*)$ and the $G_\sigma^\perp$-module Ω_μ is irreducible. Any irreducible $G_\sigma^\perp$-module is completely determined by the action of the center $Z(G_\sigma^\perp)$. Let $\zeta\colon Z(G_\sigma^\perp) \to \mathbb{C}^\times$ be the central character of Ω_μ.

We can view Ω as a G-module on which N_σ acts trivially, and similarly Ω_μ as a $G^\perp$-module with a trivial action of $N_\sigma^\perp$. Recall that, by Lemma 4.5(iii), $Z(G_\sigma^\perp) \simeq Z(G^\perp)/N_\sigma^\perp$, so we can extend ζ to a character of $Z(G^\perp)$.

If $\mu' = \mu + \pi_0\alpha$ for some $\alpha \in Q$, then $\Omega_{\mu'} = U_\alpha\Omega_\mu$ and $\Omega_{\mu+\pi_0(Q)} = \Omega_{\mu'+\pi_0(Q)}$. For $v \in \Omega_\mu$, $U_\beta \in Z(G^\perp)$, $\beta \in Q_\sigma$ (see Lemma 4.5(ii)), we have: $U_\beta U_\alpha v = C_{\alpha,\beta}^{-1} U_\alpha U_\beta v = C_{\alpha,\beta}^{-1}\zeta(U_\beta) U_\alpha v$ where $C_{\alpha,\beta}$ is given by (4.43). Hence, two pairs (μ,ζ) and (μ',ζ') correspond to the same irreducible G_σ-module if and only if they are related by:

$$\mu' = \mu + \pi_0\alpha\,, \quad \zeta'(U_\beta) = C_{\alpha,\beta}^{-1}\zeta(U_\beta)\,, \qquad \alpha\in Q\,,\ \beta\in Q_\sigma\,. \tag{4.57}$$

For $\lambda \in P_\sigma$ the element $e^{2\pi i\lambda_0} U_{(1-\sigma)\lambda} \in Z(G)$ act on Ω_μ as the scalar $e^{2\pi i(\lambda_0|\mu)}\zeta(U_{(1-\sigma)\lambda})$ (cf. Lemmas 4.4 and 4.5(ii)). Using Corollary 4.2 and Lemma 4.6, it is easy to see that the action of $Z(G)$ on Ω determines uniquely the equivalence class of (μ,ζ) under (4.57), and hence it determines the isomorphism class of the G_σ-module Ω. Conversely, different pairs (μ,ζ) give rise to different actions of $Z(G)$ on the corresponding modules $\Omega_{\mu+\pi_0(Q)}$.

This completes the proof of Proposition 4.4.

4.6. *Classification of σ-Twisted V_Q-Modules*

Combining Propositions 4.2, 4.3 and 4.4, we obtain the main result of the paper.

Theorem 4.2. *The category of σ-twisted V_Q-modules is a semisimple abelian category with finitely many isomorphism classes of simple objects, parameterized by the set Z_σ.*

Remark 4.4. The irreducible σ-twisted V_Q-module corresponding to $\lambda + Q \in Z_\sigma$ is isomorphic as an $\hat{\mathfrak{h}}_\sigma$-module to $S(\hat{\mathfrak{h}}_\sigma^+) \otimes e^{\lambda_0}\mathbb{C}[\pi_0 Q] \otimes \mathbb{C}^{d(\sigma)}$, where $\mathbb{C}$ carries the zero action and $d(\sigma)$ is the defect of σ.

Acknowledgments

We are grateful to Ivan Todorov for many stimulating discussions and for collaboration on [1]. We thank the organizers of the Varna Workshop for inviting us to present our results and for the inspiring workshop. We also acknowledge the hospitality of the Erwin Schrödinger Institute, where some of this work was done. The first author was supported in part by the Miller Institute for Basic Research in Science. The second author was supported in part by NSF grant DMS-9970007.

References

1. B. Bakalov, V.G. Kac and I.T. Todorov, "Orbifolds of lattice vertex algebras", in preparation.
2. A.A. Belavin, A.M. Polyakov and A.B. Zamolodchikov, "Infinite conformal symmetry in two-dimensional quantum field theory", *Nucl. Phys.* **B241**, 333-380 (1984).
3. R.E. Borcherds, "Vertex algebras, Kac-Moody algebras, and the Monster", *Proc. Nat. Acad. Sci. U.S.A.* **83**, no. 10, 3068-3071 (1986).
4. A. Cappelli, L.S. Georgiev and I.T. Todorov, " A unified conformal field theory description of paired quantum Hall states", *Commun. Math. Phys.* **205**, 657-689 (1999).
5. R. Dijkgraaf, C. Vafa, E. Verlinde and H. Verlinde, "The operator algebra of orbifold models", *Commun. Math. Phys.* **123**, 485-526 (1989).
6. L. Dixon, J.A. Harvey, C. Vafa and E. Witten, "String on orbifolds", *Nucl. Phys.* **B261**, 620-678 (1985); "String on orbifolds. II", *Nucl. Phys.* **B274**, 285-314 (1986).
7. C. Dong and J. Lepowsky, *Generalized vertex algebras and relative vertex operators*, Progress in Math., Vol. 112 (Birkhäuser, Boston, 1993).
8. I.B. Frenkel and V.G. Kac, "Basic representations of affine Lie algebras and dual resonance models", *Invent. Math.* **62**, no. 1, 23-66 (1980/81).

9. I.B. Frenkel, J. Lepowsky and A. Meurman, *Vertex operator algebras and the Monster*, Pure and Appl. Math., Vol. 134 (Academic Press, Boston, 1988).
10. P. Goddard, "Meromorphic conformal field theory", in: *Infinite-dimensional Lie algebras and groups* (Luminy-Marseille, 1988) Adv. Ser. Math. Phys., Vol. 7 (World Sci, Teaneck, NJ, 1989) pp. 556-587.
11. V.G. Kac, *Infinite-dimensional Lie algebras*, 3rd edition, (Cambridge University Press, Cambridge, 1990).
12. V.G. Kac, *Vertex algebras for beginners*, University Lecture Series, Vol. 10 (American Mathematical Society, Providence, RI, 1996, 2nd Edition 1998).
13. V.G. Kac and D.H. Peterson, "112 constructions of the basic representation of the loop group of E_8", in: *Symposium on anomalies, geometry, topology* (Chicago, Ill., 1985), (World Sci, Singapore, 1985) pp. 276-298.
14. V.G. Kac and I.T. Todorov, "Affine orbifolds and rational conformal field theory extensions of $W_{1+\infty}$", *Commun. Math. Phys.* **190**, 57-111 (1997).
15. M. Roitman, "On twisted representations of vertex algebras", *Adv. Math.* **176**, 53-88 (2003).

STRUCTURE THEORY OF FINITE LIE CONFORMAL SUPERALGEBRAS

DAVIDE FATTORI

Dipartimento di Matematica Pura e Applicata, Universitá di Padova, via Belzoni 7, 35131 Padova, Italy

VICTOR G. KAC

Department of Mathematics, MIT, Cambridge MA 02139, USA

ALEXANDER RETAKH

Department of Mathematics, MIT, Cambridge MA 02139, USA

1. Introduction

Lie conformal superalgebras encode the singular part of the operator product expansion of chiral fields in two-dimensional quantum field theory [7]. On the other hand, they are closely connected to the notion of a formal distribution Lie superalgebra $(\mathfrak{g}, \mathcal{F})$, i.e. a Lie superalgebra $\mathfrak{g}$ spanned by the coefficients of a family $\mathcal{F}$ of mutually local formal distributions. Namely, to a Lie conformal superalgebra R one can associate a formal distribution Lie superalgebra $(\operatorname{Lie} R, R)$ which establishes an equivalence between the category of Lie conformal superalgebras and the category of equivalence classes of formal distribution Lie superalgebras obtained as quotients of $\operatorname{Lie} R$ by irregular ideals [7].

The classification of finite simple Lie conformal superalgebras was completed in [6]. The proof relies on the methods developed in [5] for the classification of finite simple and semisimple Lie conformal algebras, and the classification of simple linearly compact Lie superalgebras [10].

The main result of the present paper is the classification of finite semisimple Lie conformal superalgebras (Theorem 6.1). Unlike in [5], we do not use the connection to formal distribution algebras in the proof of this theorem (which would require us to take care of numerous technical difficulties). We work instead entirely in the category of Lie conformal superalgebras. Our key result is the determination of finite differentiably simple Lie conformal superalgebras (Theorem 3.1). The proof of this result uses heavily the ideas of [1] and [3].

Given a finite Lie conformal superalgebra R, denote by $\operatorname{Rad} R$ the sum of all solvable ideals of R. Since the rank of a finite solvable Lie conformal (super)algebra is greater than the rank of its derived subalgebra [5], we conclude that $\operatorname{Rad} R$ is the maximal solvable ideal, hence $R/\operatorname{Rad} R$ is a finite semisimple Lie conformal superalgebra. Thus in some sense the study of general finite Lie conformal superalgebras reduces to that of semisimple and solvable superalgebras.

In the Lie conformal algebra case there is a conformal analog of Lie's theorem, stating that any non-trivial finite irreducible module over a finite solvable Lie conformal algebra is free of rank 1 [5]. However, a similar result in the "super" case is certainly false. We hope that we can develop a theory of finite irreducible modules over solvable Lie conformal superalgebras similar to that in the Lie superalgebra case [8]. We make several observations to that end in the last section of this paper.

The paper is organized as follows: we define the main objects of our study and provide some general statements in Section 2. In Section 3 we establish the structure of differentiably simple Lie conformal superalgebras. Our proof is, in fact, quite general and the result is valid for non-Lie finite conformal superalgebras as well. We list finite simple Lie conformal superalgebras in Section 4 and describe their conformal derivations. Then in Section 5 we describe conformal derivations in the differentiably simple case and thus complete the classification of finite differentiably simple Lie conformal superalgebras. This allows us to describe the structure of finite semisimple Lie conformal superalgebras in Section 6. In Section 7 we classify simple physical Virasoro pairs and, as a consequence, obtain a classification of physical Lie conformal superalgebras which generalizes that of [8]. Finally, in Section 8 we initiate the study of representations of finite solvable Lie conformal superalgebras.

2. Basic definitions and structures

2.1. *Formal distributions and conformal algebras*

Let $\mathfrak{g}$ be a Lie superalgebra. A $\mathfrak{g}$-valued *formal distribution* in one indeterminate z is a formal power series

$$a(z) = \sum_{n \in \mathbb{Z}} a_n z^{-n-1}, \quad a_n \in \mathfrak{g}.$$

The vector superspace of all formal distributions, $\mathfrak{g}[[z, z^{-1}]]$, has a natural structure of a $\mathbb{C}[\partial_z]$-module.

Let $\mathfrak{g}$ be a Lie superalgebra, and let $a(z), b(z)$ be two $\mathfrak{g}$-valued formal distributions. They are called *local* if $(z-w)^N[a(z), b(w)] = 0$ for $N \gg 0$.

Let $\mathfrak{g}$ be a Lie superalgebra, and let $\mathcal{F}$ be a family of $\mathfrak{g}$-valued mutually local formal distributions. The pair $(\mathfrak{g}, \mathcal{F})$ is called a *formal distribution Lie superalgebra* if $\mathfrak{g}$ is spanned by the coefficients of all formal distributions from $\mathcal{F}$.

The bracket of two local formal distributions is given by the formula

$$[a(z), b(w)] = \sum_j [a(w)_{(j)} b(w)] \partial_w^j \delta(z-w)/j!,$$

where $[a(w)_{(j)} b(w)] = \mathrm{Res}_z (z-w)^j [a(z), b(w)]$. Thus we get a family of operations $_{(n)}$, $n \in \mathbb{Z}_+$, on the space of formal distributions: $[a(z)_{(n)} b(z)]$ We define the λ-*bracket* on the space of formal distributions as the generating series of these operations [5, 7]:

$$[a_\lambda b] = \sum_{n \in \mathbb{Z}_+} \frac{\lambda^n}{n!} (a_{(n)} b).$$

The properties of the λ-bracket lead to the following basic definition (see [5, 7]):

A *Lie conformal superalgebra* R is a left $\mathbb{Z}/2\mathbb{Z}$-graded $\mathbb{C}[\partial]$-module endowed with a $\mathbb{C}$-linear map, called the λ-*bracket*,

$$R \otimes R \to \mathbb{C}[\lambda] \otimes R, \qquad a \otimes b \mapsto [a_\lambda b]$$

satisfying the following axioms ($a, b, c \in R$):

(sesquilinearity) $\quad [\partial a_\lambda b] = -\lambda [a_\lambda b], \;\; [a_\lambda \partial b] = (\partial + \lambda)[a_\lambda b]$,

(skew-commutativity) $\quad [b_\lambda a] = -(-1)^{p(a)p(b)} [a_{-\lambda-\partial} b]$,

(Jacobi identity) $\quad [a_\lambda [b_\mu c]] = [[a_\lambda b]_{\lambda+\mu} c] + (-1)^{p(a)p(b)} [b_\mu [a_\lambda c]]$.

(Here and further $p(a) \in \mathbb{Z}/2\mathbb{Z}$ stands for the parity of an element a.)

Similarly, one can define an *associative conformal superalgebra* by replacing the skew-commutativity and the Jacobi identity above with the following property:

$$\text{(associativity)} \qquad a_\lambda (b_\mu c) = (a_\lambda b)_{\lambda+\mu} c.$$

As a shorthand, we call R *finite* if R is finitely generated as a $\mathbb{C}[\partial]$-module.

Below follow two useful constructions of Lie conformal superalgebras (more concrete examples will be discussed in Section 4):

Example 2.1. An associative conformal superalgebra can be endowed with a λ-bracket by putting $[a_\lambda b] = a_\lambda b - (-1)^{p(a)p(b)} b_{-\lambda-\partial} a$.

Example 2.2. Let R be a Lie conformal superalgebra and let B be a commutative associative (ordinary) superalgebra. Then $R \otimes B$ carries a Lie conformal superalgebra structure defined as follows. The $\mathbb{C}[\partial]$-module structure is given by $\partial(r \otimes b) = (\partial r) \otimes b$ ($r \in R$, $b \in B$), and the λ-bracket by

$$[(r \otimes b)_\lambda (r' \otimes b')] = (-1)^{p(b)p(r')} [r_\lambda r'] \otimes (bb').$$

Notice that if R is finite and B is finite-dimensional, then $R \otimes B$ is also finite.

2.2. *Structural terminology*

We will denote by $\langle X_\lambda Y \rangle$ (or $\langle [X_\lambda Y] \rangle$, whenever appropriate) the $\mathbb{C}[\partial]$-module generated by elements of the form $x_{(n)} y$, where $x \in X, y \in Y$, and $n \in \mathbb{Z}_+$.

An *ideal* of a Lie conformal superalgebra R is a $\mathbb{C}[\partial]$-submodule I of R such that $\langle [R_\lambda I] \rangle \subset I$. An ideal I is *abelian* if $[I_\lambda I] = 0$, *central* if $[R_\lambda I] = 0$, and *nilpotent* if $\langle [\ldots \langle [I_{\lambda_1} I] \rangle_{\lambda_2} \cdots_{\lambda_k} I] \rangle = 0$ (in this case we sometimes write $I^{k+1} = 0$).

A Lie conformal superalgebra is *simple* if it is non-abelian and contains no ideals except for zero and itself.

The derived series of a Lie conformal superalgebra is built in the usual fashion: let $R' = \langle [R_\lambda R] \rangle$ and set $R^{(0)} = R$, $R^{(n+1)} = (R^{(n)})'$. Then a Lie conformal superalgebra R is *solvable* if $R^{(n)} = 0$ for some n. $\operatorname{Rad} R$ is the maximal solvable ideal of R (its existence is explained in the introduction).

A Lie conformal superalgebra is *semisimple* if it has no non-zero abelian ideals. Since the second last term of the derived series is an abelian ideal, this is equivalent to saying that R has no non-zero solvable ideals. Since the $\mathbb{C}[\partial]$-torsion is central [5], a finite semisimple Lie conformal superalgebra is free as a $\mathbb{C}[\partial]$-module.

2.3. *Conformal modules*

Let V, W be $\mathbb{Z}_2$-graded left $\mathbb{C}[\partial]$-modules. We denote by $\operatorname{End}_{\mathbb{C}[\partial]} V$ the set of all $\mathbb{C}[\partial]$-linear endomorphisms of V. Notice that $\operatorname{End}_{\mathbb{C}[\partial]} V$ has a $\mathbb{C}[\partial]$-module structure given by

$$(\partial\varphi)(v) = \partial\varphi(v) \quad \text{for any} \quad v \in V.$$

A $\mathbb{C}$-linear map $\varphi : V \to \mathbb{C}[\lambda] \otimes_{\mathbb{C}} W$ is called a *conformal linear map* if the following equation holds:

$$\varphi_\lambda(\partial v) = (\partial + \lambda)\varphi_\lambda v \quad \text{for any } v \in V.$$

The $\mathbb{C}$-vector space of all conformal linear maps from V to W is denoted by $\mathrm{Chom}(V, W)$. It has $\mathbb{C}[\partial]$-module structure if we set

$$(\partial\varphi)_\lambda(v) = -\lambda\varphi_\lambda v.$$

When $V = W$, we denote $\mathrm{Chom}(V, V)$ by $\mathrm{Cend}\, V$. When V is finite over $\mathbb{C}[\partial]$, $\mathrm{Cend}\, V$ becomes an associative conformal superalgebra with the λ-product

$$(a_\lambda b)_\mu v = a_\lambda(b_{\mu-\lambda} v), \quad a, b \in \mathrm{Cend}\, V.$$

The space $\mathrm{Cend}\, V$ endowed with a λ-bracket (see Remark 2.1) is denoted $\mathrm{gc}\, V$.

A *module* M over a Lie conformal superalgebra R is a $\mathbb{Z}_2$-graded left $\mathbb{C}[\partial]$-module M endowed with a $\mathbb{C}$-linear map $R \to \mathrm{gc}\, M$.

Alternatively, one can define a module over R by providing a map $R \to \mathbb{C}[\lambda] \otimes \mathrm{End}_{\mathbb{C}} M$, $a \mapsto a_\lambda^M$ such that

$$(\partial a)_\lambda^M m = [\partial, a_\lambda^M]m = -\lambda a_\lambda^M m,$$
$$[a_\lambda^M, b_\mu^M]v = [a_\lambda b]_{\lambda+\mu}^M v.$$

An R-module M is *simple* if it has no nontrivial R-invariant $\mathbb{C}[\partial]$-submodules.

An *endomorphism* of an R-module M is a $\mathbb{C}[\partial]$-linear map $\phi \in \mathrm{End}_{\mathbb{C}[\partial]} M$ such that for any $a \in R$ and $v \in M$ we have

$$\phi(a_\mu v) = (-1)^{p(a)p(\phi)} a_\mu \phi(v).$$

2.4. *Derivations*

A *conformal derivation* of a Lie conformal superalgebra R is a conformal endomorphism ϕ of R such that for any homogeneous $x, y \in R$

$$\phi_\lambda[x_\mu y] = [(\phi_\lambda x)_{\lambda+\mu} y] + (-1)^{p(x)p(\phi)}[x_\mu(\phi_\lambda y)].$$

We denote by $\mathrm{Cder}\, R$ (resp. $\mathrm{Cinder}\, R$) the space of all conformal derivations of R (resp. of all *inner conformal derivations*, i.e. conformal derivations of the form $\mathrm{ad}\, a$, $a \in R$, $(\mathrm{ad}\, a)_\lambda b = [a_\lambda b]$ for $b \in R$). Clearly $\mathrm{Cder}\, R$ is a subalgebra of $\mathrm{gc}\, R$.

Let D be a set of conformal derivations of R. An ideal I of R is *D-stable* if $\phi I \subseteq I$ for all $\phi \in D$. Here (and below) we use the shorthand ϕI for $\langle \phi_\lambda I \rangle$.

A Lie conformal superalgebra is *D-differentiably simple* if it contains no proper D-stable ideals. A Lie conformal superalgebra is *differentiably simple* if it is D-differentiably simple with respect to some D.

We will also work with ordinary derivations of conformal algebras. An *ordinary derivation* of a Lie conformal superalgebra R is a $\mathbb{C}[\partial]$-linear endomorphism d of R such that for any homogeneous $x, y \in R$

$$d([x_\mu y]) = [d(x)_\mu y] + (-1)^{p(x)p(d)}[x_\mu d(y)].$$

We denote the space of all ordinary derivations of R by $\operatorname{Der} R$. Remark that a conformal derivation ϕ gives rise to an ordinary derivation $\phi_{(0)}$. In particular, every element $a \in R$ gives rise to an ordinary derivation $\operatorname{ad} a_{(0)}$. We call such derivations *inner* and denote their space as $\operatorname{Inder} R$.

Remark 2.1. The operator ∂ always acts as an ordinary derivation of a conformal algebra. In some cases it is inner (e.g. when the algebra possesses a Virasoro element, see Example 4.2).

2.5. *Minimal ideals*

Lemma 2.1. *Let R be a finite Lie conformal superalgebra and J an ideal of R which contains no nonzero central elements. Then J contains a minimal ideal of R.*

Proof. Let I be a minimal rank ideal of R contained in J. Let $I_0 = \bigcap_i K_i$ be the intersection of all non-zero ideals of R contained in I. Any ideal K_i has the same rank as I, hence I/K_i is a torsion $\mathbb{C}[\partial]$-module and by [5, Proposition 3.2], R acts trivially on it. This means that for any i we have $\langle [R_\lambda I] \rangle \subseteq K_i$. Therefore $\langle [R_\lambda I] \rangle \subseteq I_0$. Note that $\langle [R_\lambda I] \rangle \neq 0$ because otherwise I would be a central ideal of R. Hence $I_0 \neq 0$ and clearly I_0 is a minimal ideal of R. □

Lemma 2.2. *Let M be a nonabelian ideal in a finite Lie conformal superalgebra R. Then*

(1) M is a minimal ideal if and only if M is $\operatorname{ad} R$*-simple.*

(2) If M is a minimal ideal, then it is differentiably simple.

(3) If M is a minimal ideal, then it is $\operatorname{Cder} R$*-invariant and* $\operatorname{Cder} R$*-simple.*

Proof. (1) is immediate. (2) follows from (1) and the fact that $\operatorname{ad} R \subseteq \operatorname{Cder} M$. As for (3), we remark that the minimality of M implies that $\langle [M_\lambda M] \rangle = M$, hence $\langle \phi_\lambda(M) \rangle \subseteq \langle [(\phi_\mu M)_\lambda M] \rangle \subseteq M$ for any $\phi \in \operatorname{Cder} R$, i.e. M is a $\operatorname{Cder} R$-invariant ideal of R. Now, let J be a nonzero $\operatorname{Cder} R$-invariant ideal of R, which is contained in M. The minimality of M implies that $J = M$. □

Note that by Lemma 2.2(3), any minimal ideal in a differentiably simple but non-simple finite Lie conformal superalgebra is abelian.

3. Differentiably simple conformal superalgebras

In this section we prove the following

Theorem 3.1. *Let R be a finite non-abelian differentiably simple Lie conformal superalgebra. Then $R \simeq S \otimes \wedge(n)$ for a simple Lie conformal superalgebra S.*

Remark 3.1. Our proof also works for any finite non-abelian differentiably simple conformal superalgebra (i.e. the one for which only sesquilinearity holds) but we do not require the result in this generality.

3.1. *Centroid*

Let R be a Lie conformal superalgebra and M an R-module. For $x \in R$, let L_x be an element of $\operatorname{Cend} R$ such that $(L_x)_\lambda y = x_\lambda y$ for any $y \in M$.

Definition 3.1. The *centroid* $C(M)$ of a module M over a Lie conformal superalgebra R is the subalgebra of the associative superalgebra $\operatorname{End}_{\mathbb{C}} M$ that consists of elements (super)commuting with $(L_x)_{(n)}$ for all $x \in R$, $n \in \mathbb{Z}_+$ and the action of ∂.

Remark 3.2. By definition $C(M)$ is a subalgebra of the associative superalgebra $\operatorname{End}_{\mathbb{C}[\partial]} M$.

Remark 3.3. We show in the proof of Lemma 3.1 below that for $a \in C(R) \subset \operatorname{End}_{\mathbb{C}[\partial]} R$, $a(x_\lambda y) = (ax)_\lambda y = (-1)^{p(a)p(x)}(x_\lambda ay)$. These conditions can be taken as the definition of the centroid of a conformal superalgebra R. (Note that here R is not necessarily Lie.)

Lemma 3.1. *If $R = \langle [R_\lambda R] \rangle$, then $C(R)$ is (super)commutative.*

Proof. The lemma follows from the equalities $a[x_\lambda y] = (-1)^{p(a)p(x)}[x_\lambda ay]$ and $a[x_\lambda y] = [ax_\lambda y]$ for any $a \in C(R), x, y \in R$. The first equality follows directly from the definition; the second is deduced from the first:

$$\begin{aligned} a[x_\lambda y] = &- (-1)^{p(x)p(y)}a[y_{-\lambda-\partial}x] = -(-1)^{p(x)p(y)+p(a)p(y)}[y_{-\lambda-\partial}ax] \\ =&(-1)^{p(x)p(y)+p(a)p(y)+(p(a)+p(x))p(y)}[ax_\lambda y] = [ax_\lambda y]. \end{aligned}$$

Then for any $a, b \in C(R)$ and any $x, y \in R$, $(ab)[x_\lambda y] = (-1)^{p(b)p(x)}[ax_\lambda by]$ and $(ba)[x_\lambda y] = (-1)^{(p(a)+p(x))p(b)}[ax_\lambda by]$, implying that a and b (super)commute. □

Since $\langle [R_\lambda R] \rangle$ is a differentiably stable ideal, it follows that $C(R)$ is (super)commutative for a differentiably simple R. We obviously have

Lemma 3.2. *For a homogeneous (i.e. even or odd) $a \in C(R)$, $\ker a$ and $\operatorname{im} a$ are ideals of R.*

We also have a version of the Schur Lemma:

Lemma 3.3. *Let M be a countable dimensional simple module over a Lie conformal superalgebra R. Then either $C(M) = \mathbb{C}\, 1_M$ or $\dim M_{\bar{0}} = \dim M_{\bar{1}}$ and*

$$C(M) = \mathbb{C}\, 1_M \oplus \mathbb{C}\, U,$$

where U is an odd operator such that $U^2 = 1_M$.

Proof. The proof follows the classical line of argument. Let $a \in C(M)$ be a non-zero even operator. The fact that M is simple implies that a is invertible. Suppose a is not a scalar. Since $\mathbb{C}$ is algebraically closed, a cannot be an algebraic element in $C(M)$. The field of rational functions in a over $\mathbb{C}$ is contained in $C(M)$, hence $C(M)$ has dimension greater than countable. On the other hand, let us fix a nonzero $x \in M$. The map $C(M) \to M$ sending a to ax is injective because $C(M)$ is a division ring. However, M is countable dimensional. The contradiction proves that $a = c1_M$ for some $c \in \mathbb{C}$.

Let $a \in C(R)$ be a non-zero odd operator. Then b is invertible (this can only happen if $\dim M_{\bar{0}} = \dim M_{\bar{1}}$). Furthermore, a^2 is an even operator, hence a scalar. Suppose a_1, a_2 are two non-zero odd operators in $C(M)$. Then $(a_1 - ca_2)^2 \in \mathbb{C}1_M$ for any $c \in \mathbb{C}$. Assume that for c and c', $(a_1 - ca_2)^2 \neq 0$ and $(a_1 - c'a_2)^2 \neq 0$. Then $(a_1 - ca_2)^2 + r^2(a_1 - c'a_2)^2 = 0$ for some $0 \neq r \in \mathbb{C}$. We obtain two non-zero odd operators b_1 and b_2 such that $b_1^2 + b_2^2 = 0$. By taking the square of $b_1 b_2 b_1^{-1} b_2^{-1}$ one easily shows that $b_1 b_2 = \pm b_1 b_2$. Thus either $(b_1 + b_2)^2 = 0$ or $(b_1 + ib_2)(b_1 - ib_2) = 0$ and it follows that a_1 is proportional to a_2. □

Corollary 3.1. *If R is a countably dimensional simple Lie conformal superalgebra, then $C(R) = \mathbb{C}1_R$.*

Proof. Assume $C(R) = \mathbb{C}1_R \oplus \mathbb{C}U$, $U^2{=}1$. Let $a = (1+U)/2$ and $\bar{a} = (1-U)/2$. R splits as $aR \oplus \bar{a}R$ and it is easy to see that $\ker a = \bar{a}R$. Using the equality $b[x_\lambda y] = [bx_\lambda y]$, we see immediately that $\ker a$ and $\operatorname{im} a$ are (non-homogeneous) ideals of R. On the other hand, $[x_\lambda ay] = \bar{a}[x_\lambda y]$ for x odd, hence $aR \cap \bar{a}R \neq \{0\}$, a contradiction. □

Remark 3.4. Lemma 3.3 and Corollary 3.1 hold for ordinary Lie superalgebras (with the same proof).

3.2. *Conformal derivations and the centroid*

Let $\phi \in \operatorname{Cder} R$, $\gamma \in \mathbb{C}$, and $n \in \mathbb{Z}_+$. We introduce the following operators acting on $\operatorname{End} R$: $\operatorname{ad}\phi_\gamma = [\phi_\gamma, \bullet]$ and $\operatorname{ad}\phi_{(n)} = [\phi_{(n)}, \bullet]$.

By direct computations, we obtain the following

Lemma 3.4. *Let $a \in C(R)$, $\phi \in \operatorname{Cder} R$.*

(1) For any $\gamma \in \mathbb{C}$, $\operatorname{ad}\phi_\gamma(a) \in C(R)$. Moreover, $\operatorname{ad}\phi_\gamma$ is a derivation of $C(R)$.

(2) For any $n \in \mathbb{Z}_+$, $\operatorname{ad}\phi_{(n)}(a) \in C(R)$. Moreover, $\operatorname{ad}\phi_{(n)}$ is a derivation of $C(R)$.

Lemma 3.5. *If $a \in C(R)$ and $\phi \in CderR$, then $a\phi \in CderR$. Also $p(a\phi) = p(a) + p(\phi)$.*

3.3. *Constructing a chain of ideals*

Let R be a finite non-abelian differentiably simple Lie conformal superalgebra. Remark that the center of R, $\{x \mid [x_\lambda R] = 0\}$ is a differentiably stable ideal, hence it is zero and by Lemma 2.1, R contains a minimal ideal I. Let D be a set of homogeneous conformal derivations of R. Our ultimate goal is to construct with the use of D a certain finite chain of ideals that starts with I (cf. [1]).

Assume now that we have constructed a chain of ideals $I = I_1 \subset I_2 \subset \cdots \subset I_q \neq R$ such that for all $2 \leq j \leq q$, $I_j/I_{j-1} \simeq I$ as R-modules. Let

$\phi \in D$ be a homogeneous conformal derivation such that $\phi I_q \not\subset I_q$. We are going to construct an ideal $I_{q+1} \supset I_q$ such that $I_{q+1}/I_q \simeq I$ as R-modules.

Remark, first that for any ideal J, a homogeneous conformal derivation ϕ induces a map $\bar{\phi} \in \mathrm{Chom}(J, R/J)$, $\bar{\phi}_\lambda x = \phi_\lambda x + \mathbb{C}[\lambda] \otimes J$. Moreover, by definition, for every $y \in R$, the following equalities holds in the R-module R/J:

$$\bar{\phi}_\lambda(y_\mu x) = (-1)^{p(\phi)p(y)}(y_\mu(\bar{\phi}_\lambda x));$$
$$\partial(\bar{\phi}_\lambda x) = \bar{\phi}_\lambda(\partial x) - \lambda\bar{\phi}_\lambda x.$$

It follows that for any $\gamma \in \mathbb{C}$, $\ker \bar{\phi}_\gamma$ is a homogeneous ideal of R and $\mathrm{im}\, \bar{\phi}_\gamma$ is an R-submodule of R/J.

Suppose now that we have constructed a chain such as above: $I_1 \subset \cdots \subset I_q \neq R$. Let j be minimal such that $\phi I_j \not\subset I_q$. Take $J = I_q$ and consider the map $\bar{\phi} \in \mathrm{Chom}(I_q, R/I_q)$ constructed as above. Restrict this map to I_j. By construction $\phi(I_{j-1}) \subseteq I_q$, so $I_{j-1} \subseteq \ker \bar{\phi}_\gamma$ for any $\gamma \in \mathbb{C}$. Thus we have a family of maps $\bar{\phi}_\gamma : I_j/I_{j-1} \to R/I_q$.

We will show that there exists γ such that $\ker \bar{\phi}_\gamma$ is zero. Indeed, if for some $\gamma \in \mathbb{C}$ there exists $x + I_{j-1} \in I_j/I_{j-1}$ such that $\bar{\phi}_\gamma x = 0$, then $\bar{\phi}_\gamma(I_j/I_{j-1}) = 0$ (it is a simple R-module by construction). Hence, if there exists such an $x + I_{j-1}$ for every γ, then $\bar{\phi}_\lambda(I_j/I_{j-1}) = 0$ and, consequently, $\phi I_j \subset I_q$.

Hence we can find $\gamma \in \mathbb{C}$ with $\ker \bar{\phi}_\gamma = 0$. Then we can define I_{q+1} as an ideal such that $I_{q+1}/I_q = \mathrm{im}\, \bar{\phi}_\gamma$.

Remark 3.5. If we start constructing a chain of ideals such as above and keep the same $\phi \in D$, we would at some point obtain I_q such that $\phi I \subset I_q$. Indeed, pick a basis $x_1, \ldots, x_n$ of I over $\mathbb{C}[\partial]$ and let d be the maximal degree in λ of all $\phi_\lambda x_i$. At every step of the above construction, we produce an ideal I_k and $\gamma_k \in \mathbb{C}$ such that $\phi_{\gamma_k} I_1 \subset I_k$. It is clear that $\gamma_1, \ldots, \gamma_d$ are pairwise distinct, thus $\phi_\lambda I_1 \subset I_d$ as required.

3.4. *Constructing the maximal ideal*

Now let D be a finite collection of homogeneous conformal derivations, $D = \{\phi_1, \ldots, \phi_m\}$, such that I is not D-stable. Let i_1 be the least index for which $\phi_{i_1} I \not\subset I$. We apply the algorithm from the previous subsection. Namely, using ϕ_{i_1} we can construct a chain of ideals $I = I_1 \subset \cdots \subset I_{r_1}$, where $\phi_{i_1} I \subset I_{r_1}$. Then use ϕ_{i_2} (with the minimal i_2) such that $\phi_{i_2} I_{r_1} \not\subset I_{r_1}$ to extend the chain to I_{r_2}, where $\phi_{i_2} I_{r_1} \subset I_{r_2}$, etc.

Either at some point we obtain $I_l = R$ or we obtain a proper ideal I_q such that $\phi_m \dots \phi_1 I \subseteq I_q$.

Suppose now that for every collection D only the second case occurs. Since R is differentiably simple, there exists a homogeneous conformal derivation ϕ such that $\phi I_q \not\subset I_q$ and, using ϕ, we can extend our chain to I_{q+1}. For any $x \in I$, let $y = \phi_{m(n_m)}(\dots \phi_{1(n_1)}x)\dots)$. Hence $y \in I_q$ and $y_\lambda I_1 \simeq y_\lambda(I_{q+1}/I_q) = 0$. But every $y \in R$ can be expressed as a $\mathbb{C}[\partial]$-linear combination of elements in the above form (for some collection $D = \{\phi_1, \dots, \phi_m\}$). Hence, $[R_\lambda I] = 0$, i.e. R has a non-trivial center, a contradiction.

Therefore, there exists a finite chain $I = I_1 \subset \dots \subset I_l = R$ such that for every $2 \le j \le l$, $I_j/I_{j-1} \simeq I$ as R-modules. Denote I_{l-1} as N. Then $R/N \simeq I$ and N is maximal.

Since $N_\lambda(R/N) = 0$, $\langle [N_\lambda I_j] \rangle \subset I_{j-1}$ for any j. Thus $N^l = 0$ and N is nilpotent.

Suppose that there exists another maximal ideal N' of R. Then $N + N' = R$ and R/N' is nilpotent and simple, a contradiction (as $R = \langle [R_\lambda R] \rangle$). We arrive at

Proposition 3.1. *Let R be a finite non-abelian differentiably simple conformal Lie superalgebra and let I be its minimal ideal. Then R possesses a chain of ideals $I = I_1 \subset I_2 \subset \dots \subset N \subset R$, where N is a unique maximal ideal of R. Each factor I_j/I_{j-1} is isomorphic to I as R-modules. Moreover, N is nilpotent and $R/N \simeq I$.*

Actually, above we only used that I is finite, this the proof implies the following aside corollary.

Corollary 3.2. *If R is a non-abelian differentiably simple Lie conformal superalgebra with a finite minimal ideal, then R itself is finite.*

Remark 3.6. Uniqueness of N implies that all minimal ideals of R are R-isomorphic.

3.5. *Centroid structure*

Here we construct an embedding $\sigma : C(R/N) \to C(R)$. For $a \in C(R/N)$, define $\sigma(a)$ as the composition of maps

$$R \twoheadrightarrow R/N \xrightarrow{a} R/N \xrightarrow{\rho} I \hookrightarrow R$$

(here the isomorphism $\rho : R/N \xrightarrow{\sim} I$ is the one constructed in the proof of Proposition 3.1).

Clearly, σ is an embedding; moreover, $\operatorname{im}\sigma(a) = I$ for all a.

For an arbitrary R-isomorphism $\theta : R/N \tilde{\rightarrow} I$, consider the map $\beta : R/N \xrightarrow{\theta} I \xrightarrow{\rho^{-1}} R/N$. Since β commutes with the action of R and ∂, we see that $\sigma(\beta) = \theta$.

Let $I = I_1 \subset I_2 \subset \ldots \subset N \subset R$ be the chain described in Proposition 3.1.

Lemma 3.6. *There exists a family of monomorphisms $\sigma_q : C(R/N) \to C(R)$, $1 \leq q \leq l$, such that for every $a \in C(R/N)$ $\sigma_q(a)R + I_{q-1} = I_q$ and $\sigma_q(a)N \subsetneq I_{q-1}$. Moreover, if $\theta : R/N \to I_q/I_{q-1}$ is an R-isomorphism, then there exists $b \in C(R/N)$ such that θ is induced by $\sigma_q(b)$.*

Proof. We put $\sigma_1 = \sigma$ constructed above. Then denote by d_q the map ϕ_γ that we used to construct I_{q+1} and let I_j be the ideal used in that construction. Put $\sigma_{q+1} = [d_q, \sigma_j]$ (a super-bracket). Modulo I_{j-1}, $\sigma_j(a)R = I_j$ for any nonzero a, hence modulo I_q, $d_q\sigma_j(a)R = I_{q+1}$. Since $\sigma_j(a)d_qR \subsetneq I_j \subset I_q$, $\sigma_{q+1}R + I_q = I_{q+1}$ as required. Similarly, $\sigma_{q+1}N \subsetneq I_q$.

Given $\theta : R/N \to I_{q+1}/I_q$, we can extend it to a map $R/N \to I_j/I_{j-1}$ that is induced by $\sigma_j(b)$ for some b. Thus $\sigma_{q+1}(b)$ induces θ. □

Using the maps σ_q constructed in the above lemma, we have

Lemma 3.7. *The map $\widetilde{\sigma} = \oplus_{q=1}^{l}\sigma_q$ is an isomorphism $\bigoplus_{l \text{ copies}} C(R/N) \tilde{\rightarrow} C(R)$. Also, $J = \sigma_1(C(R/N))$ is a minimal ideal of $C(R)$.*

Proof. Let $a \in C(R)$ and q be minimal such that $aR \subsetneq I_q$. Combined with a projection $R \to R/I_{q-1}$, a induces a map $R \to I_q/I_{q-1}$. Its kernel is a maximal ideal, i.e. N. Hence a induces an isomorphism $R/N \to I_q/I_{q-1}$. Then there exists $b \in C(R/N)$ such that $(a - \sigma_q(b))R \subsetneq I_{q-1}$. By induction, $\widetilde{\sigma}$ is onto.

Conversely, if $\widetilde{\sigma}(\oplus_1^l a_i) = 0$, then $\sigma_l(a_l)R \subsetneq N$. Thus $a_l = 0$. By induction, all $a_i = 0$.

The set $J = \sigma_1(C(R/N)) = \{a \mid aR \subset I\}$ is obviously an ideal of $C(R)$. Let $a', a'' \in J$. Then they induce isomorphisms $\theta', \theta'' : R/N \to I$. Then $a' = ba''$, where $b \in C$ induces ${\theta'}^{-1}\theta''$. □

Remark 3.7. We actually obtain a chain of ideals $J = J_1 \subset \cdots \subset J$ with properties similar to that of $I_1 \subset \cdots \subset R$.

From Lemmas 3.3 and 3.7, we have the following

Corollary 3.3. *$C(R)$ is finite-dimensional.*

Denote by D the subset of $\operatorname{Der} C(R)$ that consists of derivations $\operatorname{ad}\phi_\gamma$, $\phi \in \operatorname{Cder} R$, $\gamma \in \mathbb{C}$.

Lemma 3.8. *$C(R)$ is D-differentiably simple.*

Proof. Let H be a D-stable ideal of $C(R)$. Then $\phi_\gamma HR \subseteq HR$ for any $\phi \in \operatorname{Cder} R$, $\gamma \in \mathbb{C}$. Hence, $\phi(HR) \subseteq HR$ and $HR = R$. There exists $h \in H$ such that $hR + N = R$ (as N is maximal). Thus $0 \neq J_1 h \subseteq H \cap J_1$, i.e. $J_1 \subseteq H$. The construction of maps σ_q and $\widetilde{\sigma}$ in Lemmas 3.6 and 3.7 implies that $C \subseteq H$. □

Corollary 3.4. *Let $D' = \{\operatorname{ad}\phi_{(n)}\}$. Then $C(R)$ is D'-differentiably simple.*

Proof. A D'-stable ideal of $C(R)$ is D-stable by the definition of ϕ_γ. □

Since $C(R)$ is differentiably simple and finite-dimensional, we obtain from [3] the following

Proposition 3.2. *Let $C(R)$ be the centroid of a finite non-abelian differentiably simple conformal Lie superalgebra. Then $C(R)$ is a Grassmann superalgebra $\wedge(r)$.*

3.6. *More on differential simplicity*

At this point the non-conformal argument for a Lie superalgebra L proceeds as follows: first, establish that $C(L)$ is $\operatorname{ad} d$-differentiably simple for just one conformal derivation $d \in \operatorname{Der} L$ (with certain additional conditions on d in the super case) and then conclude that L is d-differentiably simple.

In order for this to work for a Lie conformal superalgebra R as well, we need conformal analog of d of the form $\operatorname{ad}\phi_\gamma$ for some γ. We need to tweak the proof of Theorem 5.1 in [3].

We restate the theorem itself first. Let C be a Grassmann algebra over an algebraically closed field of characteristic 0. Let C be D-differentiably simple with respect to a homogeneous set of derivations D which is both a subalgebra and a left C-module. Then there exists $d \in D$ such that C is d-differentiably simple and for the corresponding chain $J = J_1 \subset \cdots \subset C$, the map $\bar{d} : J_q \to C/J_q$ is homogeneous (i.e. $dx \equiv d_\epsilon x \mod J_q$ for all $x \in J_q$, $\epsilon = \bar{0}, \bar{1}$).

The reason for the last condition is simple: in order to build a chain of ideals we need to use a homogeneous derivation at every step. In this case we say that d is *homogeneous at every step.*

The major steps of the proof are: find a homogeneous nilpotent $m \in C = \wedge(r)$ and a derivation d_1 such that $d_1(m) = 1$. Then $D = D_0 \oplus Cd_1$, where $D_0 = \{d - d_1(m)d_1 \mid d \in D\}$ (in particular, D_0 acts as 0 on m). View C as a $\langle C, D_0\rangle$-module and use d_1 to construct the appropriate chain of ideals. The minimal ideal in this chain is isomorphic to $\wedge(r_1)$, $r_1 < r$, so we can use $d_0 \in D_0$ to refine it into the chain of ideals of C.

It follows from the proof that the requirement of D being an algebra is superfluous, thus in order to apply the construction of d_0 and d_1 to the set D' constructed in Corollary 3.4, we have to show that it is a C-module. It suffices to show that $d = \operatorname{ad}\phi_{(n)} + \operatorname{ad}\psi_{(m)} \in D'$ for $m \neq n$. Let $m < n$. Then $d = (\operatorname{ad}\phi + (-1)^{n-m}\frac{m!}{n!}\partial^{n-m}\psi)_{(n)}$.

In the same vein we can assume that d_0 and d_1 are of the form $\operatorname{ad}\phi_{(n)}$ for the same n. Hence, we can strengthen the theorem as

Lemma 3.9. *There exist $\phi \in \operatorname{Cder} R$ and n such that $C(R) = \wedge(r)$ is $\operatorname{ad}\phi_{(n)}$-differentiably simple and $\operatorname{ad}\phi_{(n)}$ is homogeneous at every step.*

Lemma 3.10. *There exist $\phi \in \operatorname{Cder} R$ and $\gamma \in \mathbb{C}$ such that $C(R)$ is $\operatorname{ad}\phi_\gamma$-differentiably simple.*

Proof. Let ϕ and n be as in the previous lemma. Assume that for each $\gamma \in \mathbb{C}$, there exists an ideal J_γ that is $\operatorname{ad}\phi_\gamma$-stable. Since $C(R) \simeq \wedge(n)$ has a unique minimal ideal, $\cap J_\gamma$ is non-empty. Thus, there exists a proper ideal J that is $\operatorname{ad}\phi_\gamma$-stable for all γ. Hence, $\operatorname{ad}\phi_\lambda$ maps J into $J[\lambda]$ and J is $\operatorname{ad}\phi_{(n)}$-stable. □

Clearly, $\operatorname{ad}\phi_\gamma$ is not necessarily homogeneous at every step. Let D be the $C(R)$-module generated by the homogeneous components of $\operatorname{ad}\phi_\gamma$. Notice that since $\operatorname{ad}\chi_\gamma + \operatorname{ad}\phi_\gamma = \operatorname{ad}(\chi + \psi)_\gamma$, every element of D arises from a conformal derivation of R.

Since $C(R)$ is D-differentiably simple, we can apply the previous argument to obtain

Lemma 3.11. *There exist ψ and γ such that $C(R)$ is $\operatorname{ad}\psi_\gamma$-differentiably simple. Moreover, $\operatorname{ad}\psi_\gamma$ is homogeneous at every step.*

Let ψ and γ be as in Lemma 3.11. We claim that R contains no ψ-stable homogeneous ideals. Indeed, let M be such an ideal. We can always assume that $I \subset M$ (e.g. let I be a minimal ideal contained in M). Moreover, $M \subset N$. Let $H = \{a \,|\, aR \subset M\}$, a proper homogeneous ideal of $C(R)$. A direct calculation shows that H is $\operatorname{ad}\psi_\gamma$-stable, a contradiction.

Thus R is ψ-stable.

We need to show that ψ_γ acts homogeneously at every step, i.e. that the map $\bar{\psi}_\gamma : I_q \to R/I_q$ is homogeneous. This will allow us to build the chain $\{I_q\}$. It suffices to show that one of the homogeneous components of $\bar{\psi}_\gamma$ acts by zero. So, let ϕ and β be such that ϕ_β is homogeneous and $\operatorname{ad}\phi_\beta : J_q \to C/J_q$ acts by zero. (We assume here that we have already built the chain $\{J_q\}$ and that the chain of ideals of R has been constructed up to I_q. We also assume that $\phi_\beta I_{q-1} \subset I_q$.) There exist $c \in J_q$ and $y \in R$ such that cy generates I_q/I_{q-1} over R. Then $\psi_\beta cy = [\psi_\beta, c]y \pm c\psi_\beta y \in I_q$. It follows that $\psi_\beta R_\lambda cy \subset I_q[\lambda]$. Summing up, we have

Lemma 3.12. *Let R be a finite non-abelian differentiably simple Lie conformal algebra. Then there exist a conformal derivation ψ and $\gamma \in \mathbb{C}$ such that R is ψ_γ-differentiably simple. Moreover, using ψ_γ we can construct a chain of homogeneous ideals $I = I_1 \subset \cdots \subset N \subset R$.*

3.7. *Splitting R*

Let R be a finite non-abelian differentiably simple Lie conformal algebra. Let ψ and γ be as in the Lemma 3.12. Let $S = \{x \,|\, \psi_\gamma x \in I\}$.

Lemma 3.13. *$R = S \oplus N$ as $\mathbb{C}[\partial]$-submodules and S is a conformal subalgebra of R.*

Proof. A direct computation shows that S is closed with respect to the ∂-action and the λ-bracket inherited from R.

Let $0 \neq x \in S \cap N$. Then $x \in I_q \backslash I_{q-1}$ for some q. Hence, $\psi_\gamma x \notin I_q$, a contradiction, and $S \cap N = 0$.

To show that $R = S + N$, observe that $I_q \subseteq I + \psi_\gamma N$ for all q. Indeed, by induction $I_{q-1} \subseteq I + \psi_\gamma N$ and $\psi_\gamma I_{q-1} \subset \psi_\gamma N$. Hence $I_q = I_{q-1} + \psi_\gamma I_{q-1} \subseteq I + \psi_\gamma N$. In particular $R \subseteq I + \psi_\gamma N$. Thus for every $x \in R$, $\psi_\gamma x = \psi_\gamma y \mod I$ for some $y \in N$ and $x - y \in S$. □

Since $S \simeq R/N$ as conformal superalgebras, we have

Corollary 3.5. *S is a finite simple Lie conformal superalgebra.*

Let $f : S \to R$ be the natural embedding. Define the map $F : \wedge(r) \otimes S \to R$ by setting

$$F(c \otimes x) = (-1)^{p(c)p(x)} c f(x), \qquad c \in \wedge(r), x \in R$$

for homogeneous elements.

Recall that we have constructed the chain $\{J_q\}$ of ideals of $C(R)$. In particular, $J_q = \sigma_1(C(S)) + \cdots + \sigma_q(C(S))$. Denote $c_q = \sigma_q(1)$ (it is homogeneous by definition of σ_q). Clearly $F(c_1 \otimes S) = I$. By induction $I_q + F(c_q \otimes S) = I_{q+1}$, hence F is surjective.

To show injectivity, assume that x is such that $F(x) = 0$ and $x = \sum_1^q c_i \otimes x_i$, $x_q \neq 0$. Since $f(x_q) \notin N$ and c_q induces an isomorphism between R/N and I_q/I_{q-1}, $c_q f(x_q) \notin I_{q-1}$. On the other hand, F maps the first $q-1$ components into I_{q-1}. Thus $c_q f(x_q) = 0$. Since $c_q f$ is a homogeneous map, we obtain $x_q = 0$.

This completes the proof of Theorem 3.1.

4. Simple Lie conformal superalgebras and their derivations

4.1. *Simple conformal superalgebras*

Example 4.1. Let $\mathfrak{g}$ be a finite-dimensional Lie superalgebra. The *loop algebra* associated to $\mathfrak{g}$ is the Lie superalgebra

$$\widetilde{\mathfrak{g}} = \mathfrak{g}[t, t^{-1}], \ \ p(at^k) = p(a) \ \text{ for } \ a \in \mathfrak{g}, \ k \in \mathbb{Z},$$

with the bracket

$$[a \otimes t^n, b \otimes t^m] = [a, b] \otimes t^{n+m} \quad (a, b \in \mathfrak{g};\ m, n \in \mathbb{Z}).$$

We introduce the family $\mathcal{F}_{\mathfrak{g}}$ of formal distributions (known as currents)

$$a(z) = \sum_{n \in \mathbb{Z}} (a \otimes t^n)\, z^{-n-1}, \quad a \in \mathfrak{g}.$$

It is easily verified that $[a(z), b(w)] = [a, b](w)\delta(z - w)$, hence $(\widetilde{\mathfrak{g}}, \mathcal{F}_{\mathfrak{g}})$ is a formal distribution Lie superalgebra. The associated Lie conformal superalgebra is $\mathbb{C}[\partial] \otimes \mathfrak{g}$, with the λ-bracket (we identify $1 \otimes \mathfrak{g}$ with $\mathfrak{g}$)

$$[a_\lambda b] = [a, b], \quad a, b \in \mathfrak{g}.$$

It is called the *current conformal algebra* associated to $\mathfrak{g}$, and is denoted $\mathrm{Cur}\, \mathfrak{g}$. The Lie conformal superalgebra $\mathrm{Cur}\, \mathfrak{g}$ is simple if and only if $\mathfrak{g}$ is a simple Lie superalgebra.

Example 4.2. We define a conformal linear map $L : \mathrm{Cur}\,\mathfrak{g} \to \mathrm{Cur}\,\mathfrak{g}$ by $L_\lambda g = (\partial + \lambda)g$ for any $g \in \mathfrak{g}$. It is immediate to verify that this is a conformal derivation of $\mathrm{Cur}\,\mathfrak{g}$. L generates the conformal algebra $\mathbb{C}[\partial]L \subset \mathrm{gc}(\mathrm{Cur}\,\mathfrak{g})$ called the *Virasoro conformal algebra* and denoted Vir. The λ-bracket is

$$[L_\lambda L] = (\partial + 2\lambda)L.$$

Thus we have constructed a Lie conformal algebra $\mathrm{Vir} \ltimes \mathrm{Cur}\,\mathfrak{g}$.

Vir can be constructed using formal distributions with coefficients in the Lie algebra $\mathbb{C}[t,t^{-1}]\partial_t$, letting $L(z) = \sum_{n\in\mathbb{Z}}(t^n\partial_t)z^{-n-1}$. Then, $[L(z), L(w)] = \partial_w L(w)\delta(z-w) + 2L(w)\partial_w\delta(z-w)$.

An element of a conformal superalgebra satisfying the above equation is called a *Virasoro element*; it is automatically even.

In all other examples below we forgo the description of conformal superalgebras in terms of formal distributions and simply provide the (conformal) generators and relations. A more detailed description can be found in [6].

Example 4.3. Recall that for the Grassmann algebra $\wedge(N)$ in the anticommuting indeterminates ξ_i, $i = 1, \ldots, N$, its Lie superalgebra of derivations is

$$W(N) = \left\{ \sum_{i=1}^{N} P_i\partial_i \mid P_i \in \wedge(N), \partial_i = \partial/\partial\xi_i \right\}.$$

The Lie conformal superalgebra W_N is defined as

$$W_N = \mathbb{C}[\partial] \otimes (W(N) \oplus \wedge(N)).$$

The λ-bracket $(a, b \in W(N); f, g \in \wedge(N))$ is as follows:

$$[a_\lambda b] = [a,b], \quad [a_\lambda f] = a(f) - \lambda(-1)^{p(a)p(f)}fa, \quad [f_\lambda g] = -(\partial + 2\lambda)fg. \tag{1}$$

The Lie conformal superalgebra W_N is simple for $N \geq 0$ and has rank $(N+1)2^N$. We also remark that $W_0 \simeq \mathrm{Vir}$.

We shall need the following representation of W_N on $\mathbb{C}[\partial] \otimes \wedge(N)$ (cf. Example 4.2):

$$a_\lambda g = a(g), \quad f_\lambda g = -(\partial + \lambda)fg, \quad a \in W(N); f, g \in \wedge(N). \tag{2}$$

As a consequence, by identifying $(\mathrm{Cur}\,\mathfrak{g}) \otimes \wedge(N)$ and $\mathfrak{g} \otimes \mathbb{C}[\partial] \otimes \wedge(N)$, we also get a representation of W_N on $\mathrm{Cur}\,\mathfrak{g} \otimes \wedge(N)$ by conformal derivations. Thus, we can construct the semi-direct product $W_N \ltimes (\mathrm{Cur}\,\mathfrak{g} \otimes \wedge(N))$.

Recall that $W(N)$ and $\wedge(N)$ are graded, $W(N) = \oplus_{j\geq -1} W(N)^j$ and $\wedge(N) = \oplus_{j\geq 0} \wedge (N)^j$, with $\deg \xi_i = 1$, $\deg \partial_i = -1$. A subalgebra L of

$W(N)$ is said to *act transitively* on $\wedge(N)$ or $\mathfrak{g}\otimes\wedge(N)$ if under the projection $W(N) \to W(N)^{-1}$, L maps onto $W(N)^{-1}$.

Similarly, a subalgebra L of W_N *acts transitively* on $\mathrm{Cur}\,\mathfrak{g}\otimes\wedge(N)$ if the projection of L to $\mathbb{C}[\partial]\otimes W(N)^{-1}$ has rank N.

Example 4.4. For an element $D = \sum_{i=1}^N P_i(\partial,\xi)\partial_i + f(\partial,\xi) \in W_N$, we define the corresponding notion of divergence:

$$\operatorname{div} D = \sum_{i=1}^{N}(-1)^{p(P_i)}\partial_i P_i - \partial f \in \mathbb{C}[\partial]\otimes\wedge(N).$$

The following identity holds in $\mathbb{C}[\partial]\otimes\wedge(N)$, where $D_1, D_2 \in W_N$ (cf. (2)):

$$\operatorname{div}[D_1{}_\lambda D_2] = (D_1)_\lambda(\operatorname{div} D_2) - (-1)^{p(D_1)p(D_2)}(D_2)_{-\lambda-\partial}(\operatorname{div} D_1). \quad (3)$$

Therefore we can define the following subalgebra of W_N:

$$S_N = \{D \in W_N \mid \operatorname{div} D = 0\}$$

The Lie conformal superalgebra S_N is simple for $N \geq 2$ and has rank $N2^N$.

Example 4.5. Let $D = \sum_{i=1}^N P_i(\partial,\xi)\partial_i + f(\partial,\xi)$ be an element of W_N. Given $a \in \mathbb{C}$, we define the deformed divergence to be

$$\operatorname{div}_a D = \operatorname{div} D + af.$$

It still satisfies formula (3), hence

$$S_{N,a} = \{D \in W_N \mid \operatorname{div}_a D = 0\}$$

is a subalgebra of W_N, which is simple for $N \geq 2$ and has rank $N2^N$.

Example 4.6. Another variation on the construction of the conformal superalgebra S_N is the following definition (N even):

$$\widetilde{S}_N = \{D \in W_N \mid \operatorname{div}((1+\xi_1\ldots\xi_N)D) = 0\}(= (1-\xi_1\ldots\xi_N)S_N).$$

We thus obtain the Lie conformal superalgebra $\widetilde{S}_N$. It is simple for $N \geq 2$ and has rank $N2^N$.

Example 4.7. We can also define the "contact" conformal superalgebra

$$K_N = \mathbb{C}[\partial]\otimes\wedge(N)$$

with the λ-bracket for $A = \xi_{i_1}\ldots\xi_{i_r}$ and $B = \xi_{j_1}\ldots\xi_{j_s}$ defined as

$$[A_\lambda B] = \left(\left(\frac{r}{2}-1\right)\partial(AB) + (-1)^r\frac{1}{2}\sum_{i=1}^{N}\partial_i A\partial_i B\right) + \lambda\left(\frac{r+s}{2}-2\right)AB.$$

K_N embeds into W_N but we will not be using this fact here.

Note also that $K_0 \simeq \text{Vir}$ and $K_2 \simeq W_1$.

The Lie conformal superalgebra K_N is simple for all $N \in \mathbb{Z}_+$, $N \neq 4$ and is a free $\mathbb{C}[\partial]$-module of rank 2^N.

Example 4.8. The Lie conformal superalgebra K_4 is not simple but its derived subalgebra K_4' is. Furthermore, $K_4 = K_4' \oplus \mathbb{C}\nu$, where $\nu = \xi_1\xi_2\xi_3\xi_4$ and, since K_4' is an ideal in K_4, the map $\operatorname{ad}\nu$ is an outer conformal derivation of K_4'.

Example 4.9. The Lie conformal superalgebra CK_6 is a simple rank 32 subalgebra of K_6, spanned over $\mathbb{C}[\partial]$ by the elements

$$-1 + \alpha\partial^3\nu, \quad \xi_i - \alpha\partial^2\xi_i^*, \quad \xi_i\xi_j - \alpha\partial(\xi_i\xi_j)^*, \quad \xi_i\xi_j\xi_k - \alpha(\xi_i\xi_j\xi_k)^*,$$

where $\alpha \in \mathbb{C}$ is a fixed number such that $\alpha^2 = -1$, $\nu = \xi_1 \dots \xi_6$, and $(\xi_{i_1}\xi_{i_2}\dots)^* = \partial_{i_1}\partial_{i_2}\dots\nu$.

The even part of CK_6 is $\text{Vir} \ltimes \text{Cur}\,\mathfrak{so}_6$ for the Virasoro element $-1 + \alpha\partial^3\nu$. For the explicit form of the commutation relations of and further details on CK_6, see [4].

The main result of [6] is the following Theorem.

Theorem 4.1. *Any finite simple Lie conformal superalgebra R is isomorphic to one of the Lie conformal superalgebras of the following list:*

(1) W_N, $(N \geq 0)$;

(2) $S_{N,a}$ $(N \geq 2,\ a \in \mathbb{C})$;

(3) $\widetilde{S}_N$ *(N even,* $N \geq 2)$;

(4) K_N $(N \geq 0,\ N \neq 4)$;

(5) K_4';

(6) CK_6;

(7) $\text{Cur}\,\mathfrak{s}$, *where* $\mathfrak{s}$ *is a simple finite-dimensional Lie superalgebra.*

We shall call the algebras (1)–(6) from the above list the Lie conformal superalgebras *of Cartan type.*

4.2. *Conformal derivations of simple Lie conformal superalgebras*

We are going describe the conformal derivations of simple Lie conformal superalgebras. With this in mind, for every finite simple Lie conformal superalgebra R we will fix a *distinguished* reductive Lie subalgebra $\mathfrak{r}$ of $(R, {}_{(0)})$.

For $R = \mathrm{Cur}\,\mathfrak{s}$, we choose as $\mathfrak{r}$ the maximal reductive subalgebra of $\mathfrak{s}$ (they are listed in [8]).

For W_N we take as $\mathfrak{r}$ the copy of $\mathfrak{gl}_N$ spanned by $\xi_i\partial_j \in W(N)$. We construct other distinguished subalgebras in a similar fashion. For S-type superalgebras, we take the subalgebras $\mathfrak{sl}_N$ of $\mathfrak{gl}_N \subset W(N)$ and for the K-type, we take $\mathfrak{r} = \mathfrak{so}_N$ spanned by $\xi_i\xi_j$. For K_4', $\mathfrak{r} = \mathfrak{so}_4 \oplus \mathbb{C}\partial\nu$ and for CK_6, the algebra $\mathfrak{so}_6$ contained in $\mathrm{Cur}\,\mathfrak{so}_6$.

Proposition 4.1. *In the following we list all cases in which a finite simple Lie conformal superalgebra has outer conformal derivations:*

(1) $\mathrm{Cder}\,K_4' = \mathrm{Cinder}\,K_4' \oplus \mathbb{C}[\partial]\nu$, *where* $\nu = \xi_1 \ldots \xi_4 \in \wedge(4)$*;*

(2) $\mathrm{Cder}(\mathrm{Cur}\,\mathfrak{s}) = \mathrm{Vir} \ltimes \mathrm{Cur}(\mathrm{Der}\,\mathfrak{s})$, *where* $\mathfrak{s}$ *is a simple finite-dimensional Lie superalgebra.*

Proof. Let R be a finite simple Lie conformal superalgebra. Then $R = \mathbb{C}[\partial] \otimes U$ and the distinguished reductive Lie algebra $\mathfrak{r}$ acts via the 0-th product completely reducibly on U. This action commutes with ∂, hence $\mathfrak{r}$ acts completely reducibly on R.

Next, $\mathrm{Cder}\,R \subset \mathrm{gc}\,R \simeq R \otimes R^*$ as an R-module, where R^* is the conformal dual of the R-module R [2, Proposition 6.4(a)]. Therefore $\mathrm{gc}\,R$ and, hence, $\mathrm{Cder}\,R$ are completely reducible as $\mathfrak{r}$-modules.

Thus, $\mathrm{Cder}\,R = \mathrm{Cinder}\,R \oplus V$ as $\mathfrak{r}$-modules. We will refer to the elements in V as *outer* conformal derivations of R.

Since $[\mathfrak{r}_{(0)}V] \subset R = \mathrm{Cinder}\,R$, we see that $\mathfrak{r}$ kills V, thus outer conformal derivations are $\mathfrak{r}$-module homomorphisms. Using this, outer conformal derivations of R can be easily computed.

In the following we provide a detailed computation in the most involved case, that of W_N. We restrict our attention to $N \geq 2$, the remaining cases being straightforward.

Recall that $W_N = \mathbb{C}[\partial] \otimes (W(N) \oplus \wedge(N))$ and that W_N is completely reducible as a $\mathfrak{gl}_N$-module. We have the following description of $W(N)^k \oplus \wedge(N)^k$ as $\mathfrak{sl}_N(\subset \mathfrak{r})$-modules. Here $R(\lambda)$ denotes the $\mathfrak{sl}_N$-module with the the highest weight λ and π_i are fundamental weights:

$$W(N)^k \oplus \wedge(N)^k = \begin{cases} R(\pi_{N-1}), & \text{if } k = -1, \\ R(\pi_{k+1} + \pi_{N-1}) \oplus R(\pi_k) \oplus R(\pi_k), & \text{if } 0 \leq k \leq N-2, \\ R(\pi_{N-1}) \oplus R(\pi_{N-1}), & \text{if } k = N-1, \\ R(0), & \text{if } k = N. \end{cases}$$

Let $\phi \in V$. Then ϕ is a $\mathfrak{gl}_N$-module homomorphism, hence it commutes with the Euler operator $E = \sum \xi_i \partial_i$ and therefore leaves each graded component invariant. The grading being consistent with parity, ϕ can only be an even conformal derivation. Let $x_k = \xi_1 \dots \xi_k E$ be the highest weight vector of $R(\pi_k) \subseteq W(N)^k$, $0 \le k \le N-1$. Let $z_k = \xi_1 \dots \xi_{k+1}\partial_N$ be the highest weight vector of $R(\pi_{k+1} + \pi_{N-1}) \subseteq W(N)^k$, $0 \le k \le N-2$. Let ∂_N be the highest weight vector of $R(\pi_{N-1}) \subseteq W(N)^{-1}$. Let $y_k = \xi_1 \dots \xi_k$ be the highest weight vector of $R(\pi_k) \subseteq \wedge(N)^k$, $0 \le k \le N$. The action of ϕ is given by

$$\begin{aligned}
&\phi_\lambda y_N = A(\partial, \lambda) y_N,\\
&\phi_\lambda x_k = P_k(\partial, \lambda) x_k + Q_k(\partial, \lambda) y_k, \quad 0 \le k \le N-1,\\
&\phi_\lambda y_k = R_k(\partial, \lambda) x_k + S_k(\partial, \lambda) y_k, \quad 0 \le k \le N-1,\\
&\phi_\lambda z_k = Z_k(\partial, \lambda) z_k, \quad 0 \le k \le N-2,\\
&\phi_\lambda \partial_N = \Omega(\partial, \lambda)\partial_N.
\end{aligned}$$

Let $g \in \mathfrak{sl}_N$. We compute $\phi_\lambda[g_\mu x_k]$:

$$\begin{aligned}
\phi_\lambda[g_\mu x_k] =& (\phi_\lambda g)_{\lambda+\mu} x_k + g_\mu(P_k(\partial, \lambda) x_k + Q_k(\partial, \lambda) y_k)\\
=& (\phi_\lambda g)_{\lambda+\mu} x_k + P_k(\partial + \mu, \lambda)(g_\mu x_k) + Q_k(\partial + \mu, \lambda)(g_\mu y_k).
\end{aligned}$$

Since for any $g \in \mathfrak{sl}_N$, there exists g' such that $g = [g', z_0]$, we have $\phi_\lambda g = \phi_\lambda(g'_{(0)} z_0) = Z_0(\partial, \lambda) g$ as ϕ_λ and $g'_{(0)}$ commute. Then using (1) we get

$$\phi_\lambda[g, x_k] = Z_0(-\lambda-\mu, \lambda)[g, x_k] + P_k(\partial+\mu, \lambda)[g, x_k] + Q_k(\partial+\mu, \lambda)(g y_k - \mu y_k g).$$

By commuting ϕ_λ and $g_{(0)}$, we get that the left-hand side of the above equality equals $P_k(\partial, \lambda)[g, x_k] + Q_k(\partial, \lambda) g y_k$.

There exists $g \in \mathfrak{sl}_N$ such that $y_k g \ne [g, x_k]$ (e.g. $g = \xi_1 \partial_1$). Hence, $Q_k(\partial + \mu, \lambda) = 0$ implying that $Q_k = 0$. Also, by comparing the left-hand and the right-hand sides, we see that $P_k(\partial, \lambda)$ is at most linear in ∂, otherwise it is impossible to cancel products of μ and ∂ that may come from $P_k(\partial + \mu, \lambda)$.

The rest of the deduction is similar and we describe it only in brief. We compute $\phi_\lambda[x_{0\mu} x_k]$, $\phi_\lambda[y_{0\mu} y_k]$ and $\phi_\lambda[g_\mu y_k]$ to show that $R_k(\partial, \lambda)$ is constant in ∂ whereas $Z_k(\partial, \lambda)$ and $S_k(\partial, \lambda)$ are at most linear in ∂. Third, we compute $\phi_\lambda[x_{0\mu} y_k]$, $\phi_\lambda[x_{0\mu} \partial_N]$, $\phi_\lambda[\partial_{N\mu} y_0]$, $\phi_\lambda[x_{k\mu} y_0]$, $\phi_\lambda[x_{0\mu} y_N]$ and $\phi_\lambda[\partial_{N\mu} \xi_N]$ to show that all the remaining polynomials $A, P_k, R_k, S_k, Z_k, \Omega_k$ are actually zero. □

Remark 4.1. Using the same arguments as in the proof of Proposition 4.1, one can calculate the ordinary derivations of finite simple Lie conformal superalgebras. Below we list all cases in which outer ordinary derivations occur (E denotes the Euler operator and ν, the highest monomial in $\wedge(N)$):

(1) $\operatorname{Der} S_{N,a} = \operatorname{Inder} S_{N,a} \oplus \mathbb{C}E_{(0)}$ $a \neq 0, N \geq 2$;

(2) $\operatorname{Der} S_{N,0} = \operatorname{Inder} S_{N,0} \oplus \mathbb{C}E_{(0)} \oplus \mathbb{C}\nu_{(0)}$, $N > 2$;

(3) $\operatorname{Der} S_{2,0} = \operatorname{Inder} S_{2,0} \oplus \mathfrak{sl}_2$, where $E_{(0)}, \nu_{(0)} \in \mathfrak{sl}_2$;

(4) $\operatorname{Der} \widetilde{S}_N = \operatorname{Inder} \widetilde{S}_N \oplus \mathbb{C}\nu_{(0)}$ $N \geq 2$, N even;

(5) $\operatorname{Der} K_4' = \operatorname{Inder} K_4' \oplus \mathbb{C}\nu_{(0)}$;

(6) $\operatorname{Der} \operatorname{Cur} \mathfrak{s} = \mathbb{C}\partial \oplus \operatorname{Der} \mathfrak{s}$, where $\mathfrak{s}$ is a simple finite-dimensional Lie superalgebra.

We will not require this result for the rest of the paper and so leave the details to the reader.

It follows that the ordinary derivations of simple finite Lie conformal superalgebras form the following Lie superalgebras:

(1) $\operatorname{Der} W_N \simeq W(N) \ltimes \wedge(N)$;

(2) $\operatorname{Der} S_{N,a} \simeq (S(N) \oplus \mathbb{C}E) \ltimes \wedge(N)'$, where $\wedge(N)'$ is the subalgebra of $\wedge(N)$ spanned by monomials of degree strictly less than N, $a \neq 0$;

(3) $\operatorname{Der} S_{N,0} \simeq (S(N) \oplus \mathbb{C}E) \ltimes \wedge(N)$, $N > 2$;

(4) $\operatorname{Der} S_{2,0} \simeq \mathfrak{so}_4 \ltimes \mathcal{H}_4$, where $\mathcal{H}_4$ is the Heisenberg algebra generated by $\xi_1, \xi_2, \partial_1, \partial_2$, and $\mathfrak{so}_4 = \mathfrak{sl}_2 \oplus \mathfrak{sl}_2$ is spanned by two $\mathfrak{sl}_2$-triples: $(\xi_2\partial_1, \xi_1\partial_1 - \xi_2\partial_2, \xi_1\partial_2)$, and (ν, E, F), where $F(\xi_1) = -\partial_2$, $F(\xi_2) = \partial_1$;

(5) $\operatorname{Der} \widetilde{S}_N \simeq ((1-\nu)S(N)) \ltimes \wedge(N)$, N even;

(6) $\operatorname{Der} K_N$, $N \neq 4$, is the natural central extension of $H(N)$ by a one-dimensional center;

(7) $\operatorname{Der} K_4' \simeq H(4)$;

(8) $\operatorname{Der} CK_6$ is the central extension of the "strange" Lie superalgebra $P(4)$ by a one-dimensional center;

(9) $\operatorname{Der} \operatorname{Cur} \mathfrak{s}$ is the direct sum of $\operatorname{Der} \mathfrak{s}$ with the one-dimensional Lie algebra.

(Cf. the calculations of derivations of corresponding formal distribution Lie superalgebras in [10] and [6].)

5. Conformal derivations of differentiably simple Lie conformal superalgebras

Here we describe conformal derivations of Lie conformal superalgebras of the form $S \otimes \wedge(n)$, where S is a finite simple Lie conformal superalgebra. By Theorem 3.1 this will take care of the conformal derivations of all non-abelian differentiably simple Lie conformal superalgebras. This allows us to obtain the complete classification of finite non-abelian differentiably simple

Lie conformal superalgebras.

5.1. *Conformal centroid*

The *conformal centroid* of a Lie conformal superalgebra R is the subalgebra $CC(R)$ of the associative conformal algebra $\mathrm{Cend}\, R$ defined by

$$CC(R) = \{\varphi \in \mathrm{Cend}\, R \mid \varphi_\lambda[x_\mu y] = [(\varphi_\lambda x)_{\lambda+\mu} y]\}.$$

Remark 5.1. A direct calculation similar to that in the proof of Lemma 3.1 shows that for $\varphi \in CC(R)$, $\varphi_\lambda[x_\mu y] = (-1)^{p(x)p(\varphi)}[x_\mu(\varphi_\lambda y)]$.

To describe conformal centroids of simple superalgebras, we need two technical lemmas:

Lemma 5.1. *Let R be a finite Lie conformal superalgebra. Suppose R contains a Virasoro element L and a $\mathbb{C}[\partial]$-basis $\{a_j\}_{j\in J}$ of R such that*

$$[L_\lambda a_j] = (\partial + \Delta_j \lambda)a_j + b_j \quad \textit{for any} \quad j \in J, \tag{4}$$

where $\Delta_j \in \mathbb{C}$, $b_j \in R$ and, whenever $\Delta_j = 0$, $b_j = 0$.
Then $CC(R) = 0$.

Proof. Let $\varphi \in CC(R)$. Suppose $\deg_\lambda \varphi_\lambda L = k$. Then $\deg_\lambda[L_\mu(\varphi_\lambda L)] = k$ and $\deg_\lambda \varphi_\lambda[L_\mu L] = \deg_\lambda(\partial+\lambda+2\mu)\varphi_\lambda L = k+1$ unless $\varphi_\lambda L = 0$. Hence the equation $\varphi_\lambda[L_\mu L] = [L_\mu(\varphi_\lambda L)]$, given in Remark 5.1 implies that $\varphi_\lambda L = 0$.

Now, for any $j \in J$, we have $\varphi_\lambda[L_\mu a_j] = \varphi_\lambda((\partial+\Delta_j\mu)a_j+d(a_j))$. On the other hand, $\varphi_\lambda[L_\mu a_j] = [(\varphi_\lambda L)_{\lambda+\mu} a_j] = 0$. Therefore, $(\partial+\lambda+\Delta_j\mu)\varphi_\lambda a_j + \varphi_\lambda b_j = 0$. Now, if $\Delta_j \neq 0$ the fact that there is no power of μ in the second summand implies that $\varphi_\lambda a_j = 0$. If $\Delta_j = 0$, then, by assumption, $b_j = 0$, hence $\varphi_\lambda a_j = 0$. □

Now we present a method of constructing a basis satisfying requirements of Lemma 5.1 which we will apply in the case of simple Lie conformal superalgebras of Cartan type.

Lemma 5.2. *Let R be a finite simple Lie conformal superalgebra, $R = \mathbb{C}[\partial] \otimes V$. Let $\mathfrak{r}$ be the distinguished reductive Lie subalgebra of R and let L be a Virasoro element of R such that*

$$[L_\lambda g] = (\partial + \lambda)g \quad \textit{for all } g \in \mathfrak{r}. \tag{5}$$

Suppose $V = \oplus_{i\in I} V_i$, where V_i is an irreducible $\mathfrak{r}$-module whose highest weight vector is v_i. Suppose that the v_i satisfy (4). Then there exists a $\mathbb{C}[\partial]$-basis $\{a_j\}_{j\in J}$ of R satisfying conditions of Lemma 5.1.

Proof. For any $i \in I$, V_i is spanned over $\mathbb{C}$ by vectors of the form $v_i^n = g^n_{(0)} \cdots g^1_{(0)} v_i$, $n \in \mathbb{Z}_+$ and $g^k \in \mathfrak{r}$ for any $k = 1, \ldots, n$.

We are going to prove by induction on n that $[L_\lambda v_i^n] = (\partial + \Delta_i \lambda) v_i^n + u_i^n$, where $u_i^n \in R$, and whenever $\Delta_i = 0$, $u_i^n = 0$ for all n. Then it would suffice to take v_i^n's as the necessary basis. We have

$$
\begin{aligned}
[L_\lambda (g^{n+1}_{(0)} v_i^n)] &= [[L_\lambda g^{n+1}]_\lambda v_i^n] + [g^{n+1}_{(0)} [L_\lambda v_i^n]] \\
&= [((\partial + \lambda) g^{n+1})_\lambda v_i^n] + [g^{n+1}_{(0)} ((\partial + \Delta_i \lambda) v_i^n + u_i^n] \\
&= (\partial + \Delta_i \lambda) [g^{n+1}_{(0)} v_i^n] + [g^{n+1}_{(0)} u_i^n].
\end{aligned}
$$

Then we take $u_i^{n+1} = [g^{n+1}_{(0)} u_i^n]$. This completes the proof. □

Proposition 5.1. *Let R be a finite simple Lie conformal superalgebra.*

(1) If R is of Cartan type, then $CC(R) = 0$;

(2) If $R = \operatorname{Cur} \mathfrak{s}$, $CC(R) = \{\varphi \in \operatorname{Cend} R \mid \varphi_\lambda s = p(\partial, \lambda) s, p(\partial, \lambda) \in \mathbb{C}[\partial, \lambda], \text{for all } s \in \mathfrak{s}\}$.

Proof. If R is of Cartan type, then it has an element L and a $\mathbb{C}[\partial]$-basis satisfying (4) and (5) (such an L is called physical in Section 7). In fact, if R is not $S_{N,a}$ or $\widetilde{S}_N$, we can choose $L = -1$ (and then all $b_j = 0$). In the case of $S_{N,a}$ or $\widetilde{S}_N$, we choose $L = -1 + \frac{1}{N}(\partial - a)E$ and $-1 + \xi_1 \ldots \xi_N + \frac{1}{N}\partial E$, respectively. The formulas that appear in the proof of [6, Proposition 4.16] show that the highest weight vectors satisfy (4) with Δ_i taking the values of $2/N, 3/N, \ldots, (N-1)/N, 1, (N+1)/N, \ldots, (2N-1)/N, 2$. Consequently, by Lemma 5.2 we can find a $\mathbb{C}[\partial]$-basis whose conformal weights are nonzero. We complete the proof of (1) by applying Lemma 5.1.

Let $\varphi \in CC(\operatorname{Cur} \mathfrak{s})$. Then for any $s, s' \in \mathfrak{s}$ we have

$$
\varphi_\lambda [s_{(0)} s'] = (-1)^{p(\varphi) p(s)} [s_{(0)} \varphi_\lambda s'].
$$

Suppose $\varphi_\lambda = \sum_{n \in \mathbb{Z}_+} \frac{\lambda^n}{n!} \varphi_{(n)}$. Then for any $n \geq 0$, $\varphi_{(n)}$ is a $\mathfrak{s}$-module homomorphism. Since $\mathfrak{s}$ is simple, if $\varphi_{(n)} \neq 0$, then $\ker \varphi_{(n)} = 0$ and $\operatorname{im} \varphi_{(n)} \simeq \mathfrak{s}$. Then $\phi_{(n)} s = \sum_i p_{n_i}(\partial) s_i(s)$. The map $s \mapsto s_i(s)$ is an $\mathfrak{s}$-automorphism. Remark 3.4 implies that $s_i(s) = c_i s$ for some $c_i \in \mathbb{C}$. Hence, $\varphi_{(n)}(s) = p_n(\partial) \otimes s$ for any $s \in \mathfrak{s}$ and $\varphi_\lambda = \sum_n \lambda^n / n! p_n(\partial) \otimes 1_{\mathfrak{s}}$. □

Remark 5.2. In fact, we have shown that $CC(\operatorname{Cur} \mathfrak{s}) \simeq \operatorname{Cend}_1$. We note that this conformal algebra is neither finite, nor commutative.

5.2. *Conformal derivations of a tensor product*

Proposition 5.2. *Let R be a finite Lie conformal superalgebra and let B be a unital commutative associative finite-dimensional superalgebra. Then*

(1) $\operatorname{Cder} R \otimes B \subseteq \operatorname{Cder}(R \otimes B)$.
(2) $CC(R) \otimes \operatorname{Der} B \subseteq \operatorname{Cder}(R \otimes B)$.
(3) *If R is simple and of Cartan type, then*

$$\operatorname{Cder}(R \otimes B) = \operatorname{Cder} R \otimes B.$$

(4) *If $R = \operatorname{Cur} \mathfrak{s}$, where $\mathfrak{s}$ is a simple finite-dimensional Lie superalgebra, then*

$$\operatorname{Cder}(\operatorname{Cur} \mathfrak{s} \otimes B) = \operatorname{Cder}(\operatorname{Cur} \mathfrak{s}) \otimes B + 1_{\mathfrak{s}} \otimes \operatorname{Cder} \operatorname{Cur} B.$$

In particular,

$$\operatorname{Cder}(\operatorname{Cur} \mathfrak{s} \otimes \wedge(N)) = \operatorname{Cur}(\operatorname{Der} \mathfrak{s}) \otimes \wedge(N) + 1_{\mathfrak{s}} \otimes W_N.$$

Proof. (1) Let $\phi \in \operatorname{Cder} R$ and $b \in B$. We set

$$(\phi \otimes b)_\lambda(r' \otimes b') = (-1)^{p(b)p(r')}(\phi_\lambda r') \otimes bb'.$$

It is easy to verify that $\phi \otimes b$ is a conformal derivation.

(2) Similarly, let $\varphi \in CC(R)$ and $d \in \operatorname{Der} B$. We set

$$(\varphi \otimes d)_\lambda(r \otimes b) = (-1)^{p(d)p(r)}(\varphi_\lambda r) \otimes d(b).$$

It is easy to verify that $\varphi \otimes d$ is a conformal derivation.

(3) Let $\phi \in \operatorname{Cder}(R \otimes B)$. For any $r \in R$ we have

$$\phi_\lambda(r \otimes 1) = \sum_{i \in I} \phi_{i\lambda}(r) \otimes b_i$$

where $\{b_i\}_{i \in I}$ is a linear basis of B. A short direct computation shows that $\phi_i \in \operatorname{Cder} R$ and $\widetilde{\phi} := \phi - \sum_{i \in I} \phi_i \otimes b_i \in \operatorname{Cder}(R \otimes B)$ is such that $\widetilde{\phi}_\lambda(r \otimes 1) = 0$. Let us fix $b \in B$. Suppose $\widetilde{\phi}_\lambda(r \otimes b) = \sum_{i \in I} \varphi_{i\lambda} r \otimes b_i$. For any $r, r' \in R$ we have $[r_\mu r'] \otimes b = [(r \otimes 1)_\mu (r' \otimes b)]$. If we apply $\widetilde{\phi}$ to both sides of this equation we see that $\varphi_i \in CC(R)$ which is zero by Proposition 5.1.

(4) Using the same argument as in (3), one can see that $\varphi_i \in CC(\operatorname{Cur} \mathfrak{s})$, so that by Proposition 5.1, $\widetilde{\phi}_\lambda(r \otimes b) = \sum_{i \in I} P_i(\partial, \lambda) r \otimes b_i$. Now, we identify $\operatorname{Cur} \mathfrak{s} \otimes B$ with $\mathfrak{s} \otimes \operatorname{Cur} B$ so that $\widetilde{\phi}_\lambda(r \otimes b) = r \otimes \sum_{i \in I} P_i(\partial, \lambda) \otimes b_i$. The map associating b to $\sum_{i \in I} P_i(\partial, \lambda) \otimes b_i$ is easily shown to be a conformal derivation of $\operatorname{Cur} B$.

In the case $B = \wedge(N)$, it remains to show that $W_N = \mathrm{Cder}(\mathrm{Cur}\wedge(N))$. We recall at first that W_N acts on $\mathrm{Cur}\wedge(N)$ by conformal derivations (see (2)). The rest is done in two steps. First, notice that if $\phi \in \mathrm{Cder}(\mathrm{Cur}\wedge(N))$, then $\phi_\lambda 1 = (\partial + \lambda)\sum_{i\in I} p_i(\lambda) f_i$, where $\{f_i\}_{i\in I}$ is a linear basis of $\wedge(N)$. Second, if we set $\widetilde{\phi} = \phi - (\sum_{i\in I} p_i(-\partial) f_i) \circ \partial$, it is immediate to see that for any $j = 1, \ldots, N$, $\widetilde{\phi}_\lambda(\xi_j) = P_j(\lambda, \xi)$, so that $\widetilde{\phi} = \sum_{j=1}^N P_j(-\partial, \xi)\partial_j$. Therefore $\phi = \widetilde{\phi} + (\sum_{i\in I} p_i(-\partial) f_i) \circ \partial \in W_N$. □

We thus obtain

Theorem 5.1. *The following is a complete list of finite non-abelian differentiably simple Lie conformal superalgebras:*

(1) $(\mathrm{Cur}\,\mathfrak{s}) \otimes \wedge(n)$, *where* $\mathfrak{s}$ *is a simple finite-dimensional Lie superalgebra;*
(2) a finite simple Lie conformal superalgebra of Cartan type.

Proof. If R is differentiably simple, then $R \simeq S \otimes \wedge(n)$ by Theorem 3.1. But if S is of Cartan type, an ideal $S \otimes I$ of R, where I is an ideal of $\wedge(n)$, is differentiably stable by Proposition 5.2(3). Part (4) of the same proposition shows that the conformal superalgebra $(\mathrm{Cur}\,\mathfrak{s}) \otimes \wedge(n)$, $\mathfrak{s}$ as above, is differentiably simple. □

6. Semisimple Lie conformal superalgebras

Here we provide a detailed description of finite semisimple Lie conformal superalgebras. In particular, we prove the following

Theorem 6.1. *Let R be a finite semisimple Lie conformal superalgebra. Then R splits into a direct sum of conformal algebras of the following types:*

(1) a finite simple Lie conformal superalgebra;
(2) a Lie conformal superalgebra L such that

$$\left(\bigoplus_{i=1}^{k}(\mathrm{Cur}\,\mathfrak{s}_i) \otimes \wedge(n_i)\right) \oplus {K_4'}^{\oplus^r} \subset L$$
$$\subset \left(\bigoplus_{i=1}^{k}(\mathrm{Cur}(\mathrm{Der}\,\mathfrak{s}_i) \otimes \wedge(n_i) + 1_{\mathfrak{s}} \otimes W_{n_i})\right) \oplus {K_4}^{\oplus^r}, \quad (6)$$

where $n_i, r \in \mathbb{Z}_+$ and $\mathfrak{s}_i$ are simple finite-dimensional Lie superalgebras, and such that for each i the projection of L onto W_{n_i} acts transitively on $(\mathrm{Cur}\,\mathfrak{s}_i) \otimes \wedge(n_i)$.

6.1. *Proof of Theorem 6.1*

Define the *socle* of a Lie conformal algebras as the sum of all its minimal ideals.

Lemma 6.1. *Let R be a finite semisimple Lie conformal superalgebra. Then its socle is a direct sum of all minimal ideals of R and there exist finitely many such ideals in R.*

Proof. Let $\{M_i\}_{i\in I}$ be the family of all minimal ideal of R. By Lemma 2.1 this family is non-empty.

Notice that for all $i,j \in I$, $[M_{i\lambda}M_j] = 0$ for $i \neq j$ by minimality. Suppose $M_i \cap \sum_{j\neq i} M_j \neq 0$ for some i. Thus by minimality of M_i, $M_i \subseteq \sum_{j\neq i} M_j$, and $[M_{i\lambda}M_i] \subset \sum_{j\neq i}[M_{i\lambda}M_j] = 0$. Hence M_i is abelian which contradicts semisimplicity of R.

Thus the sum of minimal ideals is direct. Furthermore, the rank of a direct sum is the sum of the ranks, hence the fact that R is finite implies that R contains only finitely many minimal ideals. □

We fix notations from the proof and denote the socle of R by M and the minimal ideals of R by M_i, $i = 0, 1, \ldots, l$.

The centralizer of M in R, $C_R(M) = \{x \in R \mid [x_\lambda M] = 0\}$, is zero. This follows from the fact that $C_R(M)$ is an ideal of R, but contains no minimal ideals of R, because any such ideal would be abelian contradicting the semisimplicity of R. But this contradicts Lemma 2.1.

Therefore, the homomorphism of Lie conformal superalgebras $R \to \operatorname{Cder} M$ sending x to $\operatorname{ad}_M x_\lambda$ is injective. Thus, $\operatorname{Cinder} M \subseteq R \subseteq \operatorname{Cder} M$.

The minimality of M_i implies that $\langle [M_{i\lambda}M_i] \rangle = M_i$ for any $i = 0, 1, \ldots, l$. Thus if $\phi \in \operatorname{Cder} M$ maps M_i to M_j, $\phi(M_i)$ must be zero. It follows that $\operatorname{Cder} M = \oplus_{i=0}^{l} \operatorname{Cder} M_i$.

Therefore, we have

$$\bigoplus_{i=0}^{l} \operatorname{Cinder} M_i \subseteq R \subseteq \bigoplus_{i=0}^{l} \operatorname{Cder} M_i. \tag{7}$$

Put $R_i = R \cap \operatorname{Cder} M_i$.

Lemma 6.2. *Let R be a conformal superalgebra satisfying (7). Then R is semisimple if and only if M_i is R_i-simple for all i.*

Proof. If M_i is not R_i-simple for some i, then it contains a non-trivial ideal J. By differential simplicity of M_i, J is nilpotent (see Proposition 3.1). Since J is also an ideal of R_i and hence of R, R is not semisimple.

Conversely, let R contain an abelian ideal I. Since M_i is non-abelian and R-simple for any i, $[I_\lambda M_i]$, as an R-submodule of M_i must be zero. Thus I kills M, a contradiction. □

By Lemma 2.2(2), M_i's are differentiably simple and, since they are non-abelian, we obtain a complete description of the possible form of M_i's from Theorem 5.1.

Now we can use Propositions 4.1 and 5.2 to describe possible R_i's in greater detail.

Let $M_i = S_i \otimes \wedge(n_i)$, where S_i is a simple Lie conformal superalgebra.

By Proposition 5.1, if S_i is of Cartan type, $n_i = 0$. If S_i is of Cartan type and not isomorphic to K_4', then $R_i \simeq S_i$. If $S_i \simeq K_4'$, $R_i \subset K_4$.

It remains to treat the case of a current Lie conformal algebra S_i. Let R_i be a semisimple Lie conformal superalgebra such that $\operatorname{Cur} \mathfrak{s} \otimes \wedge(n) \subset R_i \subset W_n \ltimes (\operatorname{Cur} \operatorname{Der} \mathfrak{s} \otimes \wedge(n))$, where $\mathfrak{s}$ is a finite-dimensional simple Lie superalgebra. In order to prove that the projection of R_i onto W_{n_i} acts transitively if and only if $S_i \otimes \wedge(n_i)$ is R_i-simple, we need an auxiliary lemma

Recall that a commutative associate superalgebra with the action of a Lie superalgebra $\mathfrak{a}$ by derivations is called $\mathfrak{a}$-differentiably simple if it contains no non-trivial invariant ideals.

Lemma 6.3. *$\wedge(n)$ is $\mathfrak{a}$-simple for a subalgebra $\mathfrak{a}$ of $W(n)$ if and only if $\mathfrak{a}$ acts transitively on $\wedge(n)$.*

Proof. Transitivity of the action of $\mathfrak{a}$ is equivalent to saying that $\mathfrak{a}$ contains elements $a_j = \partial_j + \sum_i p_i \partial_i$, where $p_i \in \wedge(n)$, $p_i(0) = 0$, for all j.

Then, assume that $\mathfrak{a}$ acts transitively and let I be a non-zero $\mathfrak{a}$-stable ideal of $\wedge(n)$. Then I contains the monomial $\xi_1 \ldots \xi_n$. Let $f = \xi_{k_1} \ldots \xi_{k_n}$. Since $a_j(\xi_j f) = f+$ monomials of higher degree, by induction I contains all monomials of $\wedge(n)$, i.e. $I = \wedge(n)$.

Now let $\wedge(N)$ be $\mathfrak{a}$-differentiably simple. We can assume that $\mathfrak{a}$ is closed with respect to multiplication by $\wedge(N)$. Indeed, $[a, fb] = a(f)b + (-1)^{p(a)+p(f)}[a, b]$ for $a, b \in W(N)$, a homogeneous, so $\wedge(N)\mathfrak{a}$ remains closed with respect to the Lie bracket. Also, $\mathfrak{a}$ acts transitively if and only if $\wedge(N)\mathfrak{a}$ does. Similarly, $\wedge(N)$ is $\mathfrak{a}$-differentiably simple if and only of it is $\wedge(N)$-differentiably simple.

If the projection of $\mathfrak{a}$ to $W(N)^{-1}$ is zero, then by the grading argument, $\mathfrak{a}$ leaves invariant the maximal ideal of $\wedge(N)$. Thus, possibly after a linear transformation of ξ_i's, $\mathfrak{a}$ contains an element of the form $a_N = \partial_N + \sum p_i \partial_i$, $p_i(0) = 0$.

In particular, $a_N = \partial_N + \xi_1 b + \sum q_i \partial_i$, where $b \in W(N)$, $q_i(0) = 0$ and $\partial_1 q_i = 0$. Also, we can assume that $q_N = 0$. We introduce the following change of coordinates: $\xi_i' = \xi_i - \xi_N q_i$ for $1 \le i \le N-1$, $\xi_N' = \xi_N$. Notice that $\mathbb{C}\langle \xi_1', \xi_2 \dots, \xi_N \rangle = \wedge(N)$. Indeed, let $q_1 = \xi_1 f + g$, $\partial_1 g = 0$. Then $\xi_1 = \xi_1(1 + \xi_N f) + \xi_N g$, and since $1 - f$ is invertible, this change of coordinates is valid. By induction, we obtain that $\mathbb{C}\langle \xi_1', \dots, \xi_N' \rangle$. The chain rule $\partial_i = \sum (\partial_i \xi_j')\partial_j'$ implies that in the new coordinates $a_N = \partial_N' + \xi_N c$ for some $c \in W(N)$.

Thus, we can assume that $a_N = \partial_N + \xi_N c$. Then $[a_N, a_N] = 2c$, hence $\partial_N \in \mathfrak{a}$ and we can split $\mathfrak{a}$ as $\wedge(N)\partial_N \oplus \mathfrak{a}_1$. If $\mathfrak{a}_1$ contains an element $a = \xi_N b$, then $[\partial_N, a] = b \in \mathfrak{a}_1$, i.e. we can both multiply by and cancel ξ_N in elements of $\mathfrak{a}_1$. Thus $\mathfrak{a}_1$ splits as $\mathfrak{a}_2 \oplus \xi_N \mathfrak{a}_2$, where $\mathfrak{a}_2 \subset W(N-1)$. If I is a proper $\mathfrak{a}_2$-stable ideal of $\wedge(N-1)$, then clearly $I + \xi_N I$ is a proper $\mathfrak{a}$-stable ideal. Hence, $\wedge(N-1)$ is $\mathfrak{a}_2$-differentiably simple. Moreover, $\mathfrak{a}$ acts transitively if and only if $\mathfrak{a}_2$ does.

Induction on N completes the proof. □

Remark 6.1. Along the way we proved an analog of the Frobenius theorem for $\wedge(N)$. Namely, if I is an ideal of $\wedge(N)$ and $\mathfrak{a}$ is a subalgebra of $W(N)$ which leaves I invariant, then after a change of coordinates, $\mathfrak{a} = \sum_{i=1}^{k} \wedge(N)\partial_i + \sum_{j=k+1}^{N} \wedge_k(N)\partial_j$, where $\wedge_k(N)$ is the subalgebra (without 1) of $\wedge(N)$ generated by $\xi_{k+1}, \dots, \xi_N$.

The explicit description of the action of W_n on $\operatorname{Cur} \mathfrak{s} \otimes \wedge(n)$ in the proof of Proposition 5.2(4) shows that the action of R_i is transitive if and only if the action of $R_i|_{\partial=1}$ is transitive on $\wedge(n)$. This together with Lemmas 6.2 and 6.3 completes the proof of Theorem 6.1.

6.2. *Conformal derivations*

The following is a direct analog of [3, Proposition 7.4]:

Proposition 6.1. *Let R be a finite semisimple Lie conformal superalgebra and let M be its socle. Then* $\operatorname{Cder} R = \{\phi \in \operatorname{Cder} M \mid \langle \phi_\lambda R \rangle \subset R\}$*, the normalizer of R in* $\operatorname{Cder} M$.

Proof. If I is a minimal $\operatorname{Cder} R$-stable ideal, it is either minimal or properly contains an ideal J of R. Since I is differentiably simple, J must be nilpotent, a contradiction. Hence $I \subset M$. Conversely, let M' be the sum of all minimal $\operatorname{Cder} R$-stable ideal and I a minimal ideal such that $I \not\subset M'$. Then $I \subset \operatorname{Ann} M'$ and, as $\operatorname{Ann} M'$ is an ideal of $\operatorname{Cder} R$ and contains no $\operatorname{Cder} R$-central elements, it must contain a minimal $\operatorname{Cder} R$-ideal which must be abelian. The contradiction shows that $M = M'$ and M is $\operatorname{Cder} R$-stable.

Then $\operatorname{Cder} R$ embeds into $\operatorname{Cder} M$ and the rest follows easily. □

7. Physical Virasoro pairs

In this section we compute all physical Virasoro pairs in a finite simple Lie conformal superalgebra. As a consequence, we obtain the classification of physical Lie conformal superalgebras.

7.1. *Definitions*

Let $R = \mathbb{C}[\partial] \otimes V$ be a finite simple Lie conformal superalgebra. Recall that V contains a distinguished reductive subalgebra of V; we denote it $\mathfrak{r}$.

Let R contain a Virasoro element L such that

$$[L_\lambda g] = (\partial + \lambda)g \text{ for any } g \in \mathfrak{r}. \tag{8}$$

Then (R, L) is a *simple physical Virasoro pair.*

By skew-symmetry, (8) is equivalent to $[g_\lambda L] = \lambda g$, which implies that $[g_{(0)}L] = 0$, i.e. that L is invariant with respect to the action of $\mathfrak{r}$ on R (by means of the 0th product). Also, R is completely reducible as an $\mathfrak{r}$-module because the 0th product commutes with the $\mathbb{C}[\partial]$-module structure.

7.2. *Classification results*

Theorem 7.1. *The following is a complete list of all simple physical Virasoro pairs:*

$(W_0, -1)$; $(W_N, -1 + (p_0 + p_1\partial)E)$, $N \geq 1$; $(S_{N,a}, -1 + \frac{1}{N}(\partial - a)E)$, $N \geq 2, a \in \mathbb{C}$; $(\widetilde{S}_N, -(1 - \xi_1 \dots \xi_N) + \frac{1}{N}\partial E)$, $N \geq 2$, *N even;* $(K_N, -1)$, $N \geq 1$, *N odd;* $(K_N, -1+p_1\partial\nu)$, $N > 6$, *N even;* $(K_6, -1+(p_1\partial+p_3\partial^3)\nu)$; $(K_4', -1 + (p_0 + p_1\partial)\partial\nu)$; $(CK_6, -1 + \alpha\partial^3\nu)$.

Here $E = \sum_{i=1}^N \xi_i\partial_i$ *is the Euler operator,* $\nu = \xi_1 \dots \xi_N$ *is the longest monomial in* $\wedge(N)$, $p_0, p_1, p_3 \in \mathbb{C}$ *and* $\alpha \in \mathbb{C}$, $\alpha^2 = -1$.

Proof. Recall that $W_N = \mathbb{C}[\partial] \otimes (W(N) \oplus \wedge(N))$. If $N = 0$, the equation $[L_\lambda L] = (\partial + 2\lambda)L$ implies that $L = -1$. Suppose $N \geq 1$, then $\mathfrak{r} = \mathfrak{gl}_N$.

First of all we remark that L cannot be contained in the radical of the even part of W_N, which we denote by $\mathrm{Rad}\, W_{N\,\bar{0}}$. We set

$$S = \mathbb{C}[\partial] \otimes (\oplus_{k\geq 2,\, k \text{ even}} W(N)^k), \quad T = \mathbb{C}[\partial] \otimes (\oplus_{k\geq 2,\, k \text{ even}} \wedge^k (N)).$$

Since the λ-bracket in W_N is graded, $S \oplus T \subseteq \mathrm{Rad}\, W_{N\,\bar{0}}$, hence any L must have a component in $\mathbb{C}[\partial] \otimes (W(N)^0 \oplus \mathbb{C}1)$:

$$L = Q(\partial)1 + P(\partial)E + \sum_i P_i(\partial) g_i + s + t,$$

where $\{g_i\}$ is a basis of $\mathfrak{sl}_N$, $s \in S$ and $t \in T$. Furthermore, we can compare the coefficients of the degree 0 basis elements in the equation $[L_\lambda L] = (\partial + 2\lambda)L$. In the case of $1 \in \wedge(N)$ we have $-(\partial+2\lambda)Q(-\lambda)Q(\partial+\lambda) = (\partial+2\lambda)Q(\partial)$, hence $Q(\partial) = Q$ is constant and either $Q = 0$ or $Q = -1$. Comparing the coefficients of E we get $-(\partial+\lambda)QP(\partial+\lambda) - \lambda QP(-\lambda) = (\partial+2\lambda)P(\partial)$. Comparing the terms with basis elements in $\mathfrak{sl}_N$ we get

$$(\partial+\lambda)Q \sum_i P_i(\partial+\lambda) g_i - \lambda Q \sum_i P_i(-\lambda) g_i + \sum_{i,j} P_i(-\lambda) P_j(\partial+\lambda)[g_i, g_j] = (\partial + 2\lambda) \sum_i P_i(\partial) g_i.$$

Now, $Q = 0$ implies $P = 0$ and

$$\sum_{i,j} P_i(-\lambda) P_j(\partial+\lambda)[g_i, g_j] = (\partial+2\lambda) \sum_i P_i(\partial) g_i.$$

If we examine the degree of ∂ in the last equation, we see that $P_i = 0$ for any i. It follows that if $Q = 0$, then $L = s + t \in \mathrm{Rad}\, W_{N\,\bar{0}}$, which is impossible. Hence $Q = -1$. If we examine the equation for E, which is homogeneous, we conclude that $P(\partial) = p_0 + p_1\partial$ and $L = -1 + (p_0 + p_1\partial)E + \sum_i P_i(\partial) g_i + s + t$.

Next, we impose the condition that $[L_\lambda g] = (\partial + \lambda)g$ for any $g \in \mathfrak{gl}_N$. Since $[-1_\lambda g] = (\partial+\lambda)g$, we have $[(\sum_i P_i(\partial)g_i + s + t)_\lambda g] = 0$ for any $g \in \mathfrak{gl}_N$. Consequently, in degree 0 we have $[(\sum_i P_i(\partial) g_i)_\lambda g] = 0$ for any $g \in \mathfrak{gl}_N$. In particular, it follows that $\sum_i P_i(\partial) g_i$ lies in the center of $\mathrm{Cur}\, \mathfrak{sl}_N$ which is zero. On the other hand, in degree ≥ 2 we have $[(s+t)_\lambda g] = 0$, hence $[g_{(0)}(s+t)] = 0$. By looking at the decomposition of the $\mathbb{C}[\partial]$-basis of W_N into irreducible $\mathfrak{gl}_N$-modules, we see that there are no invariants in degree ≥ 2. Consequently, $s + t = 0$ and we conclude that $L = -1 + (p_0 + p_1\partial)E$.

A similar argument, with $\mathfrak{sl}_N$ replacing $\mathfrak{gl}_N$, provides the solution for $S_{N,a}$ and $\widetilde{S}_N$.

The cases of K_N, $0 \leq N \leq 3$, K_4' and CK_6 are dealt with in a similar fashion. As for K_N, $N \geq 5$, the unique $\mathfrak{so}_N$-invariant is $\wedge^N(N)$. If N is odd, this vector is odd too, hence cannot appear in L. If N is even, from $[L_\lambda L] = (\partial + 2\lambda)L$ we obtain a homogeneous equation, which admits a linear solution for any N and a cubic solution for $N = 6$.

In the case of Cur $\mathfrak{s}$, there are no non-zero Virasoro elements. To see this, let $\{g_i\}_{i\in I}$ be a linear basis of $\mathfrak{s}$. Then $L = \sum_i P_i(\partial)g_i$ and by substituting this expression into $[L_\lambda L] = (\partial + 2\lambda)L$ and comparing the degrees in ∂, we see that $L = 0$. □

Corollary 7.1. *The following is a complete list of all pairs (R, L), where R is a finite simple Lie conformal superalgebra and L is a Virasoro element such that $R_{\bar{0}} = L \ltimes \mathrm{Cur}\,\mathfrak{g}$, where $\mathfrak{g}$ is a finite-dimensional Lie algebra and $[L_\lambda a] = (\partial + \lambda)a$ for all $a \in \mathfrak{g}$:*

$(W_1, -1 + (p_0 + p_1\partial)E)$; $(W_2, -1 + \frac{1}{2}\partial E)$; $(S_{2,a}, -1 + \frac{1}{2}(\partial - a)E)$; $(\widetilde{S}_2, -(1-\nu)+\frac{1}{2}\partial E)$; $(K_0, -1)$; $(K_1, -1)$; $(K_3, -1)$; $(K_4', -1+(p_0+p_1\partial)\partial\nu)$; $(CK_6, -1 + \alpha\partial^3\nu)$.

Proof. We have to impose the condition $[L_\lambda g] = (\partial + \lambda)g$ for any $g \in \mathfrak{g}$.

Let $w \in W(0,N)_k \subseteq W_N$ and $f \in \wedge^k(N) \subseteq W_N$. We have

$$[L_\lambda w] = (p_0 k + \partial + (1 - p_1 k)\lambda)w,$$
$$[L_\lambda f] = (p_0 k + \partial + (2 - p_1 k)\lambda)f - \lambda(p_0 - p_1\lambda)fE.$$

If $N > 2$, $W(0,N)_2 \neq 0$ and the first equation implies that $1 - 2p_1 = 1$ i.e. $p_1 = 0$. On the other hand, $\wedge^2(N) \neq 0$ too and the second equation implies that $2 - 2p_1 = 1$ i.e. $p_1 = 1/2$. The contradiction proves that $N \leq 2$. Since $W_0 \simeq K_0$ and $W_1 \simeq K_2$, we concentrate on $W_2 = \mathbb{C}[\partial] \otimes (W(2) \oplus \wedge(2))$. If we apply the second equation to $\xi_1\xi_2$, we get $p_0 = 0$ and $p_1 = 1/2$.

If $R = S_{N,a}$ or $R = \widetilde{S}_N$, then $R = \mathbb{C}[\partial] \otimes W(N)$ as $\mathfrak{sl}_N$-modules. If $N > 2$, $W(N)^2 \neq 0$ and the formulas that appear in the proof of [6, Proposition 4.16] show that the conformal weight of any vector in $W(N)^2$ is either $(N-2)/N$ or $(2N-2)/N \neq 1$. Therefore $N = 2$ is the only possibility.

If $R = K_N$, $N \geq 5$, the conformal weight of $\xi_1\xi_2\xi_3\xi_4$ with respect to any simple physical Virasoro pair (K_N, L) is 0.

As for K_4', we can check by direct computation that $K'_{4\,\bar{0}} = L \ltimes Cur\ \mathfrak{cso}_4$. Similarly in the case of K_N, $0 \leq N \leq 3$.

Finally, it was proved in [4] that $CK_{6\,\bar{0}} = L \ltimes Cur\ \mathfrak{so}_6$. □

Remark 7.1. For W_2 we have $\mathfrak{g} = \mathfrak{sl}_2 \oplus \mathfrak{b}_2$; here $\mathfrak{b}_2$ denotes the two-dimensional solvable Lie algebra. In all other cases, $\mathfrak{g} = \mathfrak{r}$.

7.3. *Physical conformal superalgebras*

A *physical* Lie conformal superalgebra is a simple physical Virasoro pair (R, L) such that the following conditions hold:

(1) $R_{\bar{0}} = L \ltimes \mathrm{Cur}\,\mathfrak{g}$, where $\mathfrak{g}$ is a finite-dimensional Lie algebra;
(2) $L_{(0)}a = \partial a$ for any $a \in R$.

We thus obtain a generalization of the main result of [9] (see also [15]).

Proposition 7.1. *The following is a complete list of physical Lie conformal superalgebras:*
$(W_1, -1+p_1\partial E)$; $(W_2, -1+\frac{1}{2}\partial E)$; $(S_{2,0}, -1+\frac{1}{2}\partial E)$; $(K_0, -1)$; $(K_1, -1)$; $(K_3, -1)$; $(K_4', -1 + (p_0 + p_1\partial)\partial\nu)$; $(CK_6, -1 + \alpha\partial^3\nu)$.

Proof. Condition (2) excludes $\widetilde{S}_2$ from the list in Corollary 7.1, implies that $a = 0$ for $S_{2,a}$ and forces $p_0 = 0$ for (W_1, L). □

The formal distribution Lie superalgebras $(\mathrm{Lie}\,R, R)$ corresponding to the Lie conformal superalgebras on the list of Proposition 7.1 are well known to physicists (except for CK_6). All of them, except for CK_6, have non-trivial central extensions. In the cases $R = K_0 = W_0, K_1, K_2 = W_1, K_3, S_{2,0}$, and K_4', these central extensions are called the Virasoro, Neveu–Schwarz, $N = 2$, $N = 3$, $N = 4$, and big $N = 4$ superconformal algebras. With the exception of the last one, they appear on the list of [12]. The last one appeared independently in [11], [13], and [14].

8. Representations of Solvable Lie Conformal Superalgebras

In this section we discuss several results towards a possible analog of Lie Theorem for finite Lie conformal superalgebras.

8.1. *Preliminaries*

First we restate the analogs of Lie theorems in the, respectively, non-conformal and conformal non-super cases:

Theorem 8.1. *[8] Let L be a finite-dimensional solvable Lie superalgebra and V its irreducible representation. Then either* $\dim V_{\bar{0}} = \dim V_{\bar{1}}$ *and* $\dim V = 2^s$, *where* $s \le \dim L_{\bar{1}}$ *or* $\dim V = 1$.

Theorem 8.2. *[5] Let R be a finite solvable Lie conformal algebra and M an R-module. Then there exists $v \in M$ such that $x_\lambda v = \varphi(a)v$, $\varphi : R \to \mathbb{C}[\lambda]$. In particular, every non-trivial simple R-module M is free and* $\operatorname{rk} M = 1$.

Remark 8.1. Let R be a Lie conformal algebra with a module M. Then $(\operatorname{Tor} R)_\lambda M = 0$, $\operatorname{Tor} M$ is a submodule of M, and $R_\lambda \operatorname{Tor} M = 0$ [5]. Since we are interested in simple modules only, below we will always assume that both the conformal algebra and its modules are free over $\mathbb{C}[\partial]$.

We will also need the following

Lemma 8.1. *[5] For a solvable Lie conformal superalgebra R, $\operatorname{rk} R > \operatorname{rk} R'$, where $R' = \langle R_\lambda R\rangle$ is the derived algebra of R.*

Another useful result is the analog of Lemma 2.1:

Lemma 8.2. *A finite module M over a finite Lie conformal superalgebra R either contains a simple R-submodule or a submodule N such that $R_\lambda N = 0$.*

8.2. *Rank* 1 *modules*

Let R be a finite solvable Lie conformal superalgebra and M an R-module, $\operatorname{rk} M = 1$. In particular, $M = \mathbb{C}[\partial]v$ for some $v \in M$ and the representation of R is given by the function $\ell : R \to \mathbb{C}[\lambda, \partial]$ such that $x_\lambda v = \ell(x)v$ for all $x \in R$. Theorem 8.2 immediately implies that either $\ell = 0$ or M is a simple $R_{\bar{0}}$-module and that ℓ as a function on $R_{\bar{0}}$ depends only on λ. A direct computation shows that $\ell(R'_{\bar{0}}) = 0$. Also, since the action of $R_{\bar{1}}$ must change parity, we conclude that $\ell(R_{\bar{1}}) = 0$.

Conversely, if ℓ satisfies all the above conditions and $\ell(\partial x) = -\lambda\ell(x)$ for all x, ℓ determines an R-module of rank one.

Let $\mathcal{L}$ be the set of functions $\ell : R \to \mathbb{C}[\lambda]$ such that $\ell(R'_{\bar{0}}) = \ell(R_{\bar{1}}) = 0$ and $\ell(\partial x) = -\lambda\ell(x)$ for all $x \in R$. Let $\mathcal{L}_0$ be a subspace of $\mathcal{L}$ consisting of ℓ for which $\ell(R') = 0$.

Remark 8.2. Let H be a subalgebra of R such that $R = H \oplus \mathbb{C}[\partial]x$, $x \in R$, $R_{\bar{0}} \subset H$. The λ-bracket in R induces an H-action on $\mathbb{C}[\partial]x$. The corresponding ℓ lies in $\mathcal{L}_0$.

Let M be an R-module and $\ell \in \mathcal{L}_0$. Let V be a $\mathbb{C}[\partial]$-generating subspace of M, i.e. $M = \mathbb{C}[\partial] \otimes V$. We introduce another R-module structure on M by putting $x_\lambda v = x^M_\lambda v + \ell(x)v$ for $x \in R$, $v \in V$ and extending the R-action to all of M in the standard way. In this way we obtain a module M' and call M and M' *$\mathcal{L}_0$-equivalent.*

8.3. *Induced modules*

In the proof of Theorem 8.1, the powers of 2 appear because, in principle, a simple module V over a Lie superalgebra L can be induced from a smaller module W over a subalgebra H of L. In particular, when $L = H \oplus \mathbb{C}g$ and $p(g) = \bar{1}$, $V = W + gW$, hence the dimension gets doubled.

In the conformal case, non-trivial induction is still possible but the rank can grow arbitrarily as the following lemma demonstrates.

Lemma 8.3. *Let R be a Lie conformal superalgebra, and let H be a (homogeneous) subalgebra of R such that $R' \subset H$ and $R = H \oplus \mathbb{C}[\partial]x$ for some homogeneous $x \in R$. Let M be a finite R-module and N an H-submodule of M that generates M. Then*

$$M = \sum_{i=1}^{l} \mathbb{C}[\partial] \left(\prod_{j=i}^{l} x_{(n_j)} \right) N \text{ for some } n_1, \ldots, n_l. \tag{9}$$

Moreover, as H-modules M and N have the same simple quotients. If N is simple, then $\operatorname{rk} M$ *is proportional to* $\operatorname{rk} N$.

Proof. Clearly, $\langle x_\lambda N \rangle + N$ is an H-submodule of M and, unless $M = N$, it contains N as a proper submodule.

Hence $U = (\langle x_\lambda N \rangle + N)/N$ is an H-module. Moreover, since $x_{(k)}(h_{(m)}w) = \pm h_{(m)}(x_{(k)}w) \mod N$ for $k, m \in \mathbb{Z}_+$ and $w \in W$, we obtain submodules $U_k = (\mathbb{C}[\partial](x_{(k)}N) + N)/N$ of U. The Leibniz rule implies that $U_k \subset U_{k+1}$. Obviously, $U = \sum U_k$ and it immediately follows that $U = U_n$ for some n. Hence, $N + \langle x_\lambda N \rangle = N + \mathbb{C}[\partial](x_{(n)}N)$. Now we can consider the latter module instead of N. Then (9) follows by induction.

We have actually obtained a filtration of M: $N \subseteq N + \mathbb{C}[\partial](x_{(0)}N) \subseteq \ldots \subseteq M$. Denote the k-th element of the filtration by M_k. Then a direct computation shows that there exists a natural surjective map $N \to M_k/M_{k-1}$ for every k (e.g. $N \to U_k/U_{k-1}$ is defined as $v \mapsto x_{(k)}v$). This proves the rest of the lemma. □

Remark 8.3. By the Jacobi identity, one can always have $n_j \leq n_{j+1}$ in (9). When x is odd the inequality is strict. The analogous lemma for ordinary algebras would then imply that $l = 1$ whenever x is odd but this is not true for conformal algebras.

Let H be a subalgebra of R, M a simple R-module and N a simple H-submodule of R such that N generates M over R. Then we say that M is *induced* from N. This does not mean that we can define an induction functor from H-modules to R-modules; the word "induced" is used here only as a shorthand.

Thus Lemma 8.3 provides a description of induced modules when $\operatorname{rk} H = \operatorname{rk} R - 1$ and $R' \subset H$.

Proposition 8.1. *Let M be a finite simple module over a finite solvable Lie conformal superalgebra R. Then all simple factors of M considered as an $R_{\bar{0}}$-module are of rank 1. The corresponding elements of $\mathcal{L}$ extended by zero to $R_{\bar{1}}$ lie in a single coset $\ell_M \in \mathcal{L}/\mathcal{L}_0$.*

Proof. We use induction on $\operatorname{rk} R$.

Let H be a subalgebra of R such that $R = H \oplus \mathbb{C}[\partial]x$, $R' \subset H$, and x is homogeneous. (Such H always exists by Lemma 8.1.) Two cases can occur: M contains a simple H-module N or there exists a submodule $N \subset M$ such that $H_\lambda N = 0$.

Consider the first case, i.e. let M be induced from N. Then by Lemma 8.3 all H-simple factors of M are isomorphic to N. Thus, if x is odd, M and N viewed as $R_{\bar{0}}$-modules have the same simple factors and we are done by induction. However, if x is even, M and N have the same simple factors only as $H_{\bar{0}}$-modules. However, let then ℓ_1, ℓ_2 be the two forms corresponding to the simple factors and let $\ell = \ell_1 - \ell_2$. By induction $\ell(R'_{\bar{1}}) = 0$, hence we are done.

Consider now the second case, i.e. let M be induced from N such that $H_\lambda N = 0$. Then all simple factors of M are killed by H. Hence, if x is odd, M has no $R_{\bar{0}}$-simple factors. If x is even and we get a simple factor that corresponds to $\ell \in \mathcal{L}$, we get $\ell(R'_{\bar{1}}) = 0$ and thus $\ell \in \mathcal{L}_0$. □

Acknowledgments

The second and the third authors were partially supported by the NSF.

References

1. R.E. Block, "Determination of the differentiably simple rings with a minimal ideal", *Ann. Math.* **90**, No. 2, 433-459 (1969).
2. C. Boyallian, V.G. Kac and J.I. Liberati, "On the classification of subalgebras of $Cend_N$ and gc_N", *J. Algebra*, **260**, 32-63 (2003).
3. S.-J. Cheng, "Differentiably simple Lie superalgebras and representations of semisimple Lie superalgebras", *J. Algebra*, **173**, 1-43 (1995).

4. S.-J. Cheng and V.G. Kac, "A new N=6 superconformal algebra", *Commun. Math. Phys.* **186**, 219-231 (1997).
5. A. D'Andrea and V.G. Kac, "Structure theory of finite conformal algebras", *Selecta Math. (N.S.)* **4**, 377-418 (1998).
6. D. Fattori and V.G. Kac, "Classification of finite simple Lie conformal superalgebras", *J. Algebra*, **258**, 23-59 (2002).
7. V.G. Kac, *Vertex algebras for beginners*, University Lecture Series, Vol. 10 (American Mathematical Society, Providence, RI, 1996, Second Edition 1998).
8. V.G. Kac, "Lie superalgebras", *Adv. Math.* **26**, 8-96 (1977).
9. V.G. Kac, "Superconformal algebras and transitive group actions on quadrics", *Commun. Math. Phys.* **186**, 233-252 (1997), *Erratum*, *Commun. Math. Phys.* **217**, 697-698 (2001).
10. V.G. Kac, "Classification of infinite-dimensional simple linearly compact Lie superalgebras", *Adv. Math.* **139**, 1-55 (1998).
11. V.G. Kac and J.W. van de Leur, "On classification of superconformal algebras", in: *Strings 88*, S.J. Gates et al., eds. (World Sci., Singapore, 1989) pp. 77-106.
12. P. Ramond and J.H. Schwarz, "Classification of dual model gauge superalgebras", *Phys. Lett.* **64B**, 75-77 (1976).
13. K. Schoutens, "$O(N)$-extended superconformal field theory in superspace", *Nucl. Phys.* **B295**, 634-652 (1988).
14. A. Severin, W. Troost and A. Van Proyen, "Superconformal algebras in two dimensions with $N = 4$", *Phys. Lett.* **208B**, 447-450 (1988).
15. G. Yamamoto, "Algebraic structures on quasiprimary states in superconformal algebras", `math.QA/0003012`.

ON CHARACTERS AND DIMENSION FORMULAS FOR REPRESENTATIONS OF THE LIE SUPERALGEBRA $\mathfrak{gl}(M|N)$

E.M. MOENS

Department of Applied Mathematics, Ghent University, Krijgslaan 281-S9, B-9000 Gent, Belgium. e-mail: ElsM.Moens@UGent.be

J. VAN DER JEUGT

Department of Applied Mathematics, Ghent University, Krijgslaan 281-S9, B-9000 Gent, Belgium. e-mail: Joris.VanderJeugt@UGent.be

We derive a new expression for the supersymmetric Schur polynomials $s_\lambda(x/y)$. The origin of this formula goes back to representation theory of the Lie superalgebra $\mathfrak{gl}(m|n)$ and gives rise to a determinantal formula for $s_\lambda(x/y)$. In the second part, we use this determinantal formula to derive new expressions for the dimension and superdimension of covariant representations V_λ of the Lie superalgebra $\mathfrak{gl}(m|n)$. In particular, we derive the t-dimension formula, giving a specialization of the character corresponding to the $\mathbb{Z}$-grading of V_λ. For a special choice of λ, the new t-dimension formula gives rise to a Hankel determinant identity.

1. Introduction

Shortly after the classification of finite-dimensional simple Lie superalgebras [1, 2], Kac considered the problem of classifying all finite-dimensional simple modules (i.e. irreducible representations) of the basic classical Lie superalgebras [3]. For a subclass of these modules, known as the 'typical' modules, Kac derived a character formula closely analogous to the Weyl character for simple modules of simple Lie algebras [3].

The problem of obtaining a character formula for the remaining 'atypical' representations has been subject of intensive investigation but is still not solved other than in various special cases. Solutions were given e.g. for covariant and contravariant tensor representations [4, 5], for so-called generic representations [6], for singly atypical representations [7, 8, 9] or for tame representations [10]. Although there is no explicit character formula, Serganova [11, 12] gave an algorithm to compute the characters of simple modules in $\mathfrak{gl}(m|n)$. This algorithm gives a method to determine the multiplicities $a_{\lambda\mu}$ of the simple modules V_μ in the composition series of a Kac module $\overline{V}_\lambda$, more specific $\mathrm{ch}\,\overline{V}_\lambda = \sum_\mu a_{\lambda,\mu}\,\mathrm{ch}\,V_\mu$ (where λ and μ refer here to the highest weight of the modules). In this algorithm, all weights μ are determined with nonzero $a_{\lambda,\mu}$ for a given weight λ; this turns out to be rather complicated. In [13] the authors conjectured a simple rule to find all nonzero $a_{\lambda,\mu}$ for any given weight μ. This conjecture was proved by Brundan [14].

This paper is concerned with covariant representations of $\mathfrak{gl}(m|n)$, i.e. representations for which the character is already known [4, 5, 15] to correspond to a supersymmetric Schur polynomial. In the following section, we show that these covariant modules are *tame*, leading to an alternative character formula. Comparing the two formulas, this gives rise to a new determinantal formula for supersymmetric Schur polynomials. A short overview of supersymmetric Schur functions is given in Section 3. Finally, in Section 4 we consider a specialization of the character, corresponding to the $\mathbb{Z}$-grading of the representation induced by the $\mathbb{Z}$-grading of $\mathfrak{gl}(m|n)$. This specialization, referred to as the t-dimension of the representation, is computed in two ways.

2. Covariant modules of the Lie superalgebra $\mathfrak{gl}(m|n)$

Let $\mathfrak{g}$ be the Lie superalgebra $\mathfrak{gl}(m|n)$. Lie superalgebras are characterized by a $\mathbb{Z}_2$-grading $\mathfrak{g} = \mathfrak{g}_{\bar{0}} \oplus \mathfrak{g}_{\bar{1}}$. Let $\mathfrak{h} \subset \mathfrak{g}$ be the Cartan subalgebra of $\mathfrak{g}$, and $\mathfrak{g} = \mathfrak{g}_{-1} \oplus \mathfrak{g}_0 \oplus \mathfrak{g}_{+1}$ the consistent $\mathbb{Z}$-grading. The dual space $\mathfrak{h}^*$ of $\mathfrak{h}$ has a natural basis $\{\epsilon_1, \ldots, \epsilon_m, \delta_1, \ldots, \delta_n\}$, and the roots of $\mathfrak{g}$ can be expressed in terms of this basis. Let Δ be the set of all roots, Δ_0 the set of even roots, and Δ_1 the set of odd roots. One can choose a set of simple roots, but note that contrary to the case of simple Lie algebras not all such choices are equivalent. The so-called *distinguished choice* [1] is given by

$$\Pi = \{\epsilon_1 - \epsilon_2, \ldots, \epsilon_{m-1} - \epsilon_m, \epsilon_m - \delta_1, \delta_1 - \delta_2, \ldots, \delta_{n-1} - \delta_n\}. \quad (1)$$

Then, with Δ_+ the set of positive roots, we have explicitly :

$$\begin{aligned} \Delta_{0,+} &= \{\epsilon_i - \epsilon_j | 1 \le i < j \le m\} \cup \{\delta_i - \delta_j | 1 \le i < j \le n\}, \\ \Delta_{1,+} &= \{\beta_{ij} = \epsilon_i - \delta_j | 1 \le i \le m,\ 1 \le j \le n\}. \end{aligned} \quad (2)$$

Thus in the distinguished basis there is only one simple root which is odd. As usual, we put

$$\rho_0 = \frac{1}{2}\Big(\sum_{\alpha\in\Delta_{0,+}} \alpha\Big), \qquad \rho_1 = \frac{1}{2}\Big(\sum_{\alpha\in\Delta_{1,+}} \alpha\Big), \qquad \rho = \rho_0 - \rho_1. \quad (3)$$

There is a symmetric form (,) on $\mathfrak{h}^*$ induced by the invariant symmetric form on $\mathfrak{g}$, and in the natural basis it takes the values $(\epsilon_i, \epsilon_j) = \delta_{ij}$, $(\epsilon_i, \delta_j) = 0$ and $(\delta_i, \delta_j) = -\delta_{ij}$. The odd roots are isotropic: $(\alpha, \alpha) = 0$ if $\alpha \in \Delta_1$. Let $\Lambda \in \mathfrak{h}^*$; the ***atypicality*** of Λ, denoted by atyp(Λ), is the maximal number of linearly independent roots β_i such that $(\beta_i, \beta_j) = 0$ and $(\Lambda, \beta_i) = 0$ for all i and j [10]. Such a set $\{\beta_i\}$ is called a Λ-maximal isotropic subset of Δ.

Given a set of positive roots Δ_+ of Δ, and a simple odd root α, one may construct a new set of positive roots [10, 16] by

$$\Delta'_+ = (\Delta_+ \cup \{-\alpha\}) \setminus \{\alpha\}. \tag{4}$$

The set Δ'_+ is called a simple reflection of Δ_+. Since we use only simple reflections with respect to simple odd roots, $\Delta_{0,+}$ remains invariant, but $\Delta_{1,+}$ will change and the new ρ is given by:

$$\rho' = \rho + \alpha. \tag{5}$$

Consider a finite-dimensional $\mathfrak{g}$-module V. Such modules are $\mathfrak{h}$-diagonalizable with weight decomposition $V = \oplus_\mu V(\mu)$, and the character is defined to be $\operatorname{ch} V = \sum_\mu \dim V(\mu)\, e^\mu$, where e^μ $(\mu \in \mathfrak{h}^*)$ is the formal exponential. If we fix a set of positive roots Δ_+, we may talk about the highest weight Λ of V and about the corresponding ρ. If Δ'_+ is obtained from Δ_+ by a simple α-reflection, where α is odd, and Λ' denotes the highest weight of V with respect to Δ'_+, then [10]

$$\Lambda' = \Lambda - \alpha \text{ if } (\Lambda, \alpha) \neq 0; \qquad \Lambda' = \Lambda \text{ if } (\Lambda, \alpha) = 0. \tag{6}$$

If α is a *simple* odd root from Δ_+ then $(\rho, \alpha) = \frac{1}{2}(\alpha, \alpha) = 0$ [10, p. 421], and therefore, following (5) and (6):

$$\begin{aligned} \Lambda' + \rho' &= \Lambda + \rho \ \text{ if } \ (\Lambda + \rho, \alpha) \neq 0, \\ \Lambda' + \rho' &= \Lambda + \rho + \alpha \ \text{ if } \ (\Lambda + \rho, \alpha) = 0. \end{aligned} \tag{7}$$

From this, one deduces that for the $\mathfrak{g}$-module V, atyp($\Lambda + \rho$) is independent of the choice of Δ_+; then atyp($\Lambda + \rho$) is referred to as the atypicality of V (if atyp($\Lambda + \rho$) $= 0$, V is typical, otherwise it is atypical). If one can choose a $(\Lambda + \rho)$-maximal isotropic subset S_Λ in Δ_+ such that $S_\Lambda \subset \Pi \subset \Delta_+$ (Π is the set of simple roots with respect to Δ_+), then the $\mathfrak{g}$-module V is called *tame*, and a character formula is known due to Kac and Wakimoto [10]. It reads:

$$\operatorname{ch} V = j_\Lambda^{-1} e^{-\rho} R^{-1} \sum_{w \in W} \varepsilon(w) w \Big(e^{\Lambda + \rho} \prod_{\beta \in S_\Lambda} (1 + e^{-\beta})^{-1} \Big), \tag{8}$$

where

$$R = \prod_{\alpha \in \Delta_{0,+}} (1 - e^{-\alpha}) \Big/ \prod_{\alpha \in \Delta_{1,+}} (1 + e^{-\alpha}) \tag{9}$$

and j_Λ is a normalization coefficient to make sure that the coefficient of e^Λ on the right hand side of (8) is 1.

Covariant modules, a particular class of finite-dimensional simple $\mathfrak{gl}(m|n)$-modules, are characterized by a partition λ inside the (m,n)-hook, i.e. such that $\lambda_{m+1} \leq n$. The representation characterized by the partition λ is denoted by V_λ. In the distinguished basis fixed by (1), the highest weight Λ_λ of V_λ in the standard ϵ-δ-basis is given by [17]

$$\Lambda_\lambda = \lambda_1\epsilon_1 + \cdots + \lambda_m\epsilon_m + \nu'_1\delta_1 + \cdots + \nu'_n\delta_n, \tag{10}$$

where $\nu'_j = \max\{0, \lambda'_j - m\}$ for $1 \leq j \leq n$ (and λ' is the partition conjugate to λ). Let us consider the atypicality of V_λ, in the distinguished basis. It is sufficient to compute the numbers $(\Lambda_\lambda + \rho, \beta_{ij})$, with $\beta_{ij} = \epsilon_i - \delta_j$, for $1 \leq i \leq m$ and $1 \leq j \leq n$, and count the number of zeros. It is convenient to put these numbers $(\Lambda_\lambda + \rho, \beta_{ij})$ in a $(m \times n)$-matrix (the atypicality matrix [8, 17]).

From the combinatorics of atypicality matrices [8], the following proposition holds.

Proposition 2.1. *The atypicality matrix of a partition λ has its zeros on row i and in column $\lambda_i + m - i + 1$ ($i = m, m-1, \ldots$) as long as these column indices are not exceeding n.*

With a partition λ we can associate the (m,n)-index k:

Definition 2.2. For λ a partition inside the (m,n)-hook, the (m,n)-index of λ is the number

$$k = \min\{i | \lambda_i + m + 1 - i \leq n\}, \qquad (1 \leq k \leq m+1). \tag{11}$$

In the rest of this paper, k will denote the (m,n)-index of λ. By Proposition 2.1, $m-k+1$ is the atypicality of V_λ. If $k = m+1$, then $S_{\Lambda_\lambda} = \emptyset$ and V_λ is typical and trivially tame. Thus in the following, we shall assume that $k \leq m$. In the initial situation, Δ_+ corresponds to the distinguished choice, and Π is the distinguished set of simple roots (1). The highest weight of V_λ is given by Λ_λ. Denote $\Lambda^{(1)} = \Lambda_\lambda$, $\rho^{(1)} = \rho$ and $\Pi^{(1)} = \Pi$. Now we perform a sequence of simple odd $\alpha^{(i)}$-reflections; each of these reflections preserve $\Delta_{0,+}$ but may change $\Lambda^{(i)}+\rho^{(i)}$ and $\Pi^{(i)}$. Denote the sequence of reflections by:

$$\Lambda^{(1)}+\rho^{(1)},\,\Pi^{(1)} \quad \overset{\alpha^{(1)}}{\longrightarrow} \quad \Lambda^{(2)}+\rho^{(2)},\,\Pi^{(2)} \quad \overset{\alpha^{(2)}}{\longrightarrow} \quad \cdots \quad \overset{\alpha^{(f)}}{\longrightarrow} \quad \Lambda'+\rho',\,\Pi'$$

where, at each stage, $\alpha^{(i)}$ is an odd root from $\Pi^{(i)}$. For given λ, consider the following sequence of odd roots (with positions on row m, row $m-1$, ..., row k):

$$\begin{array}{ll} \text{row } m: & \beta_{m,1},\beta_{m,2},\ldots,\beta_{m,\lambda_k-k+m} \\ \text{row } m-1: & \beta_{m-1,1},\beta_{m-1,2},\ldots,\beta_{m-1,\lambda_k-k+m-1} \\ \vdots & \vdots \\ \text{row } k: & \beta_{k,1},\beta_{k,2},\ldots,\beta_{k,\lambda_k} \end{array} \tag{12}$$

in this particular order (i.e. starting with $\beta_{m,1}$ and ending with β_{k,λ_k}). Then we have [18]:

Lemma 2.3. *The sequence (12) is a proper sequence of simple odd reflections for Λ_λ, i.e. $\alpha^{(i)}$ is a simple odd root from $\Pi^{(i)}$. At the end of the sequence, one finds:*

$$\begin{aligned}\Pi' = \{&\epsilon_1-\epsilon_2,\epsilon_2-\epsilon_3,\ldots,\epsilon_{k-2}-\epsilon_{k-1},\epsilon_{k-1}-\delta_1,\delta_1-\delta_2,\delta_2-\delta_3,\ldots,\\ &\delta_{\lambda_k-1}-\delta_{\lambda_k},\delta_{\lambda_k}-\epsilon_k,\epsilon_k-\delta_{\lambda_k+1},\delta_{\lambda_k+1}-\epsilon_{k+1},\epsilon_{k+1}-\delta_{\lambda_k+2},\ldots,\\ &\delta_{\lambda_k+m-k}-\epsilon_m,\epsilon_m-\delta_{\lambda_k+m+1-k},\delta_{\lambda_k+m+1-k}-\delta_{\lambda_k+m+2-k},\ldots,\delta_{n-1}-\delta_n\}.\end{aligned}$$

Furthermore,

$$\Lambda'+\rho' = \Lambda_\lambda+\rho+\sum_{i=k+1}^{m}\ \sum_{j=\lambda_i+1}^{\lambda_k-k+i}\beta_{i,j}. \tag{13}$$

The set Π' in Lemma 2.3 contains a maximal isotropic subset

$$S_{\Lambda'} = \{\epsilon_k-\delta_{\lambda_k+1},\epsilon_{k+1}-\delta_{\lambda_k+2},\ldots,\epsilon_m-\delta_{\lambda_k+m+1-k}\}.$$

So we have:

Corollary 2.4. *Every covariant module V_λ is tame.*

Thus, the character formula of Kac and Wakimoto becomes:

$$\operatorname{ch} V_\lambda = j_{\Lambda'}^{-1}e^{-\rho'}R'^{-1}\sum_{w\in W}\varepsilon(w)w\Big(e^{\Lambda'+\rho'}\prod_{\beta\in S_{\Lambda'}}(1+e^{-\beta})^{-1}\Big).$$

Putting, as usual, $e^{\epsilon_i} = x_i$ and $e^{\delta_i} = y_i$, we have shown [18] that this expression can be rewritten in a determinantal form:

$$\operatorname{ch} V_\lambda = \pm D^{-1} \det \begin{pmatrix} \left(\frac{1}{x_i+y_j}\right)_{\substack{1\le i\le m\\ 1\le j\le n}} & \left(x_i^{\lambda_j+m-n-j}\right)_{\substack{1\le i\le m\\ 1\le j\le k-1}} \\ \left(y_j^{\lambda'_i+n-m-i}\right)_{\substack{1\le i\le n-m+k-1\\ 1\le j\le n}} & 0 \end{pmatrix} \tag{14}$$

with

$$D = \frac{\prod_{i<j}(x_i - x_j)\prod_{i<j}(y_i - y_j)}{\prod_{i,j}(x_i + y_j)}.$$

3. Supersymmetric Schur polynomials

Let us now recall some notions of *supersymmetric S-functions* [4, 19, 20]. The ring of doubly symmetric polynomials in $x = (x_1, \ldots, x_m)$ and $y = (y_1, \ldots, y_n)$ is $S(x,y) = S(x) \otimes_{\mathbb{Z}} S(y)$. An element $p \in S(x,y)$ has the *cancellation property* if it satisfies the following: when the substitution $x_1 = t$, $y_1 = -t$ is made in p, the resulting polynomial is independent of t. We denote $S(x/y)$ the subring of $S(x,y)$ consisting of the elements satisfying the cancellation property. The elements of $S(x/y)$ are the supersymmetric polynomials [20].

The complete supersymmetric functions $h_r(x/y)$ belong to $S(x/y)$, and are defined by

$$h_r(x/y) = \sum_{j=0}^{r} h_{r-j}(x)e_j(y), \tag{15}$$

where $h_i(x)$ resp. $e_j(y)$ are the usual complete resp. elementary symmetric functions [21]. We can use $h_r(x/y)$ to give a first formula for the supersymmetric S-functions:

$$s_\lambda(x/y) = \det\Big(h_{\lambda_i - i + j}(x/y)\Big)_{1\le i,j\le \ell(\lambda)}, \tag{16}$$

where $\ell(\lambda)$ denotes the length of λ. The polynomials $s_\lambda(x/y)$ are identically zero when $\lambda_{m+1} > n$.

Berele and Regev [4] showed that the character of a covariant module is a supersymmetric Schur function: $\operatorname{ch} V_\lambda = s_\lambda(x/y)$. Thus, our formula (14) gives an alternative expression for $s_\lambda(x/y)$. In fact, (14) is the supersymmetric equivalent of the determinantal formula for the symmetric Schur polynomials as a quotient of two alternants.

In [21, §I.3, exercise 23], Macdonald shows that the supersymmetric S-polynomials satisfy four properties (see also [22]) which also characterize these polynomials. [a] We have shown that the polynomials defined by means of the right hand side of (14) do indeed satisfy these four properties. This provides us with an independent proof of the determinantal expression (14).

4. The t-dimension of a covariant module

Consider a finite-dimensional irreducible representation V of $\mathfrak{g}$ with weight decomposition $V = \oplus_\mu V(\mu)$. Recall that the character is $\operatorname{ch} V = \sum_\mu \dim V(\mu)\, e^\mu$. We shall consider a specialization of the character determined by

$$\begin{aligned} F(e^{\epsilon_i}) &= 1 \ (i = 1, \ldots, m) \\ F(e^{\delta_j}) &= t \ (j = 1, \ldots, n). \end{aligned} \tag{17}$$

This specialization is consistent with the $\mathbb{Z}$-grading of $\mathfrak{g}$, and the corresponding $\mathbb{Z}$-grading of V. The specialization of the character of V under F is referred to as the t-dimension of V and denoted by $\dim_t(V)$:

$$\dim_t(V) = F(\operatorname{ch} V) = \sum_\mu \dim V(\mu) F(e^\mu). \tag{18}$$

The t-dimension of V stands for the polynomial

$$F(e^\Lambda) \sum_{j \in \mathbb{Z}_+} (\dim V_{-j})\, t^j, \tag{19}$$

where $V = V_0 \oplus V_{-1} \oplus V_{-2} \oplus \cdots$ is the $\mathbb{Z}$-grading of V. Note that for the $\mathbb{Z}_2$-grading $V = V_{\bar{0}} \oplus V_{\bar{1}}$ we have $V_{\bar{0}} = V_0 \oplus V_{-2} \oplus \cdots$ and $V_{\bar{1}} = V_{-1} \oplus V_{-3} \oplus \cdots$. Therefore, the ordinary dimension of V is found by putting $t = 1$ in the expression for the t-dimension, whereas the superdimension of V is found by putting $t = -1$. So the t-dimension can also be seen as an extension of the notion of dimension and superdimension.

In order to compute the t-dimension there are two useful formulas, namely (16) and (14). Since $x_i = e^{\epsilon_i}$ and $y_j = e^{\delta_j}$, the specialization (17) corresponds to putting each $x_i = 1$ and $y_j = t$ in $s_\lambda(x/y)$. For the elementary and complete symmetric functions, such specializations are well-

---

[a] Our convention for the $s_\lambda(x/y)$ is slightly different from that of Macdonald's $s_\lambda^{\mathrm{Mac}}(x/y)$: $s_\lambda(x/y) = s_\lambda^{\mathrm{Mac}}(x/-y)$.

known:

$$h_r(x_1,\dots,x_m)\Big|_{x_i=1} = \binom{m+r-1}{r} = \binom{m+r-1}{m-1}, \tag{20}$$

$$e_r(x_1,\dots,x_m)\Big|_{x_i=1} = \binom{m}{r}. \tag{21}$$

Thus it follows from (15) and (16) that

Proposition 4.1. *The t-dimension of V_λ is given by the determinant*

$$\dim_t(V_\lambda) = \det_{1\le i,j\le \ell(\lambda)} \left(\sum_{l=0}^{\lambda_i-i+j} \binom{m+\lambda_i-i+j-l-1}{\lambda_i-i+j-l} \binom{n}{l} t^l \right). \tag{22}$$

Although this formula is simple to derive, it should be observed that in general the matrix elements in the right hand side of (22) do not have a "closed form" expression [23]: they remain polynomials of type ${}_2F_1(-t)$ in t. Even for $t=1$, the expression $\sum_{l=1}^{r}\binom{m+r-l-1}{r-l}\binom{n}{l}$ cannot be simplified in general. Only for $t=-1$, i.e. the superdimension formula $\operatorname{sdim} V_\lambda$, the expression (22) can be simplified:

$$\operatorname{sdim} V_\lambda = \det_{1\le i,j\le \ell(\lambda)} \left(\binom{m-n-1+\lambda_i-i+j}{\lambda_i-i+j} \right) \tag{23}$$

$$= \frac{\prod_{i<j}(\lambda_i-i-\lambda_j+j)}{\prod_i(\lambda_i-i+l(\lambda))} \prod_i (m-n+1-i)_{\lambda_i}. \tag{24}$$

Herein, $(a)_n = a(a+1)\cdots(a+n-1)$ is the Pochhammer symbol [24], and the determinant in (23) can be written in closed form using [25, (3.11)]. So in general the superdimension has a closed form expression (24), whereas the dimension has not.

Starting from the new determinantal formula (14) and using properties of symmetric polynomials and row and column operations, we can deduce a new t-dimension formula (see [26]):

Theorem 4.2. *The t-dimension of V_λ is* $\dim_t(V_\lambda) = \pm(1+t)^{mn}\det(R(\lambda))$, *with $R(\lambda)$ given by*

$$\begin{pmatrix} \left(\frac{(-1)^{i+j}}{(1+t)^{i+j-1}} \binom{i+j-2}{j-1} \right)_{\substack{1\le i\le m\\ 1\le j\le n}} & \left(\binom{\lambda_j+m-n-j}{i-1} \right)_{\substack{1\le i\le m\\ 1\le j\le k-1}} \\ \left(t^{\lambda'_i+n-m-i-j+1} \binom{\lambda'_i+n-m-i}{j-1} \right)_{\substack{1\le i\le n-m+k-1\\ 1\le j\le n}} & 0 \end{pmatrix}. \tag{25}$$

Compared to (22), this formula has the advantage that each matrix element is a simple binomial coefficient multiplied by a power of t or $(1+t)$, and no longer a finite series of type ${}_2F_1(-t)$. The order of the matrix in (22) is the length of the partition where as the order of the matrix in (25) is always less than or equal to $m+n$. So in general the second formula is easier to compute, see Example 4.3.

Example 4.3. In $\mathfrak{gl}(2,2)$, with $\lambda = (4,1,1,1)$, (22) resp. (25) yield:

$$\dim_t(V_{(4,1,1,1)}) =$$

$$= \det \begin{pmatrix} 5+8t+3t^2 & 6+10t+4t^2 & 7+12t+5t^2 & 8+14t+6t^2 \\ 1 & 2+2t & 3+4t+t^2 & 4+6t+2t^2 \\ 0 & 1 & 2+2t & 3+4t+t^2 \\ 0 & 0 & 1 & 2+2t \end{pmatrix}$$

$$= (1+t)^4 \det \begin{pmatrix} \frac{1}{1+t} & \frac{-1}{(1+t)^2} & 1 \\ \frac{-1}{(1+t)^2} & \frac{2}{(1+t)^3} & 3 \\ t^3 & 3t^2 & 0 \end{pmatrix} = 2t^2(6t^2+13t+6)(1+t).$$

From the equality of the two formulas (22) and (25) we can deduce certain properties of the t-dimension for particular V_λ. An interesting application follows for the special case of $\lambda = \left((n-a)^{(m-a)}\right)$, where $a = 0,1,\ldots,\min(m,n)$. For such partitions, the determinant given in (25) can be reduced and simplifies to:

$$\dim_t(V_\lambda) = (1+t)^{(m-a)(n-a)} \det_{1\le i,j\le a}\left(\binom{i+j+m+n-2a-2}{j-1} \right).$$

Taking out common factors in rows and columns in this determinant, it becomes

$$\prod_{i=1}^{a} \frac{(i+m+n-2a-1)!}{(i+n-a-1)!(i+m-a-1)!} \det_{1\le i,j\le a}\left((m+n-2a+i)_{j-1}\right).$$

This remaining determinant can easily be computed (see [26]). So we finally obtain

$$\dim_t(V_\lambda) = (1+t)^{(m-a)(n-a)} \det_{1\le i,j\le a}\left(\binom{i+j+m+n-2a-2}{j-1} \right)$$

$$= (1+t)^{(m-a)(n-a)} \prod_{i=0}^{a-1} \frac{\binom{m+n-2a+i}{n-a+i}}{\binom{m-a+i}{i}}.$$

Using this, we can prove the following result:

Corollary 4.4. *Let r and s be positive integers with $s \leq r$, u and v arbitrary variables, and*

$$A_i = \sum_{l=0}^{r} \binom{u+i-l}{r-l}\binom{u}{l} v^l = \binom{u+i}{r}\, {}_2F_1\left(\begin{matrix}-r, -u\\ -u-i\end{matrix}; -v\right).$$

Then the Hankel determinant is given by

$$\det_{0\leq i,j\leq s}(A_{i+j}) = (-1)^{s(s+1)/2}(1+v)^{(s+1)(r-s)} \prod_{i=1}^{r-s} \frac{\binom{2s+u-r+i}{s+1}}{\binom{s+i}{s+1}}.$$

References

1. V.G. Kac, *Adv. Math.* **26**, 8 (1977).
2. M. Scheunert, *The theory of Lie superalgebras* (Springer, Berlin, 1979).
3. V.G. Kac, *Lecture Notes Math.* **676**, 597 (1978).
4. A. Berele and A. Regev, *Adv. Math.* **64**, 118 (1987).
5. P.H. Dondi and P.D. Jarvis *J. Phys.* **A14**, 547 (1981).
6. I. Penkov and V. Serganova, *Int. J. Math.* **5**, 389 (1994).
7. I.N. Bernstein and D.A. Leites, *C.R. Acad. Bulg. Sci.* **33**, 1049 (1980).
8. J. Van der Jeugt, J.W.B. Hughes, R.C. King and J. Thierry-Mieg, *Comm. Algebra* **18**, 3453 (1990).
9. J. Van der Jeugt, *Comm. Algebra* **19**, 199 (1991).
10. V.G. Kac and M. Wakimoto, *Progr. Math.* **123**, 415 (1994).
11. V. Serganova, *Adv. Sov. Math.* **16**, 151 (1993).
12. V. Serganova, *Selecta Math.* **2**, 607 (1996).
13. J. Van der Jeugt and R.B. Zhang, *Lett. Math. Phys.* **47**, 49 (1999).
14. J. Brundan, *J. Amer. Math. Soc.* **16**, 185 (2003).
15. A.N. Sergeev, *Mat. Sb.* **123**, 422 (1984).
16. I. Penkov and V. Serganova, *Indag. Math.* **3**, 419 (1992).
17. J. Van der Jeugt, J.W.B. Hughes, R.C. King and J. Thierry-Mieg, *J. Math. Phys.* **31**, 2278 (1990).
18. E.M. Moens and J. Van der Jeugt, *J. Alg. Comb.* **17**, 283 (2003).
19. R.C. King, *Ars Combin.* **16A**, 269 (1983).
20. J. Stembridge, *J. Alg.* **95**, 439 (1985).
21. I.G. Macdonald, *Symmetric functions and Hall polynomials* (Oxford University Press, 2nd edition, 1995).
22. P. Pragacz and A. Thorup, *Adv. Math.* **95**, 8 (1992).
23. M. Petkovšek M, H.S. Wilf and D. Zeilberger, *A=B* (A.K. Peters, Massachusetts, 1996).
24. L.J. Slater, *Generalized Hypergeometric Functions* (Cambridge University Press, Cambridge, 1966).
25. C. Krattenthaler, *Sém. Lothar. Combin.* **42**, B42q (1999).
26. E.M. Moens and J. Van der Jeugt, *J. Lie Theory*, in press (2004).

MATCHING CONDITIONS FOR INVARIANT EIGENDISTRIBUTIONS ON SOME SEMISIMPLE SYMMETRIC SPACES

SHIGERU AOKI

Takushoku University,815-1,Tatemachi,Hachioji-shi,Tokyo 193-0985,Japan
e-mail:aoki@la.takushoku-u.ac.jp

SUEHIRO KATO

Kitasato University,1-15-1,Kitasato,Sagamihara,Kanagawa 228-8555,Japan
e-mail:kato@clas.kitasato-u.ac.jp

Let $X = G/H$ be a semisimple symmetric space whose complexification is isomorphic to $GL(p+q,\mathbf{C})/(GL(p,\mathbf{C}) \times GL(q,\mathbf{C}))$. An IED is by definition an H-invariant joint eigendistribution of the G-invariant differential operators of X. If X is Riemannian of the non-compact type, for the zonal spherical functions (i.e. IED's) on X we have a beautiful explicit formula due to Berezin-Karpelevič [6]. In this contribution, we outline a method to attack non-Riemannian cases and give an explicit form of IED's in the case of $X = GL(4,\mathbf{R})/(GL(2,\mathbf{R}) \times GL(2,\mathbf{R}))$ as an example. Our method may be considered as a generalization of that of Lie group case by Hirai and is based on another result of Berezin-Karpelevič.

1. Matching Conditions for IED's

Let $X = G/H$ be a semisimple symmetric space and $\mathcal{O}$ an H-invariant open subset of X. Let $\mathbf{D}(X)$ be the ring of invariant differential operators on X, and χ a character of $\mathbf{D}(X)$: $\chi \in \mathrm{Hom}(\mathbf{D}(X), \mathbf{C})$.

Definition 1. A distribution $\Theta \in \mathcal{D}'(\mathcal{O})$ is said to be an *Invariant Eigendistribution (IED)* with χ on $\mathcal{O}$, if it satisfies the following two conditions:

i) Θ : H-invariant

ii) $D.\Theta = \chi(D)\Theta \quad (\forall D \in \mathbf{D}(X))$

Remarks. i) If $\mathcal{O}_1 \subset \mathcal{O}_2$, then, for any IED Θ with χ on $\mathcal{O}_2$, the restriction $\Theta|_{\mathcal{O}_1}$ is obviously an IED with χ on $\mathcal{O}_1$.

ii) Let X' be the set of regular semisimple elements of X, which is open dense and H-invariant in X. Any IED on X' is necessarily a real analytic function.

iii) Let $\{J_l \,|\, l \in L\}$ be a complete system of representatives of H-conjugacy classes of Cartan subspaces of X. Put $J_l' = J_l \cap X'$. Then we have $X' = \bigsqcup_{l \in L} H.J_l'$.

In view of above remarks, for any IED Θ on X, the restriction $\Pi = \Theta|_{X'}$ is a real analytic function. In this way, putting $\Pi_l := \Theta|_{J_l'}$, we have a system of real analytic functions $\{\Pi_l\}_{l \in L}$. We study *global matching conditions*, i.e. compatibility conditions among Π_l's, to give an explicit form of IED's on $\bigsqcup_{l \in L} J_l'$.

To this end, first, for each semiregular semisimple element $x_0 \in \bigcup_{l \in L} J_l$, by the procedure described in Diag.1, we establish *local matching conditions*, i.e. compatibility conditions on a neighborhood of x_0. Then, suitably reformulating these conditions, we gather them up for all such x_0's.

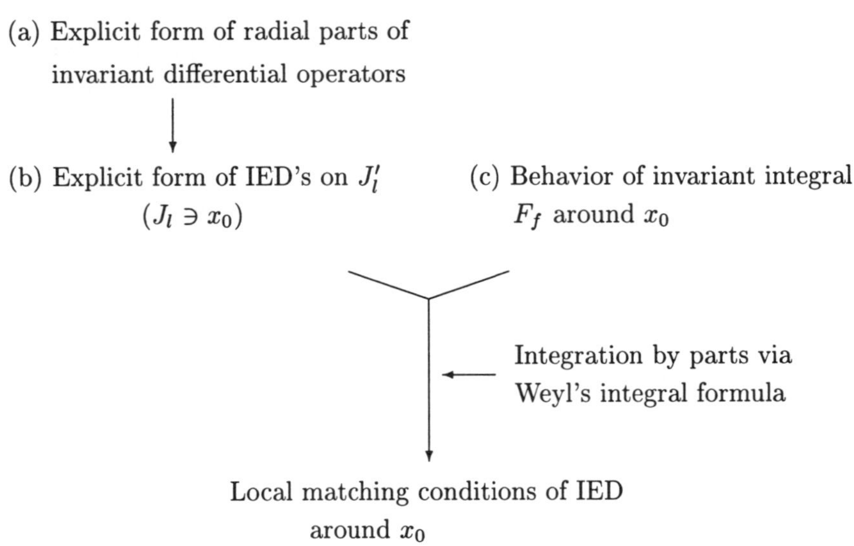

Diag.1

Definition 2. Let f be a function in $C_c^\infty(X)$. By the *invariant integral*

F_f of f, we mean a function F_f on $\bigsqcup_{l\in L} J_l'$ defined for $x \in J_l'$ by

$$F_f(x) = {}_X F_f(x) := \int_{H/Z_H(J_l)} f(h.x)\, d\tilde{h},$$

where $d\tilde{h}$ is a suitably normalized H-invariant measure on $H/Z_H(J_l)$.

Supposing that the complexification of X is locally isomorphic to $GL(p+q,\mathbf{C})/(GL(p,\mathbf{C}) \times GL(q,\mathbf{C}))$ for some integers p, q, we now explain Diag.1 in more detail:

(a) is essentially due to Berezin-Karpelevič [6] and Hoogenboom [9], because they gave the result for Riemannian symmetric spaces of the non-compact type with the complexification in question. Note that radial parts of invariant differential operators are, in our case, written using a hypergeometric type differential operator with 1 variable in a variables-separated way (cf. Lemma.).

(b) is straightforward using (a), since Π_l is a joint eigenfunction of radial parts.

(c) is investigated by using "rank reduction principle" (cf. Aoki-Kato [3]): near x_0, invariant integrals of X behave similarly to those of the attached symmetric subspace $X_{\{x_0\}}$, which is essentially of rank 1, since x_0 is semiregular. Note that the behavior of invariant integrals is well-known in rank 1 cases. Note also that for (c) we need not the supposition on the complexification.

Let J_a, J_b be (connected components of) non H-conjugate Cartan subspaces. If there is a semiregular element belonging to both of them, they are said to be *adjacent*. In this case, we denote $J_a \to J_b$ if the dimension of the split part of J_a is greater than that of J_b. (In fact, $J_a \to J_b$ means the existence of "Cayley transform" which ties up J_a with J_b (cf. Aoki-Kato [5])).

Let us mention some history of IED's on semisimple symmetric spaces (and of related topics). The rank 1 case was studied by many authors (e.g. Faraut [7], Aoki-Kato [1]). In the group case $X = (G' \times G')/\Delta G' \cong G'$, Hirai [8] studied matching conditions in the framework of his study of characters. Similar study for its C-dual $X = G'_c/G'$ was initiates by Sano [10]. For higher rank, except the Riemannian case, these two cases seemed to be main for the moment. So we studied some examples of higher rank not belonging to the cases above : $X = (U(p,q)/(U(r) \times U(p-r,q))$, $U(n,n)/GL(n,\mathbf{C})$,

$U(2,2)/(U(1,1)\times U(1,1))$, $GL(4,\mathbf{R})/(GL(2,\mathbf{R})\times GL(2,\mathbf{R}))$, each of which has the complexification X_c of the same type.

By the way, in Berezin-Karpelevič [6] treating $U(p,q)/(U(p)\times U(q))$, were stated two interesting theorems, whose proofs were given later by Hoogenboom [9]. One is on the radial parts, another on IED's. In the latter, each IED is expressed as the determinant whose elements are IED's of the rank 1 case, which turn to be hypergeometric functions.

Although our challenge in progress is to treat every symmetric spaces X with $X_c = GL(p+q,\mathbf{C})/(GL(p,\mathbf{C})\times GL(q,\mathbf{C}))$, in a systematic way (cf.Aoki-Kato [5]), we restrict ourselves to the example $GL(4,\mathbf{R})/(GL(2,\mathbf{R})\times GL(2,\mathbf{R}))$ of rank 2, in the next section, to describe the procedure in more detail and to see minutely the explicit form of IED's, which retains some flavor of the formula of Berezin-Karpelevič.

2. Case of $X = GL(4,\mathbf{R})/\,(GL(2,\mathbf{R})\times GL(2,\mathbf{R}))$

Let $G = GL(4,\mathbf{R})$, and σ an involution defined by $\sigma(g) = I_{2,2}gI_{2,2}$, where $I_{2,2} = \mathrm{diag}(1,1,-1,-1)$. Then the fixed points group $H = G^\sigma$ is isomorphic to $GL(2,\mathbf{R})\times GL(2,\mathbf{R})$. In this section, we treat $X = G/H = GL(4,\mathbf{R})/\,(GL(2,\mathbf{R})\times GL(2,\mathbf{R}))$. We identify X with the closed subset $\tilde{\sigma}(X)$ of G, where $\tilde{\sigma}(gH) = g\sigma(g^{-1})$.

2.1. *Cartan Subspaces and Weyl Groups*

$$j_0(\theta_1,\theta_2;\epsilon_1,\epsilon_2) = \begin{pmatrix} \epsilon_1\cosh\theta_1 & 0 & \epsilon_1\sinh\theta_1 & 0 \\ 0 & \epsilon_2\cosh\theta_2 & 0 & \epsilon_2\sinh\theta_2 \\ \epsilon_1\sinh\theta_1 & 0 & \epsilon_1\cosh\theta_1 & 0 \\ 0 & \epsilon_2\sinh\theta_2 & 0 & \epsilon_2\cosh\theta_2 \end{pmatrix}$$

$$j_1(\theta_1,\theta_2;\epsilon) = \begin{pmatrix} \cos\theta_1 & 0 & \sin\theta_1 & 0 \\ 0 & \epsilon\cosh\theta_2 & 0 & \epsilon\sinh\theta_2 \\ -\sin\theta_1 & 0 & \cos\theta_1 & 0 \\ 0 & \epsilon\sinh\theta_2 & 0 & \epsilon\cosh\theta_2 \end{pmatrix}$$

$$\tilde{j}_1(\theta_1,\theta_2) = \begin{pmatrix} \cos\theta_1 & 0 & 0 & \sin\theta_1 \\ 0 & \cos\theta_1 & -\sin\theta_1 & 0 \\ 0 & \sin\theta_1 & \cos\theta_1 & 0 \\ -\sin\theta_1 & 0 & 0 & \cos\theta_1 \end{pmatrix} j_0(\theta_2,\theta_2;+,+)$$

$$j_2(\theta_1,\theta_2) = \begin{pmatrix} \cos\theta_1 & 0 & \sin\theta_1 & 0 \\ 0 & \cos\theta_2 & 0 & \sin\theta_2 \\ -\sin\theta_1 & 0 & \cos\theta_1 & 0 \\ 0 & -\sin\theta_2 & 0 & \cos\theta_2 \end{pmatrix}$$

Table 1

Using the notations of Table 1, put

$$\begin{aligned} J_0^{\epsilon_1,\epsilon_2} &= \{j_0(\theta_1,\theta_2;\epsilon_1,\epsilon_2)\,|\,\theta_1,\theta_2 \in \mathbf{R}\} \qquad (\epsilon_1,\epsilon_2 = \pm), \\ J_1^{\epsilon} &= \{j_1(\theta_1,\theta_2;\epsilon)\,|\,\theta_1,\theta_2 \in \mathbf{R}\} \qquad (\epsilon = \pm), \\ J_2 &= \{j_2(\theta_1,\theta_2)\,|\,\theta_1,\theta_2 \in \mathbf{R}\}, \\ \tilde{J}_1 &= \{\tilde{j}_1(\theta_1,\theta_2)\,|\,\theta_1,\theta_2 \in \mathbf{R}\}, \\ J_0 &= J_0^{++} \sqcup J_0^{+-} \sqcup J_0^{-+} \sqcup J_0^{--}, \quad \text{and} \\ J_1 &= J_1^{+} \sqcup J_1^{-}. \end{aligned}$$

Then each of the last two equations gives the decomposition into connected components and we have a complete system $\{J_0, J_1, \tilde{J}_1, J_2\}$ of representatives of H-conjugacy classes of Cartan subspaces for $X = GL(4,\mathbf{R})/\,(GL(2,\mathbf{R}) \times GL(2,\mathbf{R}))$. We note that these Cartan subspaces are now considered as the subspaces of $\tilde{\sigma}(X)$ which is identified with X.

For an element x in J_l ($l = 0,1,2$), we denote by t_j its (j, j)-entry ($j = 1, 2$). For example, we have $t_1 = \epsilon_1 \cosh\theta_1$, $t_2 = \epsilon_2 \cosh\theta_2$ for $x = j_0(\theta_1,\theta_2;\epsilon_1,\epsilon_2) \in J_0$. Furthermore, for an element $x = \tilde{j}_1(\theta_1,\theta_2) \in \tilde{J}_1$, we put $t_1 = \cosh(-\theta_2 + i\theta_1)$ and $t_2 = \cosh(\theta_2 + i\theta_1)$. By the correspondence of $x \in J_0 \cup J_1 \cup \tilde{J}_1 \cup J_2$ to the "t-variables" (t_1, t_2), one can identify $J_0/(\mathbf{Z}_2)^2$ with $((-\infty,-1] \sqcup [1,\infty))^2$, $J_1/(\mathbf{Z}_2)^2$ with $[-1,1] \times ((-\infty,-1] \sqcup [1,\infty))$, $J_2/(\mathbf{Z}_2)^2$ with $([-1,1])^2$, and $\tilde{J}_1/\triangle\mathbf{Z}_2$ with $\{(t_1,t_2)\,|\,t_1 = \overline{t_2} \in \mathbf{C}\}$. Then, for elements $(t_1,t_2), (t_1',t_2') \in J_l/(\mathbf{Z}_2)^2$ with $l = 0,1,2$, (t_1,t_2) is mapped to (t_1',t_2') by the Weyl group $W(J_l)$ if and only if

$$\begin{cases} (t_1,t_2) = (t_1',t_2') \text{ or } (t_1,t_2) = (t_2',t_1') & (l = 0,2) \\ (t_1,t_2) = (t_1',t_2') & (l = 1). \end{cases}$$

For $\tilde{J}_1$, (t_1,t_2) is mapped to (t_1',t_2') by $W(\tilde{J}_l)$ if and only if

$$(t_1,t_2) = (t_1',t_2') \text{ or } (t_1,t_2) = (t_2',t_1').$$

2.2. *Radial Parts of Invariant Differential Operators*

Put

$$
\begin{aligned}
&\omega = t_2 - t_1,\\
&L = 4(t^2-1)\frac{d^2}{dt^2} + 8t\frac{d}{dt},\\
&L_j = 4(t_j^2-1)\frac{\partial^2}{\partial t_j^2} + 8t_j\frac{\partial}{\partial t_j} \qquad (j = 1,\, 2), \text{ and}\\
&\mathcal{S} = \{\omega^{-1}S(L_1, L_2)\omega \mid S\colon \text{symmetric polynomial with 2-variables}\}.
\end{aligned}
$$

Lemma. (Hoogenboom [9], Berezin-Karpelevič [6]). *We have the isomorphism* Φ *of* $\mathbf{D}(X)$ *to* $\mathcal{S}$ *satisfying*

$$\Phi(D)|_{J'}(f|_{J'}) = (Df)|_{J'}$$

$$\textit{for any } D \in \mathbf{D}(X),\ f \in C_H^\infty(X),\ J = J_0, J_1, \tilde{J}_1, J_2,$$

where $C_H^\infty(X)$ *is the space of all* H*-invariant* C^∞*-functions on* X.

In virtue of this fact, we denote by $\chi_{\lambda_1,\lambda_2}$ the unique character χ of $\mathbf{D}(X)$ satisfying the condition

$$\chi(\Phi^{-1}(\omega^{-1}S(L_1, L_2)\omega)) = S(\lambda_1,\, \lambda_2)$$

for any symmetric polynomial S with 2-variables. We note $\chi_{\lambda_2,\lambda_1} = \chi_{\lambda_1,\lambda_2}$. In case $\lambda_1 \neq \lambda_2$, χ is called *regular*. We call $\Phi(D)$ the *radial part* of D.

2.3. *Attached Symmetric Subspaces*

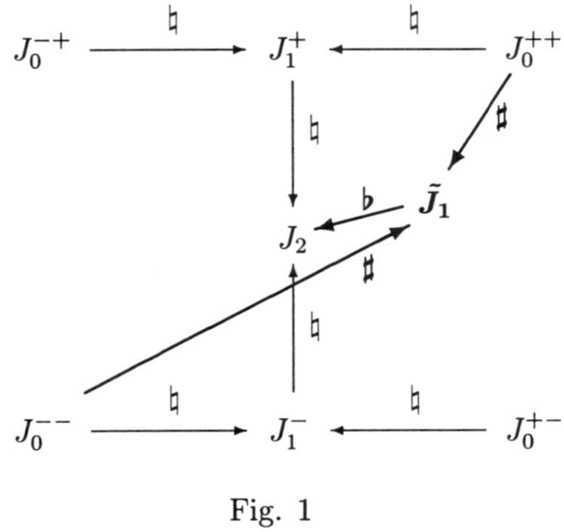

Fig. 1

In the case of $X = GL(4,\mathbf{R})/\left(GL(2,\mathbf{R}) \times GL(2,\mathbf{R})\right)$, the relation of adjacency are given in Fig.1. Any semiregular semisimple element x_0 in $J_0 \cup J_1 \cup \tilde{J}_1 \cup J_2$ corresponds to one of three symbols $\natural$, $\sharp$, $\flat$ in Fig. 1:

i) x_0 is of type $\natural$ if it belongs to $(J_0 \cap J_1) \cup (J_1 \cap J_2)$

ii) x_0 is of type $\sharp$ if it belongs to $J_0 \cap \tilde{J}_1$

iii) x_0 is of type $\flat$ if it belongs to $(hJ_2h^{-1} \cap \tilde{J}_1) \cup (J_2 \cap h^{-1}\tilde{J}_1 h)$,

where $h = h^{-1} = \begin{pmatrix} 1 & 0 & 0 & 0 \\ 0 & -1 & 0 & 0 \\ 0 & 0 & 0 & 1 \\ 0 & 0 & 1 & 0 \end{pmatrix}$. In each case, the symmetric subspace $X_{\{x_0\}} = Z_G(\tilde{\sigma}(x_0))/Z_H(\tilde{\sigma}(x_0))$ attached to x_0 is as follows:

i) $X_{\{x_0\}} \simeq GL(2,\mathbf{R})/\left(GL(1,\mathbf{R}) \times GL(1,\mathbf{R})\right) \times (\mathbf{R} \text{ or } \mathbf{T})$,

ii) $X_{\{x_0\}} \simeq \left(GL(2,\mathbf{R}) \times GL(2,\mathbf{R})\right)/\triangle GL(2,\mathbf{R}) \simeq SL(2,\mathbf{R}) \times \mathbf{R}$,

iii) $X_{\{x_0\}} \simeq GL(2,\mathbf{C})/GL(2,\mathbf{R}) \simeq SL(2,\mathbf{C})/SL(2,\mathbf{R}) \times \mathbf{T}$,

where we denote by $\simeq$ a local isomorphy.

2.4. *Local Matching Conditions*

Let x_0 be a semiregular semisimple element of type $\natural$. For simplicity, we assume moreover $x_0 \in J_1^+ \cap J_0^{++}$. Note that $t_1 = 1$ for x_0. In the case of $X_1 = GL(2,\mathbf{R})/\left(GL(1,\mathbf{R}) \times GL(1,\mathbf{R})\right)$ (a symmetric space of rank 1), we have, on a small neighborhood of the origin ($t = 1$),

$$\begin{aligned} &\{F_f(t) \mid f \in C_c^\infty(X_1)\} \\ &\quad = \{\phi(t) \mid \phi(t) = \phi_0(t) + \phi_1(t) \log|t-1| \quad (\phi_0,\, \phi_1 \in C^\infty)\}\,. \end{aligned} \qquad (*)$$

@@

Since $X_{\{x_0\}}$ is (locally) isomorphic to $GL(2,\mathbf{R})/\left(GL(1,\mathbf{R}) \times GL(1,\mathbf{R})\right) \times \mathbf{R}$, we have, from "rank reduction principle", and $(*)$, on a small neighborhood of x_0,

$$\begin{aligned} &\{F_f(t_1,t_2) \mid f \in C_c^\infty(X)\} \qquad (**) \\ &= \{\phi(t_1,t_2) \mid \phi(t_1,t_2) = \phi_0(t_1,t_2) + \phi_1(t_1,t_2) \log|t_1 - 1|.;\ \phi_0,\, \phi_1 \in C^\infty\}\,. \end{aligned}$$

@@ From $(**)$, by the procedure sketched in Diag.1, we have the local matching condition around x_0:

Let Θ be an IED defined on $X = GL(4,\mathbf{R})/\left(GL(2,\mathbf{R}) \times GL(2,\mathbf{R})\right)$, and Ψ_j's ($j = 0,\, 1$) real analytic functions defined on a neighborhood of x_0.

Then we have:

$$\Theta|_{J_0^{++'}}(t_1, t_2) = \Psi_0(t_1, t_2) + \Psi_1(t_1, t_2) \log |t_1 - 1| \quad (t_1 > 1)$$
$$\iff \Theta|_{J_1^{+'}}(t_1, t_2) = \Psi_0(t_1, t_2) + \Psi_1(t_1, t_2) \log |t_1 - 1| \quad (t_1 < 1).$$

In other words, $\Theta|_{J_0^{++'}}$ and $\Theta|_{J_1^{+'}}$ are determined mutually in a natural way.

For other semiregular semisimple elements of type $\natural$, we can prove similar local matching conditions.

We note that these local matching conditions are the same as those which appear in the case of $X = U(2,2)/GL(2, \mathbf{C})$.

For semiregular semisimple elements of type $\sharp$ or of type $\flat$, local matching conditions are essentially the same as those between two adjacent Cartan subspaces for IED's on $X = U(2,2)/(U(1,1) \times U(1,1))$.

Although the details are different and more complicated, we can also derive local matching conditions in these cases, along the lines mentioned above for elements of type $\natural$. For example, in the case of $x_0 \in J_0 \cap \tilde{J}_1$, the local matching conditions around x_0 is given as follows:

Let Θ be an IED defined on $X = GL(4, \mathbf{R}) / (GL(2, \mathbf{R}) \times GL(2, \mathbf{R}))$. Then $\omega\Theta|_{J_0'}$ is extensible as a real analytic function to a neighborhood of x_0 in J_0, and we have

$$\frac{1}{i}\frac{d}{ds}\omega\Theta|_{\tilde{J}_1'}(\tilde{j}_1(s,0)x_0)\bigg|_{s=\pm 0} = \frac{d}{ds}\omega\Theta|_{J_0'}(j_0(-s,s;+,+)x_0)\bigg|_{s=\pm 0}.$$

In case of $x_0 \in J_2 \cap \tilde{J}_1$, we have a similar local matching conditions around x_0.

By using the suitable basis ("common basis", cf. Aoki-Kato [4]) for the space of IED's with χ on J' ($J = J_0$, J_1, $\tilde{J}_1$, J_2), we first reformulate all the local matching conditions above in terms of coordinates with respect to these basis. Then gathering them up, we get a rather explicit form of IED's on $J_0' \sqcup J_1' \sqcup \tilde{J}_1' \sqcup J_2'$ (cf. Aoki-Kato [4]).

2.5. *Explicit Form of IED's*

Let λ be a complex number and put $L = 4(t^2-1)\dfrac{d^2}{dt^2} + 8t\dfrac{d}{dt}$. Any solution of the equation

$$L\phi = \lambda\phi \tag{$\star$}$$

is real analytic on $(-\infty,-1)\sqcup(-1,1)\sqcup(1,\infty)$, and a solution ϕ defined on $(-1,1)$ is of the form:

$$\begin{aligned}\phi(t) &= \phi_1(t) + \log|\,1-t\,|\phi_2(t)\\ &= \phi_3(t) + \log|\,1+t\,|\phi_4(t),\end{aligned}$$

where $\phi_1(t)$, $\phi_2(t)$ are real analytic on $(-1,\infty)$, and $\phi_3(t)$, $\phi_4(t)$ are real analytic on $(-\infty,1)$. So we can extend naturally to $(-\infty,-1)\sqcup(-1,1)\sqcup(1,\infty)$ the solution ϕ initially defined only on $(-1,1)$ as follows:

$$\phi(t) := \begin{cases}\phi_1(t) + \log|1-t|\phi_2(t) & (-1<t)\\ \phi_3(t) + \log|1+t|\phi_4(t) & (t<1).\end{cases}$$

Let α, β be complex numbers satisfying the conditions $\alpha+\beta=1$, $\alpha\beta=-\lambda$. We consider the following two solutions of the equation $(\star)$ on $(-1,1)$:

$$\begin{aligned}\phi_+(t,\lambda) &:= F\big(\alpha,\beta,1;\frac{1-t}{2}\big),\\ \phi_-(t,\lambda) &:= F\big(\alpha,\beta,1;\frac{1-t}{2}\big)\log\big|\frac{1-t}{2}\big| + F^*\big(\frac{1-t}{2}\big),\end{aligned}$$

where $F^*(z)$ is analytic on $|z|<1$,

and extend these solutions naturally to $(-\infty,-1)\sqcup(-1,1)\sqcup(1,\infty)$ in the manner described above. We denote by the same symbols the extended solutions.

Let $A(\lambda)$ be the 2×2 matrix of basis change determined by the equation:

$$(\phi_+(t,\lambda),\phi_-(t,\lambda)) = (\phi_+(-t,\lambda),\phi_-(-t,\lambda))A(\lambda).$$

Using these notations we can state:

Theorem. *Let $\chi=\chi_{\lambda_1,\lambda_2}$ be a generic character of $\mathbf{D}(X)$. Then, for an IED Θ with χ defined on X, there exist constants C_{ν_1,ν_2}'s $(\nu_1,\nu_2=\pm)$ such that we have*

$$\Theta|_{J'_0\sqcup J'_1\sqcup J'_2}(t_1,t_2) = \omega^{-1}\sum_{\nu_1,\nu_2=\pm} C_{\nu_1,\nu_2}\begin{vmatrix}\phi_{\nu_1}(t_1,\lambda_1) & \phi_{\nu_2}(t_1,\lambda_2)\\ \phi_{\nu_1}(t_2,\lambda_1) & \phi_{\nu_2}(t_2,\lambda_2)\end{vmatrix}$$

$$\begin{aligned}\Theta|_{\bar{J}'_1}(t_1,t_2) = \omega^{-1}\Bigg\{&\sum_{\nu_1,\nu_2=\pm} C_{\nu_1,\nu_2}\begin{vmatrix}\phi_{\nu_1}(t_1,\lambda_1) & \phi_{\nu_2}(t_1,\lambda_2)\\ \phi_{\nu_1}(t_2,\lambda_1) & \phi_{\nu_2}(t_2,\lambda_2)\end{vmatrix}\\ &+\operatorname{sgn}(\theta_1\theta_2)\sum_{\nu_1,\nu_2=\pm} C'_{\nu_1,\nu_2}\{\phi_{\nu_1}(t_1,\lambda_1)\phi_{\nu_2}(t_2,\lambda_2)+\phi_{\nu_2}(t_1,\lambda_2)\phi_{\nu_1}(t_2,\lambda_1)\}\Bigg\}\end{aligned}$$

where

$$C'_{--} = 0, \qquad \begin{pmatrix} C'_{++} \\ C'_{+-} \\ C'_{-+} \end{pmatrix} = B^{-1} \begin{pmatrix} \frac{\pi}{i} C_{--} \\ 0 \\ \frac{\pi}{i} C_{--} \end{pmatrix}.$$

Here B is a non-singular 3×3 matrix explicitly determined by $A(\lambda_1)$ and $A(\lambda_2)$.

As a corollary of the above theorem, we have:

Corollary. *If χ is generic, then*

$$\dim \{\Theta|_{X'};\ \Theta \textit{ is an IED on } X \textit{ with } \chi\} \leq 4.$$

Remark. On some evidence, we conjecture:

$$\dim \{\Theta|_{X'};\ \Theta \text{ is an IED on } X \text{ with } \chi\} = 4.$$

References

1. S. Aoki and S. Kato, "Connection formulas for invariant eigendistributions on certain semisimple symmetric spaces", RIMS, **570**, 111-132 (1985).
2. S. Aoki and S. Kato, "On invariant eigendistributions on $U(p,q)/(U(r) \times U(p-r,q))$", Proc. of the Japan Academy, **67** A.6, 203-207 (1991).
3. S. Aoki and S. Kato, "The behavior of invariant integrals on semisimple symmetric spaces and its application", RIMS, **1082**, 74-92 (1999).
4. S. Aoki and S. Kato, "Connection formulas for invariant eigendistributions on $GL(4,\mathbf{R})/(GL(2,\mathbf{R}) \times GL(2,\mathbf{R}))$", Bulletin of Science and Engineering Takushoku University, **7**, no.2, 15-20 (1999) (in Japanese).
5. S. Aoki and S. Kato, "Connection formulas for invariant eigendistributions on semisimple symmetric spaces: Consideration based on the signs of roots and Cayley transformations", in: Symposium on Representation Theory 2002, 126-142 (2002) (in Japanese).
6. F.A. Berezin and F.I. Karpelevič, "Zonal spherical functions and Laplace operators on some symmetric spaces", Dokl. Akad. Nauk SSSR, **118**, 9-12 (1958) (Russian).
7. J. Faraut, "Distributions sphériques sur les espaces hyperboliques", J. Math. Pures Appl. **58**, 369-444 (1979).
8. T. Hirai, "Invariant eigendistributions of Laplace operators on real simple Lie groups, II", Japan. J. Math. New Series, **2**, 27-89 (1976).
9. B. Hoogenboom "Spherical functions and differential operators on complex Grassmann manifolds", Ark. Mat. **20**, 69-85 (1982).
10. S. Sano, "Invariant spherical distributions and the Fourier inversion formula on $GL(n,\mathbf{C})/GL(n,\mathbf{R})$", J. Math. Soc. Japan, **36**, 191-219 (1984).

II. Field Theory

RATIONAL CONFORMAL CORRELATION FUNCTIONS OF GAUGE INVARIANT LOCAL FIELDS IN FOUR DIMENSIONS

N.M. NIKOLOV[1], Y.S. STANEV[1,2], I.T. TODOROV[1,3]

[1] *Institute for Nuclear Research and Nuclear Energy Tsarigradsko Chaussee 72, BG-1784 Sofia, Bulgaria*

[2] *I.N.F.N. - Sezione di Roma 2 Via della Ricerca Scientifica 1, I-00133 Roma, Italy*

[3] *Institut des Hautes Etudes Scientifiques, Le Bois-Marie, 35 route de Chartres, F-91440 Bures-sur-Yvette, France*

e-mail addresses: mitov@inrne.bas.bg, stanev@roma2.infn.it, todorov@inrne.bas.bg

Global conformal invariance in Minkowski space and the Wightman axioms imply strong locality (Huygens principle) and rationality of correlation functions thus providing an extension of the concept of a vertex algebra to higher (even) dimensions D. We (p)review current work on a model of a hermitian scalar field $\mathcal{L}$ of scale dimension 4 (for $D = 4$) which can be interpreted as the Lagrangian of a gauge field theory that generates the algebra of gauge invariant local observables in a conformally invariant renormalization group fixed point.

1. Introduction

The present paper provides a concise review and a continuation of our work [16] [14] [15] [13] aimed at constructing a non-trivial *globally conformal invariant* (GCI) 4-dimensional *quantum field theory* (QFT) model. Our attempt to build such a model is based on the following results of [16].

Invariance under finite conformal transformations in Minkowski space-time M and local commutativity imply the *Huygens principle*: the commutator of two local Bose fields vanishes for non-isotropic separations. The Huygens principle and energy positivity yield rationality of correlation functions (Theorem 3.1 of [16]). These results allow to extend any GCI QFT to compactified Minkowski space $\bar{M}$, which admits the following convenient complex variable realization ([20] [14] [15] [13] [17]):

$$\bar{M} = \{z_\mu = e^{2\pi i\zeta} u_\mu, \mu = 1, \ldots, 4; \zeta \in \mathbb{R}, \\ u \in \mathbb{S}^3 = \{u \in \mathbb{R}^4; u^2 = \mathbf{u}^2 + u_4^2 = 1\}\} = \frac{\mathbb{S}^1 \times \mathbb{S}^3}{\mathbb{Z}_2}. \tag{1.1}$$

Fields $\phi(z)$ are expressed as formal power series of the form

$$\phi(z) = \sum_{\nu \in \mathbb{Z}} \sum_{m \geq 0} (z^2)^\nu \, \phi_{\{\nu,m\}}(z), \ \Delta \, \phi_{\{\nu,m\}}(z) = 0,$$

$$\phi_{\{\nu,m\}}(\lambda z) = \lambda^m \, \phi_{\{\nu,m\}}(z), \ z^2 := \sum_{\mu=1}^4 z_\mu^2, \ \Delta = \sum_{\mu=1}^4 \frac{\partial^2}{\partial z_\mu^2}, \tag{1.2}$$

$\phi_{\{\nu,m\}}(z)$ being (in general, multicomponent) operator valued (homogeneous, harmonic) polynomial in z. The expansion (1.2) singles the eigenvalues of the *conformal energy operator* H defined up to an additive constant (the vacuum energy) by the commutation rules

$$[H, \phi(z)] = \left(d + z\frac{\partial}{\partial z}\right) \phi(z), \\ \text{implying } [H, \phi_{\{\nu,m\}}(z)] = (d + 2\nu + m)\, \phi_{\{\nu,m\}}(z), \tag{1.3}$$

where d is the *scale* (or *conformal*) *dimension* of the field ϕ. The Minkowski space *spectral conditions* (including *energy positivity*) imply analyticity of the vector valued function $\phi(z)\,|0\rangle$ (where $|0\rangle$ is the conformally invariant vacuum vector) in the *z-picture image*

$$T_+ = \left\{ z \in \mathbb{C}^4 : |z^2| < 1, \ |z|^2 \left(:= \sum_{\mu=1}^4 |z^\mu|^2 \right) < \frac{1 + |z^2|^2}{2} \right\} \tag{1.4}$$

of the *forward tube*

$$\mathcal{T}_+ = \left\{ \xi = x + iy : x, y \in M, y^0 > |\mathbf{y}| := \sqrt{y_1^2 + y_2^2 + y_3^2} \right\}$$

under the complex conformal transformation (with singularities)

$$h : M_{\mathbb{C}}(\ni \xi) \to E_{\mathbb{C}}(\ni z), \mathbf{z} = \frac{\boldsymbol{\xi}}{\omega_\xi}, \ z_4 = \frac{1 - \xi^2}{\omega_\xi}, \\ \xi^2 = \boldsymbol{\xi}^2 - (\xi^0)^2, \ \omega_\xi = \frac{1}{2}(1 + \xi^2) - i\xi^0, \ z^2 = \frac{1 + \xi^2 + 2i\xi^0}{1 + \xi^2 - 2i\xi^0}. \tag{1.5}$$

(Note that h maps the real Minkowski space M onto the open dense subset of $\bar{M}$ (1.1) such that $\mathbf{z}^2 + (z_4 + 1)^2 = 2e^{2\pi i\zeta}(\cos 2\pi\zeta + u_4) \neq 0$.) It follows, in particular, that no negative powers of z^2 appear in $\phi(z)\,|0\rangle$:

$$\phi_{\{\nu,m\}}(z)\,|0\rangle = 0 \qquad \text{for } \nu < 0. \tag{1.6}$$

This is unambiguous [13], because, as it is well known, every homogeneous polynomial $p(z)$ of degree m has a *unique decomposition* $p(z) = h(z)+z^2q(z)$ where h is harmonic (of degree m) and q is homogeneous of degree $m-2$.

The resulting (*analytic*) *z-picture* provides a higher dimensional generalization [13] of *chiral vertex algebras* (which have been an outgrow of physicists' work on conformal field theory and dual resonance models, formalized by R.E. Borcherds [1] and since subject of numerous studies, including several books – see, *e.g.* [11] [7] and references therein).

The 4-dimensional vertex algebra of GCI fields with rational correlation functions corresponds to the *algebra of local observables* in Haag's approach [10] to QFT. Its *isotypical* (or *factorial*) *representations* (*i.e.* multiples with a finite multiplicity of an irreducible representation) give rise to the *superselection sectors* of the theory. The intertwiners between the vacuum and other superselection sectors are higher dimensional counterparts of primary fields (which typically have fractional dimensions and multivalued n-point distributions).

We shall add to the traditional assumption that the conserved (symmetric, traceless) *stress-energy tensor* $T_{\mu\nu}(z)$ is a local observable the requirement that so is the scalar, gauge invariant *Lagrangian density* $\mathcal{L}(z)$ (of dimension $d = 4$). Then the construction of a GCI QFT model becomes a rather concrete program of writing down rational (conformally invariant and "crossing symmetric") correlation functions and studying the associated *operator product expansions* (OPE).

After a brief review (in Sec. 2.1) of the results of [16] reformulated in the above z-picture we outline (in Sec. 2.2) the general truncated GCI 4-point function w_4^t of a neutral scalar field of (integer) dimension d. This is a homogeneous rational function of degree $-2d$ in the (complex) Euclidean invariant variables

$$\rho_{ij} = z_{ij}^2\,,\ z_{ij} = z_i - z_j\,,\ z^2 = \mathbf{z}^2 + z_4^2 \tag{1.7}$$

of denominator $(\rho_{12}\rho_{13}\rho_{14}\rho_{23}\rho_{24}\rho_{34})^{d-1}$ and numerator, a homogeneous polynomial of degree $4d-6$ (for $d \geq 2$) depending (linearly) on $\left[\!\left[\frac{d^2}{3}\right]\!\right]$ ($[\![a]\!]$ standing for the integer part of the real a) real parameters. It is just $c(\rho_{13}\rho_{24} + \rho_{12}\rho_{34} + \rho_{14}\rho_{23})$ for the simplest candidate for a non-trivial, $d = 2$, model, and involves 5 parameters for the physically interesting case of a $d = 4$ Lagrangian density. As the model of a $d = 2$ scalar field was proven in [14] to correspond to normal products of free (massless) scalar fields we concentrate in the rest of the paper on the $d = 4$ case. We study in Sec. 3 OPE organized in bilocal fields of fixed twist which provide what could be called a *conformal partial wave expansion* of the 4-point function (a concept, introduced in [2], see also [3], and recently revisited in [5]). The bilocal field $V_1(z_1, z_2)$ of (lowest) dimension $(1,1)$, which admits a Taylor expansion in z_{12} involving only twist 2 symmetric traceless tensors, is harmonic in each argument allowing to compute (in Sec. 3.2) its (rational) 4-point function. The corresponding (crossing) symmetrized contribution to w_4^t gives rise to a 3-parameter sub-family of the original 5-parameter family of GCI 4-point functions. This provides what we call (in Sec. 3.3) a *minimal model* corresponding to a given V_1.

We argue, in Sec. 4.1 (summarizing results of [15]), that the Lagrangian $\mathcal{L}(z)$ of a gauge field theory should have vanishing odd point functions and should not involve a $d = 2$ scalar field in its OPE. This reduces to 3 the 5 parameters in w_4^t. One of the remaining parameters corresponds to the Lagrangian (*i.e.*, the contracted normal square) of a free Maxwell field. Another corresponds to a twist 4 contribution that is not recovered in the minimal version of the theory. We are thus led to consider a 1-parameter family of 4-point functions that appears as the simplest candidate for a non-trivial GCI QFT consistent with Wightman axioms [18], which may correspond to a (conformally invariant) renormalization group fixed point of a gauge field theory. As a first step in studying this distinguished simple model we point out, in Sec. 4.2, that the resulting special bilocal field $V_1 = v(z_1, z_2)$ has the properties of a bilinear combination of a free Weyl spinor and its conjugate. This permits the computation of higher point functions as displayed in Sec. 4.3.

The results and the challenging open problems are discussed in Sec. 4.4.

2. Implications of GCI. General 4-point function

2.1. *Strong locality and energy positivity imply rationality*

It follows from local commutativity and GCI that for any pair of conjugate (Bose or Fermi) fields $\psi(z_1)$ and $\psi^*(z_2)$ there is a positive integer N_ψ such that

$$\rho_{12}^N \{\psi(z_1)\,\psi^*(z_2) - \varepsilon_\psi\,\psi^*(z_2)\,\psi(z_1)\} = 0 \text{ for } N \geq N_\psi\,,\ \rho_{12} := z_{12}^2 \qquad (2.1)$$

where $\varepsilon_\psi (= \pm 1)$ is the fermion parity of ψ. If ψ transforms under an *elementary local field representation* of the spinor conformal group $SU(2,2)$ (see [12, 19]) – *i.e.* one induced by a $(2j_1 + 1)(2j_2 + 1)$ dimensional representation $(d; j_1, j_2)$ of the maximal compact subgroup $S(U(2) \times U(2))$ of $SU(2,2)$ (d being the $U(1)$ character coinciding with the scale dimension), then

$$N_\psi = d + j_1 + j_2\,,\ \varepsilon_\psi = (-1)^{2j_1+2j_2} = (-1)^{2d}\,. \tag{2.2}$$

It follows that for any n-point function of GCI local fields and for large enough $N \in \mathbb{N}$ the product

$$F_{1\dots n}(z_1,\dots,z_n) := \left(\prod_{1\le i<j\le n} \rho_{ij}\right)^N \langle 0|\,\phi_1(z_1)\dots\phi_n(z_n)\,|0\rangle \tag{2.3}$$

($\rho_{ij} = z_{ij}^2 \equiv (z_i - z_j)^2$), is $\mathbb{Z}_2$ symmetric under any permutation of the factors within the vacuum expectation value. Energy positivity, on the other hand, implies that $\langle 0|\,\phi_1(z_1)\dots\phi_n(z_n)\,|0\rangle$, and hence $F_{1\dots n}(z_1,\dots,z_n)$ do not contain negative powers of z_n^2. It then follows from the symmetry and the homogeneity of $F_{1\dots n}$ that it is a polynomial in all z_i^μ. Thus the (Wightman) correlation functions are rational functions of the coordinate differences. (See for more detail [17]; an equivalent Minkowski space argument based on the support properties of the Fourier transform of (the x-space counterpart of) (2.3) is given in [16].)

Rationality of correlation functions implies that all dimensions of GCI fields should be integer or half integer depending on their spin, more precisely, that sums like N_ψ (2.2) should be integer. This condition is, however, not sufficient for rationality even of 3-point functions.

Observation 2.1. *The necessary and sufficient condition for the existence of a GCI 3-point function $\langle 0|\,\phi_1(z_1)\,\phi_2(z_2)\,\phi_3(z_3)\,|0\rangle$ of elementary conformal fields $\phi_i(z)$ of $S(U(2)\times U(2))$ weight $(d_i; j_{i1}, j_{i2})$ is:*

$$N_i := d_i + j_{i1} + j_{i2} \in \mathbb{N}\,,\ \frac{1}{2}\sum_{i=1}^{3} N_i \in \mathbb{N}, \sum_{i=1}^{3} d_i \in \mathbb{N}\,. \tag{2.4}$$

The statement follows from the explicit knowledge of 3-point functions (see for reviews [21] [19] [9]).

In particular, there is no Yukawa type rational conformal 3-point function of a pair of conjugate canonical $\left(d = \frac{3}{2}\right)$ spinor fields and a canonical $(d = 1)$ scalar field. Similarly, one observes that 2-point functions of free massless fields in odd space-time dimensions are not rational and hence cannot be GCI.

It is important for the feasibility of constructing a GCI model that the singularities of n-point functions (the integer N in (2.3)) is majorized by that of the 2-point function (N_ψ in (2.2)) whenever Wightman positivity is satisfied.

2.2. *General truncated 4-point function of a GCI scalar field*

Infinitesimal (or Euclidean) conformal invariance is sufficient to determine 2- and 3-point functions (see, *e.g.* [21]). One can construct, however, two independent conformally invariant cross-ratios out of four points,

$$s = \frac{\rho_{12}\rho_{34}}{\rho_{13}\rho_{24}}\,, \qquad t = \frac{\rho_{14}\rho_{23}}{\rho_{13}\rho_{24}}\,, \tag{2.5}$$

so that a simple minded symmetry argument does not determine the 4-point functions. GCI, on the other hand, combined with Wightman axioms, yields rationality and thus allows to construct higher point correlation functions involving just a finite number of free parameters. In particular, the truncated 4-point function of a hermitian scalar field ϕ of (integer) dimension d can be written in the form ([14] Sec. 1):

$$\begin{aligned} w_4^t \equiv w^t(z_1, z_2, z_3, z_4) &:= \langle 1234\rangle - \langle 12\rangle\langle 34\rangle - \langle 13\rangle\langle 24\rangle - \langle 14\rangle\langle 23\rangle \\ &= \frac{(\rho_{13}\rho_{24})^{d-2}}{(\rho_{12}\rho_{23}\rho_{34}\rho_{14})^{d-1}}\, P_d(s,t)\,, \quad P_d(s,t) = \sum_{i\geq 0, j\geq 0}^{i+j\leq 2d-3} c_{ij}\, s^i t^j\,, \end{aligned} \tag{2.6}$$

where $\langle 1 \ldots n\rangle$ is a short-hand for the n-point function of ϕ:

$$\langle 1 \ldots n\rangle = \langle 0|\, \phi(z_1) \ldots \phi(z_n)\,|0\rangle\,, \quad \langle 12\rangle = B_\phi\, \rho_{12}^{-d}\,. \tag{2.7}$$

In writing down (2.6) we have used the fact that for space-time dimensions $D > 2$ Hilbert space positivity implies that the degree of singularities of the truncated n-point function ($n \geq 4$) is strictly smaller than the degree of the 2-point function.

Furthermore, crossing symmetry (which is a manifestation of local commutativity) implies an $\mathcal{S}_3$ symmetry of P_d:

$$\begin{aligned} &s_{i i+1} P_d(s,t) = P_d(s,t),\ i = 1, 2, \\ &s_{12}\, P_d(s,t) := t^{2d-3} P_d\left(\frac{s}{t}, \frac{1}{t}\right), \quad s_{23}\, P_d(s,t) := s^{2d-3} P_d\left(\frac{1}{s}, \frac{t}{s}\right), \end{aligned} \tag{2.8}$$

s_{ij} being the substitution exchanging the arguments z_i and z_j. The number of independent crossing symmetric polynomials P_d is $\left[\!\left[\frac{d^2}{3}\right]\!\right]$ (the integer part of $d^2/3$: $\left[\!\left[\frac{d^2}{3}\right]\!\right] = n(2d-3n)$ for $3n - 1 \leq d \leq 3n + 1$, $n = 0, 1, 2, \ldots$).

The 1-parameter family of crossing symmetric polynomials for $d = 2$ is $P_2(s,t) = c\,(1 + s + t)$, *i.e.*

$$w_4^t = c\left\{(\rho_{12}\rho_{23}\rho_{34}\rho_{14})^{-1} + (\rho_{13}\rho_{23}\rho_{24}\rho_{14})^{-1} + (\rho_{12}\rho_{13}\rho_{24}\rho_{34})^{-1}\right\}, \tag{2.9}$$

thus corresponding to the sum of three 1-loop diagrams for a sum of normal products of free massless fields:

$$\phi(z) = \frac{1}{2}\sum_{i=1}^{N} :\varphi_i^2(z): \; ([\varphi_i(z_1), \varphi_j(z_2)] = 0 \text{ for } i \neq j, \; \Delta\,\varphi_i(z) = 0)\,. \quad (2.10)$$

Indeed, it was proven in [14] that $\phi(z)$ generates under commutations a central extension of the infinite symplectic algebra $\mathrm{sp}(\infty, \mathbb{R})$ for $d = 2$ and that the unitary vacuum representations of this algebra correspond to integer central charge $c = N (\in \mathbb{N})$. Thus, (2.10) is the general form of a $d = 2$ GCI field satisfying Wightman axioms (including Hilbert space positivity) and involving a unique rank 2 symmetric traceless tensor of dimension 4 in its OPE algebra.

The physically most attractive example, corresponding to a $d = 4$ scalar field $\mathcal{L}(z)$ that can be interpreted as a QFT Lagrangian density, gives rise to a 5-parameter truncated 4-point function [15] of type (2.6) with

$$P_4(s,t) = \sum_{\nu=0}^{2} a_\nu J_\nu(s,t) + st\,[b(Q_1(s,t) - 2Q_2(s,t)) + c\,Q_2(s,t)]\,. \quad (2.11)$$

As we shall see in Sec. 3, J_ν correspond to the symmetrized twist 2 contribution to the OPE of two $\mathcal{L}$'s:

$$J_0(s,t) = s^2(1+s) + t^2(1+t) + s^2t^2(s+t) \quad (2.12)$$

is the only one among the 5 polynomials in the right hand side of (2.11) which does not vanish for $s = 0$, $t = 1$,

$$J_1(s,t) = s(1-s)(1-s^2) + t(1-t)(1-t^2) + st\,[(s-t)(s^2-t^2) - 2Q_1]\,, \quad (2.13)$$

$$J_2(s,t) = (1+t^3)\,[(1+s-t)^2 - s] - 3s\,(1-t) + s^3\,[(1+t-s)^2 - t]\,; \quad (2.14)$$

Q_i are $\mathcal{S}_3$ symmetric polynomials of degree 2:

$$Q_1 = 1 + s^2 + t^2\,, \; Q_2 = s + t + st\,,$$
$$Q_1(s,t) - 2Q_2(s,t) = (1-s-t)^2 - 4st\,, \quad (2.15)$$

and the contributions $st\,Q_j$ to P_4 correspond to twist 4 and higher in the OPE.

The above choice of basic $\mathcal{S}_3$ symmetric polynomials is not accidental: it is essentially determined by its relation to the "partial wave" expansion of

w_4^t to be displayed in the next section. We only note at this point that the term proportional to $J_2(s,t)$ is reproduced by the free Maxwell Lagrangian

$$\mathcal{L}_0(x) = -\frac{1}{4} : F_{\mu\nu}(z)\, F^{\mu\nu}(z) : ; \tag{2.16}$$

here $F_{\mu\nu}$ is characterized by its 2-point function

$$\langle 0|\, F_{\mu_1\nu_1}(z_1)\, F_{\mu_2\nu_2}(z_2)\, |0\rangle = R_{\mu_1\mu_2}(z_{12})\, R_{\nu_1\nu_2}(z_{12}) - R_{\mu_1\nu_2}(z_{12})\, R_{\nu_1\mu_2}(z_{12}) \tag{2.17}$$

where $R_{\mu\nu}$ is related to the vector representation of the conformal inversion:

$$R_{\mu\nu}(z) = \frac{r_{\mu\nu}(z)}{z^2}\,,\ r_{\mu\nu}(z) = \delta_{\mu\nu} - 2\,\frac{z_\mu z_\nu}{z^2}\,,\ (r^2)_{\mu\nu} = \delta_{\mu\nu}\,. \tag{2.18}$$

(The calculation proving that Eqs. (2.16–2.18) yield a w_4^t proportional to $\rho_{13}^2\rho_{24}^2(\rho_{12}\rho_{23}\rho_{34}\rho_{14})^{-3}\, J_2(s,t)$ is given in Appendix B to [14].)

3. OPE in terms of bilocal fields

3.1. *Fixed twist fields. Conformal partial wave expansion*

The infinite series of integrals of local tensor fields appearing in the singular part of the OPE of $\mathcal{L}(z_1)\,\mathcal{L}(z_2)$ can be organized into a finite sum of bilocal (scalar) fields:

$$\begin{aligned} \mathcal{L}(z_1)\,\mathcal{L}(z_2) &= \frac{B}{\rho_{12}^4} + \sum_{\kappa=1}^{3} \rho_{12}^{\kappa-4}\, V_\kappa(z_1,z_2) + :\mathcal{L}(z_1)\,\mathcal{L}(z_2): \,, \\ V_\kappa(z_1,z_2) &= V_\kappa(z_2,z_1)\,. \end{aligned} \tag{3.1}$$

By definition, V_κ is a bilocal conformal field of dimension (κ,κ) which only involves (derivatives of) *twist* 2κ *tensor fields* in its Taylor expansion in z_{12}. More precisely, it can be written in the form (*cf.* [8] [2] [21] [4] [14])

$$V_\kappa(z_1,z_2) = \sum_{\ell=0}^{\infty} C_{\kappa\ell} \int_0^1 K_{\kappa\ell}(\alpha, \rho_{12}\,\Delta_2)\, O_{2\kappa,\ell}(z_2 + \alpha\, z_{12}; z_{12})\, d\alpha\,, \tag{3.2}$$

where $O_{2\kappa\ell}(z;w)$ are (contracted) symmetric traceless tensor fields,

$$\begin{aligned} &O_{2\kappa,\ell}(z;w) = O_{2\kappa}^{\mu_1\ldots\mu_\ell}(z)\, w_{\mu_1}\ldots w_{\mu_\ell} \\ &(\text{tracelessness} \Leftrightarrow \Delta_w\, O_{2\kappa,\ell}(z;w) = 0)\,, \end{aligned} \tag{3.3}$$

of scale dimension $d_{\kappa,\ell} = 2\kappa + \ell$ (*i.e.* of *fixed twist* $d_{\kappa\ell} - \ell = 2\kappa$). The (4-dimensional) Laplacian Δ_w acts on the 4-vector w while Δ_2 acts on z_2 (for fixed z_{12}). The integrodifferential operator with kernel $K_{k\ell}$ in the right hand side of (3.2) is defined to transform the 2-point function of $O_{2\kappa,\ell}$ into the 3-point function

$$\langle 0|\, V_\kappa(z_1,z_2)\, O_{2\kappa,\ell}(z_3;w)\,|0\rangle = A_{\kappa\ell}\,\frac{(X_{12}^3\,w)^\ell}{(\rho_{13}\rho_{23})^\kappa}\ \text{for } w^2=0\,, \tag{3.4}$$

$$X_{12}^3 = \frac{z_{13}}{\rho_{13}} - \frac{z_{23}}{\rho_{23}} \quad \left((X_{12}^3)^2 = \frac{\rho_{12}}{\rho_{13}\rho_{23}} \right) \tag{3.5}$$

(see [4] and [15] Sec. 3, where more general OPE – for any scale dimension d and for complex fields are considered, and the kernel $K_{\kappa\ell}$ which implements (3.4) is written down). For real $\mathcal{L}$, due to the symmetry property (3.1) of V_κ, only even rank tensors (even ℓ) appear in the expansion (3.2).

The "normal product" $:\mathcal{L}(z_1)\,\mathcal{L}(z_2):$ is defined by (3.1) and can in turn be expanded in higher twist V_κ ($\kappa \geq 4$), accompanied by positive powers of ρ_{12}.

The usefulness of the expansion (3.1) is enhanced by the fact that fields of different twists are mutually orthogonal under vacuum expectation values:

$$\langle 0|\, V_\kappa(z_1,z_2)\,|0\rangle = 0 = \langle 0|\, V_\kappa(z_1,z_2)\, V_\lambda(z_3,z_4)\,|0\rangle \ \text{ for } 0<\kappa\neq\lambda\,. \tag{3.6}$$

For equal dimensions we can write

$$\langle 0|\, V_\kappa(z_1,z_2)\, V_\kappa(z_3,z_4)\,|0\rangle = (\rho_{13}\rho_{24})^{-\kappa} f_\kappa(s,t)$$
$$\left(s_{12}\, f_\kappa(s,t) = t^{-\kappa} f_\kappa\left(\frac{s}{t},\frac{1}{t}\right) = f_\kappa(s,t)\right)\,. \tag{3.7}$$

As a consequence of the OPE (3.2) and of (3.4) we find the following (light-cone) *conformal partial wave expansion:*

$$f_\kappa(0,t) = \sum_{\ell=0}^{\infty} B_{\kappa\ell}(1-t)^{2\ell}\, F(2\ell+\kappa,\, 2\ell+\kappa;\, 4\ell+2\kappa;\, 1-t)\,, \tag{3.8}$$

$F(\alpha,\beta;\gamma;x) = \sum_{n=0}^{\infty} \frac{(\alpha)_n(\beta)_n}{n!(\gamma)_n}\, x^n = 1 + \frac{\alpha\beta}{\gamma}\, x + \ldots$ being Gauss' hypergeometric function. The *structure constants* $B_{\kappa\ell}$, unlike those appearing in (3.2) and (3.4) are invariant under rescaling of $O_{2\kappa 2\ell}$:

$$B_{\kappa\ell} := A_{\kappa 2\ell}\, C_{\kappa 2\ell} \left(O_{2\kappa 2\ell} \to \lambda\, O_{2\kappa 2\ell} \Rightarrow A_\kappa \to \lambda\, A_{\kappa 2\ell},\, C_{\kappa 2\ell} \to \frac{1}{\lambda}\, C_{\kappa 2\ell}\right)\,, \tag{3.9}$$

The fixed twist condition allows to restore the s-dependence of f_κ. This is particularly transparent in the $\kappa = 1$ case, in which $O_{2\ell}(z;w)$ are conserved symmetric tensor fields (including, for $\ell = 2$, the stress-energy tensor).

Writing the expression in the right hand side of (3.4) for $\kappa = 1$, $w^2 = 0$ as $\frac{1}{\ell!}\left(\frac{w}{2}\left(\frac{\partial}{\partial z_2} - \frac{\partial}{\partial z_1}\right)\right)^\ell \frac{1}{\rho_{13}\rho_{23}}$ we observe that it is harmonic in both z_1 and z_2. It then follows from (3.1) that V_1 is harmonic in each argument:

$$\Delta_1 V_1(z_1, z_2) = 0 = \Delta_2 V_1(z_1, z_2) \quad \left(\Delta_j = \sum_{\mu=1}^{4} \frac{\partial^2}{(\partial z_j^\mu)^2}\right). \tag{3.10}$$

This implies what may be called the *conformal Laplace equation*[a] for f_1:

$$\Delta_{st} f_1(s,t) = 0, \; \Delta_{st} := s\,\frac{\partial^2}{\partial s^2} + t\,\frac{\partial^2}{\partial t^2} + (s+t-1)\,\frac{\partial^2}{\partial s\,\partial t} + 2\left(\frac{\partial}{\partial s} + \frac{\partial}{\partial t}\right). \tag{3.11}$$

Its general solution is expressed in terms of the "chiral variables" u, v of [4] (see also [6])

$$f_1(s,t) = \frac{g(u) - g(v)}{u - v} \text{ for } s = uv, \; t = (1-u)(1-v). \tag{3.12}$$

The function g is expressed in terms of $f_1(0,t)$ as $g(u) = u f_1(0, 1-u)$ (for $s = 0 = v$). Furthermore, as demonstrated in [4], the following extension of (3.12) is valid for $\kappa > 1$:

$$\begin{aligned} f_\kappa(s,t) = \frac{1}{u-v} \{ & F(\kappa-1,\, \kappa-1;\, 2\kappa-2;\, v)\, g_\kappa(u) \\ & - F(\kappa-1,\, \kappa-1;\, 2\kappa-2;\, u)\, g_\kappa(v)\}. \end{aligned} \tag{3.13}$$

(We shall view (3.13) as a generalization of (3.12) which can be included as a special case if we set $F(0,0;0;u) \equiv 1$.) Then the light cone expansion (3.8) (appearing as an "initial condition") implies

$$g_\kappa(u) = u\, f_\kappa(0, 1-u) = u \sum_{\ell=0}^{\infty} B_{\kappa\ell}\, u^{2\ell}\, F(2\ell+\kappa, 2\ell+\kappa; 4\ell+2\kappa; u) \tag{3.14}$$

($g_\kappa(0) = 0$). The functions $f_\kappa(0,t)$, on the other hand, can be determined (for a given P_4 (2.11)) inductively in κ as follows:

$$\begin{aligned} & f_1(0,t) = t^{-3} P_4(0,t), \\ & f_\kappa(0,t) = \lim_{s \to 0} \left\{ s^{1-\kappa} \left[t^{-3} P_4(s,t) - \sum_{\nu=1}^{\kappa-1} s^{\nu-1} f_\nu(s,t) \right] \right\}. \end{aligned} \tag{3.15}$$

It is important to realize that the kernels $K_{\kappa\ell}$, the functional form of the 3-point function (3.4), and the hypergeometric functions defining the conformal partial waves (3.8) are *universal*; only the structure constants $B_{\kappa\ell}$ depend on $\mathcal{L}(z)$ and are, in fact, determined by its 4-point function.

[a]The operator (3.11) has appeared in various contexts in [6] [4] and [15].

3.2. *Symmetrized contribution of twist 2 (conserved) tensors*

We now proceed to writing down the general rational solution of (3.11) for which $t^3 f_1(0,t)$ is a polynomial of degree not exceeding 5 and which satisfies the symmetry condition $s_{12}\, f_1(s,t) = f_1(s,t)$, in particular (according to (3.7)), it obeys

$$f_1\left(0, \frac{1}{t}\right) = t\, f_1(0,t)\,. \tag{3.16}$$

The first requirement follows from the comparison of the homogeneous factor multiplying P_d in (2.6) for $d = 4$ with the factor $(\rho_{12}\rho_{34})^{-3}(\rho_{13}\rho_{24})^{-1}$ (resulting from (3.1) and (3.7)) which multiplies $f_1(s,t)$ in the expression for w_4^t. By (crossing) symmetrizing the contribution to w_4^t so obtained we shall reproduce the basic polynomials J_ν (2.12–2.14) in (2.11).

Anticipating the result of (Secs. 3.2 and 4.2 of) [15] we choose the following complete set of initial conditions, $f_1(0,t) = j_\nu(0,t)$, $\nu = 0,1,2,$ consistent with the symmetry property (3.16):

$$\begin{aligned} j_0(0,t) &= 1 + t^{-1},\ j_1(0,t) = j_0(0,t)\,\frac{(1-t)^2}{t}\,, \\ j_2(0,t) &= j_1(0,t)\left[\frac{(1-t)^2}{t} + 1\right]. \end{aligned} \tag{3.17}$$

Note that the rational function $I(t) := \frac{(1-t)^2}{t}$ which enters the expressions for j_1 and j_2 is the minimal degree polynomial in t and t^{-1} that is non-identically zero and satisfies $I\left(\frac{1}{t}\right) = I(t)$ and $I(1) = 0$.

The solutions of (3.11) with these initial conditions are given by

$$\begin{aligned} j_0(s,t) &= j_0(0,t) = 1 + t^{-1}\,, \\ j_1(s,t) &= \left(\frac{1-t}{t}\right)^2 (1 + t - s) - 2\frac{s}{t} = j_1(0,t) - s(1 + t^{-2})\,, \\ j_2(s,t) &= (1 + t^{-3})[(1 + s - t)^2 - s] - 3s(1-t)\,t^{-3}\,. \end{aligned} \tag{3.18}$$

As none of the above solutions is symmetric under the exchange s_{23} (2.8), locality implies that such twist 2 contributions to the partial wave expansion should be necessarily accompanied by higher twist terms present in the symmetrized expressions

$$\begin{aligned} J_\alpha(s,t) &= (1 + s_{23} + s_{13})\, t^3\, j_\alpha(s,t),\ \alpha = 0,1 \\ J_2(s,t) &= \frac{1}{2}\,(1 + s_{23} + s_{13})\, t^3\, j_2(s,t)\,. \end{aligned} \tag{3.19}$$

The presence of different factors (1 and $\frac{1}{2}$) in front of the symmetrizer in the two formulae calls for a precise definition of what is meant by saying that "J_ν is a *symmetrized expression* of $t^3 j_\nu$". We note that for any value of λ_ν the expression $J_\nu = \lambda_\nu(1 + s_{23} + s_{13})\, t^3\, j_\nu$ is $\mathcal{S}_3$ symmetric. We shall say that such a J_ν is a symmetrization of $t^3 j_\nu$ if $J_\nu - t^3\, j_\nu$ tends to zero for $s \to 0$. It then in fact follows that the possible values of λ_ν are 1 and 1/2.

It should be stressed that not every solution of (3.11) admits a symmetrized version in this sense. (The reader can verify that *e.g.* $j_2 - j_1$ does not.) Thus the above requirement restricts the choice of basis. In fact, it follows from (2.12–2.14) and from (3.18) that the differences $J_\nu - t^3\, j_\nu$ have different orders for small s:

$$\begin{aligned}
J_0(s,t) - t^3 j_0(s,t) &= s^2(1 + t^3 + s(1+t^2)),\\
J_1(s,t) - t^3 j_1(s,t) &= s(1-t)(1-t^3) - s^2(1+t^3) - s^3(1+t)^2 + s^4(1+t),\\
J_2(s,t) - t^3 j_2(s,t) &= s^3(1+t+t^2) - 2s^4(1+t) + s^5.
\end{aligned} \tag{3.20}$$

Comparing (3.20) with (3.15) we deduce that the differences $J_1 - t^3\, j_1$, $J_0 - t^3\, j_0$ and $J_2 - t^3\, j_2$ carry twist 4 and higher, twist 6 and higher and twist 8 and higher, respectively. The above basis is clearly unique with such a separation property.

Putting everything together we can, in principle, determine all structure constants $B_{\kappa\ell}$. It follows from (3.4) (3.9) and from the relation $A_{\kappa\ell} = N_{\kappa\ell}\, C_{\kappa\ell}$ where $N_{\kappa\ell}(> 0)$ stands for the normalization of the 2-point function of $O_{2\kappa\ell}$ that $B_{\kappa\ell} = N_{\kappa 2\ell}\, C^2_{\kappa 2\ell}$ should be positive if Hilbert space (or Wightman) positivity holds. (The full argument uses the classification [12] of unitary positive energy representations of $SU(2,2)$ according to which the state spaces spanned by $O_{2\kappa 2\ell}(z,w)|0\rangle$, for $\kappa = 1, 2, \ldots$, $\ell \geq 0$, belong to the unitary series.) Thus, such a calculation will restrict the admissible values of the parameters a_ν, b, c and B in (2.11) and (3.1), providing a non-trivial positivity check for the 4-point function of $\mathcal{L}$. We shall display the corresponding equations and their solution for $\kappa = 1, 2, 3$ (the twists for which the 2-point normalization B does not contribute).

Inserting in the left hand side of (3.8) for $\kappa = 1$ the expression $f(0,t) = \sum_{\nu=0}^{2} a_\nu\, j_\nu(0,t)$ we can solve with respect to $B_{1\ell}$ with the result

$$B_{1\ell} = \frac{1}{\binom{4\ell}{2\ell}} \left\{2a_0 + 2\ell(2\ell+1)[2a_1 + (2\ell-1)(\ell+1)\, a_2]\right\}. \tag{3.21}$$

The equation for $\kappa = 2$ involves a_1, b and c:

$$f_2(0, 1-u) = a_1 u \left(\frac{1}{(1-u)^3} - 1 \right) + \frac{bu^2}{(1-u)^2} + \frac{c}{1-u}$$
$$= \sum_{\ell=0}^{\infty} B_{2\ell}\, u^{2\ell} F(2\ell+2, 2\ell+2; 4\ell+4; u) . \qquad (3.22)$$

Its solution is

$$B_{2\ell} = \frac{1}{\binom{4\ell+1}{2\ell}} \{\ell(2\ell+3)[(\ell+1)(2\ell+1)\, a_1 + 2b] + c\} . \qquad (3.23)$$

For $\kappa = 3$ we have to use the expression (3.13) for $f_2(s,t)$ which involves a log term as $F(1,1;2;v) = \sum_{n=1}^{\infty} \frac{v^{n-1}}{n} = \frac{1}{v} \log \frac{1}{1-v}$. The result is

$$f_3(0, 1-u) = \left(a_0 + \frac{a_1}{2}\right)(1 + (1-u)^{-3}) - \frac{3}{2} \frac{b}{1-u} \left(1 + \frac{1}{1-u}\right)$$
$$+ \frac{c}{2} \left\{ \frac{2u-1}{u(1-u)} \left(1 + \frac{2}{u(1-u)}\right) - 2 \frac{\log(1-u)}{u^3} \right\}$$
$$= \sum_{\ell=0}^{\infty} B_{3\ell}\, u^{2\ell}\, F(2\ell+3, 2\ell+3; 4\ell+6; u) . \qquad (3.24)$$

A computer aided calculation (using Maple) gives in this case

$$B_{3\ell} = \frac{1}{2} \binom{4\ell+3}{2\ell+1}^{-1} \{(\ell+1)(2\ell+3)\,[(\ell+2)(2\ell+1)(2a_0+a_1) - 6b + 4c] - c\}. \qquad (3.25)$$

The positivity of $B_{j\ell}$, $j = 1, 2, 3$ implies

$$a_0 \geq 0\, ,\ a_1 + a_2 > 0\, ;\ 3a_1 + b \geq 0\, ,\ c \geq 0\, ;\ 6(2a_0 + a_1 - 3b) + 11c \geq 0\, . \quad (3.26)$$

This leaves a nonempty domain in the space of (4-point function) parameters in which positivity holds.

3.3. *The concept of a minimal model*

We observe that the twist two contribution to the $2n$-point function,

$$w_1(1, 2; 3, 4; \ldots; 2n-1, 2n) :=$$
$$\left(\prod_{i=1}^{n} \rho_{2i-1,2i}^{-3} \right) \langle 0|\, V_1(z_1, z_2)\, V_1(z_3, z_4) \ldots V_1(z_{2n-1}, z_{2n}) |0\rangle \ , \quad (3.27)$$

combined with locality, implies the existence of higher twist terms. The question arises whether one can define a "minimal model" obtained by an appropriate symmetrization of such a bilocal field contribution. The difficulty in making this idea precise comes from the fact that there are two distinct notions of symmetrization – as displayed in (3.19). We shall therefore restrict our attention to symmetrizable V_1 which we proceed to define.

We begin by extending the notion of truncated $2n$-point function to bilocal fields, setting

$$w_1^t(1,2;\ldots;2n-1,2n) = w_1(1,2;\ldots;2n-1,2n) \quad \text{for} \quad n<4\,,$$

$$\begin{aligned} w_1^t(1,2;\ldots;7,8) = w_1(1,2;\ldots;7,8) &- w_1(1,2;3,4)\,w_1(5,6;7,8) \\ &- w_1(1,2;5,6)\,w_1(3,4;7,8) \\ &- w_1(1,2;7,8)\,w_1(3,4;5,6) \end{aligned} \tag{3.28}$$

(and similar expressions involving symmetric subtractions for $n > 4$). We say that V_1 is *symmetrizable* if for any $n = 2, 3, \ldots$, there is a λ_n such that the function

$$w^t(1,2,\ldots,2n) = \lambda_n \sum_{i_{2k-1}<i_{2k}\leq 2n} w_1^t(1,i_2;\ldots;i_{2n-1},i_{2n}) \tag{3.29}$$

where the sum is spread over all $(2n-1)!!$ permutations $(1,2,\ldots,2n) \to (1,i_2,\ldots,i_{2n})$ whose entries satisfy the above inequalities, involves the same twist two contribution with respect to the pair of arguments $(2i-1,2i)$ for $i = 1,\ldots,n$ as $w_1^t(1,2;\ldots;2n-1,2n)$:

$$\lim_{\rho_{2i-1,2i}\to 0}\{\rho_{2i-1,2i}^3(w^t(z_1,\ldots,z_{2n}) - w_1^t(1,2;\ldots;2n-1,2n))\} = 0\,. \tag{3.30}$$

We then define a *minimal model* by identifying the truncated $2n$-point correlation function with $w^t(1,\ldots,2n)$ (3.29).

Remark 3.1. The symmetrizability of V_1 is only a sufficient condition enabling us to construct a minimal model. For instance, the truncated 4-point function (2.6) (2.11) for $b = c = 0$ is "minimal" although V_1 is not symmetrizable according the above definition, but is a sum of symmetric expressions associated with different factors λ in (3.29). It follows from the results summed up in Sec. 2 that both J_0 and J_2 have free field real

$$V_1^{(0)}(z_1,z_2) = \sum_{i=1}^{N} :\varphi_i(z_1)\,\varphi_i(z_2): \quad (\Delta\,\varphi_i(z) = 0) \tag{3.31}$$

and to the expression (4.21) of [15],

$$V_1^{(2)}(z_1,z_2) = C\left\{\frac{1}{4}\,z_{12}^2 : F^{\sigma\tau}(z_1)F_{\sigma\tau}(z_2) : -\delta^{\sigma\tau} z_{12}^{\mu} z_{12}^{\nu} : F_{\sigma\mu}(z_1)F_{\tau\nu}(z_2) :\right\}$$

$$(dF(z) = 0 = d^*F(z))\,, \tag{3.32}$$

respectively. As we shall see (in Sec. 4.2, below), $J_1(s,t)$ also admits such a realization. This allows to extend the construction of

$$w_4^t = \frac{\rho_{13}^2\rho_{24}^2}{(\rho_{12}\rho_{23}\rho_{34}\rho_{14})^3}\,(a_0\,J_0(s,t) + a_1\,J_1(s,t) + a_2\,J_2(s,t)) \tag{3.33}$$

to higher point correlation functions, thus obtaining a minimal model of a more general type than the above considered case of a symmetrizable V_1.

4. Is there a non-trivial gauge field theory model?

4.1. *Restrictions on the parameters in the 4-point function*

We now address the question how to characterize the local gauge invariant Lagrangian, which gives rise to a 4-form

$$\mathcal{L}(z)\,dz_1 \wedge dz_2 \wedge dz_3 \wedge dz_4 = \mathrm{tr}\,({}^*F(z) \wedge F(z))\,, \tag{4.1}$$

where F is the (Maxwell, Yang-Mills) curvature 2-form and ${}^*F$ is its Hodge dual:

$$F(z) = \frac{1}{2}\,F_{\mu\nu}(z)\,dz^{\mu} \wedge dz^{\nu}\,,\quad {}^*F(z) = \frac{1}{4}\,\varepsilon_{\kappa\lambda\mu\nu}\,F^{\kappa\lambda}\,dz^{\mu} \wedge dz^{\nu}\,, \tag{4.2}$$

without introducing gauge dependent quantities like F (in the non-abelian case) or the connection 1-form (of the gauge potential) A. We first note that a pure gauge Lagrangian of type (4.1) (*i.e.* a Lagrangian without matter fields) should not allow for a scalar of dimension (= twist) 2 in the OPE of $\mathcal{L}(z_1)\,\mathcal{L}(z_2)$. In view of (3.21) this implies

$$a_0 = 0\,. \tag{4.3}$$

Furthermore, assuming invariance of the theory under "electric-magnetic" (or Hodge) duality[b], and noting that ${}^*({}^*F) = -F$ in Minkowski space, we deduce that the theory should be invariant under a change of sign of $\mathcal{L}$. Hence all odd-point functions of $\mathcal{L}$ should vanish. We make the stronger

---

[b]We thank Dirk Kreimer for a discussion on this point.

assumption that *no* scalar field of dimension 4 should appear in the OPE of two $\mathcal{L}$'s. According to (3.23) this implies

$$c = 0 \,. \tag{4.4}$$

(The vanishing of the 3-point function of the Maxwell Lagrangian is verified by a direct calculation.) We are thus left with the 3 parameters a_1, a_2, and b in the truncated 4-point function, the positivity restrictions (3.26) implying

$$a_1 \geq 0 \,, \; a_1 + a_2 > 0 \,, \; -3a_1 \leq b \leq \frac{1}{3}\, a_1 \,. \tag{4.5}$$

Clearly for $a_1 = 0$ we shall also have $b = 0$ and the truncated 4-point function will be a multiple of that of the free electromagnetic Lagrangian (2.16) (Sec. 2.2). In order to go beyond the free field theory we shall assume $a_1 > 0$ and will study the properties of the contribution J_1 (2.13) to w_4^t.

4.2. *A composite bilocal field reproducing J_1*

We denote by $v(z_1, z_2)$ the field V_1 which reproduces the term $j_1(s,t)$ (3.18) in the expansion of f_1 (defined by (3.7) for $\kappa = 1$); in other words, we set

$$\begin{aligned} \langle 0|\, v(z_1,z_2)\, v(z_3,z_4)\,|0\rangle &= \frac{a}{\rho_{13}\rho_{24}}\, j_1(s,t) \\ &= a\left(\frac{t^{-1} - st^{-1} - 1}{\rho_{14}\rho_{23}} + \frac{t - s - 1}{\rho_{13}\rho_{24}}\right) . \end{aligned} \tag{4.6}$$

The two terms in the right hand side are obtained from one another by the substitution $s_{12}(z_1 \rightleftarrows z_2)$ suggesting that there exists a free field realization of $v(z_1, z_2)$ which reproduces (4.6) (for an appropriate choice of the constant a). Indeed one can check that the CP even projection of the product of two conjugate Weyl fields does the job.

In order to write it down in the above z-picture we introduce the 2×2 matrix realization of the quaternion algebra setting

$$\not{z} := \sum_{\mu=1}^{4} z_\mu\, Q_\mu = \begin{pmatrix} z_4 - iz_3 & -z_2 - iz_1 \\ z_2 - iz_1 & z_4 + iz_3 \end{pmatrix} , \; \not{z}^+ := \sum_{\mu=1}^{4} z_\mu\, Q_\mu^+ = z_4 - \mathbf{z}\,\mathbf{Q} \tag{4.7}$$

(where, in the last equation, we have omitted the 2×2 unit matrix Q_4). It is characterized by the basic anticommutation relations

$$\not{z}\,\not{w}^+ + \not{w}\,\not{z}^+ = 2zw \; (= 2(\mathbf{z}\mathbf{w} + z_4 w_4)) \,. \tag{4.8}$$

A *free Weyl field* ψ is a complex 2-component field of $S(U(2)\times U(2))$ weight $(\frac{3}{2};\frac{1}{2},0)$ satisfying the *Weyl equation*

$$\not\partial\,\psi(z) := Q_\mu \frac{\partial}{\partial z_\mu}\,\psi(z) = 0 \quad \left(= \frac{\partial}{\partial z_\mu}\,\psi^+(z)\,Q_\mu\right); \qquad (4.9)$$

for an appropriate normalization its 2-point function is given by

$$\langle 0|\,\psi(z_1)\,\psi^+(z_2)\,|0\rangle = \frac{\not z_{12}^{+}}{\rho_{12}^2}\,. \qquad (4.10)$$

A straightforward computation (see Appendix A) shows that the bilocal field

$$v(z_1,z_2) =: \psi^+(z_1)\,\not z_{12}\,\psi(z_2) : - : \psi^+(z_2)\,\not z_{12}\,\psi(z_1) := v(z_2,z_1) \qquad (4.11)$$

(where :: stands for the standard free field normal product) indeed reproduces (4.6) for $a=2$.

The bilocal field (4.11), being symmetric under the exchange of arguments, only involves even rank (conserved) tensors in its expansion in local fields. The product $v(z_1,z_2)\,v(z_3,z_4)$, however, gives also rise to odd rank tensors, in particular, a conserved current, which will thus contribute to the 8-point function of $\mathcal{L}$. A gauge invariant conserved current, on the other hand, can only appear in an abelian, say $U(1)$, gauge theory. This looks puzzling: are not we dealing with a free Weyl theory in disguise? The answer is no: in the theory of a free Weyl field there is no room for a dimension 4 scalar. In particular, the Lagrangian of a free Weyl field vanishes, while in the model at hand $\mathcal{L}$ is coupled to v since the OPE of two $\mathcal{L}$'s involves v.

The basic question in the title of this section can then be sharpened as follows: *is there a minimal model* (in the sense of Sec. 3.3) *for a scalar field* $\mathcal{L}(z)$ *with the reduced OPE* (in which we have set $a_2=b=0$)

$$\mathcal{L}(z_1)\,\mathcal{L}(z_2) = \frac{B}{\rho_{12}^4} + \frac{v(z_1,z_2)}{\rho_{12}^3} + O(\rho_{12}^{-2})\,, \qquad (4.12)$$

where $v(z_1,z_2)$ has correlation functions corresponding to the ansatz (4.11)?

We have no definitive answer to this simple question. We know that the 4 point function of $\mathcal{L}$ with truncated part

$$w_4^t = \frac{a\rho_{13}^2\rho_{24}^2}{(\rho_{12}\rho_{23}\rho_{34}\rho_{14})^3} \qquad (4.13)$$
$$\{s(1-s)(1-s^2)+t(1-t)(1-t^2)+st[(s-t)(s^2-t^2)-2-2s^2-2t^2]\}$$

satisfies all Wightman axioms – including positivity (for a non-empty set of positive "a" and B). Happily, as we shall see in the next subsection, the ansatz (4.11) also allows to construct higher (even point) correlation functions of $\mathcal{L}$. Then one should be able to attack the positivity problem in its full generality (all other Wightman axioms being obviously satisfied by rational symmetric correlation functions). Albeit we are ultimately interested in the theory with a truncated 4-point function depending on all 3 admissible parameters, a_1, a_2, and b, it should be easier to first handle the simpler model defined by (4.11), (4.12).

4.3. *Elementary contributions to w^t_{2n}, symmetric under $\mathbb{Z}_2 \times \mathbb{Z}_n$*

We observe that the expression (3.19) for J_1 is a sum of 3 terms each of which splits into two, as displayed in the right hand side of (4.6). This observation can be understood as follows: each of the 6 contributions, say

$$\mathcal{W}(12;34) = \frac{2}{\rho_{12}^3 \rho_{34}^3} \frac{\rho_{13}\rho_{24} - \rho_{12}\rho_{34} - \rho_{14}\rho_{23}}{\rho_{14}^2 \rho_{23}^2}, \tag{4.14}$$

(where we have set $a = 2$) has a $\mathbb{Z}_2 \times \mathbb{Z}_2$ symmetry generated by the pair of involutive transformations $(12;34) \to (21;43)$ and $(12;34) \to (43;21)$. The number of such *elementary contributions* is $|\mathcal{S}_4|/|\mathbb{Z}_2 \times \mathbb{Z}_2| = \frac{4!}{4} = 3!$ (where we denote by $|G|$ the number of elements of the finite group G). Such a counting readily extends to the general case of a $2n$-point function.

The $2n$-point function

$$\langle 0|\, v(z_1, z_2)\, v(z_3, z_4) \ldots v(z_{2n-1}, z_{2n})\, |0\rangle \tag{4.15}$$

of the composite field (4.11) has a $(\mathbb{Z}_2)^{\times n} \times \mathcal{S}_n$ symmetry (consisting of exchanging the arguments of each individual factor and of permuting the v's). There are, hence, $\frac{|\mathcal{S}_{2n}|}{2^n |\mathcal{S}_n|} = (2n-1)!!$ different $2n$-point functions $\langle 0|\, v(z_1, z_{i_2}) \ldots\, v(z_{i_{2n-1}}, z_{i_{2n}})\, |0\rangle$. The expression for each such $2n$-point function has $(2n-2)!! = 2^{n-1}(n-1)!$ elementary contributions labeled by the set of pairs (ij) for which the corresponding rational function has a pole in ρ_{ij}. Each elementary contribution to w^t_{2n}, say $\mathcal{W}(12; 34; \ldots; 2n-1\, 2n)$, possesses a $\mathbb{Z}_2 \times \mathbb{Z}_n$ symmetry, where $\mathbb{Z}_n$ is the group of cyclic permutations of ordered pairs of arguments generated by $(12, 34, \ldots, 2n-1\, 2n) \to (2n-1\, 2n, 12, \ldots, 2n-3\, 2n-2)$, while the non-trivial element of $\mathbb{Z}_2$ is the involutive permutation $(12, \ldots, 2n-1\, 2n) \to (2n\, 2n-1, \ldots, 21)$.

For the 6-point function we have $5!! = 15$ vacuum expectation values of type (4.15) each consisting of $2^2 \cdot 2! = 8$ elementary $\mathbb{Z}_2 \times \mathbb{Z}_3$ $(= \mathbb{Z}_6)$

symmetric contributions. One of the resulting $8.15 = 120$ elementary contributions to w_6^t is $\mathcal{W}(12;34;56)$; for the corresponding term in (4.15) we find (see Appendix A):

$$\rho_{12}^3\rho_{34}^3\rho_{56}^3\,\mathcal{W}(12;34;56) = (\rho_{16}\rho_{23}\rho_{45})^{-2}\{\rho_{12}(\rho_{34}\rho_{56} - \rho_{35}\rho_{46} + \rho_{36}\rho_{45})$$
$$- \rho_{13}(\rho_{24}\rho_{56} - \rho_{25}\rho_{46} + \rho_{26}\rho_{45}) + \rho_{14}(\rho_{23}\rho_{56} - \rho_{25}\rho_{36} + \rho_{26}\rho_{35})$$
$$- \rho_{15}(\rho_{23}\rho_{46} - \rho_{24}\rho_{36} + \rho_{26}\rho_{34}) + \rho_{16}(\rho_{23}\rho_{45} - \rho_{24}\rho_{35} + \rho_{25}\rho_{34})\}\,, \quad (4.16)$$

thus exhibiting the general pattern: it displays a Wick structure of a fermionic correlation function with "propagators" ρ_{ij}. In the perspective of studying positivity property of a sum of 120 terms of type (4.16) (plus 15 products of 2-point functions $\langle 1\, i_2\rangle \langle i_3\, i_4\rangle \langle i_5\, i_6\rangle$ with $i_{2k-1} < i_{2k} \le 6$) it seems more promising to try to use the OPE (4.12) for the minimal model at hand together with the fact that the theory of a free Weyl field (and hence the theory of the bilocal field (4.11)) does satisfy Wightman positivity.

4.4. *Concluding remarks*

Global conformal invariance [16] opens the way of constructing 4- (or higher) dimensional QFT models satisfying all Wightman axioms (except for asymptotic completeness). Experience with gauge field theory suggests that the simplest local gauge invariant observable is the Lagrangian density $\mathcal{L}$. The present update of our effort to construct a non-perturbative GCI gauge QFT [14] [15] displays some new features and suggests new questions (or new ways of approaching old ones).

– We emphasize that the main tool for attacking the difficult problem of Wightman positivity are the *conformal partial wave expansions* of 4-point functions. They should be extended to 4-point functions of composite (tensor) fields or, alternatively, to higher point functions of $\mathcal{L}(z)$. OPE provide just a means to derive such expansions with invariant under rescaling structure constants (like (3.9)).

– The notion of a *minimal model* is introduced that is fully determined by a (twist 2) symmetrizable bilocal field $V_1(z_1, z_2)$ which is harmonic in each argument.

– It is demonstrated that the non-trivial twist 2 contribution to the 4-point function (which is *not* reproduced by an $\mathcal{L}(z)$ that is a normal product of free fields) requires a bilocal field $V_1 = v(z_1, z_2)$ that can be realized by such a normal product (involving free Weyl spinors ψ and ψ^+ – (see (4.11)).

The resulting rather special minimal model appears to be sufficiently simple to enable one to decide whether it is consistent with Wightman positivity and to find out whether its properties (in particular, the existence of a conserved $U(1)$ current in the OPE of four $\mathcal{L}$'s) agree with the conjecture that we are dealing with a non-perturbative renormalization group fixed point of a gauge field theory.

Acknowledgments

I.T. acknowledges the hospitality of l'Institut des Hautes Etudes Scientifiques (Bures-sur-Yvette) where this paper was written. The research of Ya.S. has been supported in part by EC contracts HPRN-CT-2000-00122 and -00148. All three authors acknowledge partial support by the Research Training Network within Framework Programme 5 of the European Commission under contract HPRN-CT-2002-00325.

Appendix A. Computing correlation functions of $v(z_1, z_2)$

The 4-point function (4.6) for v given by (4.11) and the 2-point function of ψ normalized according to (4.10) is expressed as a symmetric combination of traces:

$$\langle 0|\, v(z_1, z_2)\, v(z_3, z_4)\, |0\rangle = \frac{\mathrm{tr}\{\not z_{12}(\not z_{23}^{+}\not z_{34}\not z_{14}^{+} + \not z_{14}^{+}\not z_{34}\not z_{23}^{+})\}}{\rho_{14}^2\,\rho_{23}^2} - \frac{\mathrm{tr}\{\not z_{12}(\not z_{24}^{+}\not z_{34}\not z_{13}^{+} + \not z_{13}^{+}\not z_{34}\not z_{24}^{+})\}}{\rho_{13}^2\,\rho_{24}^2}\,. \tag{A.1}$$

It is sufficient to compute the first term since the second can be obtained from it by the substitution $z_3 \rightleftarrows z_4$. To do that we shall use the following trace formula for the product of any four 4-vectors a, b, c, d written as quaternions:

$$\mathrm{tr}(\not a \not b^{+} \not c \not d^{+}) = 2[(ab)(cd) - (ac)(bd) + (ad)(bc) + \det(a, b, c, d)]\,,$$
$$2(ab) = \mathrm{tr}\, \not a \not b^{+}\,, \tag{A.2}$$

$\det(a, b, c, d)$, the determinant of the 4×4 matrix of the components of the four (column) vectors, changing sign under transposition of any two arguments. It follows that

$$\mathrm{tr}(\not z_{12}\not z_{23}^{+}\not z_{34}\not z_{14}^{+} + \not z_{12}\not z_{14}^{+}\not z_{34}\not z_{23}^{+})$$
$$= 4[(z_{12}z_{23})(z_{34}z_{14}) - (z_{12}z_{34})(z_{14}z_{23}) + (z_{12}z_{14})(z_{23}z_{34})]\,. \tag{A.3}$$

To reproduce (4.14) one uses the relations

$$2\,z_{12}z_{23} = \rho_{13} - \rho_{12} - \rho_{23}\,,\ 2\,z_{34}z_{14} = \rho_{34} + \rho_{14} - \rho_{13}\,,\ \text{etc.}$$

$$2\,z_{12}z_{34} = \rho_{14} + \rho_{23} - \rho_{13} - \rho_{24}\,,\ 2\,z_{14}z_{23} = \rho_{13} + \rho_{24} - \rho_{12} - \rho_{34}\ \text{etc.} \quad (A.4)$$

Similarly, the (polynomial) *elementary contribution* (4.16) to the 6-point function

$$\rho_{16}^2\rho_{23}^2\rho_{45}^2\ \langle 0|\, v(z_1, z_2)\, v(z_3, z_4)\, v(z_5, z_6)\, |0\rangle$$

is given by

$$P(12;34;56) := \mathrm{tr}\{\not z_{12}(\not z_{23}^{+}\not z_{34}\not z_{45}^{+}\not z_{56}\not z_{16}^{+} + \not z_{16}^{+}\not z_{56}\not z_{45}^{+}\not z_{34}\not z_{23}^{+})\}$$
$$= 4\{(z_{12}z_{23})[(z_{34}z_{45})(z_{56}z_{16}) - (z_{34}z_{56})(z_{45}z_{16}) + (z_{34}z_{16})(z_{45}z_{56})]$$
$$-(z_{12}z_{34})[(z_{23}z_{45})(z_{56}z_{16}) - (z_{23}z_{56})(z_{45}z_{16}) + (z_{23}z_{16})(z_{45}z_{56})] + \ldots$$
$$+(z_{12}z_{16})[(z_{23}z_{39})(z_{45}z_{56}) - (z_{23}z_{45})(z_{34}z_{56}) + (z_{23}z_{56})(z_{34}z_{45})]\} \quad (A.5)$$

(5×3 terms). Applying to this expression the relations of type (A.4) (and using Maple to simplify the result) we recover (4.16).

References

1. R.E. Borcherds, "Vertex algebras, Kac-Moody algebras, and the Monster", *Proc. Nat. Acad. Sci. USA*, **83**, 3068-3071 (1983).
2. V.K. Dobrev, G. Mack, V.B. Petkova, S.G. Petrova and I.T. Todorov, *Harmonic Analysis of the n-dimensional Lorentz Group and Its Application to Conformal Quantum Field Theory*, (Springer, Berlin et al, 1977).
3. V.K. Dobrev, V.B. Petkova, S.G. Petrova and I.T. Todorov, "Dynamical derivation of vacuum operator product expansion in conformal field theory", *Phys. Rev.* **D13**, 887-912 (1976).
4. F.A. Dolan and H. Osborn, "Conformal four point functions and operator product expansion", *Nucl. Phys.* **B599**, 459-496 (2001), hep-th/0011040.
5. F.A. Dolan and H. Osborn, "Conformal partial waves and operator product expansion", *Nucl. Phys.* **B678**, 491-507 (2004), hep-th/0309180.
6. B. Eden, A.C. Petkou, C. Schubert and E. Sokatchev, "Partial nonrenormalization of the stress-tensor four-point function in $N = 4$SYM and AdS/CFT", *Nucl. Phys.* **B607**, 191-212 (2001), hep-th/0009106.
7. E. Frenkel and D. Ben-Zvi, *Vertex Algebras and Algebraic Curves*, (AMS 2001).
8. S. Ferrara, R. Gatto, A. Grillo and G. Parisi, "The shadow operator formalism for conformal algebra vacuum expectation values and operator products", *Nuovo Cim. Lett.* **4**, 115-120 (1972).
9. E.S. Fradkin and M.Ya. Palchik, "New developments in d-dimensional conformal quantum field theory", *Phys. Rep.* **300**, 1-112 (1998); *Conformal Quantum Field Theory in D-Dimensions* (Kluwer, Dordrecht et al, 1996).

10. R. Haag, *Local Quantum Physics, Fields, Particles, Algebras*, (Springer, Berlin et al, 1992).
11. V.G. Kac, *Vertex Algebras for Beginners*, 2nd ed., ULS Vol. 10, (AMS, Providence, R.I., 1998).
12. G. Mack, "All unitary representations of the conformal group $SU(2,2)$ with positive energy", *Commun. Math. Phys.* **55**, 1–28 (1977).
13. N.M. Nikolov, "Vertex Algebras in Higher Dimensions and Globally Conformal Invariant Quantum Field Theory", Commun. Math. Phys., to appear; hep-th/0307235.
14. N.M. Nikolov, Ya.S. Stanev and I.T. Todorov, "Four dimensional CFT models with rational correlation functions", *J. Phys. A: Math. Gen.* **35**, 2985-3007 (2002), hep-th/0110230.
15. N.M. Nikolov, Ya.S. Stanev and I.T. Todorov, "Globally conformal invariant gauge field theory with rational correlation functions", Nucl. Phys. **B670** [FS] 373-400 (2003), hep-th/0305200.
16. N.M. Nikolov and I.T. Todorov, "Rationality of conformally invariant local correlation functions on compactified Minkowski space", *Commun. Math. Phys.* **218**, 417-436 (2001), hep-th/0009004.
17. N.M. Nikolov and I.T. Todorov, "Finite temperature correlation functions and modular forms in a globally conformal invariant QFT", ICTP preprint IC/2003/165.
18. R.F. Streater and A.S. Wightman, *PCT, Spin and Statistics, and All That*, (W.A. Benjamin, 1964; Princeton Univ. Press, Princeton, N.J., 2000).
19. I.T. Todorov, "Local field representations of the conformal group and their applications", in: L. Streit (ed.) *Mathematics and Physics. Lectures on Recent Results* Vol. 1, (World Scientific, Singapore 1985) pp. 195-338.
20. I.T. Todorov, "Infinite dimensional Lie algebras in conformal QFT models", in: A.O. Barut and H.-D. Doebner (eds.) *Conformal Groups and Related Symmetries. Physical Results and Mathematical Background*, *Lecture Notes in Physics* Vol. 261, (Springer, Berlin et al, 1986) pp. 387-443.
21. I.T. Todorov, M.C. Mintchev and V.B. Petkova, *Conformal Invariance in Quantum Field Theory*, (Scuola Normale Superiore, Pisa, 1978).

RENORMALISATION OF NONCOMMUTATIVE SCALAR FIELD THEORIES

HARALD GROSSE

Institut für Theoretische Physik der Universität Wien
Boltzmanngasse 5, A-1090 Wien, Austria
e-mail: harald.grosse@univie.ac.at

RAIMAR WULKENHAAR

Max-Planck-Institut für Mathematik in den Naturwissenschaften
Inselstraße 22–26, D-04103 Leipzig, Germany
e-mail: raimar.wulkenhaar@mis.mpg.de

We present the main ideas of our proof that the real ϕ^4-models on noncommutative $\mathbb{R}^2$ and noncommutative $\mathbb{R}^4$ are renormalisable to all orders when working in the matrix base. The proof uses renormalisation group techniques, power-counting theorems for ribbon graphs as well as orthogonal polynomials.

1. Introduction

Quantum field theory on Euclidean or Minkowski space is extremely successful. For suitably chosen action functionals one achieves a remarkable agreement of up to 10^{-11} between theoretical predictions and experimental data. However, combining the fundamental principles of both general relativity and quantum mechanics one concludes that space(-time) cannot be a differentiable manifold [1]. To the best of our knowledge, such a possibility was first discussed in [2].

To make this transparent, let us ask how we explore technically the geometry of space(-time). The building blocks of a manifold are the *points* labelled by coordinates $\{x^\mu\}$ in a given chart. Points enter quantum field theory via the *values* of the fields at the point labelled by $\{x^\mu\}$. This observation provides a way to "visualise" the points: we have to prepare a distribution of matter which is sharply localised around $\{x^\mu\}$. For a perfect visualisation we need a δ-distribution of the matter field. This is physically not possible, but one would think that a δ-distribution could be arbitrarily well approximated. However, that is not the case, there are limits of localisability long before the δ-distribution is reached.

Let us assume that there is a matter distribution which is believed to have two separated peaks within a space-time region R of diameter d. How do we test this conjecture? We perform a scattering experiment in the hope to find interferences which tell us about the internal structure in the region R. We clearly need test particles of de Broglie wave length $\lambda = \frac{\hbar c}{E} \lesssim \mathrm{d}$, otherwise we can only resolve a single peak. For $\lambda \to 0$ the gravitational field of the test particles becomes important. The gravitational field created by an energy E can be measured in terms of the Schwarzschild radius

$$r_s = \frac{2G_N E}{c^4} = \frac{2G_N \hbar}{\lambda c^3} \gtrsim \frac{2G_N \hbar}{\mathrm{d}c^3}\,, \tag{1}$$

where G_N is Newton's constant. If the Schwarzschild radius r_s becomes larger than the radius $\frac{\mathrm{d}}{2}$, the inner structure of the region R can no longer be resolved (it is behind the horizon). Thus, $\frac{\mathrm{d}}{2} \geq r_s$ leads to the condition

$$\frac{\mathrm{d}}{2} \gtrsim \ell_P := \sqrt{\frac{G_N \hbar}{c^3}}\,, \tag{2}$$

which means that the Planck length ℓ_P is the fundamental length scale below of which length measurements become meaningless. Space-time cannot be a manifold.

Since geometric concepts are indispensable in physics, we need a replacement for the space-time manifold which still has a geometric interpretation. Quantum physics tells us that whenever there are measurement limits we have to describe the situation by non-commuting operators on a Hilbert space. Fortunately for physics, mathematicians have developed a generalisation of geometry, baptised noncommutative geometry [3], which is perfectly designed for our purpose. However, in physics we need more than just a better geometry: We need renormalisable quantum field theories modelled on such a noncommutative geometry.

Remarkably, it turned out to be very difficult to renormalise quantum field theories even on the simplest noncommutative spaces [4]. It would be a wrong conclusion, however, that this problem singles out the standard commutative geometry as the only one compatible with quantum field theory. The problem tells us that we are still at the very beginning of *understanding* quantum field theory. Thus, apart from curing the contradiction between gravity and quantum physics, in doing quantum field theory on noncommutative geometries we learn a lot about quantum field theory itself.

2. Field theory on noncommutative $\mathbb{R}^D$ in momentum space

The simplest noncommutative generalisation of Euclidean space is the so-called noncommutative $\mathbb{R}^D$. Although this space arises naturally in a certain limit of string theory [5], we should not expect that it is a good model for nature. In particular, the noncommutative $\mathbb{R}^D$ does not allow for gravity. For us the main purpose of this space is to develop an understanding of quantum field theory which has a broader range of applicability.

The noncommutative $\mathbb{R}^D$, $D = 2, 4, 6, \dots$, is defined as the algebra $\mathbb{R}^D_\theta$ which as a vector space is given by the space $\mathcal{S}(\mathbb{R}^D)$ of (complex-valued) Schwartz class functions of rapid decay, equipped with the multiplication rule

$$(a \star b)(x) = \int \frac{d^D k}{(2\pi)^D} \int d^D y \; a(x+\tfrac{1}{2}\theta\cdot k)\, b(x+y)\, \mathrm{e}^{\mathrm{i}k\cdot y} , \qquad (3)$$

$$(\theta\cdot k)^\mu = \theta^{\mu\nu} k_\nu , \quad k\cdot y = k_\mu y^\mu , \quad \theta^{\mu\nu} = -\theta^{\nu\mu} .$$

The entries $\theta^{\mu\nu}$ in (3) have the dimension of an area. The physical interpretation is $||\theta|| \approx \ell_P^2$. Much information about the noncommutative $\mathbb{R}^D$ can be found in [6].

A field theory is defined by an action functional. We obtain action functionals on $\mathbb{R}^D_\theta$ by replacing in standard action functionals the ordinary product of functions by the $\star$-product. For example, the noncommutative ϕ^4-action is given by

$$S[\phi] := \int d^D x \Big(\frac{1}{2} \partial_\mu \phi \star \partial^\mu \phi + \frac{1}{2} m^2 \phi \star \phi + \frac{\lambda}{4!} \phi \star \phi \star \phi \star \phi \Big) . \qquad (4)$$

The action (4) is then inserted into the partition function

$$Z[j] := \int \mathcal{D}\phi \; \mathrm{e}^{-\frac{1}{\hbar}(S[\phi] - \int d^4 x\, \phi(x) j(x))} , \qquad (5)$$

which gives rise to the correlation functions (= expectation values)

$$\langle \phi(x_1) \dots \phi(x_n) \rangle = Z[0]^{-1} \hbar^n \frac{\delta^n Z[j]}{\delta j(x_1) \dots \delta j(x_n)} \Big|_{j(x)=0} . \qquad (6)$$

As usual we solve (5) perturbatively by Feynman graphs. Due to $\int d^D x \; (a \star b)(x) = \int d^D x \; a(x) b(x)$, the propagator in momentum space is unchanged. For later purpose it is, however, convenient to write it as a double line, $\overset{p}{=\!=\!=\!=}$ $= (p^2 + m^2)^{-1}$. The novelty are phase factors in the vertices, which we also write in double line notation,

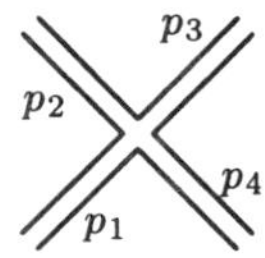

$$= \frac{\lambda}{4!}\, e^{-\frac{i}{2}\sum_{i<j} p_i^\mu p_j^\nu \theta_{\mu\nu}} \,. \tag{7}$$

The double line notation reflects the fact that the vertex (7) is invariant only under cyclic permutations of the legs (using momentum conservation). The resulting Feynman graphs are *ribbon graphs* which depend crucially on how the valences of the vertices are connected. For *planar graphs* the total phase factor of the integrand is independent of internal momenta, whereas *non-planar graphs* have a total phase factor which involves internal momenta. For example, the one-loop contribution to the two-point function splits as follows into a planar part

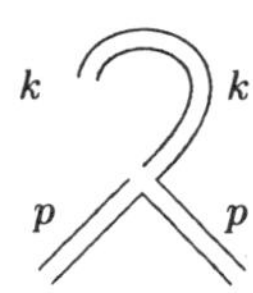

$$= \frac{\lambda}{6}\int \frac{d^4k}{(2\pi)^4}\frac{1}{k^2+m^2} \tag{8}$$

and a non-planar part

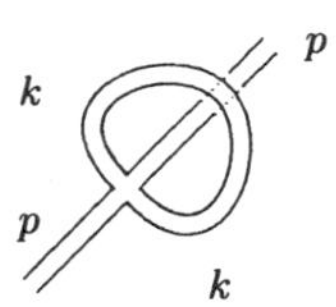

$$= \frac{\lambda}{12}\int \frac{d^4k}{(2\pi)^4}\frac{e^{ip^\mu k^\nu \theta_{\mu\nu}}}{k^2+m^2} = \frac{\lambda}{48\pi^2}\sqrt{\frac{m^2}{\tilde{p}^2}}\, K_1\Big(\sqrt{m^2\tilde{p}^2}\Big)\,, \tag{9}$$

where $\tilde{p}_\mu := \theta_{\mu\nu}p^\nu$.

Planar graphs are treated as usual. The resulting phase factor is precisely of the form of the original two-point function or vertex (7) so that the divergence can be removed via the normalisation conditions. Here, the contribution (8) can entirely be removed by a suitable normalisation condition for the physical mass. The contribution from the non-planar graph (9) is—at first sight—finite, which is a relic of the original motivation that noncommutativity would serve as regulator. The finiteness is important, because the momentum dependence (9) does not appear in the original action (4), which means that a divergence of the form (9) cannot be absorbed by multiplicative renormalisation.

However, the expansion of the modified Bessel function K_1 shows that the contribution (9) behaves $\sim \tilde{p}^{-2}$ for small momenta. If we insert the graph (9) declared as finite as a subgraph into a bigger graph, one easily builds examples (with an arbitrary number of external legs) which lead to non-integrable integrals at small inner momenta. This is the so-called UV/IR-mixing problem [4].

The heuristic argumentation can be made exact: Chepelev and Roiban have proven a power-counting theorem [7, 8] which relates the power-counting degree of divergence to the topology of the ribbon graph. The rough summary of the power-counting theorem is that noncommutative field theories with quadratic divergences become meaningless beyond a certain loop order. The situation is better for field theories with logarithmic UV/IR-divergences, e.g. supersymmetric models. These can be formulated to any loop order. However, the logarithmic IR-divergences at exceptional external momenta are still present so that the correlation functions are unbounded: For every $\delta > 0$ one finds non-exceptional momenta such that $\big|\langle\phi(p_1)\dots\phi(p_n)\rangle\big| > \frac{1}{\delta}$. In the remainder of this article we present an approach which solves these problems.

3. Renormalisation group approach to noncommutative scalar models

We have seen that quantum field theories on noncommutative $\mathbb{R}^D$ are not renormalisable by standard Feynman graph evaluations. One may speculate that the origin of this problem is the too naïve way one performs the continuum limit. A way to treat the limit more carefully is the use of flow equations. The idea goes back to Wilson [9]. It was then used by Polchinski [10] to give a very efficient renormalisability proof of commutative ϕ^4-theory. Applying Polchinski's method to the noncommutative ϕ^4-model, we can hope to be able to prove renormalisability to all orders, too. There is, however, a serious problem of the momentum space proof. We have to guarantee that planar graphs only appear in the distinguished interaction coefficients for which we fix the boundary condition at the renormalisation scale Λ_R. Non-planar graphs have phase factors which involve inner momenta. Polchinski's method consists in taking norms of the interaction coefficients, and these norms ignore possible phase factors. Thus, we would find that boundary conditions for non-planar graphs at Λ_R are required. Since there is an infinite number of different non-planar structures, the model is not renormalisable in this way. A more careful examination of the phase factors is also not possible because the cut-off integrals prevent the Gaußian integration required for the parametric integral representation [7, 8].

Fortunately, there is a matrix representation of the noncommutative $\mathbb{R}^D$ where the $\star$-product becomes a simple product of infinite matrices. The price for this simplification is that the propagator becomes complicated, but the difficulties can be overcome.

3.1. *Matrix representation*

For simplicity we restrict ourselves to the noncommutative $\mathbb{R}^2$. There exists a matrix base $\{f_{mn}(x)\}_{m,n\in\mathbb{N}}$ of the noncommutative $\mathbb{R}^2$ which satisfies

$$(f_{mn}\star f_{kl})(x)=\delta_{nk}f_{ml}(x)\;,\qquad \int d^2x\; f_{mn}(x)=2\pi\theta_1\;, \tag{10}$$

where $\theta_1:=\theta_{12}=-\theta_{21}$. In terms of radial coordinates $x_1=\rho\cos\varphi$, $x_2=\rho\sin\varphi$ one has

$$f_{mn}(\rho,\varphi)=2(-1)^m \mathrm{e}^{\mathrm{i}(n-m)\varphi}\sqrt{\tfrac{m!}{n!}}\Big(\sqrt{\tfrac{2\rho^2}{\theta_1}}\Big)^{n-m}L_m^{n-m}\Big(\tfrac{2\rho^2}{\theta_1}\Big)\,\mathrm{e}^{-\frac{\rho^2}{\theta_1}}\;, \tag{11}$$

where $L_n^\alpha(z)$ are the Laguerre polynomials. See also [6]. The matrix representation was also used to obtain exactly solvable noncommutative quantum field theories [11, 12].

Now we can write down the noncommutative ϕ^4-action in the matrix base by expanding the field as $\phi(x)=\sum_{m,n\in\mathbb{N}}\phi_{mn}f_{mn}(x)$. It turns out, however, that in order to prove renormalisability we have to consider a more general action than (4) at the initial scale Λ_0. This action is obtained by adding a harmonic oscillator potential to the standard noncommutative ϕ^4-action:

$$\begin{aligned} S[\phi] &:= \int d^2x\Big(\frac{1}{2}\partial_\mu\phi\star\partial^\mu\phi+2\Omega^2(\tilde{x}^\mu\phi)\star(\tilde{x}_\mu\phi)+\frac{1}{2}\mu_0^2\phi\star\phi \\ &\qquad\qquad +\frac{\lambda}{4!}\phi\star\phi\star\phi\star\phi\Big)(x) \\ &= 2\pi\theta_1\sum_{m,n,k,l}\Big(\frac{1}{2}G_{mn;kl}\phi_{mn}\phi_{kl}+\frac{\lambda}{4!}\phi_{mn}\phi_{nk}\phi_{kl}\phi_{lm}\Big)\;, \end{aligned} \tag{12}$$

where $\tilde{x}^\mu:=\theta^{\mu\nu}x_\nu$ and

$$G_{mn;kl}:=\int\frac{d^2x}{2\pi\theta_1}\,\Big(\partial_\mu f_{mn}\star\partial^\mu f_{kl}+4\Omega^2(\tilde{x}^\mu f_{mn})\star(\tilde{x}_\mu f_{kl})+\mu_0^2 f_{mn}\star f_{kl}\Big)\,. \tag{13}$$

We view Ω as a regulator and refer to the action (12) as describing a regularised ϕ^4-model. The action (12) could also be obtained by restricting a complex ϕ^4-model with magnetic field [11, 12] to the real part. One finds

$$G_{mn;kl} = \Big(\mu_0^2+\frac{2}{\theta_1}(1+\Omega^2)(n+m+1)\Big)\delta_{nk}\delta_{ml} \\ -\frac{2}{\theta_1}(1-\Omega^2)\big(\sqrt{(n+1)(m+1)}\,\delta_{n+1,k}\delta_{m+1,l} - \sqrt{nm}\,\delta_{n-1,k}\delta_{m-1,l}\big)\,. \tag{14}$$

The kinetic matrix $G_{mn;kl}$ has the important property that $G_{mn;kl} = 0$ unless $m + k = n + l$. The same relation is induced for the propagator $\Delta_{nm;lk}$ defined by $\sum_{k,l=0}^{\infty} G_{mn;kl}\Delta_{lk;sr} = \sum_{k,l=0}^{\infty} \Delta_{nm;lk}G_{kl;rs} = \delta_{mr}\delta_{ns}$. In order to evaluate the propagator we first diagonalise the kinetic matrix $G_{mn;kl}$:

$$G_{m,m+\alpha;l+\alpha,l} = \sum_{y\in\mathbb{N}} U_{my}^{(\alpha)}\big(\mu_0^2+\tfrac{4\Omega}{\theta}(2y+\alpha+1)\big)U_{yl}^{(\alpha)}\,, \tag{15}$$

$$U_{ny}^{(\alpha)} = \sqrt{\binom{\alpha+n}{n}\binom{\alpha+y}{y}\Big(\frac{1-\Omega}{1+\Omega}\Big)^{2n+2y+\alpha+1}\Big(\frac{4\Omega}{1-\Omega^2}\Big)^{\alpha+1}} \\ \times M_n\Big(y; 1+\alpha, \frac{(1-\Omega)^2}{(1+\Omega)^2}\Big)\,, \tag{16}$$

where $M_n(y;\beta,c) = {}_2F_1\Big(\genfrac{}{}{0pt}{}{-n,-y}{\beta}\Big|1-c\Big)$ are the (orthogonal) Meixner polynomials [13]. A lengthy calculation gives

$$\Delta_{mn;kl} = \frac{\theta_1}{2(1+\Omega^2)}\delta_{m+k,n+l} \sum_{v=\frac{|m-l|}{2}}^{\frac{\min(m+l,k+n)}{2}} B\big(\tfrac{1}{2}+\tfrac{\mu_0^2\theta_1}{8\Omega}+\tfrac{1}{2}(m+k)-v, 1+2v\big) \\ \times\sqrt{\binom{n}{v+\frac{n-k}{2}}\binom{k}{v+\frac{k-n}{2}}\binom{m}{v+\frac{m-l}{2}}\binom{l}{v+\frac{l-m}{2}}}\Big(\frac{1-\Omega}{1+\Omega}\Big)^{2v} \\ \times {}_2F_1\left(\genfrac{}{}{0pt}{}{1+2v\,,\ \frac{1}{2}+\frac{\mu_0^2\theta_1}{8\Omega}-\frac{1}{2}(m+k)+v}{\frac{3}{2}+\frac{\mu_0^2}{2\sqrt{1-\omega}\,\mu^2}+\frac{1}{2}(m+k)+v}\middle|\frac{(1-\Omega)^2}{(1+\Omega)^2}\right). \tag{17}$$

Here, $B(a,b)$ is the Beta-function and $F(\genfrac{}{}{0pt}{}{a,b}{c};z)$ the hypergeometric function. We recall that in the momentum space version of the ϕ^4-model, the interactions contain oscillating phase factors which make a renormalisation by flow equations impossible. Here we use an adapted base which eliminates the phase factors from the interaction. We see from (17) that these oscillations do not reappear in the propagator. Note that all matrix elements $\Delta_{nm;lk}$ are non-zero for $m + k = n + l$.

3.2. *The Polchinski equation for matrix models*

We summarise here our derivation [14] of the Polchinski equation for the noncommutative ϕ^4-theory in the matrix base. According to Polchinski's derivation of the exact renormalisation group equation [10] we now consider a (at first sight) different problem than the matrix version of (5):

$$\begin{aligned} Z[J,\Lambda] &= \int \Big(\prod_{a,b} d\phi_{ab}\Big) \exp\big(-S[\phi,J,\Lambda]\big)\,, \\ S[\phi,J,\Lambda] &= (2\pi\theta_1)\Big(\sum_{m,n,k,l} \frac{1}{2}\phi_{mn}G^K_{mn;kl}(\Lambda)\,\phi_{kl} + \sum_{m,n,k,l} \phi_{mn}F_{mn;kl}[\Lambda]J_{kl} \\ &\qquad + \sum_{m,n,k,l} \frac{1}{2}J_{mn}E_{mn;kl}[\Lambda]J_{kl} + L[\phi,\Lambda] + C[\Lambda]\Big)\,, \\ G^K_{mn;kl}(\Lambda) &= \prod_{i\in\{m,n,k,l\}} K\big[\tfrac{i}{\Lambda^2\theta_1}\big]^{-1} G_{mn;kl}\,. \end{aligned} \tag{18}$$

with $L[0,\Lambda]=0$. The cut-off function $K(x)$ is a smooth decreasing function with $K(x)=1$ for $0\leq x\leq 1$ and $K(x)=0$ for $x\geq 2$. Accordingly, we define

$$\Delta^K_{nm;lk}(\Lambda) = \prod_{i\in\{m,n,k,l\}} K\big[\tfrac{i}{\Lambda^2\theta_1}\big]\Delta_{nm;lk}\,. \tag{19}$$

The function $C[\Lambda]$ is the vacuum energy and the matrices E and F, which are not necessary in the commutative case, must be introduced because the propagator Δ is non-local. It is in general not possible to separate the support of the sources J from the support of the Λ-variation of K. We would obtain the original problem for the choice

$$\begin{aligned} L[\phi,\infty] &= \sum_{m,n,k,l} \frac{\lambda}{4!}\phi_{mn}\phi_{nk}\phi_{kl}\phi_{lm}\,, \\ C[\infty] = 0\,, &\qquad E_{mn;kl}[\infty]=0\,, \qquad F_{mn;kl}[\infty] = \delta_{ml}\delta_{nk}\,. \end{aligned} \tag{20}$$

However, we shall expect divergences in the partition function which require a renormalisation, i.e. additional (divergent) counterterms in $L[\phi,\infty]$. In the Feynman graph solution of the partition function one carefully adapts these counterterms so that all divergences disappear. If such an adaptation is possible and all counterterms are local, the model is considered as perturbatively renormalisable.

Following Polchinski [10] we proceed differently to prove renormalisability. We first ask ourselves how we have to choose L, C, E, F in order to make $Z[J,\Lambda]$ *independent* of Λ. After straightforward calculation one finds the answer

$$\Lambda\frac{\partial L[\phi,\Lambda]}{\partial\Lambda} = \sum_{m,n,k,l}\frac{1}{2}\Lambda\frac{\partial\Delta^K_{nm;lk}(\Lambda)}{\partial\Lambda}\left(\frac{\partial L[\phi,\Lambda]}{\partial\phi_{mn}}\frac{\partial L[\phi,\Lambda]}{\partial\phi_{kl}} - \frac{1}{2\pi\theta_1}\left[\frac{\partial^2 L[\phi,\Lambda]}{\partial\phi_{mn}\,\partial\phi_{kl}}\right]_\phi\right), \qquad (21)$$

where $\big[f[\phi]\big]_\phi := f[\phi]-f[0]$. The corresponding differential equations for C,E,F are easy to integrate [14]. Now, instead of computing Green's functions from $Z[J,\infty]$ we can equally well start from $Z[J,\Lambda_R]$, where it leads to Feynman graphs with vertices given by the Taylor expansion coefficients $A^{(V)}_{m_1n_1;\dots;m_Nn_N}$ in

$$L[\phi,\Lambda] = \lambda\sum_{V=1}^{\infty}\big(2\pi\theta_1\lambda\big)^{V-1}\sum_{N=2}^{\infty}\frac{1}{N!}\sum_{m_i,n_i}A^{(V)}_{m_1n_1;\dots;m_Nn_N}[\Lambda]\phi_{m_1n_1}\cdots\phi_{m_Nn_N}\,. \qquad (22)$$

These vertices are connected with each other by internal lines $\Delta^K_{nm;lk}(\Lambda)$ and to sources J_{kl} by external lines $\Delta^K_{nm;lk}(\Lambda_0)$. Since the summation variables are cut-off in the propagator (19), loop summations are finite, provided that the interaction coefficients $A^{(V)}_{m_1n_1;\dots;m_Nn_N}[\Lambda]$ are bounded. Thus, renormalisability amounts to prove that for certain initial conditions (parametrised by finitely many parameters!) the evolution of L according to (21) does not produce any divergences.

Inserting the expansion (22) into (21) and restricting to the part with N external legs we get the graphical expression

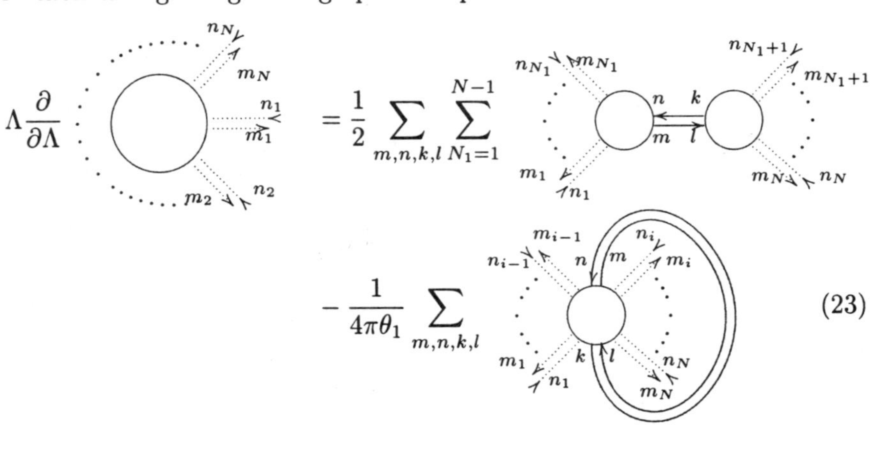

(23)

Combinatorial factors are not shown and symmetrisation in all indices $m_i n_i$ has to be performed. On the RHS of (23) the two valences mn and kl of subgraphs are connected to the ends of a *ribbon* which symbolises the differentiated propagator $\underset{m \quad l}{\overset{n \quad k}{\rightleftarrows}} = \Lambda\frac{\partial}{\partial\Lambda}\Delta^{K}_{nm;lk}$. We see that for the simple fact that the fields ϕ_{mn} carry two indices, the effective action is expanded into ribbon graphs.

In the expansion of L there will occur very complicated ribbon graphs with crossings of lines which cannot be drawn any more in a plane. A general ribbon graph can, however, be drawn on a *Riemann surface* of some *genus* g. In fact, a ribbon graph *defines* the Riemann surfaces topologically through the *Euler characteristic* χ. We have to regard here the external lines of the ribbon graph as amputated (or closed), which means to directly connect the single lines m_i with n_i for each external leg $m_i n_i$. A few examples may help to understand this procedure:

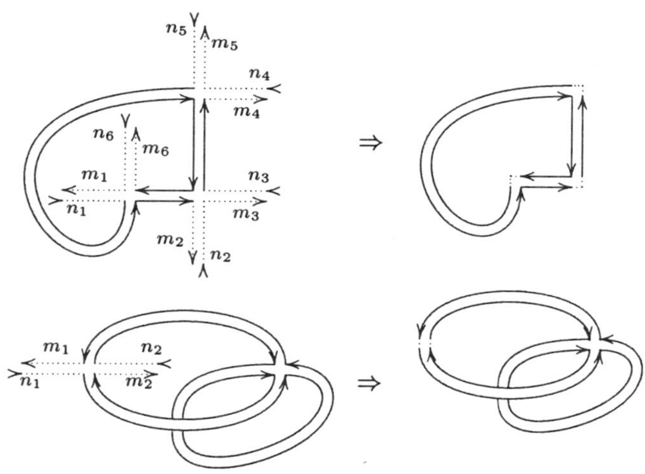

$\tilde{L} = 2$, $I = 3$, $V = 3$, $g = 0$, $B = 2$, $N = 6$

$\tilde{L} = 1$, $I = 3$, $V = 2$, $g = 1$, $B = 1$, $N = 2$ (24)

The genus is computed from the number $\tilde{L}$ of single-line loops, the number I of internal (double) lines and the number V of vertices of the graph according to Euler's formula $\chi = 2 - 2g = \tilde{L} - I + V$. The number B of boundary components of a ribbon graph is the number of those loops which carry at least one external leg. There can be several possibilities to draw the graph and its Riemann surface, but $\tilde{L}, I, V, B$ and thus g remain unchanged. Indeed, the Polchinski equation (21) interpreted as in (23) tells us which external legs of the vertices are connected. It is completely irrelevant how the ribbons are drawn between these legs. In particular, there is no distinction between overcrossings and undercrossings.

We expect that non-planar ribbon graphs with $g > 0$ and/or $B > 1$ behave differently under the renormalisation flow than planar graphs having $B = 1$ and $g = 0$. This suggests to introduce a further grading in g, B in the interactions coefficients $A^{(V,B,g)}_{m_1n_1;\dots;m_Nn_N}$. Technically, our strategy is to apply the summations in (23) either to the propagator or the subgraph only and to maximise the other object over the summation indices. For that purpose one has to introduce further characterisations of a ribbon graph which disappear at the end, see [14].

3.3. *ϕ^4-theory on noncommutative* $\mathbb{R}^2$

First one estimates the A-functions by integrating (21) perturbatively between an initial scale Λ_0 to be sent to ∞ later on and the renormalisation scale Λ_R:

Lemma 3.1. *The homogeneous parts $A^{(V,B,g)}_{m_1n_1;\dots;m_Nn_N}$ of the coefficients of the effective action describing a regularised ϕ^4-theory on $\mathbb{R}^2_\theta$ in the matrix base are for $2 \le N \le 2V+2$ and $\sum_{i=1}^N(m_i - n_i) = 0$ bounded by*

$$\begin{aligned}
&\left|A^{(V,B,g)}_{m_1n_1;\dots;m_Nn_N}[\Lambda, \Lambda_0, \Omega, \rho_0]\right| \\
&\quad \le \left(\Lambda^2\theta_1\right)^{2-V-B-2g}\left(\frac{1}{\Omega}\right)^{3V-\frac{N}{2}+B+2g-2} P^{2V-\frac{N}{2}}\left[\ln\frac{\Lambda_0}{\Lambda_R}\right]. \qquad (25)
\end{aligned}$$

We have $A^{(V,B,g)}_{m_1n_1;\dots;m_Nn_N} \equiv 0$ for $N > 2V+2$ or $\sum_{i=1}^N(m_i - n_i) \neq 0$. By $P^q[x]$ we denote a polynomial in x of degree q.

The proof of (25) for general matrix models by induction goes over 20 pages! The formula specific for the ϕ^4-model on $\mathbb{R}^2_\theta$ follows from the asymptotic behaviour of the cut-off propagator (19), (17) and a certain index summation, see [14, 15].

We see from (25) that the only divergent function is

$$\begin{aligned}
A^{(1,1,0)}_{m_1n_1;m_2n_2} &= A^{(1,1,0)}_{00;00}\delta_{m_1n_2}\delta_{m_2n_1} \\
&\quad + \left(A^{(1,1,0)}_{m_1n_1;m_2n_2}[\Lambda, \Lambda_0, \rho^0] - A^{(1,1,0)}_{00;00}\delta_{m_1n_2}\delta_{m_2n_1}\right), \qquad (26)
\end{aligned}$$

which is split into the distinguished divergent function

$$\rho[\Lambda, \Lambda_0, \Omega, \rho^0] := A^{(1,1,0)}_{00;00}[\Lambda, \Lambda_0, \Omega, \rho^0] \qquad (27)$$

for which we impose the boundary condition $\rho_R := \rho[\Lambda_R, \Lambda_0, \Omega, \rho^0] = 0$ and a convergent part with boundary condition at Λ_0.

One remarks that the limit $\Omega \to 0$ in (25) is singular. In fact the estimation for $\Omega = 0$ with an optimal choice of the ρ-coefficients (different than

(27)!) would grow with $\big(\Lambda\sqrt{\theta_1}\big)^{V-\frac{N}{2}-B-2g+2}$. Since the exponent of Λ can be arbitrarily large, there would be an infinite number of divergent interaction coefficients, which means that the ϕ^4-model is not renormalisable when keeping $\Omega = 0$.

In order to pass to the limit $\Lambda_0 \to \infty$ one has to control the total Λ_0-dependence of the functions $A^{(V,B,g)}_{m_1n_1;\dots;m_Nn_N}[\Lambda, \Lambda_0, \Omega[\Lambda_0], \rho^0[\Lambda_R, \Lambda_0, \rho_R]]$. This leads again to a differential equation in Λ, see [15]. It is then not difficult to see that the regularised ϕ^4-model with $\Omega > 0$ is renormalisable. It turns out that one can even prove more [15]: On can endow the parameter Ω for the oscillator frequency with an Λ_0-dependence so that in the limit $\Lambda_0 \to \infty$ one obtains a standard ϕ^4-model without the oscillator term:

Theorem 3.1. *The ϕ^4-model on $\mathbb{R}^2_\theta$ is (order by order in the coupling constant) renormalisable in the matrix base by adjusting the bare mass $\Lambda_0^2\rho[\Lambda_0]$ to give $A^{(1,1,0)}_{00;00}[\Lambda_R] = 0$ and by performing the limit $\Lambda_0 \to \infty$ along the path of regulated models characterised by $\Omega[\Lambda_0] = \big(1+\ln\frac{\Lambda_0}{\Lambda_R}\big)^{-1}$. The limit $A^{(V,B,g)}_{m_1n_1;\dots;m_Nn_N}[\Lambda_R, \infty] := \lim_{\Lambda_0\to\infty} A^{(V,B,g)}_{m_1n_1;\dots;m_Nn_N}[\Lambda_R, \Lambda_0, \Omega[\Lambda_0], \rho^0[\Lambda_0]]$ of the expansion coefficients of the effective action $L[\phi, \Lambda_R, \Lambda_0, \Omega[\Lambda_0], \rho^0[\Lambda_0]]$ exists and satisfies*

$$\begin{aligned}&\Big|\lambda\big(2\pi\theta_1\lambda\big)^{V-1} A^{(V,B,g)}_{m_1n_1;\dots;m_Nn_N}[\Lambda_R, \infty] \\ &\qquad - \big(2\pi\theta_1\lambda\big)^{V-1} A^{(V,V^e,B,g,\iota)}_{m_1n_1;\dots;m_Nn_N}\big[\Lambda_R, \Lambda_0, \tfrac{1}{(1+\ln\frac{\Lambda_0}{\Lambda_R})}, \rho^0[\Lambda_0]\big]\Big| \\ &\leq \frac{\Lambda_R^4}{\Lambda_0^2}\Big(\frac{\lambda}{\Lambda_R^2}\Big)^V \Big(\frac{(1+\ln\frac{\Lambda_0}{\Lambda_R})}{\Lambda_R^2\theta_1}\Big)^{B+2g-1} P^{5V-N-1}\Big[\ln\frac{\Lambda_0}{\Lambda_R}\Big]\,. \end{aligned} \tag{28}$$

In this way we have proven that the real ϕ^4-model on $\mathbb{R}^2_\theta$ is perturbatively renormalisable when formulated in the matrix base. This proof was not simply a generalisation of Polchinski's original proof to the noncommutative case. The naïve procedure would be to take the standard ϕ^4-action at the initial scale Λ_0, with Λ_0-dependent bare mass to be adjusted such that at Λ_R it is scaled down to the renormalised mass. Unfortunately, this does not work. In the limit $\Lambda_0 \to \infty$ one obtains an unbounded power-counting degree of divergence for the ribbon graphs. The solution is the observation that the cut-off action at Λ_0 is (due to the cut-off) not translation-invariant. We are therefore free to break the translational symmetry of the action at Λ_0 even more by adding a harmonic oscillator potential for the fields ϕ. There exists a Λ_0-dependence of the oscillator frequency Ω with $\lim_{\Lambda_0\to\infty}\Omega = 0$ such that the effective action at Λ_R is convergent (and thus bounded) order by order in the coupling constant in the limit $\Lambda_0 \to \infty$. This means that the partition function of the original (translation-invariant) ϕ^4-model without cut-off and with suitable divergent bare mass can equally well be solved by Feynman graphs with propagators cut-off at Λ_R and vertices given by the bounded expansion coefficients of the effective action at Λ_R. Hence, this model is renormalisable, and in contrast to the naïve Feynman graph approach in momentum space [8] there is no problem with exceptional configurations. Whereas the treatment of the oscillator potential is easy in the matrix base, a similar procedure in momentum space will face enormous difficulties. This makes clear that the adaptation of Polchinski's renormalisation programme is the preferred method for noncommutative field theories.

3.4. ϕ^4-*theory on noncommutative* $\mathbb{R}^4$

The renormalisation of ϕ^4-theory on $\mathbb{R}^4_\theta$ in the matrix base is performed in an analogous way. We choose a coordinate system in which $\theta_1 = \theta_{12} = -\theta_{21}$ and $\theta_2 = \theta_{34} = -\theta_{43}$ are the only non-vanishing components of θ. Moreover, we assume $\theta_1 = \theta_2$ for simplicity. Then we expand the scalar field according to $\phi(x) = \sum_{m_1,n_1,m_2,n_2 \in \mathbb{N}} \phi_{\substack{m_1 \; n_1 \\ m_2 \; n_2}} f_{m_1 n_1}(x_1,x_2) f_{m_2 n_2}(x_3,x_4)$. The action (12) with integration over $\mathbb{R}^4$ leads then to a kinetic term generalising (14) and a propagator generalising (17). Using estimates on the asymptotic behaviour of that propagator one proves the four-dimensional generalisation of Lemma 3.1 on the power-counting degree of the N-point functions. For $\Omega > 0$ one finds that all non-planar graphs ($B > 1$ and/or $g > 0$) and all graphs with $N \geq 6$ external legs are convergent.

The remaining infinitely many planar two- and four-point functions have to be split into a divergent ρ-part and a convergent complement. Using some sort of locality for the propagator (17), which is a consequence of its derivation from Meixner polynomials, one proves that

$$
\begin{aligned}
A^{\text{planar}}_{\substack{m_1 \; n_1 \\ m_2 \; n_2};\substack{k_1 \; l_1 \\ k_2 \; l_2}} - \delta_{m_1 l_1}\delta_{n_1 k_1}\delta_{m_2 l_2}\delta_{n_2 k_2}\Bigg(& A^{\text{planar}}_{\substack{0\,0 \\ 0\,0};\substack{0\,0 \\ 0\,0}} \\
& + m_1\Big(A^{\text{planar}}_{\substack{1\,0 \\ 0\,0};\substack{0\,1 \\ 0\,0}} - A^{\text{planar}}_{\substack{0\,0 \\ 0\,0};\substack{0\,0 \\ 0\,0}}\Big) + n_1\Big(A^{\text{planar}}_{\substack{0\,1 \\ 0\,0};\substack{1\,0 \\ 0\,0}} - A^{\text{planar}}_{\substack{0\,0 \\ 0\,0};\substack{0\,0 \\ 0\,0}}\Big) \\
& + m_2\Big(A^{\text{planar}}_{\substack{0\,0 \\ 1\,0};\substack{0\,0 \\ 0\,1}} - A^{\text{planar}}_{\substack{0\,0 \\ 0\,0};\substack{0\,0 \\ 0\,0}}\Big) + n_2\Big(A^{\text{planar}}_{\substack{0\,0 \\ 0\,1};\substack{0\,0 \\ 1\,0}} - A^{\text{planar}}_{\substack{0\,0 \\ 0\,0};\substack{0\,0 \\ 0\,0}}\Big)\Bigg)
\end{aligned}
$$

$$
\begin{aligned}
&- \Big(\sqrt{(m_1{+}1)(n_1{+}1)}\delta_{m_1+1,l_1}\delta_{n_1+1,k_1}\delta_{m_2l_2}\delta_{n_2k_2} \\
&\quad + \sqrt{m_1n_1}\delta_{m_1-1,l_1}\delta_{n_1-1,k_1}\delta_{m_2l_2}\delta_{n_2k_2}\Big) A^{\text{planar}}_{\substack{1\,1\,;\,0\,0\\0\,0\,;\,0\,0}} \\
&- \Big(\sqrt{(m_2{+}1)(n_2{+}1)}\delta_{m_2+1,l_2}\delta_{n_2+1,k_2}\delta_{m_1l_1}\delta_{n_1k_1} \\
&\quad + \sqrt{m_2n_2}\delta_{m_2-1,2_1}\delta_{n_2-1,k_2}\delta_{m_1l_1}\delta_{n_1k_1}\Big) A^{\text{planar}}_{\substack{0\,0\,;\,0\,0\\1\,1\,;\,0\,0}} \,, \qquad (29)
\end{aligned}
$$

$$
A^{\text{planar}}_{\substack{m_1\,n_1\,;\dots;\,m_4\,n_4\\m'_1\,n'_1\,;\dots;\,m'_4\,n'_4}} - \Big(\tfrac{1}{6}\delta_{\substack{n_1\,m_2\\n'_1\,m'_2}}\delta_{\substack{n_2\,m_3\\n'_2\,m'_4}}\delta_{\substack{n_3\,m_4\\n'_3\,m'_4}}\delta_{\substack{n_4\,m_1\\n'_4\,m'_1}} + 5\text{ perm's}\Big) A^{\text{planar}}_{\substack{0\,0\,;\dots;\,0\,0\\0\,0\,;\dots;\,0\,0}} \,, \qquad (30)
$$

are convergent functions, thus identifying

$$
\begin{aligned}
\rho_1 &:= A^{\text{planar}}_{\substack{0\,0\,;\,0\,0\\0\,0\,;\,0\,0}} \,, \\
\rho_2 &:= A^{\text{planar}}_{\substack{1\,0\,;\,0\,1\\0\,0\,;\,0\,0}} - A^{\text{planar}}_{\substack{0\,0\,;\,0\,0\\0\,0\,;\,0\,0}} = A^{\text{planar}}_{\substack{0\,0\,;\,0\,0\\1\,0\,;\,0\,1}} - A^{\text{planar}}_{\substack{0\,0\,;\,0\,0\\0\,0\,;\,0\,0}} \,, \\
\rho_3 &:= A^{\text{planar}}_{\substack{1\,1\,;\,0\,0\\0\,0\,;\,0\,0}} = A^{\text{planar}}_{\substack{0\,0\,;\,0\,0\\1\,1\,;\,0\,0}} \\
\rho_4 &:= A^{\text{planar}}_{\substack{0\,0\,;\,0\,0\,;\,0\,0\,;\,0\,0\\0\,0\,;\,0\,0\,;\,0\,0\,;\,0\,0}} \qquad (31)
\end{aligned}
$$

as the distinguished divergent ρ-functions for which we impose boundary conditions at Λ_R. Details will be given in [16].

The function ρ_3 has no commutative analogue. Due to (29) it corresponds to a normalisation condition for the frequency parameter Ω in (14). This means that in contrast to the two-dimensional case we cannot remove the oscillator potential with the limit $\Lambda_0 \to \infty$. In other words, the oscillator potential in (12) is a necessary companionship to the $\star$-product interaction. This observation is in agreement with the UV/IR-entanglement first observed in [4]. Whereas the UV/IR-problem prevents the renormalisation of ϕ^4-theory on $\mathbb{R}^4_\theta$ in momentum space [8], we have found a self-consistent solution of the problem by providing the unique (due to properties of the Meixner polynomials) renormalisable extension of the action. We remark that the diagonalisation of the free action via the Meixner polynomials leads to discrete momenta as the only difference to the commutative case. The inverse of such a momentum quantum can be interpreted as the size of the (finite!) universe, as it is seriously discussed in cosmology [17].

Acknowledgement

Harald Grosse thanks Vladimir Dobrev for the invitation and kind hospitality at the conference.

References

1. S. Doplicher, K. Fredenhagen and J.E. Roberts, "The Quantum structure of space-time at the Planck scale and quantum fields", *Commun. Math. Phys.* **172**, 187 (1995) [hep-th/0303037].
2. E. Schrödinger, "Über die Unanwendbarkeit der Geometrie im Kleinen", *Naturwiss.* **31**, 342 (1934).
3. A. Connes, *Noncommutative geometry* (Academic Press, 1994).
4. S. Minwalla, M. Van Raamsdonk and N. Seiberg, "Noncommutative perturbative dynamics", *J. High En. Phys.* 02 (2000) 020 [hep-th/9912072].
5. N. Seiberg and E. Witten, "String theory and noncommutative geometry", *J. High En. Phys.* 09 (1999) 032 [hep-th/9908142].
6. V. Gayral, J.M. Gracia-Bondía, B. Iochum, T. Schücker and J.C. Várilly, "Moyal planes are spectral triples", hep-th/0307241.
7. I. Chepelev and R. Roiban, "Renormalization of quantum field theories on noncommutative $\mathbb{R}^d$. I: Scalars", *J. High En. Phys.* 05 (2000) 037 [hep-th/9911098].
8. I. Chepelev and R. Roiban, "Convergence theorem for non-commutative Feynman graphs and renormalization", *J. High En. Phys.* 03 (2001) 001 [hep-th/0008090].
9. K.G. Wilson and J.B. Kogut, "The Renormalization Group And The Epsilon Expansion", *Phys. Rep.* **12**, 75 (1974).
10. J. Polchinski, "Renormalization And Effective Lagrangians", *Nucl. Phys.* **B231**, 269 (1984).
11. E. Langmann, R.J. Szabo and K. Zarembo, "Exact solution of noncommutative field theory in background magnetic fields", *Phys. Lett.* **B569**, 95 (2003) [hep-th/0303082].
12. E. Langmann, R.J. Szabo and K. Zarembo, "Exact solution of quantum field theory on noncommutative phase spaces", hep-th/0308043.
13. R. Koekoek and R.F. Swarttouw, "The Askey-scheme of hypergeometric orthogonal polynomials and its q-analogue", math.CA/9602214.
14. H. Grosse and R. Wulkenhaar, "Power-counting theorem for non-local matrix models and renormalisation", hep-th/0305066.
15. H. Grosse and R. Wulkenhaar, "Renormalisation of ϕ^4 theory on noncommutative $\mathbb{R}^2$ in the matrix base", *J. High En. Phys.* 12 (2003) 019 [hep-th/0307017].
16. H. Grosse and R. Wulkenhaar, "Renormalisation of ϕ^4 theory on noncommutative $\mathbb{R}^4$ in the matrix base", in preparation.
17. J.P.M. Luminet, J. Weeks, A. Riazuelo, R. Lehoucq and J.P. Uzan, "Dodecahedral space topology as an explanation for weak wide-angle temperature correlations in the cosmic microwave background", *Nature*, **425**, 593 (2003) [astro-ph/0310253].

ON THE INSERTION-ELIMINATION LIE ALGEBRA OF FEYNMAN GRAPHS

KURUSCH EBRAHIMI-FARD
I.H.É.S.
Le Bois-Marie, 35, Route de Chartres,
F-91440 Bures-sur-Yvette, France
kurusch@ihes.fr
and
Universität Bonn - Physikalisches Institut
Nussallee 12, D-53115 Bonn, Germany

DIRK KREIMER
C.N.R.S.-I.H.É.S.
Le Bois-Marie, 35, Route de Chartres,
F-91440 Bures-sur-Yvette, France
kreimer@ihes.fr

IGOR MENCATTINI
Boston University - Center for Mathematical Physics
111 Cummington Street, Boston, MA 02115, U.S.A.
igorre@math.bu.edu

In this brief note we would like to review part of the recent work on the Hopf algebraic approach to renormalization in perturbative Quantum Field Theory. We hereby focus on the insertion and elimination operation defined on Feynman graphs. The simple operation of inserting graphical objects like Feynman graphs one into another carries a pre-Lie algebra structure. Two natural representations of the underlying Lie algebra in terms of derivations acting on the commutative (Hopf) algebra of Feynman graphs are given. The special case of ladder graphs is further explored.

1. Introduction

The perturbative approach to Quantum Field Theory (QFT) is the most successful description of the fundamental subatomic processes to be found in nature. Kreimer [1] and later Connes and Kreimer [2, 3] revealed a rich algebraic structure underlying this theory. It gives rise to commutative

Hopf algebras being the dual of the universal enveloping algebra of the corresponding Lie algebras[a].

Perturbation theory essentially is given by series expansions over so-called one particle irreducible Feynman graphs. These expansions are organized by the order of a coupling constant measuring the strength of an interaction. The higher the order in the expansion the more complex the Feynman graphs become. The latter can be characterized by the fact that Feynman graphs contain Feynman graphs as substructures. Loosely speaking we can say that Feynman graphs are made out of Feynman graphs.

Renormalization theory of perturbative QFT relies on the iterated structure of Feynman graphs. The Hopf algebraic approach to the process of renormalization highlights this fact, mainly by organizing the combinatorics of this process in terms of a coproduct which disentangles a Feynman graph into its building blocks, i.e. Feynman graphs. The physical model given in terms of a Lagrangian provides the basic elements out of which Feynman graphs are made, i.e. propagators represented by edges, and vertices.

From a more abstract point of view the use of the notion of coalgebras with respect to graphical objects like Feynman diagrams seems to be a very natural one [5, 6]. In this note nevertheless we will reverse this idea and show that the two simple operations of insertion and elimination of subgraphs in Feynman graphs gives new and rich structures. The above mentioned Lie algebras can be represented by derivations defined naturally in terms of these two operations on the commutative Hopf algebra of Feynman diagrams. The underlying Lie algebra of graphs comes from a pre-Lie algebra. The former has a seemingly simple but unexpected rich structure. In [7] the first steps of its exploration were recently undertaken and will be reported here. Let us mention here that our presentation follows mainly the original work of Connes and Kreimer in [8].

In the next section we will give a more precise definition of graphs and introduce the notion of n-particle irreducibility and the residue of a graph. We then define the two main operations, insertion and elimination of subgraphs giving rise to a pre-Lie structure. In section three we define two representations of the Lie algebra in terms of derivations, given by the insertion and elimination operation. Section four reviews the recent progress made for a special class of diagrams, the so called ladder graphs. We end this work with some conclusions and an outlook for further work in this direction.

---

[a]See [4] for a very recent and readable short introduction and review.

2. Pre-Lie Algebra of Feynman Graphs

We start with the

Definition: *A $n-$particle irreducible graph Γ is made out of vertices $v \in \Gamma^{[0]}$ and edges $f \in \Gamma^{[1]}$, such that when removing $\mathbb{N} \ni k \leq n$ of its edges it is still connected.*

Of main interest for us in what follows is the class of $1-$PI graphs.

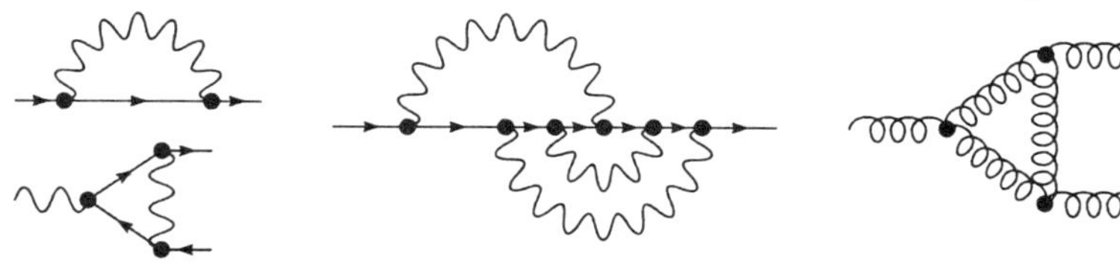

Fig.1 1-PI graphs from Quantum Electro- and Chromodynamics (QED, QCD).

The set of edges $\Gamma^{[1]}$ splits into the subset of external and internal edges, $\Gamma^{[1]} = \Gamma^{[1]}_{ext} \cup \Gamma^{[1]}_{int}$.

Definition: *An edge $f \in \Gamma^{[1]}$ of a graph Γ is called external (internal) if $v_f := f \cap \Gamma^{[0]}$ is a set of one (two) element(s).*

Here and for the rest of the paper we exclude self-loops: no edge starts and ends at the same vertex. Edges sometimes are called propagators and they can be of different type, mainly symbolized by the way they are drawn, for instance: (un-)oriented straight, curly or dashed lines. The notion of type for edges implies the

Definition: *The type of a vertex $v \in \Gamma^{[0]}$ is defined by the set $f_v := \{f \in \Gamma^{[1]} \mid f \cap v \neq \emptyset\}$.*

Edges and vertices may be of different type, and it might be of interest to compare them. We therefore introduce the

Definition: *We call two sets of edges E_1 and E_2 compatible, i.e. $E_1 \sim E_2$ iff they contain the same number of edges, of the same type.*

This equivalence relation allows us to compare vertices v_1, $v_2 \in \Gamma^{[0]}$ by their types, i.e. they are of the same type if $f_{v_1} \sim f_{v_2}$. We already mentioned the exclusion of self-loops, and we will also not consider so called vacuum graphs for which $\Gamma^{[1]}_{ext} = \emptyset$. Graphs for which $\Gamma^{[1]}_{ext}$ consists of only two elements are called self-energy graphs. If there are more than two external edges we call them interaction graphs. These notions are mainly motivated by applications in elementary particle physics. Here we suppose the type of a vertex to be of cardinality bigger than one. Vertices of cardinality two may be associated with edges.

An important notion is the shrinking of a graph to its residue graph, we denote this operation by $\Gamma \to res(\Gamma)$.

Definition: *The residue $res(\Gamma)$ of a graph Γ is defined to be the result of identifying $\Gamma^{[0]} \cup \Gamma^{[1]}_{int}$ with a point in Γ.*

The graph Γ is reduced to a vertex v of type $f_v = \Gamma^{[1]}_{ext}$. We can therefore compare the residue of two graphs. The following example shows an interaction graph in QED shrinked to its residue, the QED vertex.

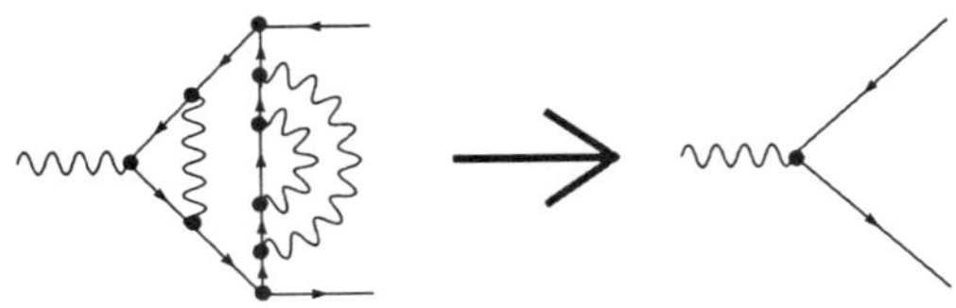

Fig.2 Shrinking of a 4-loop QED interaction graph to its residue.

Obviously, $res(\Gamma)^{[1]} \sim \Gamma^{[1]}_{ext}$ and graphs with a compatible external edge structure all map to the same residue graph.
We now introduce the insertion or gluing operation for graphs. It is the opposite of the shrinking of graphs to its residue. Essentially this means replacing a certain vertex $v \in \Gamma^{[0]}$ of type f_v by a graph γ for which $\gamma^{[1]}_{ext} \sim f_v$. This composition of graphs, denoted by $\Gamma * \gamma$ gives a new graph Γ'.
In general there are several ways to link the two sets of edges and one has to sum over all bijections. Also we sum over all places where the graph γ can be glued into Γ. It is important to note that in the case where there is no $\gamma^{[1]}_{ext}$ compatible vertex $v \in \Gamma^{[0]}$ the result is defined to vanish, $\Gamma * \gamma = 0$. Also we impose a normalization with respect to the symmetry of graphs. Topologically different graphs are supposed to appear with unit multiplicity. Here we give as an example the gluing operation for the so called ϕ^3-theory.

Fig.3 Gluing operation.

The first graph on the r.h.s. should appear twice but due to the normalization we have to multiply by a factor 1/2. Note though that if we keep such multiplicities, we simply obtain a slightly modified (pre-)Lie algebra. We denote the $\mathbb{K}$-vector space of 1−PI Feynman graphs by $\mathcal{H}^l_{FG}$. Beside the fact that the insertion composition can be zero for non-zero elements, it is important to realize that it is not associative, but of (right) pre-Lie type.

Proposition 1: *$\mathcal{H}^l_{FG}$ equipped with the insertion composition of graphs is of pre-Lie type, i.e. holds the following (right) pre-Lie relation:*

$$(\Gamma_1 * \Gamma_2) * \Gamma_3 - \Gamma_1 * (\Gamma_2 * \Gamma_3) = (\Gamma_1 * \Gamma_3) * \Gamma_2 - \Gamma_1 * (\Gamma_3 * \Gamma_2). \qquad (1)$$

The proof of this follows from the fact that we sum over all ways to insert a graph into another, and the l.h.s. (r.h.s.) just results in all ways where Γ_2 and Γ_3 can go in different places into Γ_1.

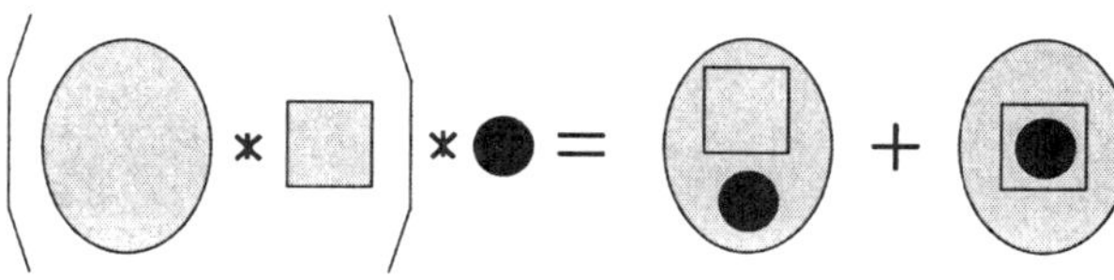

Fig.4 Insertion possibilities for Gluing operation.

We see here that this pre-Lie structure is of a very general type. The fact that we use here a Quantum Field theoretical language mainly comes from the applications of this structure we have in mind. But it applies in a much more general way to graph theoretical objects. We see also that this simple gluing operation reverses the general idea to operate in terms of coalgebras on graphs. It somehow inverts the coproduct and puts graphs together in a nontrivial way.
Antisymmetrizing this pre-Lie product on $\mathcal{H}^l_{FG}$ results in a Lie bracket $[\Gamma_1, \Gamma_2] := \Gamma_1 * \Gamma_2 - \Gamma_2 * \Gamma_1$. In the following we will give two representations of this Lie algebra $\mathcal{L}_{FG}$ of Feynman graphs.

3. Two Representations as Derivations on the Hopf Algebra of Feynman Graphs

In the last section we introduced the free $\mathbb{K}$-module $\mathcal{H}^l_{FG}$. Equipped with the insertion operation we found a (pre-)Lie algebra $\mathcal{L}_{FG}$ of Feynman graphs. We will now refine our notation a little bit and denote the generators of $\mathcal{H}^l_{FG}$ by ξ_Γ, Γ a 1$-$PI Feynman graph. The generators of the dual of $\mathcal{H}^l_{FG}$ are denoted by Z_Γ and both are extended by linearity. We have the following Kronecker pairing:

$$\langle Z_{\Gamma_1}, \xi_{\Gamma_2} \rangle := \delta_{\Gamma_1 \Gamma_2}. \qquad (2)$$

Later, we want to refer to the indices, i.e. the graphs, we therefore introduce the following notation:

$$\bar{Z}_\Gamma = \bar{\xi}_\Gamma := \Gamma. \qquad (3)$$

Using the concatenation product on $\mathcal{H}^l_{FG}$, i.e. union of graphs, we get a connected graded commutative polynomial $\mathbb{K}$-algebra which we denote by $\mathcal{H}_{FG}$. Following [2] we may equip $\mathcal{H}_{FG}$ with a coproduct

$$\Delta(\xi_\Gamma) := \xi_\Gamma \otimes 1 + 1 \otimes \xi_\Gamma + \sum_i \xi^{(i)}_{\Gamma'} \otimes \xi^{(i)}_{\Gamma''} \tag{4}$$

which gives the Hopf algebra of Feynman graphs, the linear part of which we will denote by $\mathcal{H}^l_{FG}$. For later convenience we will use the Sweedler notation $\Delta(\xi_\Gamma) := \sum_i \xi^{(i)}_{\Gamma'} \otimes \xi^{(i)}_{\Gamma''}$. The coproduct disentangles an element in $\mathcal{H}_{FG}$, i.e. a graph, into the elements $\xi^{(i)}_{\Gamma'}$ and what is left over, $\xi^{(i)}_{\Gamma''}$, after the subgraph $\overline{\xi^{(i)}_{\Gamma'}}$ has been replaced by its $res\big(\overline{\xi^{(i)}_{\Gamma'}}\big)$. It determines thereby the gluing data $G_{(i)}$ for reinserting the (sub)graph $\overline{\xi^{(i)}_{\Gamma'}}$ into the graph $\overline{\xi^{(i)}_{\Gamma''}}$. We now define the following two representations of the Feynman (pre-)Lie algebra $\mathcal{L}_{FG}$ on $\mathcal{H}^l_{FG}$.

Definition: *The maps* $\phi^\pm : \mathcal{H}^l_{FG} \to End(\mathcal{H}^l_{FG})$,

$$\phi^+(\xi_{\Gamma_1}) =: Z^+_{\Gamma_1}, \quad Z^+_{\Gamma_1}(\xi_{\Gamma_2}) := \xi_{\Gamma_2 * \Gamma_1} \tag{5}$$

$$\phi^-(\xi_{\Gamma_1}) =: Z^-_{\Gamma_1}, \quad Z^-_{\Gamma_1}(\xi_{\Gamma_2}) := \sum_i \langle Z_{\Gamma_1}, \xi^{(i)}_{\Gamma'_2}\rangle\, \xi^{(i)}_{\Gamma''_2} \tag{6}$$

define an anti-representation and representation, respectively, of $\mathcal{L}_{FG}$ *on* $\mathcal{H}^l_{FG}$, *i.e.:*

$$\phi^\pm([\xi_{\Gamma_1}, \xi_{\Gamma_2}]) = \mp[\phi^\pm(\xi_{\Gamma_1}), \phi^\pm(\xi_{\Gamma_2})]. \tag{7}$$

The proof is straightforward and relies on the pre-Lie relation (1). Both representations operate as derivations on $\mathcal{H}_{FG}$. For ϕ^+ this follows from the sum over all possibilities to insert Γ_1 into Γ_2 and the fact that the product is the union of graphs. For ϕ^- it follow from the Kronecker pairing and from the form of the coproduct (4).

As a first step in the investigation of these two representations we would like to understand the commutator $[Z^+_{\Gamma_1}, Z^-_{\Gamma_2}]$. We therefore introduce the following third operator on $\mathcal{H}_{FG}$. For $\Gamma^{[1]}_{1\,ext} \sim \Gamma^{[1]}_{2\,ext}$ we define

$$Z_{(\Gamma_1,\Gamma_2)}(\xi_{\Gamma_3}) := \sum_i \langle Z_{\Gamma_2}, \xi^{(i)}_{\Gamma'_3}\rangle\, \xi^{(i)}_{\Gamma''_3} *_{G_{(i)}} \xi_{\Gamma_1}. \tag{8}$$

The gluing data $G_{(i)}$ specify the place where Γ_1 has to be inserted and what the normalization is. This operator replaces Γ_2 by Γ_1 in Γ_3 wherever possible. Obviously, the operators $Z^+_{\Gamma_1}$ and $Z^-_{\Gamma_2}$ can be written as:

$$Z^+_{\Gamma_1} = Z_{(\Gamma_1, res(\Gamma_1))} \tag{9}$$

$$Z^-_{\Gamma_1} = Z_{(res(\Gamma_1), \Gamma_1)}. \tag{10}$$

The following theorem gives a general formula for the commutator of two operators of the form (8).

Theorem 1[8]: *For 1−PI Feynman graphs* Γ_1, Γ_2, Γ_3, Γ_4, *for which* $res(\Gamma_1) \sim res(\Gamma_2)$ *and* $res(\Gamma_3) \sim res(\Gamma_4)$ *the commutator:*

$$[Z_{(\Gamma_1,\Gamma_2)}, Z_{(\Gamma_3,\Gamma_4)}] = Z_{\left(\overline{Z_{(\Gamma_1,\Gamma_2)}(\xi_{\Gamma_3})},\Gamma_4\right)} - Z_{\left(\Gamma_3,\overline{Z_{(\Gamma_2,\Gamma_1)}(\xi_{\Gamma_4})}\right)} \tag{11}$$

$$-Z_{\left(\overline{Z_{(\Gamma_3,\Gamma_4)}(\xi_{\Gamma_1})},\Gamma_2\right)} + Z_{\left(\Gamma_1,\overline{Z_{(\Gamma_4,\Gamma_3)}(\xi_{\Gamma_2})}\right)} \tag{12}$$

$$-\delta_{\Gamma_2\Gamma_3} Z_{(\Gamma_1,\Gamma_4)} + \delta_{\Gamma_1\Gamma_4} Z_{(\Gamma_3,\Gamma_2)}. \tag{13}$$

Therefore the operator $Z_{(\Gamma_1,\Gamma_2)}$ *defines a closed Lie subalgebra of derivations on* $\mathcal{H}_{FG}$.

The two δ-functions take care of a possible overcounting. The proof of this theorem is easy and only needs the following point to be emphasized. When acting with $Z_{(\Gamma_1,\Gamma_2)}Z_{(\Gamma_3,\Gamma_4)}$ on ξ_Γ we have to distinguish 3 cases:
1) Γ_2 and Γ_4 are replaced by Γ_1 and Γ_3 respectively in different places.
2) Γ_4 is replaced by Γ_3 and then Γ_2 is replaced by Γ_1 in Γ_3.
3) Replacing the graph Γ_4 by Γ_3 in Γ creates the graph Γ_2 which is then replaced by Γ_1.
In the first case the order of operators commutes, therefore this case does not contribute in the above commutator. Case 2) is represented by $Z_{\left(\overline{Z_{(\Gamma_1,\Gamma_2)}(\xi_{\Gamma_3})},\Gamma_4\right)}$. The last case can be expressed by $Z_{\left(\Gamma_1,\overline{Z_{(\Gamma_4,\Gamma_3)}(\xi_{\Gamma_2})}\right)}$, since a subgraph with external edge structure compatible with $\Gamma^{[1]}_{2\,ext}$ appears in Γ but with Γ_4 instead of Γ_3 as sub-substructure. We therefore replace first Γ_3 by Γ_4 in Γ_2, i.e. $Z_{(\Gamma_4,\Gamma_3)}(\xi_{\Gamma_2})$. Since this process did not alter $\Gamma^{[1]}_{2\,ext}$, i.e. $\Gamma^{[1]}_{2\,ext} \sim \overline{Z_{(\Gamma_4,\Gamma_3)}(\xi_{\Gamma_2})}^{[1]}_{ext}$, we replace the graph $\overline{Z_{(\Gamma_4,\Gamma_3)}(\xi_{\Gamma_2})}$ by Γ_1.
Inverting the order of indices in $Z_{(\Gamma_1,\Gamma_2)} \to Z_{(\Gamma_2,\Gamma_1)}$ defines an anti-involution. The following two corollaries follow immediately from the above theorem.

Corollary 1: $[Z^-_{\Gamma_1}, Z^-_{\Gamma_2}] = -Z^-_{[Z^+_{\Gamma_1},Z^+_{\Gamma_2}]}$, furthermore

$$[Z^+_{\Gamma_1}, Z^-_{\Gamma_2}] = [Z_{(\Gamma_1,res(\Gamma_1))}, Z_{(res(\Gamma_2),\Gamma_2)}] \tag{14}$$

$$= -\delta_{res(\Gamma_1)res(\Gamma_2)} Z_{(\Gamma_1,\Gamma_2)} + \delta_{\Gamma_1\Gamma_2} Z_{(res(\Gamma_2),res(\Gamma_1))}$$

$$-Z^-_{Z^-_{\Gamma_1}(\xi_{\Gamma_2})} - Z^+_{Z^-_{\Gamma_2}(\xi_{\Gamma_1})}. \tag{15}$$

We will now report recent results for the case where the graphs have a certain ladder structure, i.e. for the sub-Hopf algebra of so called ladder graphs.

4. The Ladder case

In this section we will describe some features of the Insertion-Elimination Lie algebra when we restrict ourself to $\mathcal{H}_L$, a ladder sub-Hopf algebra of the Hopf algebra of Feynman graphs.

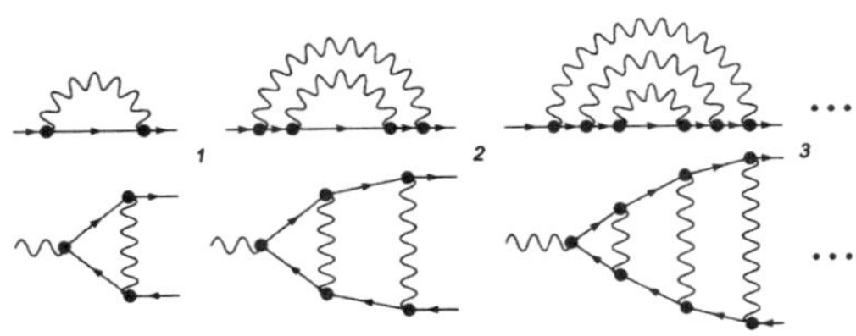

Fig.5 Example of ladder QED graphs of order 1, 2, and 3.

In what follows we will consider only ladder sub-algebras of $\mathcal{H}_{FG}$ isomorphic to the ladder Hopf algebra of rooted trees without decorations. The generators of a such given Hopf algebra will be denoted by ξ_{Γ_k} with k non negative integer number. The restriction of the co-product to such sub-Hopf algebra is given on the generators ξ_{Γ_k} by the following formula:

$$\Delta(\xi_{\Gamma_k}) = \xi_{\Gamma_k} \otimes 1 + 1 \otimes \xi_{\Gamma_k} + \sum_{j=1}^{k-1} \xi_{\Gamma_j} \otimes \xi_{\Gamma_{k-j}}.$$

We also remark that $\mathcal{H}_L$ is graded by $n \in \mathbb{Z}_{\geq 0}$, i.e for each ξ_{Γ_n}, $\deg \xi_{\Gamma_n} = n$ and that such integer can be interpreted as the loop number of the graph Γ_n. In particular the homogeneous components of $\mathcal{H}_L$ are 1 dimensional.

For each couple (n, m) of non negative integers we can consider the couple of ladder graphs $(\xi_{\Gamma_n}, \xi_{\Gamma_m})$ to which we associate the symbol $Z_{(n,m)}$. Let us now consider the vector space $\mathcal{L}_L = span_{\mathbb{K}}\{Z_{(n,m)} |\ n, m \in \mathbb{Z}_{\geq 0}\}$. Now the **theorem 1** of the previous section becomes:

Theorem 2: *$\mathcal{L}_L$ is a Lie algebra whose bracket is given by the following formula:*

$$\begin{aligned}\left[Z_{(n,m)}, Z_{(l,s)}\right] = &\ \Theta(l-m)Z_{(l-m+n,s)} - \Theta(s-n)Z_{(l,s-n+m)} \\ &-\Theta(n-s)Z_{(n-s+l,m)} + \Theta(m-l)Z_{(n,m-l+s)} \\ &-\delta_{ml}Z_{(n,s)} + \delta_{ns}Z_{(l,m)}, \end{aligned} \tag{16}$$

where $\Theta(n-m) = 1$ if $n \geq m$, 0 otherwise, and δ_{nm} is the usual Kronecker delta.

Now we can state the following proposition, [7]:

Proposition 2: *1) $\mathcal{L}_L$ is a $\mathbb{Z}$-graded Lie algebra, i.e $\mathcal{L}_L = \oplus_{n\in\mathbb{Z}} l_n$, whose homogeneous components $l_n = span_{\mathbb{K}}\{Z_{(k,m)} | k-m = n\}$ are infinite*

dimensional;
2)

$$\mathcal{L}_L = L^+ \oplus L^0 \oplus L^-,$$

where $L^+ = \oplus_{n\geq 0} l_n$, $L^- = \oplus_{n\leq 0} l_n$, $L^0 = l_0$ *and* $[L^+, L^0] \subset L^+$, $[L^-, L^0] \subset L^-$.

Let us now consider the following $\mathbb{Z}_{\geq 0}$-graded $\mathbb{K}$-vector space:

$$\mathbb{S} = \bigoplus_{n\geq 0} \mathbb{K}t_n,$$

where we assign $deg\, t_n = n$. We have the following, [7]:

Proposition 3: $\mathbb{S}$ *is a* $\mathcal{L}_L$*-module via the following action:*

$$Z_{(n,m)} t_k = t_{k-n+m}\ if\ m \leq k,\ 0\ otherwise.$$

In particular we can show that, [7]:

Proposition 4: $\mathbb{S}$ *is irreducible highest weight* $\mathcal{L}_L$*-module with highest weight vector* t_0.

Using **Proposition 3** we can relate the Lie algebra $\mathcal{L}_L$ to some well known infinite dimensional Lie algebras. In what follows we will describe such relations with the algebra of infinite matrices of finite rank that we are going to introduce; for the relation of $\mathcal{L}_L$ with the Heisenberg algebra we refer the reader to the reference [7].

Let us now introduce the following Lie algebra:

Definition: $\mathfrak{gl}_+(\infty) = span_{\mathbb{K}}\{E_{i,j}, i, j \in \mathbb{Z}_{\geq 0} | [E_{i,j}, E_{n,m}] = E_{i,m}\delta_{j,n} - E_{n,j}\delta_{i,m}\}$

Remark: *The Lie algebra* $\mathfrak{gl}_+(\infty)$ *has a natural representation in terms of infinite matrices of finite ranks, i.e infinite matrices having only a finite number of non zero entries.*

Let us define the following map:

$$\phi : \mathfrak{gl}_+(\infty) \longrightarrow \mathcal{L}_L,$$

where $\phi(E_{i,j}) = Z_{(i,j)} - Z_{(i+1,j+1)}$. Then:

Proposition 5, [7]: ϕ *is a morphism of Lie algebras.*

Following the previous remark it makes sense to consider $\mathfrak{sl}_+(\infty) = \{a \in \mathfrak{gl}_+(\infty) | \, tr\, a = 0\}$. We can now state the following:

Proposition 6, [7]: *The set* $\{f_i = Z_{(i+1,i)} - Z_{(i+2,i+1)}, e_i = Z_{(i,i+1)} - Z_{(i+1,i+2)} | \ i \geq 0\}$ *is a system of Chevalley's generators, while the set*

$\{\check{\alpha}_i = Z_{(i,i)} - 2Z_{(i+1,i+1)} + Z_{(i+2,i+2)} |\, i \geq 0\}$ *is a system of simple co-roots for* $\mathfrak{sl}_+(\infty)$.

We now recall that given a Lie algebra $\mathfrak{g}$ with a set of Chevalley's generators e_i, f_i and simple co-roots $\check{\alpha}_i$, we can define the so called *Chevalley's involution*:

$$\omega : \mathfrak{g} \longrightarrow \mathfrak{g}$$

via the following:

$$\omega(e_i) = -f_i,\ \omega(f_i) = -e_i,\ \omega(\check{\alpha}_i) = -\check{\alpha}_i.$$

We conclude this section with the following:

Proposition 7, [7]: *The homomorphism of Lie algebras* $C : \mathcal{L}_L \longrightarrow \mathcal{L}_L$ *defined on the generators by* $C(Z_{(n,m)}) = -Z_{(m,n)}$ *restricts on* $\mathfrak{sl}_+(\infty)$ *to the Chevalley's involution.*

5. Conclusions and Outlook

Insertion and elimination of subgraphs are the two main operations necessary to deal with perturbative Quantum Field Theory: insertion is initiated by the quantum equations of motions, the Dyson–Schwinger equations, which tell us that at any instance of propagation or interaction the whole amplitude can contribute, and thus we get graphs built out of graphs. This self-similarity falls onto itself then in the renormalization process which eliminates subgraphs in favour of their counterterms. The balance we are left with of these two operations is then the renormalized amplitude prescribing the observable world around us. An efficient understanding of the interplay of insertion and elimination also for non-ladder graphs is thus a crucial task for the future.

Acknowledgments

The first author would like to thank the German Academic Exchange Service (DAAD) for financial support and the IHÉS for support and hospitality.

References

1. D. Kreimer, "On the Hopf algebra structure of perturbative quantum field theories", *Adv. Theor. Math. Phys.* **2**, 303 (1998).
2. A. Connes and D. Kreimer, "Renormalization in quantum field theory and the Riemann-Hilbert problem. I. The Hopf algebra structure of graphs and the main theorem", *Commun. Math. Phys.* **210**, 249 (2000).

3. A. Connes and D. Kreimer, "Renormalization in quantum field theory and the Riemann-Hilbert problem. II. The β-function, diffeomorphisms and the renormalization group", *Commun. Math. Phys.* **216**, 215 (2001).
4. S. Weinzierl, "Hopf algebra structures in particle physics", preprint hep-th/0310124 (Oct. 2003).
5. W. Schmitt, "Hopf algebras in enumerative combinatorics", in: "Gian-Carlo Rota on combinatorics", ed. J.P.S. Kung, Contemp. Mathematicians, (Birkhäuser Boston, Boston, MA, 1995) pp. 345-356.
6. S.A. Joni and G.-C. Rota, "Coalgebras and bialgebras in combinatorics. Umbral calculus and Hopf algebras", (Norman, Okla., 1978), Contemp. Math. vol. 6 (Amer. Math. Soc., Providence, R.I., 1982) pp. 1-47.
7. I. Mencattini and D. Kreimer, "Insertion and Elimination Lie Algebra: The Ladder Case", *Lett. Math. Phys.*, in print (2003) (math.QA/0308042).
8. A. Connes and D. Kreimer, "Insertion and elimination: the doubly infinite Lie algebra of Feynman graphs", *Ann. Henri Poincaré*, **3**, no. 3, 411 (2002).

SUPERCONFORMAL KINEMATICS AND DYNAMICS IN THE ADS/CFT CORRESPONDENCE

EMERY SOKATCHEV

Laboratoire d'Annecy-le-Vieux de Physique Théorique LAPTH
B.P 110, F-74941 Annecy-le-Vieux, France
Emery.Sokatchev@cern.ch

We consider correlation functions of BPS operators in the context of the AdS/CFT correspondence. The consequences of the superconformal Ward identities as well as some dynamical restrictions on the amplitudes are discussed. These results are compared to explicit calculations in AdS supergravity.

1. The rôle of BPS operators in the AdS/CFT correspondence

The AdS/CFT conjecture [1]-[3], in its weakest form, predicts the correspondence between Kaluza-Klein states in Type IIB SG on $AdS_5 \times S^5$ and 1/2 BPS short gauge invariant operators in the boundary $SCFT_4$ of $N = 4$ super-Yang-Mills (SYM) in the 't Hooft limit.

BPS operators are very special representations of the superconformal symmetry underlying both the $AdS_5 \times S^5$ and $SCFT_4$ pictures. Their main property is that the symmetry fixes their conformal weight. For this reason they are called protected operators. The simplest example is the stress-tensor multiplet in $N = 4$ SYM.

A lot of effort has been made to test this correspondence by computing and comparing correlation functions of BPS operators in SCFT to their SG counterparts [4]-[41].

Most of the results so far take the form of non-renormalization theorems for various correlators of 1/2 BPS operators:

- Non-renormalization of 2- and 3-point functions
- Non-renormalization of n–point functions of extremal and subextremal types
- Partial non-renormalization of 4-point functions:

– of stress-tensor multiplets, corresponding to the massless AdS supergravity multiplet
– of 1/2 BPS operators of weight 3 and 4, corresponding to massive KK states

These results have been obtained by different methods: explicit perturbative and instanton calculations in SCFT, analysis of the superconformal invariants undergoing BPS shortening, SG calculations.

A lot of what we know about correlators of 1/2 BPS operators is just due to superconformal kinematics. However, using a well-known QFT tool called the insertion formula combined with CSUSY, we can obtain further, dynamical restrictions on the correlators in SCFT, in particular, on 4-point functions. These take the form of partial non-renormalization, meaning that the quantum corrections to the free amplitude are given by a smaller number of conformal invariant functions, compared to the general solution of the SCWI.

Remarkably, all the available explicit SG calculations (to this date for weights 2, 3 and 4) using the Arutyunov-Frolov $AdS_5 \times S^5$ SG action, as well as the instanton calculation of the Roman group, show complete agreement with these dynamical predictions of SCFT. These results constitute some of the rare concrete evidence for the AdS/CFT correspondence and are the main subject of this talk.

2. BPS short multiplets

2.1. *The superconformal algebra PSU(2,2/4)*

Here is the relevant odd part of the superalgebra:

$$\begin{aligned}
\{Q^i_\alpha, \bar{Q}_{\dot\alpha j}\} &= 2\delta^i_j(\sigma^\mu)_{\alpha\dot\alpha}P_\mu \ , \\
\{S_{\alpha j}, \bar{S}^i_{\dot\alpha}\} &= 2\delta^i_j\,(\sigma^\mu)_{\alpha\dot\alpha}K_\mu \ , \\
\{Q^i_\alpha, S^\beta_j\} &= -\delta^i_j(\sigma^{\mu\nu})_\alpha{}^\beta M_{\mu\nu} - 4\delta^\beta_\alpha T^i_j - 2i\delta^\beta_\alpha\delta^i_j D
\end{aligned}$$

with generators: Q of Poincaré supersymmetry and S of special conformal supersymmetry; P of translations, K of conformal boosts, M of the Lorentz group, D of dilatations, T of SU(4) (no R charge in PSU(2,2/4))

2.2. *Realization on superfields in real superspace*

The standard (real) superspace is defined as the coset

$$\frac{\mathrm{PSU}(2,2/4)}{\{K,S,\bar{S},M,D,T\}} = (x^\mu, \theta^\alpha_i, \bar\theta^{\dot\alpha i})$$

Superfields $\Phi(x^\mu, \theta_i^\alpha, \bar\theta^{\dot\alpha i})$ are defined as superconformal primaries:

$$s_{\alpha i}\Phi = \bar s^i_{\dot\alpha}\Phi = k_\mu \Phi = 0$$

The Unitary Irreducible Representations (UIR) are labeled by

$$D(\ell; j_1, j_2; a_1, a_2, a_3)$$

where ℓ is the conformal weight, j_1, j_2 are the two Lorentz ($SL(2,\mathbb{C})$) "spins" and a_1, a_2, a_3 are the SU(4) Dynkin labels.

2.3. *Chiral superspace*

Question: Can we realize PSU(2,2/4) in a smaller superspace? In other words, can we use *shorter superfields*?

The answer is well known in $N = 1$ SUSY: Use *chiral superfields.* They are obtained by adding all the right- (or left-)handed Q generators to the coset denominator. Let us try the same in $N = 4$:

$$\frac{\mathrm{PSU}(2,2/4)}{\{K, \dots, T, \bar Q\}} = (x^\mu, \theta_i^\alpha)$$

This imposes a new constraint on the superconformal primary Φ:

$$\bar q_i^{\dot\alpha}\Phi = 0$$

whose compatibility condition is

$$\{\bar q_i^{\dot\alpha}, \bar s^j_{\dot\beta}\}\Phi = \left[-\delta_i^j (\sigma^{\mu\nu})^{\dot\alpha}{}_{\dot\beta} m_{\mu\nu} - 2\delta^{\dot\alpha}_{\dot\beta}\left(2t^i_j + \delta^j_i \ell\right)\right]\Phi = 0$$

It implies $\ell = 0$ - too strong! There are no chiral superfields in $N = 4$ (with vanishing R or central charge).

Watch out for the improper use of the words "chiral primary operators"!

2.4. *Grassmann-analytic or BPS superfields*

New possibility in extended ($N > 1$) SUSY: instead of putting all $\bar Q$s (or all Qs) in the coset denominator, put, e.g., only the first 2 of the Qs and the last 2 of the $\bar Q$s:

$$\frac{\mathrm{PSU}(2,2/4)}{\{K, \dots, T, Q^1, Q^2, \bar Q_3, \bar Q_4\}} = (x^\mu, \theta_3^\alpha, \theta_4^\alpha, \bar\theta^1_{\dot\alpha}, \bar\theta^2_{\dot\alpha})$$

1/2 BPS defining constraints

$$q^1_\alpha \Phi = q^2_\alpha \Phi = \bar q_3^{\dot\alpha}\Phi = \bar q_4^{\dot\alpha}\Phi = 0$$

Compatibility conditions in the left- and right-handed sectors:

$$j_1 = j_2 = 0 \; ;$$
$$\ell = a_2 \; ;$$
$$a_1 = a_3 = 0 \; ,$$

where a_i are the SU(4) Dynkin labels.

The 1/2 BPS superfields realize the UIRs

$$D(a_2; 0, 0; 0, a_2, 0)$$

We call the single non-vanishing Dynkin label a_2 "charge".

The BPS multiplets are *protected* because they have *fixed ("quantized") weight (dimension).*

3. $N = 4$ SYM and 1/2 BPS operators

3.1. $N = 4$ *SYM on shell*

The $N = 4$ SYM on-shell multiplet is described by a real scalar superfield (field-strength)

$$W^{ij}(x, \theta, \bar{\theta}) = -W^{ji} = \frac{1}{2}\epsilon^{ijkl}\overline{W}_{kl}$$

in the **6** of SU(4) (Dynkin labels $[0, 1, 0]$) satisfying the following on-shell constraints (massless field equations):

$$D^{(k}_{\alpha} W^{i)j} = 0 \; , \qquad \bar{D}_{\dot{\alpha}\{k} W^{i\}j} = 0$$

where $() \equiv$ symmetrization, $\{\} \equiv$ traceless part.

The on-shell constraints can be rewritten as G-analyticity (BPS shortness conditions) with the help of *harmonic variables.* The harmonics u^I_i (and $u^i_I = (u^I_i)^*$) form an SU(4) matrix:

$$u \in \mathrm{SU}(4): \quad \begin{aligned} & u^I_i u^i_J = \delta^I_J \; , \\ & u^I_i u^j_I = \delta^j_i \; , \\ & \varepsilon^{i_1 \ldots i_4} u^1_{i_1} \ldots u^4_{i_4} = 1 \end{aligned}$$

Here i is an index in the fundamental representation of SU(4) and $I = 1, \ldots, 4$ are the projections of the second index onto the subgroup $[\mathrm{U}(1)]^3$.

The harmonics allow us to project SU(4) indices onto $[\mathrm{U}(1)]^3$ without breaking SU(4). The *harmonic functions* are $\mathrm{SU}(4)_L$ invariant but $[\mathrm{U}(1)]^3_R$

covariant. Thus, they effectively live on the 6-dimensional complex compact space:

$$f(u) \text{ is defined on } \frac{\mathrm{SU}(4)}{[\mathrm{U}(1)]^3}$$

Now, use harmonics to project the SYM constraints. The projected field-strength

$$W^{12} = W^{ij}(x,\theta,\bar\theta)u_i^1 u_j^2$$

satisfies the constraints

$$D^1_\alpha W^{12} = D^2_\alpha W^{12} = \bar D_{\dot\alpha\,3} W^{12} = \bar D_{\dot\alpha\,4} W^{12} = 0$$

where

$$D^I_\alpha = D^i_\alpha u^I_i\ , \quad \bar D^{\dot\alpha}_J = \bar D^{\dot\alpha}_i u^i_J$$

These constraints can be solved in an appropriate G-analytic basis in superspace (the analog of the chiral basis). There

$$W^{12}(x_A, \theta_3, \theta_4, \bar\theta^1, \bar\theta^2, u)$$

becomes a manifestly SU(4) covariant 1/2 BPS superfield of weight (charge) 1.

In fact, the θ expansion of W^{12} is ultrashort:

$$W^{12} = u_i^{[1}u_j^{2]}\ \phi^{ij}(x) + \theta^\alpha_{[3}u^i_{4]}\ \psi_{\alpha\ i}(x) + \theta_{[3}\sigma^{\mu\nu}\theta_{4]}\ F_{\mu\nu}(x)$$
$$+\text{c. c.} + \text{derivative terms}$$

where ϕ, ψ and F are the on-shell scalars, spinors and YM field strength of $N = 4$ SYM.

3.2. *Composite BPS operators*

The remarkable property of Grassmann analytic superfields is that they form a ring under multiplication.

1/2 BPS gauge invariant operators of weight k are obtained as composites (single or multitrace):

$$1/2 \text{ BPS:} \qquad O^k_{1/2} = \mathrm{Tr}\ [W^{12}(\theta_{3,4}, \bar\theta^{1,2})]^k$$

An important particular case is $k = 2$ corresponding to the *stress-tensor multiplet*:

$$
\begin{aligned}
T^{1122} &= \mathrm{Tr}\,[W^{12}W^{12}] \\
&= u^1_i u^1_k u^2_j u^2_l \; \phi^{ij}(x)\phi^{kl}(x) + \cdots \\
&+ \theta_{[3}\sigma^\mu \bar\theta^{[1} u^i_{4]} u^{2]}_j \; R^j_{\mu\;i}(x) + \cdots \\
&+ \theta_3 \sigma^{\mu\lambda}\theta_4 \bar\theta^1 \sigma_{\lambda\nu}\bar\theta^2 \; T^\nu_\mu(x) + \cdots \\
&+ [\bar\theta^1 \cdot \bar\theta^1 \bar\theta^2 \cdot \bar\theta^2 + \text{c. c.}]\; L(x) \\
&+ \text{derivative terms}
\end{aligned}
$$

where $R^j_{\mu\;i}$ is the SU(4) R current, T^ν_μ is the stress tensor and L is the on-shell Lagrangian of $N = 4$ SYM.

4. Correlators of 1/2 BPS operators

4.1. *General properties of two-, three- and four-point functions*

Counting the θs:
Consider the n–point correlation function of 1/2 BPS composites

$$\langle O^{k_1}_{1/2}(1)\cdots O^{k_n}_{1/2}(n)\rangle$$

At each point it depends on $1/2\times 4\times 4 = 8$ Grassmann variables (θ components); altogether, this makes $8n$ Grassmann variables. On the other hand, the correlator should be covariant under two supersymmetries (Q and S) with $2 \times 4 \times 4 = 32$ Grassmann parameters. It is then clear that we can form $8(n-4)$ invariant combinations of θ's if $n > 4$.

If $n \leq 4$, no such combinations exist. Consequently, the 2-, 3- and 4-point supercorrelators are uniquely determined from their lowest ($\theta = 0$) components.

4.2. *Two- and three-point functions*

Let us start with the elementary 2-point function (propagator) for the SYM field strength itself. Its lowest ($\theta = 0$) component is determined by conformal and SU(4) covariance:

$$\langle W^{12}(x,0,u)W^{12}(0,0,v)\rangle = C\frac{(1^{12}2^{12})}{x^2}$$

where C is a normalization constant,

$$(1^{12}2^{12}) = \epsilon^{ijkl}u^1_i u^2_j v^1_k v^2_l$$

and u, v are two sets of harmonic variables. The θ dependence at point 1 can be restored by making the nilpotent shift

$$\hat{x}^\mu = x^\mu - 2i\frac{(1^{23}2^{12})}{(1^{12}2^{12})}\theta_3\sigma^\mu\bar{\theta}^1 - 2i\frac{(1^{24}2^{12})}{(1^{12}2^{12})}\theta_4\sigma^\mu\bar{\theta}^1 - 1 \leftrightarrow 2$$

so that

$$\langle W^{12}(x,\theta,u)W^{12}(0,0,v)\rangle = C\frac{(1^{12}2^{12})}{\hat{x}^2}$$

For arbitrary $O^k_{1/2} = \mathrm{Tr}[W^{12}]^k$ the 2-point function is just the k-th power:

$$\langle O^k_{1/2}(1)O^k_{1/2}(2)\rangle = C_k\frac{(1^{12}2^{12})^k}{\hat{x}^{2k}_{12}} \equiv C_k\frac{(12)^k}{\hat{x}^{2k}_{12}}$$

Three-point functions are obtained by multiplication of 2-point functions and have the form

$$\langle O^a_{1/2}(1)O^b_{1/2}(2)O^c_{1/2}(3)\rangle = C_{abc}\frac{(12)^a}{\hat{x}^{2a}_{12}}\frac{(23)^b}{\hat{x}^{2b}_{23}}\frac{(31)^c}{\hat{x}^{2c}_{31}}$$

So, CSUSY kinematics predicts the form of the 2- and 3-point functions up to the normalization C which is given by the dynamics.

4.3. *Four-point functions*

As explained earlier, the 4-point correlators

$$\langle O^{k_1}_{1/2}O^{k_2}_{1/2}O^{k_3}_{1/2}O^{k_4}_{1/2}\rangle$$

can be reconstructed from their lowest component. The latter is determined by conformal and R symmetry but now it involves arbitrary functions of the conformal cross-ratios

$$s = \frac{x^2_{12}x^2_{34}}{x^2_{13}x^2_{24}}, \qquad t = \frac{x^2_{14}x^2_{23}}{x^2_{13}x^2_{24}}$$

Take the simplest example $k_i = 2$ (four stress-tensor multiplets). The lowest component is made out of 6 propagator structures with arbitrary coefficients:

$$\langle O^2_{1/2}O^2_{1/2}O^2_{1/2}O^2_{1/2}\rangle_{\theta=0}$$

$$= a_1(s,t)\frac{(12)^2(34)^2}{x_{12}^4 x_{34}^4} + a_2(s,t)\frac{(13)^2(24)^2}{x_{13}^4 x_{24}^4} + a_3(s,t)\frac{(14)^2(23)^2}{x_{14}^4 x_{23}^4}$$
$$+b_1(s,t)\frac{(13)(14)(23)(24)}{x_{13}^2 x_{14}^2 x_{23}^2 x_{24}^2} + b_2(s,t)\frac{(12)(14)(23)(34)}{x_{12}^2 x_{14}^2 x_{23}^2 x_{34}^2}$$
$$+b_3(s,t)\frac{(12)(13)(24)(34)}{x_{12}^2 x_{13}^2 x_{24}^2 x_{34}^2}$$

These 6 terms come in two crossing-equivalence classes, so that crossing symmetry reduces the 6 coefficients to only *two independent functions*:

$$a_1(s,t) = a_3(t,s) = a_1(s/t, 1/t)$$
$$a_2(s,t) = a_2(t,s) = a_3(s/t, 1/t)$$
$$b_1(s,t) = b_3(t,s) = b_1(s/t, 1/t)$$
$$b_2(s,t) = b_2(t,s) = b_3(s/t, 1/t)$$

Similarly, for $k = 3$ we obtain that the correlator is determined by 10 propagator structures in 3 crossing-equivalence classes, so that the crossing symmetric amplitude involves 3 independent functions.

For $k = 4$ we find 15 structures with 4 independent functions.

4.4. *Superconformal Ward identities*

We are not finished with the 4-point kinematics yet! It can tell us even more about the coefficient functions above, in the form of superconformal Ward identities. Their origin is related to the completion of the lowest component to a full superconformal covariant. Each propagator factor is just promoted to a full 2-point superfunction by replacing $x \rightarrow \hat{x}$. However, this is not so simple with the cross-ratios. Their completion to full superconformal covariants,

$$\hat{s} = s + \theta \text{ terms}, \quad \hat{t} = t + \theta \text{ terms}$$

involves intrinsic harmonic singularities. The resulting harmonic functions are not polynomial, i.e. they do not correspond to finite dimensional UIRs of SU(4).

To suppress these singularities we need to restrict the coefficient functions by differential constraints or SCWI. Their solution is given by holomorphic functions. For example, for $k = 2$

$$h(z) = \frac{1}{z}a_1(z,\bar{z}) + z a_3(z,\bar{z}) + b_2(z,\bar{z}), \qquad \partial_{\bar{z}} h(z) = 0$$

where the complex variables $z, \bar{z}$ are defined by

$$s = \frac{z\bar{z}}{(1+z)(1+\bar{z})}, \qquad t = \frac{1}{(1+z)(1+\bar{z})}$$

Note that the holomorphic functions are present in the free amplitude:

$$h_{\text{free}}(z) = \frac{1}{z} + z + \frac{4}{N^2-1}$$

In fact, later on we shall see that the dynamics restricts the holomorphic functions to stay free $\rightarrow$ partial non-renormalization.

So, the SCWI effectively reduce the functional freedom.

5. Dynamical input from the insertion formula

5.1. *The insertion formula*

Recall that the $N = 4$ SYM on-shell Lagrangian is a component of the stress-tensor multiplet:

$$O^2_{1/2} \equiv \text{Tr }[W^{12}W^{12}] = \cdots + \bar{\theta}^1 \cdot \bar{\theta}^1 \bar{\theta}^2 \cdot \bar{\theta}^2 \; L_{N=4}(x) + \text{c. c.} + \cdots$$

The on-shell action can be written down as an integral over 4 $\bar{\theta}$:

$$S_{N=4} = \frac{1}{g^2} \int d^4x \; d\bar{\theta}^1 \cdot d\bar{\theta}^1 d\bar{\theta}^2 \cdot d\bar{\theta}^2 \; \text{Tr }[W^{12}W^{12}]$$

This action is supersymmetric only because Tr $[W^{12}W^{12}]$ is an ultrashort superfield on shell.

Now, in QFT correlators are given by the path integral

$$\langle O^{k_1}_{1/2}(1) \cdots O^{k_n}_{1/2}(n) \rangle = \int e^{\frac{i}{g^2} \int \text{Tr }[W^{12}W^{12}]} O^{k_1}_{1/2}(1) \cdots O^{k_n}_{1/2}(n)$$

Differentiating w.r.t. the coupling g^2 we obtain the *insertion formula*

$$\frac{\partial}{\partial g^2} \langle O^{k_1}_{1/2}(1) \cdots O^{k_n}_{1/2}(n) \rangle \sim \int d^4x_0 d^4\bar{\theta}_0 \langle O^2_{1/2}(0) O^{k_1}_{1/2}(1) \cdots O^{k_n}_{1/2}(n) \rangle$$

In other words, the quantum corrections to the n-point correlator are determined by the $n+1$-point correlator obtained by inserting the $N = 4$ SYM on-shell Lagrangian.

5.2. *(4 + 1)-point nilpotent invariants*

Take $n = 4$, so the r.h.s. of the insertion formula involves a $(4+1)$-point superconformal invariant. Surprisingly, CSUSY puts more powerful restrictions on it than on the 4-point invariant in the l.h.s. The reason is that the $(4+1)$-point invariant must be nilpotent.

Indeed, let us use the entire CSUSY to shift away the θs at the external points 1 to 4. We are left with θ_0 and $\bar\theta_0$ at the insertion point 0. The Grassmann measure in the action involves all the four $\bar\theta_0$, so we need the nilpotent invariant

$$\langle O^2_{1/2}(0)O^{k_1}_{1/2}(1)\cdots O^{k_4}_{1/2}(4)\rangle|_{\theta_{1,2,3,4}=0} \sim \bar\theta^1_0\cdot\bar\theta^1_0\bar\theta^2_0\cdot\bar\theta^2_0$$

(recall that such invariants only exist for 5 or more points)

Restoring the θ dependence at points 1 to 4, we obtain

$$\langle O^2_{1/2}(0)O^{k_1}_{1/2}(1)\cdots O^{k_4}_{1/2}(4)\rangle$$

$$= [\bar\theta^1_0\cdot\bar\theta^1_0\bar\theta^2_0\cdot\bar\theta^2_0\; R^{2,2,2,2} + \text{terms with } \theta_{1,2,3,4}] \times F^{k_1-2,k_2-2,k_3-2,k_4-2}(x_0,\cdots,x_4;u_1,\cdots,u_4)$$

where

$$\begin{aligned} R^{2,2,2,2} &= \frac{(12)^2(34)^2}{x^4_{12}x^4_{34}}\,s + \frac{(13)^2(24)^2}{x^4_{13}x^4_{24}} + \frac{(14)^2(23)^2}{x^4_{14}x^4_{23}}\,t \\ &+\frac{(12)(14)(23)(34)}{x^2_{12}x^2_{14}x^2_{23}x^2_{34}}\,(1-s-t) + \frac{(12)(13)(24)(34)}{x^2_{12}x^2_{13}x^2_{24}x^2_{34}}\,(t-s-1) \\ &+\frac{(13)(14)(23)(24)}{x^2_{13}x^2_{14}x^2_{23}x^2_{24}}\,(s-t-1) \end{aligned}$$

is a particular harmonic polynomial carrying charges (Dynkin labels) 2 at points 1 to 4. The entire charge at point 0 is absorbed by the nilpotent factor. The remaining freedom is encoded in the harmonic and space-time function $F^{k_1-2,k_2-2,k_3-2,k_4-2}$ whose charges at points 1 to 4 are reduced by two units.

Integrating over the insertion point (the $\bar\theta_0$ integral is trivial), we obtain the quantum part of the 4-point amplitude:

$$\frac{\partial}{\partial g^2}\langle O^{k_1}_{1/2}(1)\cdots O^{k_4}_{1/2}(4)\rangle =$$

$$R^{2,2,2,2}\; F^{k_1-2,k_2-2,k_3-2,k_4-2}(x_1,\cdots,x_4;u_1,\cdots,u_4)$$

where F is a conformally covariant 4-point function with charges k_i-2 at each point. Its harmonic dependence must be polynomial, and this determines the number of independent functions in the amplitude.

The crucial prediction of the insertion formula is the *factorization* $R^{2,2,2,2}\times F^{k_1-2,k_2-2,k_3-2,k_4-2}$:

→ this amplitude automatically satisfies the SCWI
→ it gives no contribution to the holomorphic functions
→ the holomorphic functions stay free (partial non-renormalization)

5.3. *Restrictions on the quantum part of the amplitude*

All the results below are completely general, i.e. they apply to both the perturbative and non-perturbative sectors.

5.3.1. *Three points*

The $(3+1)$-point nilpotent invariant simply does not exist (not enough θs), hence

$$\frac{\partial}{\partial g^2}\langle O^{k_1}_{1/2}(1)\cdots O^{k_3}_{1/2}(3)\rangle = 0$$

This explains why 3-point functions of 1/2 BPS operators are not renormalized.

5.3.2. *Four points, extremal and subextremal*

With 4 points the question is how many propagator structures can enter the function $F^{k_1-2,k_2-2,k_3-2,k_4-2}$. It is easy to see that if

$$k_1 = k_2 + k_3 + k_4 \quad \text{extremal correlator}$$

or

$$k_1 = k_2 + k_3 + k_4 - 2 \quad \text{subextremal correlator}$$

the function $F^{k_1-2,k_2-2,k_3-2,k_4-2}$ cannot be constructed. Indeed,

$$(k_1 - 2) > (k_2 - 2) + (k_3 - 2) + (k_4 - 2)$$

and it is impossible to connect point 1 to points 2, 3 and 4 with all the $k_1 - 2$ propagators. This explains why extremal and subextremal 4-point functions of 1/2 BPS operators are not renormalized.

5.3.3. *Four points, $k_i = 2$*

For $k_i = 2$ (stress-tensor multiplets) we have a harmonic-independent (charge 0) and conformal invariant function

$$F^{0,0,0,0} = F(s,t)$$

So, the quantum part of the amplitude is determined by a single function of 2 variables.

Recall that the SCWI predict an additional holomorphic function. We find it hidden in the free part of the amplitude.

Here are the 6 coefficients of the complete amplitude (free + quantum parts):

$$\begin{aligned}
a_1 &= 1 + s\,F(s,t)\\
a_2 &= 1 + F(s,t)\\
a_3 &= 1 + t\,F(s,t)\\
b_1 &= \tfrac{4}{N^2-1} + (s-t-1)\,F(s,t)\\
b_2 &= \tfrac{4}{N^2-1} + (1-s-t)\,F(s,t)\\
b_3 &= \tfrac{4}{N^2-1} + (t-s-1)\,F(s,t)
\end{aligned}$$

5.3.4. *Four points, $k_i = 3$ and $k_i = 4$*

For $k_i = 3$ we have three possible propagator structures:

$$F^{1,1,1,1} = \frac{(12)(34)}{x_{12}^2 x_{34}^2}\,\alpha(s,t) + \frac{(14)(23)}{x_{14}^2 x_{23}^2}\,\beta(s,t) + \frac{(13)(24)}{x_{13}^2 x_{24}^2}\,\gamma(s,t)$$

Crossing symmetry reduces them to only one independent function of two variables $\gamma(s,t)$:

$$\alpha(s,t) = 1/s\,\gamma(t/s, 1/s)\,,\ \beta(s,t) = 1/t\,\gamma(s/t, 1/t)\,,\ \gamma(s,t) = \gamma(t,s)$$

Again, the quantum part of the amplitude is given by 1 function of 2 variables. The 2 holomorphic functions allowed by the SCWI are only seen in the free amplitude.

For $k_i = 4$ $F^{2,2,2,2}$ contains 6 structures reduced to just 2 independent by crossing symmetry. So, the insertion formula predicts 2 functions of 2 variables in the quantum part, leaving the 2 holomorphic functions to the free part.

General partial non-renormalization theorem: for any choice of the weights k the dynamics fixes all the holomorphic functions allowed by the SCWI at their free values.

This phenomenon has an alternative explanation from the OPE point of view. The holomorphic functions are expanded in terms of protected multiplets (BPS short or semishort) only. The dynamics forbids such OPE exchanges beyond the free OPE.

5.4. *Large N degeneracy in perturbation theory*

Explicit low-order perturbative calculations show that in the large N limit the number of independent functions in the amplitude is limited and does not increase with the weight k: *(i)* At one loop the complete amplitude for $k = 2, 3, 4$ is determined by the same single function (up to a numerical factor), the one-loop scalar box $\Phi^{(1)}(s,t)$. It is likely that this also holds for arbitrary k (so far only checked for a subset of the crossing-equivalence classes). *(ii)* At two loops the degeneracy is lifted for $k = 3$ (the single function is not the same as for $k = 2$) and for $k = 4$ (we find two distinct functions instead of one). However, the degeneracy reappears at $k \geq 5$: all such amplitudes are determined by the same two functions, combinations of the one- and two-loop scalar boxes $\Phi^{(1)}(s,t)$ and $\Phi^{(2)}(s,t)$.

This phenomenon has a simple explanation in terms of Feynman diagrams. The diagram has a frame made out of a large number of matter lines, some of which are connected with a limited number of gluon lines, the rest remaining free. In the large N limit the color combinatorics is trivial, so the actual number of independent functions is determined by the limited number of different topologies.

This perturbative degeneracy means that entire classes of operators in the OPE are indistinguishable at the given loop level (have the same anomalous dimension), but start drifting apart at the next level.

It would be very interesting to examine the instanton sector for $k = 3$ and 4, to see if the partial non-renormalization pattern is respected and if there are any traces of the perturbative degeneracy.

6. Confronting CFT with AdS

6.1. *Effective action on $AdS_5 \times S^5$*

According to the AdS/CFT conjecture, 1/2 BPS operators of charge $k = 2$ are dual to the massless KK modes in AdS SG, and those with $k > 2$ to massive modes. The effective 5-d action for scalars s_k needed to compute the 4-point correlator of ANY 1/2 BPS operators in $N = 4$ SYM_4 is:

$$S(s) = \frac{N^2}{8\pi^2} \int d^5x \sqrt{g_a} \, (L_2 + L_3 + L_4)$$

The scalar s_k, $k \geq 2$ in the irrep $[0, k, 0]$ of $SU(4)$ is the superconformal primary of a BPS short supermultiplet with top spin 2:

$$[s, A_\mu, \varphi_{\mu\nu}, \phi, C_\mu, t, ...]_k \,.$$

The dots stand for a number of other fields which do not appear in the cubic couplings involving two fields s_k.to know the SG effective action up to quartic terms.

Working out this effective action is a formidable problem which was solved by Arutyunov and Frolov. The equations of motion that follow from this action are related to the original non-Lagrangian equations of IIB supergravity on the $AdS_5 \times S^5$ background by of a huge number of field redefinitions and by hidden relations among different kinds of spherical harmonics.

Take for example $k = 2 \rightarrow [0,2,0] = \mathbf{20}$. The KK mode s_2 can be described as a traceless symmetric 6×6 matrix Λ. The general effective Lagrangian is reduced to

$$\begin{aligned} L = R &- \frac{1}{4}\mathrm{tr}\big(\nabla_\mu \Lambda \nabla^\mu \Lambda - 4\Lambda^2\big) + \frac{1}{3}\mathrm{tr}\Lambda^3 \\ &+ \frac{1}{2^4}\mathrm{tr}\big(\nabla_\mu \Lambda^2 \nabla^\mu \Lambda^2\big) - \frac{3}{2^2}\mathrm{tr}\Lambda^4 + \frac{1}{2^3}\left(\mathrm{tr}\Lambda^2\right)^2 \\ &+ \frac{1}{2}\mathrm{tr}F_{\mu\nu}F^{\mu\nu} - 2\mathrm{tr}(\nabla^\mu \Lambda\Lambda A_\mu). \end{aligned}$$

One sees that except for self-interactions the scalars from the massless multiplet interact only via exchanges of the massless graviton and of the massless vector field. It has been shown that this action coincides with the action of gauged $N = 8$ supergravity on the AdS_5 background. For higher values of k, $k = 3, 4, \ldots$ the Lagrangians become quite complicated due to the increasing number of fields mediating the interaction of two scalars s_k.

6.2. *Main results*

- The supergravity-induced four-point amplitude splits into a "free" and a "quantum" parts, precisely reproducing the SCFT pattern.
- The "quantum" part contains the set of independent functions of 2 variables predicted by the insertion procedure.
- The numerical coefficients in the "free" part are exactly those of the free SCFT amplitude. In the case $k = 4$ the possible admixtures of double-trace operators are suppressed.

What makes this so unexpected is that contrary to the SYM theory, the supergravity action has no coupling constant and therefore no natural separation of the free part from the interacting part! We see this as concrete evidence that the AdS/CFT correspondence is not just kinematics!

Here is a summary of the results of the supergravity calculations for $k = 2, 3, 4$:

- Weight 2

$$F_2^{\infty}(s,t) = -\frac{4}{N^2} D_{2224}\,,$$

- Weight 3

$$F_3^{\infty}(s,t) = -\frac{9}{N^2}\left(D_{2235} + sD_{3335}\right)\,,$$

- Weight 4

$$\begin{aligned} F_4^{\infty}(s,t) &= -\frac{4}{N^2}\left(2D_{2246} + 2sD_{3346} + s^2 D_{4446}\right)\,, \\ G_4^{\infty}(s,t) &= -\frac{16}{N^2} s\Big(D_{4222} + (D_{4233} + D_{4323}) \\ &\quad + (D_{4244} + D_{4424} + 5D_{4334}) - D_{5335} + D_{4446}\Big)\,. \end{aligned}$$

Here ∞ refers to the supergravity approximation (infinite value of the 't Hooft coupling) and the conformal invariant D-functions are defined by the AdS integral

$$D_{\Delta_1\Delta_2\Delta_3\Delta_4} = \int \frac{d^5w}{w_0^5} \prod_{i=1}^{4} \left(\frac{w_0}{w_0^2 + (\vec{w} - x_i)^2}\right)^{\Delta_i}$$

7. Conclusions

- The general field-theoretical predictions about the dynamical structure of the correlators of 1/2 BPS operators in $N = 4$ SYM are remarkably confirmed by explicit amplitudes following from the supergravity effective action.
- The QFT insertion procedure and the effective 5d supergravity action comprising all KK modes are at the heart of this study.

Many open questions:

- In all the cases we have dealt with the SG action is of the sigma-model type (all 4-derivative vertices vanish). Is there a sigma model extending gauged $N = 8$ supergravity?

- Is there an analogue of the insertion procedure in supergravity? A systematic superspace procedure for constructing the supergravity effective action is needed.
- What makes the perturbative and the supergravity amplitudes so different and do they have anything in common ?
- Generic unprotected operators (e.g., Konishi) are expected to decouple in the supergravity limit. What is the mechanism for this?
- Is the perturbation series convergent, Borel summable ... ?
- Non-perturbative (instanton) contributions vs the insertion formula.
- ...

Acknowledgments

The author acknowledges a longstanding collaboration with G. Arutyunov, F. Dolan, B. Eden, S. Ferrara, P. Howe, H. Osborn, S. Penati, A. Petkou, A. Santambrogio, C. Schubert and P. West.

References

1. J. Maldacena, "The large N limit of superconformal field theories and supergravity", *Adv. Theor. Math. Phys.* **2**, 231 (1998).
2. G.G. Gubser, I.R. Klebanov and A.M. Polyakov, "Gauge theory correlators from noncritical string theory", *Phys. Lett.* **B428**, 105 (1998), hep-th/9802109.
3. E. Witten, "Anti-de Sitter space and holography", *Adv. Theor. Math. Phys.* **2**, 253 (1998), hep-th/9802150.
4. L. Andrianopoli and S. Ferrara, "Short and long SU(2,2/4) multiplets in the AdS/CFT correspondence", *Lett. Math. Phys.* **48**, 145 (1999), hep-th/9812067; "K-K excitations on AdS(5) x S(5) as N = 4 *primary* superfields", *Phys. Lett.* **B430**, 248 (1998), hep-th/9803171.
5. L. Andrianopoli, S. Ferrara, E. Sokatchev and B. Zupnik, "Shortening of primary operators in N-extended SCFT(4) and harmonic-superspace analyticity", *Adv. Theor. Math. Phys.* **3**, 1149 (1999), hep-th/9912007.
6. G. Arutyunov and S. Frolov, "Scalar quartic couplings in type IIB supergravity on AdS(5) x S(5)", *Nucl. Phys.* **B579**, 117 (2000), hep-th/9912210; "Four-point functions of lowest weight CPOs in N = 4 SYM(4) in supergravity approximation", *Phys. Rev.* **D62**, 064016 (2000); hep-th/0002170; "On the correspondence between gravity fields and CFT operators", *J. High En. Phys.* 04 (2000) 017, hep-th/0003038.
7. G. Arutyunov, B. Eden and E. Sokatchev, "On non-renormalization and OPE in superconformal field theories", *Nucl. Phys.* **B619**, 359 (2001), hep-th/0105254.

8. G. Arutyunov, S. Frolov and A.C. Petkou, "Operator product expansion of the lowest weight CPOs in N = 4 SYM(4) at strong coupling", *Nucl. Phys.* **B586**, 547 (2000), hep-th/0005182; "Perturbative and instanton corrections to the OPE of CPOs in N = 4 SYM(4)", *Nucl. Phys.* **B602**, 238 (2001), hep-th/0010137.
9. G. Arutyunov, B. Eden, A.C. Petkou and E. Sokatchev, "Exceptional non-renormalization properties and OPE analysis of chiral four-point functions in N = 4 SYM(4)", hep-th/0103230.
10. G. Arutyunov and E. Sokatchev, "Implications of superconformal symmetry for interacting (2,0) tensor multiplets", *Nucl. Phys.* **B635**, 3 (2002), hep-th/0201145; "On a large N degeneracy in N = 4 SYM and the AdS/CFT correspondence", *Nucl. Phys.* **B663**, 163 (2003), hep-th/0301058.
11. G. Arutyunov, F.A. Dolan, H. Osborn and E. Sokatchev, "Correlation functions and massive Kaluza-Klein modes in the AdS/CFT correspondence", *Nucl. Phys.* **B665**, 273 (2003), hep-th/0212116.
12. G. Arutyunov, S. Penati, A. Santambrogio and E. Sokatchev, "Four-point correlators of BPS operators in N = 4 SYM at order g**4", *Nucl. Phys.* **B670**, 103 (2003), hep-th/0305060.
13. M. Bianchi, M.B. Green, S. Kovacs and G. Rossi, "Instantons in supersymmetric Yang-Mills and D-instantons in IIB superstring theory", *J. High En. Phys.* 08 (1998) 013, hep-th/9807033.
14. M. Bianchi, S. Kovacs, G. Rossi and Y.S. Stanev, "On the logarithmic behavior in N = 4 SYM theory", *J. High En. Phys.* 08 (1999) 020, hep-th/9906188; "Anomalous dimensions in N = 4 SYM theory at order g^4", *Nucl. Phys.* **B584**, 216 (2000), hep-th/0003203; "Properties of the Konishi multiplet in N = 4 SYM theory", *J. High En. Phys.* 05 (2001) 042, hep-th/0104016.
15. M. Bianchi, B. Eden, G. Rossi and Y.S. Stanev, "On operator mixing in N = 4 SYM", *Nucl. Phys.* **B646**, 69 (2002), hep-th/0205321.
16. M. Bianchi, G. Rossi and Y.S. Stanev, "Surprises from the resolution of operator mixing in N = 4 SYM", hep-th/0312228.
17. F.A. Dolan and H. Osborn, "Conformal four point functions and the operator product expansion", *Nucl. Phys.* **B599**, 459 (2001), hep-th/0011040; "Superconformal symmetry, correlation functions and the operator product expansion", *Nucl. Phys.* **B629**, 3 (2002), hep-th/0112251.
18. V.K. Dobrev and V.B. Petkova, "All Positive Energy Unitary Irreducible Representations Of Extended Conformal Supersymmetry", *Phys. Lett.* **B162**, 127 (1985); "On The Group Theoretical Approach To Extended Conformal Supersymmetry: Classification Of Multiplets", *Lett. Math. Phys.* **9**, 287 (1985).
19. B. Eden, P.S. Howe, C. Schubert, E. Sokatchev and P.C. West, "Four-point functions in N = 4 supersymmetric Yang-Mills theory at two loops", *Nucl. Phys.* **B557**, 355 (1999), hep-th/9811172; "Simplifications of four-point functions in N = 4 supersymmetric Yang-Mills theory at two loops", *Phys. Lett.* **B466**, 20 (1999), hep-th/9906051; "Extremal correlators in four-dimensional SCFT", *Phys. Lett.* **B472**, 323 (2000), hep-th/9910150.

20. B. Eden, P.S. Howe and P.C. West, "Nilpotent invariants in N=4 SYM", *Phys. Lett.* **B463**, 19-26 (1999), hep-th/9905085.
21. B. Eden, A.C. Petkou, C. Schubert and E. Sokatchev, "Partial non-renormalisation of the stress-tensor four-point function in N = 4 SYM and AdS/CFT", *Nucl. Phys.* **B607**, 191 (2001), hep-th/0009106.
22. B.U. Eden, P.S. Howe, E. Sokatchev and P.C. West, "Extremal and next-to-extremal n-point correlators in four-dimensional SCFT", *Phys. Lett.* **B494**, 141 (2000), hep-th/0004102.
23. B.U. Eden, P.S. Howe, A. Pickering, E. Sokatchev and P.C. West, "Four-point functions in N = 2 superconformal field theories", *Nucl. Phys.* **B581**, 523 (2000), hep-th/0001138.
24. B. Eden, C. Schubert and E. Sokatchev, "Three-loop four-point correlator in N = 4 SYM", *Phys. Lett.* **B482**, 309 (2000), hep-th/0003096; "Four-point functions of chiral primary operators in N = 4 SYM", hep-th/0010005.
25. B. Eden and E. Sokatchev, "On the OPE of 1/2 BPS short operators in N = 4 SCFT(4)", *Nucl. Phys.* **B618**, 259 (2001), hep-th/0106249.
26. S. Ferrara and E. Sokatchev, "Superconformal interpretation of BPS states in AdS geometries", *Int. J. Theor. Phys.* **40**, 935 (2001), hep-th/0005151.
27. A. Galperin, E. Ivanov, S. Kalitsyn, V. Ogievetsky and E. Sokatchev, "Unconstrained N=2 Matter, Yang-Mills And Supergravity Theories In Harmonic Superspace", *Class. Quant. Grav.* **1**, 469 (1984).
28. A.S. Galperin, E.A. Ivanov, V.I. Ogievetsky and E.S. Sokatchev, *Harmonic Superspace* (Cambridge Univ. Press, 2001).
29. G.G. Hartwell and P.S. Howe, *Class. Quant. Grav.* **12**, 1823 (1995); "(N, p, q) harmonic superspace", *Int. J. Mod. Phys.* **A10**, 3901 (1995), hep-th/9412147.
30. P.J. Heslop and P.S. Howe, "A note on composite operators in N = 4 SYM", *Phys. Lett.* **B516**, 367 (2001), hep-th/0106238; "OPEs and 3-point correlators of protected operators in N = 4 SYM", *Nucl. Phys.* **B626**, 265 (2002), hep-th/0107212; "Four-point functions in N = 4 SYM", *J. High En. Phys.* 01 (2003) 043, hep-th/0211252.
31. E. D'Hoker, D.Z. Freedman, S.D. Mathur, A. Matusis and L. Rastelli, "Graviton exchange and complete four-point functions in the AdS/CFT correspondence", *Nucl. Phys.* **B562**, 353 (1999), hep-th/9903196.
32. E. D'Hoker, D.Z. Freedman and L. Rastelli, "AdS/CFT four-point functions: How to succeed at z-integrals without really trying", *Nucl. Phys.* **B562**, 395 (1999); hep-th/9905049.
33. E. D'Hoker, D.Z. Freedman, S.D. Mathur, A. Matusis and L. Rastelli, "Extremal correlators in the AdS/CFT correspondence", hep-th/9908160.
34. E. D'Hoker, S.D. Mathur, A. Matusis and L. Rastelli, "The operator product expansion of N = 4 SYM and the four-point functions of supergravity", *Nucl. Phys.* **B589**, 38 (2000), hep-th/9911222.
35. E. D'Hoker, P. Heslop, P. Howe and A.V. Ryzhov, "Systematics of quarter BPS operators in N = 4 SYM", *J. High En. Phys.* 04 (2003) 038, hep-th/0301104.
36. P.S. Howe and P.C. West, *Phys. Lett.* **B400**, 307 (1997), hep-th/9611075; "Is

N = 4 Yang-Mills theory soluble?", hep-th/9611074.
37. P.S. Howe, E. Sokatchev and P.C. West, "3-point functions in N = 4 Yang-Mills", *Phys. Lett.* **B444**, 341 (1998), hep-th/9808162.
38. K.A. Intriligator, "Bonus symmetries of N = 4 super-Yang-Mills correlation functions via AdS duality", *Nucl. Phys.* **B551**, 575 (1999), hep-th/9811047.
39. S.M. Lee, S. Minwalla, M. Rangamani and N. Seiberg, "Three-point functions of chiral operators in D = 4, N = 4 SYM at large N", *Adv. Theor. Math. Phys.* **2**, 697 (1998), hep-th/9806074.
40. S.M. Lee, "AdS(5)/CFT(4) four-point functions of chiral primary operators: Cubic vertices", *Nucl. Phys.* **B563**, 349 (1999), hep-th/9907108.
41. S. Minwalla, "Restrictions imposed by superconformal invariance on quantum field theories", *Adv. Theor. Math. Phys.* **2**, 781 (1998), hep-th/9712074.

RENORMALONS AND FRACTIONAL INSTANTONS

CESAR GOMEZ
Instituto de Fisica Teorica UAM/CSIC, Madrid, Spain

1. Introduction

What follows is a brief summary of my talk in the Varna Workshop last June. It is part of a work in progress in collaboration with Sumit Das. I will concentrate this lecture in the interplay between renormalon divergences in quantum field theory and fractional instanton effects.

2. Divergences in Quantum Field Theory: Renormalons and Fractional Instantons

Physical amplitudes in QFT are defined in terms of perturbative expansions in the coupling constant. Generically we have

$$G(g^2) = \sum_n a_n (g^2)^n \tag{1}$$

where g is the expansion parameter or coupling constant. There are two main sources of divergences for this expansion. The first one comes from the fact that the number of Feynman diagrams contributing at order n grows as $n!$. This kind of divergences in normally associated with instantons. The other type of divergences is associated with the contribution of Feynman diagrams that depend on the number n of loops as $n!$. It is this type of divergences that are known under the generic name of renormalons [1].

A typical way to try to define the sum of the perturbative expansion is using a Borel transform. In fact we can easily prove that

$$G(g^2) = \int_0^\infty dz \sum_n \frac{a_n}{n!} z^n e^{-\frac{z}{g^2}} \tag{2}$$

The Borel transform $F(z)$ is defined by

$$F(z) = \sum_n \frac{a_n}{n!} z^n \tag{3}$$

Clearly if

$$a_n = a^n n! \tag{4}$$

for some coefficient a, we observe that the Borel transform $F(z)$ is divergent from $z = \frac{1}{a}$.

A simple way to relate singularities of the Borel transform with the stationary points of the Euclidean action i.e with instantons, is as follows. Let us define

$$Z(g^2) = \int dA e^{-\frac{1}{g^2}S(A)} \tag{5}$$

where we assume that we have fixed the gauge. Let us now introduce a real variable z and look for the solution $A(z)$ of

$$S(A) = z \tag{6}$$

i.e the gauge configuration A such that $S(A) = z$. Changing variables we get

$$Z(g^2) = \int_0^\infty dz (\frac{\partial S}{\partial A})_z^{-1} e^{-\frac{z}{g^2}} \tag{7}$$

which clearly implies that the stationary point configuration $\frac{\partial S}{\partial A} = 0$ produces a singularity in the Borel transform. This singularity is at $z = S(A_{inst}) = 8\pi^2$, therefore we can associate instantons with a perturbative expansion of type

$$Z(g^2) = \sum_n a^n n! (g^2)^n \tag{8}$$

with

$$a = \frac{1}{8\pi^2} \tag{9}$$

Of course equation 7 is a bit too simple since we should consider the moduli M_z of gauge inequivalent configurations A such that $S(A) = z$. The correct expression would be

$$Z(g^2) = \int_{M_z} \int_0^\infty dz (\frac{\partial S}{\partial A})_z^{-1} e^{-\frac{z}{g^2}} \tag{10}$$

In the instanton case we know that the moduli $M(z)$ is isomorphic to $R^4.R^+$ where R^4 parametrizes the translations and R^+ the dilatations. Assuming that the Borel integral is dominated by the instanton we get the well known result

$$L_{eff}(g^2) dx^4 = dx^4 \int d\rho \frac{1}{\rho^5} e^{-\frac{8\pi^2}{g^2}} \tag{11}$$

where we are considering the case of pure gauge theory.

In the large N planar limit is known that the number of Feynmann diagrams at level n grow only geometrically therefore singularities of instanton type are absent. Thus we can in the large N planar limit concentrate our attention on renormalon singularities.

Let us first review briefly the type of n loop diagrams that contribute as $n!$. These are diagrams where we integrate over some internal momentum k a chain of self energy insertions with the following typical k dependence

$$k^2 ln(k^2) \tag{12}$$

In this case we find integrals of the type

$$\int d^4k^2 C^n \frac{(ln(\frac{k^2}{\Lambda^2}))^n}{k^{2\alpha}} \tag{13}$$

After performing the integral we get

$$\Lambda^d n! a^n \tag{14}$$

with

$$a = \frac{\beta_1}{d} \tag{15}$$

where

$$d = 2(\alpha - 2) \tag{16}$$

and β_1 is the one loop coefficient of the beta function. Therefore renormalon contributions are associated with perturbative expansions

$$G_\alpha(g^2) = \sum_n n! a^n (g^2)^n \tag{17}$$

with a given by 15. Using Borel transform we can perform the following approximation. From 17 we get

$$G(g^2) = \int dz e^{f(z) - \frac{z}{g^2}} \tag{18}$$

where

$$e^{f(z)} = \sum_n a^n z^n \tag{19}$$

Now we approximate 18 by the saddle point. The result for small g is

$$G(g^2) \sim e^{-\frac{8\pi^2}{g^2 a}} \tag{20}$$

From here we can define an effective action

$$L_{eff}(g^2)dx^4 = dx^4 \Lambda^d e^{-\frac{8\pi^2}{g^2 a}} \qquad (21)$$

This is dimensionally correct only if $d = 4$. We can perform the previous computation in the case of $N = 1$ super Yang Mills. In this case we get

$$L_{eff}(g^2)dx^4 d\theta^2 = dx^4 d\theta^2 \Lambda^d e^{-\frac{8\pi^2}{g^2 a}} \qquad (22)$$

which defines an effective action only if $d = 3$. Using now 15 and taking into account that

$$\beta_1 = 3N \qquad (23)$$

we get

$$L_{eff}(g^2)dx^4 d\theta^2 = dx^4 d\theta^2 \Lambda^3 e^{-\frac{8\pi^2}{g^2 N}} \qquad (24)$$

i.e the fractional instanton effect. In this sense we can tentatively conclude that renormalons in $N = 1$ super Yang Mills are just what normally is known under the name of fractional instantons.

3. The Effective Action for $N = 1$ super Yang Mills

Let us consider the following function

$$f(x) = N\Lambda^3 e^{\frac{x}{N}} \qquad (25)$$

and let us perform the Legendre transform

$$Lf(S) = Sx(S) - f(x(S)) \qquad (26)$$

where $x(S)$ is the solution to

$$f'(x) = S \qquad (27)$$

we get

$$Lf(S) = NSln(\frac{S}{\Lambda^3}) - NS \qquad (28)$$

We realize that the effective superpotential for $N = 1$ super Yang Mills $W(S)$ [2] where S is the composite operator TrW^2 is

$$W(S) = -Lf(S) \qquad (29)$$

Thus the effective superpotential is just the Legendre transform of $f(x)$. Denoting $x = \frac{1}{g^2}$ we observe that $W(S)$ is just the Legendre transform of the fractional instanton contribution we have derived in previous section as the saddle point approximation to the renormalon contribution.

Notice also the following general structure. If you start with the renormalization group equation

$$\frac{1}{g^2}(S) = Nln(\frac{S}{\Lambda^3}) \tag{30}$$

and you look for a function $f(x = \frac{1}{g^2})$ such that the solution $x(S)$ to $f'(x) = S$ is the renormalization group equation, then the effective action $W(S)$ is just the Legendre transform of f.

4. Some Comments on the Mass Gap

The proof of the existence of a mass gap in non abelian gauge theories is a one million dollar question. From equation 22 it looks that in the case of $N = 1$ super Yang Mills we only need to work out the contribution of renormalons and to consider $d = 3$. However the reader should notice that in the instanton case the integration dx^4 has a natural meaning as the integration over the moduli of translations of the instanton. However in the renormalon case we have not any classical solution and we cannot ,at least directly, interpret dx^4 in 22 as any moduli integration. In the $N = 1$ case since $\beta_1 = 3N$ for $d = 3$ we get a fractional instanton action and we can think of this as a sort of classical configuration with moduli of translations but without moduli of dilatations. To prove that this is in fact the case is part of the intrinsic difficulty of proving the existence of a mass gap in $N = 1$ supersymmetric Yang Mills.

Acknowledgments

It is a pleasure to thank the organizers of the Varna Workshop and specially Vlado Dobrev for the hospitality during such a nice and interesting meeting.

References

1. G. 't Hooft, "Planar Diagram Field Theory", in: *Progress in Gauge Field Theory*, NATO Adv. Study Inst., Series. B: Physics, Vol. 115 (Plenum Press, 1984) p. 271.
2. G. Veneziano and S. Yankielowicz, *Phys. Lett.* **B113**, 231 (1982).

III. String Theory

THE HAGEDORN/DECONFINEMENT PHASE TRANSITION IN WEAKLY COUPLED LARGE N GAUGE THEORIES

OFER AHARONY[1], JOSEPH MARSANO[2], SHIRAZ MINWALLA[2], KYRIAKOS PAPADODIMAS[2], MARK VAN RAAMSDONK[3,4]

[1] *Department of Particle Physics, Weizmann Institute of Science, Rehovot 76100, Israel*

[2] *Jefferson Physical Laboratory, Harvard University, Cambridge, MA 02138, USA*

[3] *Department of Physics, Stanford University, Stanford, CA 94305, USA*

[4] *Department of Physics and Astronomy, University of British Columbia, Vancouver, BC, V6T 1Z1, Canada*

We demonstrate that weakly coupled, large N, d-dimensional $SU(N)$ gauge theories on a class of compact spatial manifolds (including $S^{d-1}\times$ time) undergo deconfinement phase transitions at temperatures proportional to the inverse length scale of the manifold in question. The low temperature phase has a free energy of order one, and is characterized by a stringy (Hagedorn) growth in its density of states. The high temperature phase has a free energy of order N^2. These phases are separated either by a single first order transition that generically occurs below the Hagedorn temperature or by two continuous phase transitions, the first of which occurs at the Hagedorn temperature. These phase transitions may be continuously connected to the usual flat space deconfinement transition in the case of confining gauge theories, and to the Hawking-Page nucleation of AdS_5 black holes in the case of the $\mathcal{N} = 4$ supersymmetric Yang-Mills theory. We suggest that deconfinement transitions may generally be interpreted in terms of black hole formation in a dual string theory. Our analysis proceeds by first reducing the Yang-Mills partition function to a $(0+0)$-dimensional integral over a unitary matrix U, which is the holonomy (Wilson loop) of the gauge field around the thermal time circle in Euclidean space; deconfinement transitions are large N transitions in this matrix integral.

1. Introduction

The thermodynamics of large N $SU(N)$ gauge theories is interesting for at least two different reasons. On the one hand, these theories are believed to share many qualitative features with the finite N non-Abelian gauge theories which are relevant to real-world QCD (or GUTs), including confinement at zero temperature and a deconfinement transition as the temperature is increased. On the other hand, large N gauge theories are believed [1, 2] to be dual to weakly coupled string theories, which display Hagedorn behavior (reviewed below) [3] associated with singularities in various thermodynamic quantities. Thus, by studying large N gauge theory thermodynamics, one may hope to achieve a better understanding of both deconfinement transitions in gauge theory and Hagedorn behavior in string theory and, possibly, some relationship between the two.

Unfortunately, asymptotically free gauge theories in Minkowski space are strongly coupled except at very high temperatures, so direct analysis of the thermodynamic behavior usually relies on numerical or lattice techniques. However, by placing the theory on a compact space, one obtains a tunable dimensionless parameter $R\Lambda_{QCD}$ (where R is the size of the compact space and Λ_{QCD} is the dynamically generated scale of the gauge theory[1]) which gives back the Minkowski space theory when it is large, but gives a weakly coupled theory when it is small.[2] In this weak coupling limit the thermodynamics can be studied at all temperatures in perturbation theory, and this analysis is the goal of the present paper. Specifically, we investigate the thermodynamics of d-dimensional $U(N)$ (or $SU(N)$) gauge theories with arbitrary adjoint-valued matter fields [3] on compact spatial manifolds, such as S^{d-1}, in the limit of weak coupling. This theory is simple enough to analyze exactly in the large N limit; we will find that it has a surprisingly rich structure with many of the expected features of strongly coupled gauge theory thermodynamics, including stringy Hagedorn behavior and a deconfinement transition.

---

[1] We use here, and throughout the paper, notations which are appropriate for four dimensional gauge theories. For lower dimensional gauge theories, Λ_{QCD} should be replaced by a scale which is an appropriate power of the 't Hooft coupling constant $g_{YM}^2 N$.

[2] A similar dimensionless parameter appears also in conformal gauge theories like the $d = 4$ $\mathcal{N} = 4$ supersymmetric Yang-Mills theory, which can also be studied at weak coupling.

[3] Our methods can be easily generalized to include also matter fields in other representations.

We should start by emphasizing that even at very weak coupling, $SU(N)$ Yang-Mills theory on a compact manifold behaves very differently from (N^2-1) copies of the $U(1)$ theory. The excitations of the $U(1)$ theory are arbitrary numbers of photons; in contrast (for example) a single $SU(N)$ gluon is not an allowed excitation on a compact manifold, as gluons source lines of color electric flux. Due to Gauss' law, the only allowed excitations of an $SU(N)$ theory on any compact manifold are combinations of gluons (and any other charged particles in the theory) that are grouped into $SU(N)$ singlets.

As we will see below, the projection onto the singlet sector introduces effective interactions between the gluons. It will turn out that these interactions are negligible at high energies, but they dominate the dynamics at low energies. As a consequence, in the $N \to \infty$ limit, the gauge theories under study in this paper have at least two distinct phases. The lowest temperature phase (dominated by states with $E/N^2 \ll 1/R$) has a strongly stringy flavor; it is characterized (see [4, 5] for the free theory) by a Hagedorn-like density of states $\rho(E)$ that grows exponentially with energy, $\rho(E) \propto e^{E/T_H}$ with $T_H = C/R$, where C is a constant depending on the matter content of the theory and on the shape of the compactification manifold. On the other hand, the high temperature phase (dominated by states with $E/N^2 \gg 1/R$) behaves qualitatively like a gas of free particles; the free energy F takes the form $F = N^2 f(T)$, where $f(T) \propto T^d$ for $T \gg 1/R$. In some cases these two phases may be separated by a more mysterious intermediate temperature phase.

The free gauge theory undergoes precisely one phase transition as a function of the temperature; this transition is of first order and it occurs precisely at the Hagedorn temperature T_H. As far as we know, this result was first found by [6] in a specific case, and was derived more generally in a beautiful paper by Sundborg [4].[4] For finite N, this phase transition is smoothed out, as it must be for any theory with a finite number of degrees of freedom at finite volume.

Turning on a small 't Hooft coupling $\lambda \equiv g_{YM}^2 N$ qualitatively alters the behavior of the free gauge theory. It turns out that the interacting theory displays one of two possible behaviors, depending on the sign of a coefficient determined by two-loop and three-loop vacuum diagrams. When

[4] Sundborg's results do not seem to have received wide attention; in particular, we learned of his paper only after we had independently rederived all of its results, which have substantial overlap with section 3 and parts of section 5 of our paper.

varying the temperature, this theory either undergoes a single first order transition *below* the Hagedorn temperature, or it undergoes two continuous phase transitions, one at the Hagedorn temperature and the second one slightly above it. Note that, in the first case, the phase transition shields the Hagedorn spectrum of the theory in the sense that there is no temperature at which string excitations of arbitrarily high energy dominate the partition function. In the second case the first of the two phase transitions is truly Hagedorn-like; the singularities in the neighborhood of the phase transition directly encode the high-energy density of states of stringy oscillators.

We derive the results described in the previous paragraphs by reducing the thermal partition function to an integral over a single unitary $U(N)$ (or $SU(N)$) matrix

$$Z(\beta) = \int [dU] \exp[-S_{eff}(U)], \tag{1.1}$$

where $U = e^{i\beta\alpha}$ (α is the zero mode of A_0 on $\mathcal{M} \times S^1$, $\mathcal{M}$ is the spatial manifold in question, and $\beta \equiv 1/T$). The mode α is the lightest mode [5] of the gauge theory on $\mathcal{M} \times S^1$; $S_{eff}(U)$, the quantum effective action for this light mode, may be computed by integrating out all other fields in the theory. This procedure (integrating out the heavy modes) may be explicitly performed in perturbation theory, generating an expression for $S_{eff}(U)$ as a power series in the 't Hooft coupling λ. The lowest term in this power series, corresponding to the free gauge theory, is particularly easy to compute, either by evaluating one-loop vacuum diagrams or by explicitly counting gauge-invariant states in free Yang-Mills theory using a projection onto the singlet sector of the theory.[6] For the free theory we obtain an expression (of the form (1.1)) for $Z(\beta)$ which is exact, even at finite N, and applies to any gauge theory with any matter content. The computation of higher order terms is straightforward in principle but tedious in practice.

Having computed S_{eff} we proceed to evaluate (1.1) in the usual manner, by changing the integration variable to the eigenvalues $e^{i\theta}$ of U. For the adjoint theories we consider, the resulting effective action in the free limit is simply the sum over a pairwise potential between the eigenvalues with a temperature-independent repulsive term and an attractive term that increases from zero to infinite strength as the temperature increases from zero to infinity. In the large N limit, the integral is determined by a saddle point characterized by a density of eigenvalues $\rho(\theta)$ that minimizes the effective

[5] We assume that there are no additional zero modes for any of the fields.
[6] This second method was already used to derive similar expressions in [4,6].

action. At sufficiently low temperatures, the repulsive term dominates and $\rho(\theta)$ is constant on the circle. At high enough temperatures, the attractive term forces the eigenvalues to bunch together and $\rho(\theta)$ vanishes outside a narrow interval on the circle [4]. In the free theory, these two regimes are separated by a first order transition at the Hagedorn temperature in which the eigenvalue distribution jumps discontinuously from the uniform distribution to a sinusoidal distribution. However, it turns out that depending on the details of the interactions in S_{eff}, the nature of this transition may be modified at arbitrarily weak non-zero coupling.

Based on the general form of S_{eff}, we show that two classes of behavior are possible at weak coupling, as noted earlier. In the first case, we have a single first order phase transition at which the uniform low temperature distribution jumps discontinuously to a non-uniform distribution. This is similar to the behavior in the free theory, but the transition happens strictly below the Hagedorn temperature for non-zero coupling. The second possibility is that the eigenvalue distribution evolves continuously as a function of temperature, changing from a uniform to a non-uniform (but nowhere vanishing) distribution at a first critical temperature, and then developing a gap (on which the distribution vanishes) at a second critical temperature. Consequently, the corresponding theory undergoes two phase transitions as a function of temperature; the first of these is a second order transition at the Hagedorn temperature, while the second is a third order transition closely related to the Gross-Witten phase transition [7] of two dimensional lattice gauge theories.

Which of these two possibilities is realized depends on the sign of a particular coefficient in S_{eff} which appears to depend on the details of the field content and interactions of the theory in question. We are currently involved in a computation of this coefficient for the $d = 4$ pure Yang-Mills theory and for the $d = 4$ $\mathcal{N} = 4$ supersymmetric Yang-Mills (SYM) theory; we hope to report the result of this computation soon [8].

The phase transition at (or near) the Hagedorn temperature has a natural interpretation in the gauge theory,[7] it is simply a deconfining transition. The low-temperature phase can be thought of as a gas of singlet 'glueballs', with the free energy scaling as N^0, while the high temperature phase, with the free energy scaling as N^2, is a plasma of gluons (and other particles).

[7] As we have described above, in some situations the theory undergoes two phase transitions upon raising the temperature. The discussion in this paragraph applies to the first of these.

Indeed, the traditional deconfinement order parameter, the Polyakov loop in the fundamental representation $\frac{1}{N}\langle \mathrm{tr}(Pe^{i\oint A_0})\rangle \simeq \frac{1}{N}\langle \mathrm{tr}(U)\rangle$, is a good order parameter for the phase transitions described in this paper. It is zero in the confining phase, and non-zero in the deconfining phase (see Subsec 4.5 for caveats to this statement).

The phase transitions we discuss in this paper appear to be continuously related to several other interesting phase transitions. In the case of the 3+1-dimensional $\mathcal{N} = 4$ $SU(N)$ SYM theory compactified on a sphere, our analysis applies to the deconfinement transition at weak 't Hooft coupling; the deconfinement transition at strong coupling is related [9] by the AdS/CFT correspondence [10] to the Hawking-Page transition [11] of gravitational theories on asymptotically anti-de Sitter (AdS) spaces. For asymptotically free gauge theories, our weakly coupled results apply when the scale R of the compactification manifold is much smaller than the strong coupling scale $1/\Lambda_{QCD}$. However, when we take $R\Lambda_{QCD} \gg 1$, our phase transitions should turn into the infinite volume deconfinement transition.

We propose a set of phase diagrams for gauge theories as a function of coupling (λ or $R\Lambda_{QCD}$, respectively) and temperature that interpolate between our weakly coupled results and the known strong coupling behavior. We are led to speculate on a dual interpretation of deconfinement transitions; in particular, we conjecture that deconfinement transitions are always associated with black hole formation in a dual string theory. We argue that the mysterious intermediate temperature phase associated with second order deconfinement transitions (if such a phase is realized in some theory) would be dual to a string theory in a background dominated by a strange new type of stable black hole.

The contribution is organized as follows. In §2, we show that large N free gauge theories have a Hagedorn-like behavior of their spectrum and provide a first derivation of the matrix model expression for the thermal partition function of compactified gauge theories by explicitly summing over gauge-invariant states. In §3, we derive the same expression by using a path integral formalism. In §4, we describe the solution of this matrix model, exhibit the phase transition, and both compute and discuss the relevant order parameters. In §5, we generalize the path integral derivation of (1.1) to interacting gauge theories and determine the general form of S_{eff} in perturbation theory. Utilizing these results, we discuss the different possibilities for the phase structure of the theory at weak coupling. In §6, we suggest extrapolations of our results to strong coupling. We also speculate on the interpretation of deconfinement transitions in terms of black hole formation in a dual string theory.

2. Partition function for free Yang-Mills theory on a compact space

We now proceed to directly analyze the thermodynamics of weakly coupled Yang-Mills theory on a compact space, beginning in this section (and the next two) with the free theory. For simplicity, we restrict here to spaces (such as S^{d-1}) for which all modes of the various fields are massive. In most of this section we will discuss the specific case of a $U(N)$ gauge theory with fields in the adjoint representation, but it is easy to generalize our arguments to more general theories. In particular, our exact results for the partition function in §2.3 will be completely general. As usual, the behavior of thermodynamic quantities in the canonical ensemble is governed by the partition function

$$Z(\beta) = \sum_{physical\ states} x^{E_i} = \int \rho(E) x^E dE \,, \tag{2.1}$$

where $\rho(E)$ is the density of states and we define $x \equiv e^{-\beta} = e^{-1/T}$. As we discussed earlier, it is essential to keep in mind that the constraint from Gauss' law for a gauge theory on a compact space implies that the physical states over which we sum must be gauge-invariant.

2.1. *Two matrix harmonic oscillators give Hagedorn behavior for large N*

By expanding the fields into modes on the compact space (e.g. spherical harmonics on S^3), our field theory may be viewed as a quantum mechanical system with infinitely many degrees of freedom (one for each mode). In the free theory, these degrees of freedom are decoupled harmonic oscillators, each in some representation of the gauge group. Before embarking on our detailed analysis of this theory, we present a simple demonstration that even restricting to two such oscillators in the adjoint representation of $U(N)$ yields a theory that displays Hagedorn-like behavior in the large N limit.

Thus, we consider two such modes, each with unit energy, created by operators $A^\dagger$ and $B^\dagger$ in the adjoint of $U(N)$. Physical states are gauge-invariant and correspond to traces of products of these operators acting on the Fock space vacuum. Single-trace states of energy E are specified by a series of E $A^\dagger$'s and $B^\dagger$'s inside the trace. The number of such states satisfies

$$2^E/E < n(E) \leq 2^E\,, \tag{2.2}$$

where the upper bound arises since for each of the E positions in the trace we may choose $A^\dagger$ or $B^\dagger$, and the lower bound comes because cyclicity of the trace equates a given state to at most $E-1$ others. In the large N theory all of these states are independent, so the density of independent single-trace states is Hagedorn-like :

$$\rho(E) = \frac{\partial n}{\partial E} \sim E^\alpha e^{\beta_H E}\,, \tag{2.3}$$

where the Hagedorn temperature in this case is $T_H = 1/\beta_H = 1/\ln(2)$.

For finite N, we cannot have Hagedorn behavior at all energies, since the high temperature behavior must be that of a field theory (or quantum mechanics in the toy example described above) with finitely many fields. The departure from Hagedorn behavior at high energies comes about because the oscillators have only finitely many gauge-invariant degrees of freedom (of order N^2), so that only a number of traces of order N^2 give independent states. Thus, at energies of order N^2, trace relations will cut off the exponential growth of states, resulting in a cross-over to field theory (or quantum mechanics) behavior. In the limit of large N, this cross-over becomes a sharp transition, as we will see explicitly in the next sections.

2.2. *Exact partition function for* $N = \infty$

We now show that the exact partition function of our free gauge theory in the strict $N = \infty$ limit (where no trace relations exist) may be obtained by simple counting arguments, following [4, 5].

We consider a model with $m \geq 2$ matrix-valued bosonic harmonic oscillators; the number m can be finite or infinite, as in the example of a compactified d-dimensional gauge theory. Let the ith oscillator have energy E_i. We encode the spectrum of oscillators in a "single-particle partition function" $z(x) \equiv \sum_i x^{E_i}$ where the sum goes over all oscillators.[8] Examples of such single-particle partition functions for compactified $3+1$ dimensional theories will be provided in §2.4 below.

[8] To be precise, this is the single-particle partition function for the $U(1)$ theory, since in general the single oscillator states are non-physical.

Following the discussion above, the partition function of single-trace states with k oscillators is given by

$$Z_k = \frac{z(x)^k}{k} + (positive) \ , \tag{2.4}$$

where the factor of $1/k$ compensates for cyclicity in the trace, and the $+(positive)$ acknowledges that this is an over-compensation in cases where the trace breaks up into repeated sequences of oscillators. Since in the large N limit a single-trace state can have any number of oscillators, we need to sum this result over k to find a partition function for single-trace states,

$$Z_{ST} = \sum_{k=1}^{\infty} Z_k = -\ln(1 - z(x)) + (positive) \ . \tag{2.5}$$

Actually, it is not too difficult to promote (2.5) to an exact expression which correctly accounts for repetitions inside the trace. This careful counting utilizes Polya theory and was already performed in [4,5] - see Appendix B of [12] for a review. The exact result in the large N limit is

$$Z_{ST} = -\sum_{q=1}^{\infty} \frac{\varphi(q)}{q} \ln(1 - z(x^q)) \, , \tag{2.6}$$

where $\varphi(q)$ represents the number of positive integers which are not larger than q and are relatively prime to q. Note that the first term in (2.6) is precisely the term we explicitly wrote down in (2.5).

Using (2.6), we may now write the full partition function of our model, summing over states with arbitrarily many traces. The space of multi-trace states is simply the Fock space of single-trace states, so the full multi-trace partition function Z is easily obtained from Z_{ST} (taking into account the fact that the single-trace states behave as identical bosonic particles),

$$\begin{aligned} \ln(Z) &= \sum_{n=1}^{\infty} \frac{1}{n} Z_{ST}(x^n) = -\sum_{n=1}^{\infty} \frac{1}{n} \sum_{q=1}^{\infty} \frac{\varphi(q)}{q} \ln(1 - z(x^{qn})) = \\ &= -\sum_{k=1}^{\infty} \frac{1}{k} (\sum_{q|k} \varphi(q)) \ln(1 - z(x^k)) = -\sum_{k=1}^{\infty} \ln(1 - z(x^k)) \, , \end{aligned} \tag{2.7}$$

where we have used the formula $\sum_{q|k} \varphi(q) = k$. It is easy to generalize this result to include also fermionic oscillators; denoting by z_B the single-particle partition function of the bosonic oscillators, and by z_F the single-particle partition function of the fermionic oscillators, we obtain

$$\ln(Z) = -\sum_{k=1}^{\infty} \ln(1 - z_B(x^k) + (-1)^k z_F(x^k)) \ . \tag{2.8}$$

The inclusion of fermions does not change the qualitative behavior, so it is sufficient to assume that all oscillators are bosonic for the purposes of qualitative discussions.

Equation (2.8) gives the exact partition function in the strict $N = \infty$ theory. It is not correct for the finite N theory because we have assumed all traces to be independent.[9] However, as long as the thermodynamics is dominated by states for which the number of oscillators in a trace is small compared to N^2 (which will certainly be true for low enough temperatures), trace relations will be unimportant and (2.8) should give a very good approximation to the correct behavior.

In order to understand the dynamics of (2.7) we note some obvious properties of the "single-particle partition function" $z(x)$, namely that $z(0) = 0$, $z(x)$ is a monotonically increasing function of x, and that (since $m \geq 2$) $z(1) > 1$. It follows that the equation $z(x) = 1$ has a unique solution (which we will denote by $x = x_H$) for $0 < x < 1$. If we denote the corresponding inverse temperature by β_H ($e^{-\beta_H} = x_H$), then $\ln(Z(\beta))$ is well defined for $\beta > \beta_H$, but diverges like $-\ln(\beta - \beta_H)$ as β tends to β_H from above. For $\beta < \beta_H$ (high temperatures) $Z(\beta)$ in (2.7) is ill defined. All these features are reproduced by the formula where $\rho(E) \approx e^{\beta_H E}$ at high energies. Consequently, our system of oscillators has Hagedorn-like thermodynamics with the inverse Hagedorn temperature β_H. Note that $z(x) = 2x$ for the two oscillator model of the previous subsection, yielding $x_H = 1/2$ and $\beta_H = \ln(2)$ in agreement with the discussion in that subsection.

The fact that $Z(\beta)$ diverges for $T > T_H = \beta_H^{-1}$ indicates that T_H is a limiting temperature if N is set strictly equal to infinity. Of course, this cannot be correct at any finite N, no matter how large. Indeed, a system of mN^2 harmonic oscillators (or even a compactified field theory for which m is infinite but there is a finite number of oscillators below any fixed energy) can certainly be heated to any temperature. It must instead be the case that at high temperatures ($T > T_H$), the condition for the validity of (2.8) (namely, that the number of fields contained in the typical state is much smaller than N^2) fails; we will see in §4 that this is indeed the case. Thus, in the high temperature phase, trace relations invalidate the analysis above, and traces no longer provide a useful basis for $U(N)$ singlets in the high temperature phase. This is true at any finite N and so it is also true in

[9] When the number of fields in a trace is of order N^2 trace identities relate various states that are assumed to be different in (2.7). For example, any 2×2 Hermitian matrix M obeys $\mathrm{tr}(M^3) = \frac{1}{2}(3\mathrm{tr}(M)\mathrm{tr}(M^2) - \mathrm{tr}(M)^3)$.

the large N limit of principal interest to us. In the next subsection we therefore turn to a more generally applicable method of dealing with the $U(N)$ singlet condition.

2.3. *Exact partition function for free Yang-Mills theory*

In this section, we will derive an exact expression for the partition function of free Yang-Mills theory with arbitrary gauge group and matter content, on any compact space for which all modes of the various fields are massive. As we have discussed, the basic excitations of such a theory are single-particle states, corresponding to oscillators which arise by expanding the various fields into physical modes on the space. The general physical state is obtained by acting with arbitrary collections of these oscillators on the Fock space vacuum, with the constraint that the total state should be a singlet of the gauge group.

To obtain an explicit expression for the partition function, suppose we have bosonic modes with energies E_i in representations R_i of the gauge group and fermionic modes with energies E'_i in representations R'_i of the gauge group. Then, the partition function may be expressed as a sum over the occupation numbers of all modes, with a Boltzmann factor $e^{-\beta E}$ corresponding to the total energy, and a numerical factor counting the number of singlets (physically allowed states) in the corresponding product of representations,

$$\begin{aligned} Z(x) = \sum_{n_1=0}^{\infty} x^{n_1 E_1} \sum_{n_2=0}^{\infty} x^{n_2 E_2} \cdots \sum_{n'_1=0}^{\infty} x^{n'_1 E'_1} \sum_{n'_2=0}^{\infty} x^{n'_2 E'_2} \cdots \times \\ \{\#\text{ of singlets in } sym^{n_1}(R_1) \otimes sym^{n_2}(R_2) \otimes \cdots \otimes \\ \otimes\, anti^{n'_1}(R'_1) \otimes anti^{n'_2}(R'_2) \otimes \cdots \}\,. \end{aligned} \tag{2.9}$$

Note that in order to correctly account for particle statistics, we must symmetrize (antisymmetrize) representations corresponding to identical bosonic (fermionic) modes. An explicit expression for the group theory factor may be obtained by using the fact that the number of singlets in a product of representations is simply the integral over the group manifold of the product of characters for these representations, as reviewed in appendix A of [12]. Thus, we find

$$Z(x) = \int [dU] \prod_i \left\{ \sum_{n_i=0}^{\infty} x^{n_i E_i} \chi_{sym^{n_i}(R_i)}(U) \right\} \times$$

$$\times \prod_i \left\{ \sum_{n'_i=0}^{\infty} x^{n'_i E'_i} \chi_{anti^{n'_i}(R'_i)}(U) \right\} . \qquad (2.10)$$

The sums in brackets are generating functions for the characters of symmetrized (antisymmetrized) products of arbitrarily many copies of a given representation R_i (R'_i). As is shown in appendix A of [12], these turn out to have simple expressions given by equation (A.8). Using this equation, we find that

$$Z(x) = \int [dU] \exp \left\{ \sum_{i;m=1}^{\infty} \frac{1}{m} x^{mE_i} \chi_{R_i}(U^m) \right\} \times$$
$$\times \exp \left\{ \sum_{i;m=1}^{\infty} \frac{(-1)^{m+1}}{m} x^{mE'_i} \chi_{R'_i}(U^m) \right\} . \qquad (2.11)$$

Finally, since the sums over i in the exponentials are simply sums over the single-particle modes, it is convenient to define bosonic and fermionic single-particle partition functions for each representation R, counting the single-particle states in this representation without the degeneracy coming from the dimension of the representation and without any gauge-invariance constraints,

$$z_B^R(x) = \sum_{R_i=R} x^{E_i} , \qquad z_F^R(x) = \sum_{R'_i=R} x^{E'_i} . \qquad (2.12)$$

With these definitions, we may express our final result for the partition function as [4]

$$Z(x) = \int [dU] \exp \left\{ \sum_R \sum_{m=1}^{\infty} \frac{1}{m} \left[z_B^R(x^m) + (-1)^{m+1} z_F^R(x^m) \right] \chi_R(U^m) \right\} . \qquad (2.13)$$

Thus, the partition function for an arbitrary free field theory (with no zero modes) on a compact space may be expressed as a single-matrix integral, with the action determined in a simple way by the single-particle modes of the theory.

2.4. $U(N)$ gauge theories with adjoint matter on $S^3 \times \mathbb{R}$

For most of the remainder of this talk, we will focus on $U(N)$ gauge theories on a sphere with adjoint matter. In this case, using $\chi_{adj}(U) = \mathrm{tr}(U)\mathrm{tr}(U^\dagger)$

(where we denote by tr without a subscript the trace in the fundamental representation), we obtain the unitary matrix model

$$Z(x) = \int [dU] \exp\left\{ \sum_{m=1}^{\infty} \frac{1}{m}(z_B(x^m) + (-1)^{m+1} z_F(x^m)) \mathrm{tr}(U^m)\mathrm{tr}((U^\dagger)^m) \right\} . \tag{2.14}$$

It is straightforward to work out explicit expressions for the single particle partition functions for various types of fields on a sphere of unit radius. In appendix B of [12] (see also [4, 5]) we find that for (3 + 1)-dimensional scalars, vectors, and chiral fermions on $S^3 \times \mathbb{R}$,

$$z_{S4}(x) = \frac{x + x^2}{(1-x)^3}, \qquad z_{V4}(x) = \frac{6x^2 - 2x^3}{(1-x)^3}, \qquad z_{F4}(x) = \frac{4x^{\frac{3}{2}}}{(1-x)^3} . \tag{2.15}$$

In a 3 + 1 dimensional free $U(N)$ gauge theory with n_S scalar fields, n_V vector fields and n_F chiral fermions (all in the adjoint representation), we should use in (2.14) $z_B(x) = n_S z_{S4}(x) + n_V z_{V4}(x)$, $z_F(x) = n_F z_{F4}(x)$. Note that all of these functions (and the single-particle partition functions in general) increase monotonically from $z = 0$ at $x = 0$ (zero temperature), and in all dimensions above 0 + 1 they diverge at $x = 1$ (infinite temperature).[10]

For $SU(N)$ gauge theories, the only difference (beyond the fact that U must have determinant one) is that every $\mathrm{tr}(U)\mathrm{tr}(U^\dagger)$ in (2.14) should be replaced everywhere by $(\mathrm{tr}(U)\mathrm{tr}(U^\dagger) - 1)$, the character of the adjoint representation in $SU(N)$.

3. Path integral derivation of the matrix integral and an order parameter

In the previous section we computed the partition function of free Yang-Mills theory on a compact space $\mathcal{M}$ by counting gauge-invariant states. However, $Z(T)$ may also be computed by the vacuum path integral of Euclidean Yang-Mills theory [11] on $\mathcal{M} \times S^1_T$ where the circumference of the thermal circle S^1_T is $\beta = 1/T$. In this section we will rederive the expression (2.14) for the partition function by computing this path integral for gauge theories on an S^3 of unit radius.

[10] For a d dimensional field theory compactified on S^{d-1}, $z(x)$ diverges as $2\mathcal{N}^{dof}/(1-x)^{d-1}$, where $\mathcal{N}^{dof}$ is the number of single-particle degrees of freedom.

[11] We thank Nima Arkani Hamed for emphasizing the utility of this approach to us, and for several extremely enjoyable and productive conversations on related issues.

Though it gives an identical final result for the partition function, the path integral derivation provides a physical interpretation for the unitary matrix U as the Wilson loop of the gauge field (averaged over the compact space) around the thermal circle. Consequently, $\frac{1}{N}\mathrm{tr}(U)$ is precisely the Polyakov loop operator; a standard order parameter for deconfinement. We will see in the next section that $\frac{1}{N}\mathrm{tr}(U)$ is also the natural order parameter for a large N phase transition of the matrix model (2.14).

A further advantage of the path integral derivation is that it generalizes easily to the calculation of the large N partition function $Z(T,\lambda)$ at weak 't Hooft coupling $\lambda \equiv g_{YM}^2 N$ when the theory is no longer free. In §5.2 we analyze the general structure of $S_{eff}(U,\lambda)$ at small λ. An explicit calculation of the weak-coupling partition function, in specific gauge theories, is postponed to a future paper [8].

3.1. *Basic set-up*

We work in the gauge

$$\partial_i A^i = 0\,, \tag{3.1}$$

where $i = 1, 2, 3$ runs over the sphere coordinates, and ∂_i are covariant derivatives. The choice (3.1) fixes the gauge freedom only partially; it leaves spatially independent but time dependent gauge transformations unfixed. We fix this residual gauge-invariance with the condition

$$\partial_t \alpha(t) = 0\,, \tag{3.2}$$

where

$$\alpha \equiv \frac{1}{\omega_3}\int_{S^3} A_0\,, \tag{3.3}$$

where ω_3 is the volume of S^3.

The mode α will play a special role in what follows because it is the only zero mode (mode whose action vanishes at quadratic order) in the decomposition of Yang-Mills theory into Kaluza-Klein modes on $S^3 \times S^1$. Consequently, α fluctuations are strongly coupled at every value of λ, including in the limit $\lambda \to 0$. In particular, they cannot directly be integrated out in perturbation theory.

In order to proceed with the perturbative evaluation of the partition function we will adopt a two step procedure. In the first step, discussed in this section, we integrate out all non-zero modes [12] and generate an effective

[12] As α is a zero mode this is a standard Wilsonian procedure.

action $S_{eff}(\alpha)$ for α. $S_{eff}(\alpha)$ is non-trivial even at zero coupling and, as we will discuss in §5, it is further modified perturbatively in λ. Once we have obtained $S_{eff}(\alpha)$, we may then proceed to perform the integral over α in order to obtain $Z(T)$, cf. [12].

3.2. *The integration measure*

In this subsection we will define $S_{eff}(\alpha)$ more carefully. Recall that the Yang-Mills free energy may be written as

$$e^{-\beta F} = \int d\alpha \int \mathcal{D}A\Delta_1\Delta_2 e^{-S_{YM}(A,\alpha)}\,, \tag{3.4}$$

where Δ_1 is the Faddeev-Popov determinant conjugate to (3.1), Δ_2 is the Faddeev-Popov determinant conjugate to (3.2), and S_{YM} is the Yang-Mills action. It is not difficult to explicitly evaluate Δ_2 (see later in this subsection) and verify that it is independent of A. Consequently, (3.4) may be rewritten as

$$e^{-\beta F} = \int d\alpha\Delta_2 \exp[-S_{eff}(\alpha)]\,, \tag{3.5}$$

where

$$\exp[-S_{eff}(\alpha)] = \int \mathcal{D}A\Delta_1 \exp[-S_{YM}(A,\alpha)]\ . \tag{3.6}$$

In the rest of this subsection, we will explicitly evaluate Δ_2 and so determine the effective measure of integration (3.5).

It follows from (3.2) that

$$\Delta_2 = \det{}'\left(\partial_0 - i\,[\alpha, *]\right)\,, \tag{3.7}$$

where the prime asserts that the determinant is over non-zero modes of A_0. Denoting by λ_i $(i = 1, \cdots, n)$ the eigenvalues of α, and choosing a convenient basis of matrix functions whose time-dependence is given by $\exp(2\pi int/\beta)$, the determinant is easily evaluated as the product

$$\begin{aligned}\Delta_2 &= \prod_{n\neq 0}\prod_{i,j}\left[\frac{2\pi in}{\beta} - i\,(\lambda_i - \lambda_j)\right] = \\ &= \left(\prod_{m\neq 0}\frac{2\pi im}{\beta}\right)\prod_{i,j}\frac{2}{\beta\,(\lambda_i-\lambda_j)}\sin\left(\frac{\beta\,(\lambda_i-\lambda_j)}{2}\right)\ .\end{aligned} \tag{3.8}$$

Notice that up to an overall constant,

$$d\alpha\Delta_2 = [dU]\,, \tag{3.9}$$

where

$$d\alpha = \prod_i d\lambda_i \prod_{i<j} (\lambda_i - \lambda_j)^2 \tag{3.10}$$

is the left-right invariant integration measure over Hermitian matrices α, and

$$[dU] = \prod_i d\lambda_i \prod_{i<j} \sin^2\left(\frac{\beta(\lambda_i - \lambda_j)}{2}\right) \tag{3.11}$$

is the left-right invariant integration measure in the integral over the unitary matrices

$$U \equiv e^{i\beta\alpha} . \tag{3.12}$$

We will see in the next section at one loop order and argue generally in §5 that S_{eff} may be regarded as a function of U rather than α, so that (3.5) may be written as

$$e^{-\beta F} = \int [dU] e^{-S_{eff}(U)} , \tag{3.13}$$

where $S_{eff}(U)$ is defined in (3.6).

3.3. *Evaluation of S_{eff} at one-loop*

The path integral in (3.5) may be evaluated diagrammatically, generating an expansion of $S_{eff}(U)$ in powers of the gauge coupling. In this subsection we use this method to evaluate $S_{eff}(U)$ to lowest order in λ ($\mathcal{O}(\lambda^0)$). This evaluation is rather simple as only one-loop graphs in the expansion of (3.5) contribute to S_{eff} at this order.

We first consider the pure gauge theory. The Faddeev-Popov determinant for gauge-fixing (3.1) is

$$\det \partial_i D^i = \int \mathcal{D}c\mathcal{D}\bar{c} e^{-\bar{c}\partial_i D^i c} , \tag{3.14}$$

where D^i denotes a gauge covariant derivative

$$D_i c = \partial_i - i[A_i, c] , \tag{3.15}$$

and c and $\bar{c}$ are complex ghosts. Consequently, the action of the free gauge theory may be written as

$$e^{-S_{eff}(U)} = \int \mathcal{D}A_i \mathcal{D}A_0 \mathcal{D}c\mathcal{D}\bar{c}\delta(\partial_i A^i) \times$$

$$\times \ \exp\left[-\int \mathrm{Tr}\left(\frac{1}{2}A_i(\widetilde{D}_0^2+\partial^2)A^i+\frac{1}{2}A_0\partial^2 A_0+\bar{c}\partial^2 c\right)\right], \quad (3.16)$$

where

$$\widetilde{D}_0 X \equiv \partial_0 X - i[\alpha, X]\ . \quad (3.17)$$

To proceed further we note that any vector field on the sphere may be decomposed as

$$A_i = \partial_i\varphi + B_i\,, \quad (3.18)$$

where $\partial_i B^i = 0$. The integral over φ in (3.16) is easily performed using the δ function, yielding $1/\sqrt{\det'(\partial_i\partial^i)}$ where the derivatives act on scalar functions on S^3 and the prime denotes omission of the zero mode. The integral over A_0 yields the identical factor (the zero mode is α which is not integrated over). The integral over the ghosts, on the other hand, evaluates to $\det'(\partial_i\partial^i)$. These three factors cancel nicely, so that (3.16) simplifies to

$$e^{-S_{eff}(U)} = \int \mathcal{D}B_i \exp\left[-\frac{1}{2}\int \mathrm{Tr}\left(B_i(\widetilde{D}_0^2+\partial^2)B^i\right)\right]. \quad (3.19)$$

Thus,

$$S_{eff} = \frac{1}{2}\ln(\det(-\widetilde{D}_0^2-\partial^2))\,, \quad (3.20)$$

where the operator acts on the space of divergenceless vector functions on the sphere, i.e. the space of vector functions spanned by the vector spherical harmonics. Consequently

$$S_{eff} = \frac{1}{2}\sum_{\Delta} n(\Delta)\ln\det(-\widetilde{D}_0^2+\Delta^2)\,, \quad (3.21)$$

where Δ^2 are the eigenvalues of the Laplacian acting on vector spherical harmonics on the compact manifold and $n(\Delta)$ is the degeneracy of each eigenvalue. When the compact space is an S^3 of unit radius we have $\Delta = h+1$ (for integer $h \geq 0$) and $n(\Delta) = 2h(h+2)$ (see the counting of vector spherical harmonics in appendix B of [12]).

The determinant of $(-\widetilde{D}_0^2+\Delta^2)$ is easily evaluated by passing to Fourier space in the time direction, yielding the infinite product

$$\det_{U(N)}\left[\prod_{n=-\infty}^{\infty}\left(\frac{4\pi^2 n^2}{\beta^2}+\frac{4\pi n}{\beta}\alpha+\alpha^2+\Delta^2\right)\right]. \quad (3.22)$$

The infinite product of matrices may be rewritten as

$$
\begin{aligned}
(\alpha^2+\Delta^2)\ &\left[\prod_{m\neq 0}\frac{\beta}{2\pi m}\right]^{-2}\left[\prod_{n\neq 0}\left(1+\frac{\beta\alpha}{\pi n}+\frac{(\alpha^2+\Delta^2)\,\beta^2}{4\pi^2 n^2}\right)\right]= \\
&= (\alpha^2+\Delta^2)\left[\prod_{m\neq 0}\frac{\beta}{2\pi m}\right]^{-2}\times \\
&\times\left[\prod_{n=1}^{\infty}\left(1-\frac{\beta^2(\alpha+i\Delta)^2}{4\pi^2 n^2}\right)\left(1-\frac{\beta^2(\alpha-i\Delta)^2}{4\pi^2 n^2}\right)\right]= \\
&= \left[\prod_{m\neq 0}\frac{\beta}{2\pi m}\right]^{-2}\left(\frac{4}{\beta^2}\right)\sin\left[\frac{\beta(\alpha+i\Delta)}{2}\right]\sin\left[\frac{\beta(\alpha-i\Delta)}{2}\right]= \\
&= \frac{2}{\beta^2}\left[\prod_{m\neq 0}\frac{\beta}{2\pi m}\right]^{-2}[\cosh(\beta\Delta)-\cos(\beta\alpha)]= \\
&= \mathcal{N}e^{\beta\Delta}(1-e^{-\beta\Delta+i\beta\alpha})(1-e^{-\beta\Delta-i\beta\alpha})\,,
\end{aligned}
\tag{3.23}
$$

where

$$
\mathcal{N}=\left(\frac{1}{\beta^2}\left[\prod_{m\neq 0}\frac{\beta}{2\pi m}\right]^{-2}\right)\,. \tag{3.24}
$$

The divergent factor $\mathcal{N}$ is a constant independent of both Δ and α, and we set it to unity to reproduce the free energy of the harmonic oscillator. Thus

$$
\ln(\det(-\widetilde{D}_0^2+\Delta^2))=\mathrm{Tr}\left(\beta\Delta+\ln(1-e^{-\beta\Delta+i\beta\alpha})+\ln(1-e^{-\beta\Delta-i\beta\alpha})\right), \tag{3.25}
$$

where the trace is over $N^2\times N^2$ dimensional matrices and α acts in the adjoint representation. Expanding the logarithms in a power series, summing over Δ, and passing from the adjoint to the fundamental ($\mathrm{Tr}_{adj}(e^{in\beta\alpha})\to \mathrm{tr}(U^n)\mathrm{tr}(U^{-n})$) and using (3.20) we obtain

$$
S_{eff}=\frac{1}{2}\beta N^2\sum_{\Delta}\Delta n(\Delta)-\sum_{n=1}^{\infty}\frac{z_V(x^n)}{n}\mathrm{tr}(U^n)\mathrm{tr}(U^{-n})\,. \tag{3.26}
$$

For the case of an S^3 of unit radius the first term (appropriately regularized) is equal to $\frac{11}{120}\beta N^2$, where $\frac{11}{120}$ is the Casimir energy $\sum_0^\infty h(h+1)(h+2)$ for a vector field on the unit sphere.

It is a simple matter to generalize this calculation to include the contributions from free conformally coupled scalar and spinor fields. When the compact manifold is S^3, the contribution to S_{eff} from a single additional scalar field is

$$\delta S_{eff} = \frac{1}{2}\ln(\det(-\widetilde{D}_0^2 - \partial^2 + 1)), \tag{3.27}$$

where the operator acts on scalar fields on $S^3 \times S^1$ and the constant piece of the operator is a consequence of the $\mathcal{R}\varphi^2$ term in the Lagrangian for conformally coupled scalars. As above, this determinant is easily evaluated to yield

$$S_{eff} = \frac{1}{240}\beta N^2 - \sum_{n=1}^{\infty} \frac{z_S(x^n)}{n}\mathrm{tr}(U^n)\mathrm{tr}(U^{-n}), \tag{3.28}$$

where $\frac{1}{240} = \frac{1}{2}\sum_{n=0}^{\infty}(n+1)^3$. A Weyl spinor field on S^3 contributes [13]

$$\begin{aligned} S_{eff} &= -\ln(\det(-\widetilde{\partial}_0\gamma^0 - \not{\partial})) = \\ &= \frac{17}{960}\beta N^2 - \sum_{n=1}^{\infty} \frac{(-1)^{n+1} z_F(x^n)}{n}\mathrm{tr}(U^n)\mathrm{tr}(U^{-n}), \end{aligned} \tag{3.29}$$

where $\frac{17}{960} = -\sum_{n=1}^{\infty} n(n+1)(n+\frac{1}{2})$.

Thus, we have rederived the expression for the free partition function (2.14), but this time with a physical interpretation for the matrix U as the *holonomy around the time circle*, and for $\frac{1}{N}\mathrm{tr}(U)$ as a Wilson loop around the time circle, or a Polyakov loop (in the fundamental representation).

4. Solution of the free Yang-Mills matrix model

In this section we proceed to directly analyze the unitary matrix model (2.14) for the exact partition function of free $U(N)$ Yang-Mills theory in order to extract the thermodynamic behavior.

We begin by recalling that any unitary matrix model with gauge-invariant action and measure may be rewritten entirely in terms of the eigenvalues of the unitary matrix, which must lie on the unit circle. Denoting these eigenvalues by $\{e^{i\alpha_i}\}$ (with $-\pi < \alpha_i \le \pi$), we may rewrite the partition function (2.14) in terms of the eigenvalues by the replacements

$$\int [dU] \to \prod_i \int_{-\pi}^{\pi} [d\alpha_i] \prod_{i<j} \sin^2\left(\frac{\alpha_i - \alpha_j}{2}\right); \qquad \mathrm{tr}(U^n) \to \sum_j e^{in\alpha_j}. \tag{4.1}$$

---

[13] We take the spinor field to be anti-periodic on the S^1 in order to compute the trace of $e^{-\beta H}$ rather than the trace of $(-1)^F e^{-\beta H}$.

With only adjoint matter, we find from (2.14) that the effective theory for the eigenvalues is governed entirely by a pairwise potential,

$$Z(x) = \int [d\alpha_i] e^{-\sum_{i\neq j} V(\alpha_i - \alpha_j)}, \tag{4.2}$$

where

$$\begin{aligned} V(\theta) &= -\ln|\sin(\theta/2)| - \sum_{n=1}^{\infty} \frac{1}{n}\left[z_B(x^n) + (-1)^{n+1} z_F(x^n)\right]\cos(n\theta) = \\ &= \ln(2) + \sum_{n=1}^{\infty} \frac{1}{n}(1 - z_B(x^n) - (-1)^{n+1} z_F(x^n))\cos(n\theta)\,. \end{aligned} \tag{4.3}$$

In the first line, the first term coming from the measure is a temperature-independent repulsive potential, while the remaining terms provide an attractive potential which increases from zero to infinite strength as the temperature increases from zero to infinity. This suggests that at low temperatures, the minimum action configurations will correspond to eigenvalues distributed evenly around the circle, while at high temperatures, the eigenvalues will tend to bunch up.

For finite N, the partition function receives contributions from all configurations of eigenvalues and depends smoothly on the temperature. On the other hand, it is well known that in the large N limit, the matrix model partition function is dominated by the minimum action configurations, with the exact leading and subleading terms in the $1/N$ expansion of the free energy given, respectively, by the minimum of the action and by the Gaussian integral about the minimum action configuration. When this minimum action configuration (or its derivatives) changes abruptly as a function of temperature, we may have a phase transition in the large N limit (see, for example, [7]). We will now see that exactly this behavior occurs and leads to a phase transition for our model.

We begin in §4.1 by analyzing the low temperature phase. In §4.2 we analyze the behavior near the phase transition temperature, and in §4.3 we analyze the high temperature limit. In §4.4 we summarize the thermodynamical behavior of the theory. In §4.5 we discuss how the Polyakov loop may be used as an order parameter at finite volume.

4.1. *Low temperature behavior*

Consider first the theory at low temperatures, where the effective potential between the eigenvalues is dominated by the repulsive term. In this case, we expect that the pairwise repulsion drives the eigenvalues to spread uniformly around the circle. In fact, since there is no difference between displacing any individual eigenvalue in the uniform distribution to the left and displacing it to the right, the uniform eigenvalue distribution will always be a stationary point of the action for a pairwise potential. To see when this stationary point is a minimum, it is convenient to introduce an eigenvalue distribution $\rho(\theta)$ proportional to the density of eigenvalues $e^{i\theta}$ of U at the point θ. Note that ρ must be everywhere non-negative, and we may choose its normalization so that

$$\int_{-\pi}^{\pi} d\theta \rho(\theta) = 1 . \tag{4.4}$$

With this definition, the effective action for the eigenvalues becomes

$$\begin{aligned} S[\rho(\theta)] &= N^2 \int d\theta_1 \int d\theta_2 \rho(\theta_1)\rho(\theta_2) V(\theta_1 - \theta_2) = \\ &= \frac{N^2}{2\pi} \sum_{n=1}^{\infty} |\rho_n|^2 V_n(T) , \end{aligned} \tag{4.5}$$

where in the second line, we have defined $\rho_n \equiv \int d\theta \rho(\theta) \cos(n\theta)$ and $V_n \equiv \int d\theta V(\theta) \cos(n\theta)$ to be the Fourier modes on the circle of ρ and V, respectively, and we assume without loss of generality that the eigenvalue distribution is symmetric around $\theta = 0$. From the latter expression, it is clear that the uniform distribution ($\rho_{n\geq 1} = 0$) will be an absolute minimum of the potential as long as all V_n are positive. From (4.3) we see that

$$V_n = \frac{2\pi}{n}(1 - z_B(x^n) - (-1)^{n+1} z_F(x^n)) , \tag{4.6}$$

so the uniform distribution is an absolute minimum if and only if

$$z_B(x^n) + (-1)^{n+1} z_F(x^n) < 1 \tag{4.7}$$

for all n. But, since the single-particle partition functions are monotonically increasing, and $0 \leq x < 1$, the $n = 1$ condition is always strongest, and the uniform distribution will be stable for temperatures $T < T_H = -1/\ln(x_H)$ where x_H is the solution to

$$z(x_H) \equiv z_B(x_H) + z_F(x_H) = 1 ; \tag{4.8}$$

note that this is precisely the Hagedorn temperature we discussed in §2.2. As long as we have more than a single quantum-mechanical mode, equation (4.8) always has a unique solution with $0 < x < 1$, so the uniform distribution becomes an unstable extremum beyond some finite temperature T_H determined by the single-particle partition function.

For $T < T_H$, we may now evaluate the free energy at leading and subleading order in N. Since $\mathrm{tr}(U^n)$ vanishes for any $n \geq 1$ for the uniform distribution, the classical value for the action (2.14) (and thus the leading $\mathcal{O}(N^2)$ contribution to the free energy) vanishes. The first non-zero contribution to the free energy thus arises from the Gaussian integral around this configuration. From the quadratic action (4.5),(4.6), we see that the leading contribution to the partition function is [14]

$$Z(x) = \prod_{n=1}^{\infty} \frac{1}{1 - z_B(x^n) - (-1)^{n+1} z_F(x^n)}, \tag{4.9}$$

where the normalization has been arbitrarily fixed by choosing the vacuum free energy to vanish, $Z(x = 0) = 1$ (more physically, we could choose $Z(0)$ to account for the Casimir energy of the vacuum, see the previous subsection). This is precisely the same as the result (2.8) that we found above by counting the states with $E \ll N^2$; we see that this is indeed the leading large N behavior as long as $T < T_H$.

This partition function diverges at $T = T_H$, as should be expected, since this is the point where the quadratic action (4.5) develops an unstable direction. The corresponding divergence in the free energy $F = -T\ln(Z)$ goes like

$$F \to T_H \ln(T_H - T) \tag{4.10}$$

as T approaches T_H from below. By the general discussion in §2, we may conclude that this divergence is associated with a Hagedorn density of states,

$$\rho(E) \propto E^0 e^{\beta_H E}, \tag{4.11}$$

as we also found by counting states in §2. Thus, just as for perturbative string theory, we have a Hagedorn divergence which is associated with an unstable mode in the Euclidean path integral; note that the power of the energy is also the same as in string theory at finite volume. In this case, the "tachyon" causing the divergence is the lowest "momentum" fluctuation mode ρ_1 of the eigenvalues. In the next subsection, we will be able to see explicitly what the endpoint of this tachyon condensation brings as we raise the temperature beyond the Hagedorn temperature.

---

[14] The 'Jacobian' for the 'change of variables' from λ_i to ρ_n appears to be irrelevant at every order in the $1/N$ expansion, see [13].

4.2. *Behavior near the transition*

At the temperature $T = T_H$, V_1 vanishes and the mode ρ_1 (the lowest momentum fluctuation in the eigenvalue distribution) becomes massless. Since the action is quadratic, this corresponds to an exactly flat direction in the potential, and the corresponding family of minimum action configurations are given (up to overall translations of θ) by

$$\rho(\theta) = \frac{1}{2\pi}(1 + t\cos(\theta)), \qquad 0 \leq t \leq 1\,, \tag{4.12}$$

which interpolate between the uniform distribution and a sinusoid vanishing at a single point. For $T > T_H$, the quadratic form S has negative coefficients, so all minimum action configurations must lie at the boundary of configuration space, at a point where a hyperboloid of constant S lies tangent to this boundary. Since this boundary is provided by the positivity condition $\rho(\theta) \geq 0$, these minimum action eigenvalue distributions necessarily vanish on a subset of the circle.

In the limit of small, positive $\Delta T = T - T_H$, the action contour $S = 0$ is a cone with opening angle going to zero. The contours of smaller S are hyperboloids inside this cone, so the minimum action configuration lies on the boundary of the configuration space inside the cone. Finally, it is easy to see that the region of allowed ρ's is convex (if points $\{\rho_n\}$ and $\{\rho'_n\}$ are in the region then all points $\{t\rho_n + (1-t)\rho'_n\}$ are in the region for $0 \leq t \leq 1$). Thus, the boundary region interior to the cone is a simply-connected neighborhood of the $t = 1$ configuration in (4.12) whose size goes to zero as $T \to T_H$. We conclude that the minimizing configuration changes continuously from the $t = 1$ configuration in (4.12) for T in some interval above T_H.

The leading behavior of the free energy in this vicinity of T_H is then given by evaluating the V_1 term in the action (4.5) on the $t = 1$ configuration in (4.12), since the effects of changing the eigenvalue distribution come in at higher orders.

Thus, the leading result for the minimum action is

$$S^{min}_{T>T_H}(x) = \frac{N^2}{8\pi}(T - T_H)\, V_1'(T_H) + \ldots \tag{4.13}$$

from which we may obtain the free energy as $F = TS^{min}$. The leading behavior near the transition is therefore given by

$$\lim_{N\to\infty} \frac{1}{N^2} F_{T\to T_H}(T) = \begin{cases} 0 & T < T_H \\ -\frac{1}{4}(T - T_H) z'(x_H) \frac{x_H}{T_H} & T > T_H \end{cases} \tag{4.14}$$

so we have a first order phase transition at the Hagedorn temperature. Note also that the behavior is characteristic of a deconfinement transition, since the free energy is $\mathcal{O}(1)$ below the transition and $\mathcal{O}(N^2)$ above the transition. This will be verified in §4.5 below when we discuss the behavior of a second order parameter for confinement, related to the Polyakov loop.

4.3. *High temperature behavior*

For general values of $T > T_H$, we will not be able to write down an exact expression for the free energy, though we describe in the next subsection a formal solution that may be used to evaluate the free energy to arbitrary accuracy. However, the theory simplifies again in the large temperature limit, where the potential becomes strongly attractive.[15] In this case, the action is minimized by a tightly clustered configuration of eigenvalues that approaches a delta function in the limit of infinite temperature. Thus, we should have $\rho_n = 1$ to leading order, and

$$S = \frac{N^2}{2\pi} \sum_{n=1}^{\infty} V_n(T \gg 1) \ . \tag{4.15}$$

The high temperature limit of the single-particle partition functions depends only on the space-time dimension d and the number of physical polarizations $\mathcal{N}^{dof}$ of the fields in the theory,

$$z_i(x) \to 2\mathcal{N}_i^{dof} T^{d-1} + \mathcal{O}(T^{d-2}) \ . \tag{4.16}$$

Using (4.6), we find that the high temperature limit of the free energy is

$$F(x \to 1) = -2N^2 T^d \zeta(d) \left[\mathcal{N}_B^{dof} + \left(1 - \frac{1}{2^{d-1}}\right) \mathcal{N}_F^{dof} \right] + \mathcal{O}(T^{d-1}) \ . \tag{4.17}$$

The limiting free energy density here coincides with that of the flat space theory, which should be expected since in the free field analysis we are dealing with a scale-invariant theory, so taking the dimensionless temperature TR to infinity is equivalent to taking the limit of large volume at fixed temperature (see [16–18] among others for the high temperature expansion of the free energy for the $U(1)$ theory).

See Section 5 of [12] for more details on the solution of this matrix model, including an evaluation of the Free Energy in a powerseries expansion about the phase transition point.

---

[15] The high temperature regime and its relation to string theory was discussed for uncompactified theories in [14] in a very similar language to the one we are using here, and this discussion was applied to two dimensional QCD in [15].

4.4. *Summary of thermodynamic behavior*

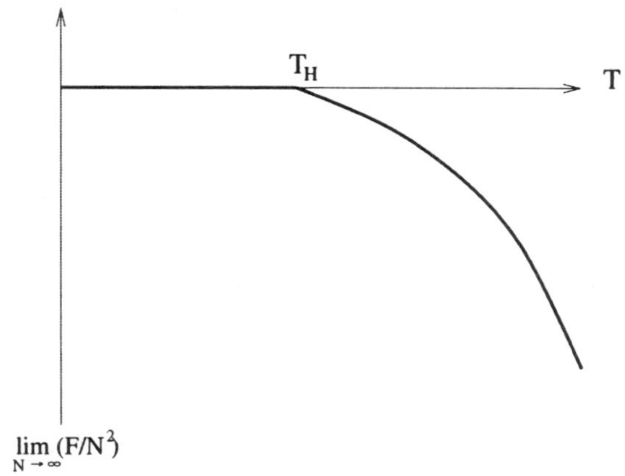

Fig. 1. Free energy as a function of temperature in free Yang-Mills theories.

We are now in a position to present a reasonably detailed picture of the behavior of the free energy of the free gauge theory at all temperatures, depicted in figure 1.

We find that the uniform eigenvalue distribution provides an absolute minimum of the matrix model potential for all temperatures below some T_H determined by the single-particle partition function as

$$z(T_H) = 1 \,. \tag{4.18}$$

The free energy in this regime, given by (4.9), is $\mathcal{O}(1)$ and decreases monotonically to a logarithmic divergence (4.10) at $T = T_H$, characteristic of a Hagedorn density of states (4.11).

As T increases past T_H, the uniform eigenvalue distribution develops an unstable mode which condenses to give a pure sinusoid distribution (4.12) for T just above T_H, and a distribution with a gap for $T > T_H$. For small $(T - T_H)$, the $\mathcal{O}(N^2)$ free energy decreases linearly in $T - T_H$, so the first derivative of the free energy is discontinuous and we have a first order phase transition. The free energy as a function of temperature may be written as a perturbation expansion in $T - T_H$.

In the high temperature limit, the pairwise potential between eigenvalues becomes strongly attractive, and the eigenvalue distribution approaches a delta function. In this regime, the free energy asymptotes to (4.17) which corresponds to the expected flat-space free-energy density. Using the formal exact solution of the previous section, it should be possible to work out the high temperature behavior as a perturbation expansion in $1/T$, as we did explicitly for the behavior near the transition.

4.5. *The Polyakov loop as an order parameter at finite volume*

In the theory we analyzed above, the Polyakov loop $\langle \frac{1}{N}\mathrm{tr}(U)\rangle$ vanishes in both the high temperature and the low temperature phases. However, the reason for this vanishing is rather different in the two phases. The low temperature phase is governed by a single saddle point (uniform distribution of eigenvalues); the Polyakov loop vanishes on this saddle point. On the other hand the high temperature phase is governed by a one parameter set of saddle points (related to each other by the $U(1)$ rotation $U \to e^{i\alpha}U$, which acts on the eigenvalue distribution by shifting θ periodically). The value of $\langle \frac{1}{N}\mathrm{tr}(U)\rangle$ is non-zero when evaluated on each saddle point individually but vanishes upon integrating over α (averaging over all saddle points [16]), since the phase varies over a complete circle.[17] To obtain an order parameter which correctly distinguishes the two phases, it is therefore sufficient to use the squared magnitude of the Polyakov loop, $\langle|\mathcal{P}|^2\rangle = \lim_{N\to\infty}\langle \frac{1}{N^2}|\mathrm{tr}(U)|^2\rangle$.

This order parameter may be given a nice physical interpretation by the following alternative definition. Define $\langle \frac{1}{N}\mathrm{tr}(U)\rangle_\epsilon$ as the expectation value of $\langle \frac{1}{N}\mathrm{tr}(U)\rangle$ upon perturbing the effective action in (2.14) by the infinitesimal term $N\epsilon(\mathrm{tr}(U) + \mathrm{tr}(U^\dagger))$. Note that this perturbation breaks the $U(1)$ symmetry (or the $\mathbb{Z}_N$ symmetry for $SU(N)$); in the presence of this perturbation, the high temperature phase is also governed by a single saddle point. Consequently, the order parameter

$$\mathcal{P}^{eff} = \lim_{\epsilon\to 0}\lim_{N\to\infty}\langle \frac{1}{N}\mathrm{tr}(U)\rangle_\epsilon \tag{4.19}$$

is zero in the low temperature phase but non-zero in the high temperature phase.[18] It is easy to show that in the large N limit, $\langle|\mathcal{P}|^2\rangle = |\mathcal{P}^{eff}|^2$, so the two definitions of the order parameter are equivalent.

---

[16] Witten [9] has described a rather similar phenomenon in the bulk dual to the strongly coupled $\mathcal{N}=4$ Yang Mills.

[17] For $SU(N)$ theories this is not an integral but a sum over N discrete saddle points, but it has the same effect.

[18] Note that this construction is rather similar to the definition of the order parameter of a ferromagnet in 3 dimensions. Naively, the magnetization of the ferromagnet vanishes due to $SO(3)$ symmetry in both the high and low temperature phases. However if we define $|M|_\epsilon$ to be the magnetization in the presence of a small magnetic field in the z direction, then $\lim_{\epsilon\to 0}\lim_{V\to\infty}|M|_\epsilon$ (where V is the volume) is the standard order parameter for the ferromagnet.

The construction presented in the previous paragraph may be understood physically as follows. As discussed above $\langle \mathrm{tr}(U) \rangle$ vanishes in both phases for purely kinematical reasons; Gauss' law does not allow the introduction of a quark on a compact space. However, the perturbation described in the previous paragraph effectively adds a condensate (or plasma) of classical quarks and anti-quarks to the theory; in the presence of such a condensate the constraint from Gauss' law is circumvented (recall that flux lines decay in a Higgs phase). In the high temperature phase the additional classical quark has finite free energy even in the limit that the classical Higgsing is taken to zero. On the other hand, in the low temperature phase the Polyakov loop vanishes as this Higgsing is taken to zero.

The analysis above shows that $\mathcal{P}^{eff} = 1/2$ immediately after the phase transition, and it goes up to $\mathcal{P}^{eff} \to 1$ at high temperatures.

5. Phase structure at weak coupling

In this section we turn to an analysis of the phase transition at non-zero 't Hooft coupling.[19] We will see that an arbitrarily small non-zero coupling qualitatively changes the behavior of the transition. In §5.1 we deduce the most general form of $S_{eff}(U, \lambda)$ allowed by gauge invariance. In §5.2 we analyze the structure of $S_{eff}(U, \lambda)$ in perturbation theory. In §5.3 we study the structure of the deconfinement transition (discussed in detail for the free theory in the previous section) at weak coupling. We end in §5.4 with a description of the implications of our results for the microcanonical ensemble.

5.1. *General properties of the effective action*

For any value of the coupling constant, gauge invariance imposes tight constraints on the form of the effective action $S_{eff}(\alpha)$ (3.6) for the zero mode of the gauge field; $S_{eff}(\alpha)$ should be invariant under all space-time gauge transformations that:

1) Are single valued (up to an element of the center of $U(N)$) on $\mathcal{M} \times S^1$;
2) Preserve the gauge-fixing conditions (3.1) and (3.2).

We will restrict attention to gauge transformations $U(t)$ that are inde-

[19] Some of the key ideas underlying this section and the next one arose in discussions with R. Gopakumar. We thank him in particular for emphasizing the qualitative difference between the two scenarios depicted in figure 4, and for forcing us to understand the order parameter in a clearer fashion.

pendent of the position on the compact space. Under such a transformation

$$A_0 \to V(t)A_0V^\dagger(t) - i\partial_t VV^\dagger\,, \tag{5.1}$$

so that

$$\alpha \to V(t)\alpha V^\dagger(t) - i\partial_t VV^\dagger\,. \tag{5.2}$$

$V(t)$ obeys the condition (2) above (and (5.2) makes sense) only when the right-hand side of (5.2) is independent of time.

Constant (time independent) gauge transformations clearly satisfy the requirements (1) and (2) above. Consequently, $S_{eff}(\alpha)$ is invariant under $\alpha \to V\alpha V^\dagger$. We may use this invariance to diagonalize α. Once this has been done, we consider the further gauge transformations

$$V(t) = e^{itD}\,, \tag{5.3}$$

where D is a diagonal matrix, whose eigenvalues are all integral multiples of $2\pi/\beta$. α transforms under the gauge transformation generated by (5.3) as $\alpha \to \alpha + D$. This implies that $S_{eff}(\alpha)$ is invariant under separate shifts of any of the eigenvalues of α by multiples of $2\pi/\beta$. It follows from these invariances that S_{eff} is really a function only of $\mathrm{tr}(U^n)$ (for all n) where $U = e^{i\beta\alpha}$ (see (3.12)) is the zero mode holonomy around the time circle.

Finally, consider $V(t) = e^{\frac{2\pi ikt}{\beta N}}$ where k is an integer if the gauge group is $SU(N)$, and arbitrary for $U(N)$. $V(t)$ obeys the single-valuedness condition (1), as $e^{\frac{2\pi ik}{N}}$ belongs to the center of the gauge group. Under the gauge transformation generated by $V(t)$, $\alpha \to \alpha + \frac{2\pi k}{N\beta}$. Consequently, S_{eff} should be invariant under $U \to e^{\frac{2\pi ik}{N}}U$, for all integers k if the gauge group is $SU(N)$, and for any k for $U(N)$. In the limit $N \to \infty$ the two cases coincide, and $S_{eff}(U)$ must be invariant under $U \to e^{i\theta}U$ for arbitrary θ. Putting everything together, we conclude that S_{eff} may depend on U only in combinations of the form

$$\mathrm{tr}(U^{n_1})\cdots\mathrm{tr}(U^{n_k})\mathrm{tr}(U^{-n_1-\cdots-n_k})\,. \tag{5.4}$$

5.2. *The general form of the effective action in perturbation theory*

In large N perturbation theory, S_{eff} is generated by planar vacuum diagrams obtained by integrating out the massive modes. In this calculation, α is a background field, which appears diagrammatically via external line insertions on the index loops of diagrams. Planar diagrams at order λ^{k-1} include $(k+1)$ index loops, each of which leads to a trace in the final result (which may contain arbitrarily many factors of α). From the discussion of the previous subsection, the only terms with $(k+1)$ traces allowed by gauge invariance take the form (5.4). It follows that the most general form of the planar contribution to S_{eff} in perturbation theory is

$$S_{eff}(U) = N^2[\sum_n m_n^2(x)|u_n|^2 + \lambda \sum_{m,n} F_{m,n}(x)(u_m u_n u_{-n-m} + c.c.) + \\ +\lambda^2 \sum_{m,n,p} F_{m,n,p}(x)(u_m u_n u_p u_{-m-n-p} + c.c.) + \cdots], \qquad (5.5)$$

where we have defined $u_n \equiv \mathrm{tr}(U^n)/N$ (note that $u_n^* = u_{-n}$ and $u_0 = 1$). This agrees, of course, with the quadratic form of the one-loop effective action that we found in sections 3 and 4. Also, the u_n coincide with the variables ρ_n used in §4 for eigenvalue distributions which are symmetric about $\theta = 0$, so the values of m_n^2 in the free field theory may be read off from equation (4.6).

5.3. *Possible phase structures at weak coupling*

Recall from §4 that, at $\lambda = 0$, u_1 is massless at the phase transition temperature T_H, while the other u_n's are all massive. Consequently, u_1 will continue to be the lightest mode in the vicinity of the phase transition also at weak coupling. Thus, for analyzing the phase transition at weak coupling, it is useful to obtain an effective action for u_1 by integrating out the u_n's with $n > 1$. [20] It follows from (5.5) that the leading large N terms in the resulting effective action are of the form

$$S_{eff}(u_1) = N^2 \left(m_1^2(x,\lambda)|u_1|^2 + \sum_{n=2}^{\infty} \lambda^{2n-2} B_n(x,\lambda)|u_1|^{2n} \right), \qquad (5.6)$$

where m_1^2 and the B_n's are functions of the temperature and power series in λ starting (generically) from a λ^0 term.

In particular, to $\mathcal{O}(\lambda^2)$ we have

$$S_{eff}(u_1) = N^2 \left(m_1^2|u_1|^2 + b|u_1|^4 \right), \qquad (5.7)$$

where $b(x) = B_2(x,0)\lambda^2$ is a function of temperature which is generically non-zero. It is not difficult to compute $B_2(0)$ starting from (5.5). The only

[20] As we are interested in the large N limit, it is sufficient to integrate out these modes classically.

terms in (5.5) that contribute are

$$S'_{eff}(U) = N^2 \left(m_1^2 |u_1|^2 + m_2^2 |u_2|^2 + \lambda I (u_{-2} u_1^2 + u_2 u_{-1}^2) + \lambda^2 A |u_1|^4 \right). \tag{5.8}$$

At this order, u_2 may be integrated out from (5.8) by setting it to be equal to its classical value $u_2 = -I\lambda u_1^2/m_2^2$, [21] yielding (5.7) with $b = (A - I^2/m_2^2)\lambda^2$. Note that because the eigenvalue density (5.9) has to be non-negative everywhere, our expressions for the effective action (5.6) and (5.7) are valid only for u_1 such that $|u_1| \leq \frac{1}{2} + \mathcal{O}(\lambda)$.

As in §4, the one-loop contribution to the free energy coming from integrating over u_1 diverges at the temperature $T_H(\lambda)$ at which $m_1^2(x, \lambda)$ goes to zero and u_1 becomes massless. For any value of the coupling, at leading order in the distance from this temperature, we have $m_1^2 \simeq K(T_H(\lambda) - T)$ for some positive constant K. This divergence signals a Hagedorn behavior of the single-particle spectrum of the theory with Hagedorn temperature T_H (computable in perturbation theory), so this behavior persists (at least in the microcanonical ensemble) even at non-zero 't Hooft coupling.

As the saddle point at $u_1 = 0$ is unstable for $T > T_H$, the theory described by (5.7) clearly undergoes a phase transition (to another saddle point) at some $T \leq T_H$. Whether this phase transition occurs at $T < T_H$ or $T = T_H$ depends on whether the value of b (defined in (5.7)) at the Hagedorn temperature T_H is positive or negative, as we now argue in detail.[22]

First, consider the case $b > 0$. For $T < T_H$, $u_1 = 0$, corresponding to the uniform eigenvalue distribution, is clearly a global minimum of the effective action (5.7). For $T > T_H$, however, $u_1 = 0$ is unstable and S_{eff} is minimized at $|u_1|^2 = |m_1|^2/2b$. The value of the effective action at this minimum is $S_{eff} = -N^2|m_1|^4/4b$, which is of order $(T - T_H)^2$, and so the phase transition at $T = T_H$ is of second order. As the temperature is raised above T_H, the eigenvalue distribution smoothly becomes non-uniform until we reach $|u_1|^2 = 1/4$; this occurs at $m_1^2 = -b/2$ (at leading order in λ). At this point, $\rho(\theta)$ vanishes at some θ and we have reached the boundary of the space of eigenvalue distributions and the edge of the validity of (5.7). As the temperature is further raised the eigenvalue distribution develops a gap on the circle, and the theory undergoes a further phase transition

[21] Thus setting the eigenvalue distribution to

$$\rho(\theta) = \frac{1}{2\pi} \left(1 + 2|u_1| \cos(\theta + \alpha) - \frac{2\lambda I}{m_2^2} |u_1|^2 \cos(2\theta + 2\alpha) \right) \tag{5.9}$$

for some α.

[22] See [19] for a similar discussion in a related context.

similar to the Gross-Witten transition [7]. This second phase transition (at $T = T_H + \frac{b}{2K}$) is of third order. The behavior above this second transition temperature is no longer captured by the effective action (5.6), and the full action is required to analyze it.

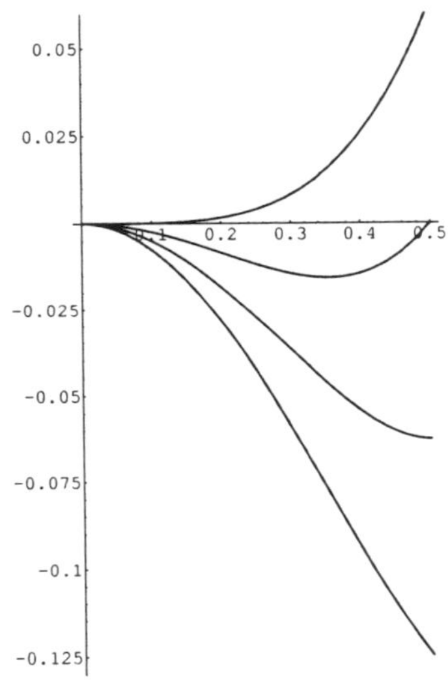

Fig. 2. Plots of $S_{eff}(u_1)$ as a function of u_1 for $b = 1$ at several values of m_1^2 (from top to bottom) : $m_1^2 = 0$ (Hagedorn temperature and first phase transition temperature), $m_1^2 = -\frac{1}{4}$, $m_1^2 = -\frac{1}{2}$ (second phase transition temperature), and $m_1^2 = -\frac{3}{4}$.

In the case where b is negative, S_{eff} develops a new local minimum at $|u_1| = \frac{1}{2}$ (the boundary of our order parameter space) when $m_1^2 < |b|/2$; note this happens below the Hagedorn temperature. When $m_1^2 > |b|/4$, the free energy at this new local minimum is positive, and the saddle point $|u_1| = \frac{1}{2}$ is disfavored compared to $u_1 = 0$. However, when we raise the temperature to $m_1^2 < |b|/4$, the free energy at $|u_1| = \frac{1}{2}$ becomes negative, and so this saddle dominates over the $u_1 = 0$ saddle point. Consequently, at $T = T_H - \frac{|b|}{2K}$, $|u_1|$ jumps discontinuously from zero to $\frac{1}{2}$ and the theory undergoes a first order phase transition. When the temperature is further raised, the eigenvalue distribution develops a break on the circle and the theory is no longer described by (5.6). This behavior is qualitatively similar to that of the free theory which we analyzed in §4, except that the phase transition now happens below the Hagedorn temperature T_H.

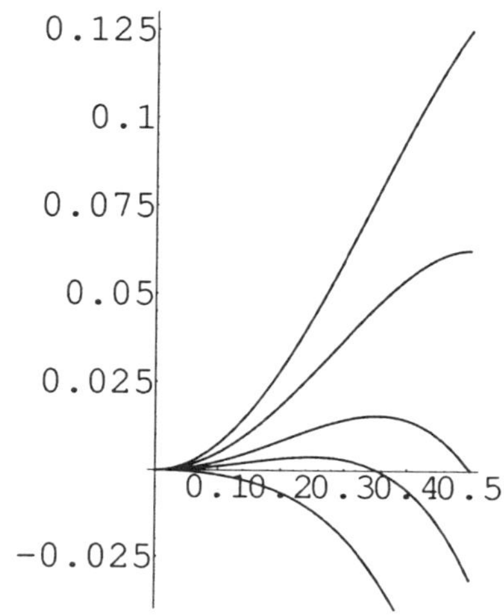

Fig. 3. Plots of $S_{eff}(u_1)$ as a function of u_1 for $b = -1$ at several values of m_1^2 (from top to bottom) : $m_1^2 = \frac{3}{4}$, $m_1^2 = \frac{1}{2}$ (new phase nucleated), $m_1^2 = \frac{1}{4}$ (first order phase transition temperature), $m_1^2 = \frac{1}{8}$, and $m_1^2 = -\frac{1}{8}$.

If b vanishes exactly then the higher order terms in (5.6) are required for analyzing the phase transition, but generically b will not vanish at the Hagedorn temperature. In our discussions above we assumed that b was constant in a range of temperatures near the Hagedorn temperature; this is consistent at weak coupling since the range of temperatures which is relevant for the discussions above is of order λ^2, so the changes in b within this range are of higher order in λ.

5.4. *Density of states as a function of energy*

In the previous subsections we discussed the canonical partition function as a function of temperature for Yang-Mills theories at weak coupling. We found three qualitatively different classes of $Z(\beta)$ depending on whether b (see (5.7)) is negative, zero or positive. In this subsection we will qualitatively describe the corresponding Yang-Mills theories in the microcanonical ensemble, seeing how the density of states depends on the energy in each of these three different classes of models.

In order to make contact with the canonical ensemble discussed in previous subsections, we will find it convenient to characterize the microcanonical density of states $S(E) = \ln(\rho(E))$ in a rather peculiar way. We will find it convenient to plot the logarithm of the effective temperature $\ln(T(E)) = -\ln(\partial S(E)/\partial E)$ as a function of $\ln(E)$ [23]; such plots were used for very similar purposes in [20].

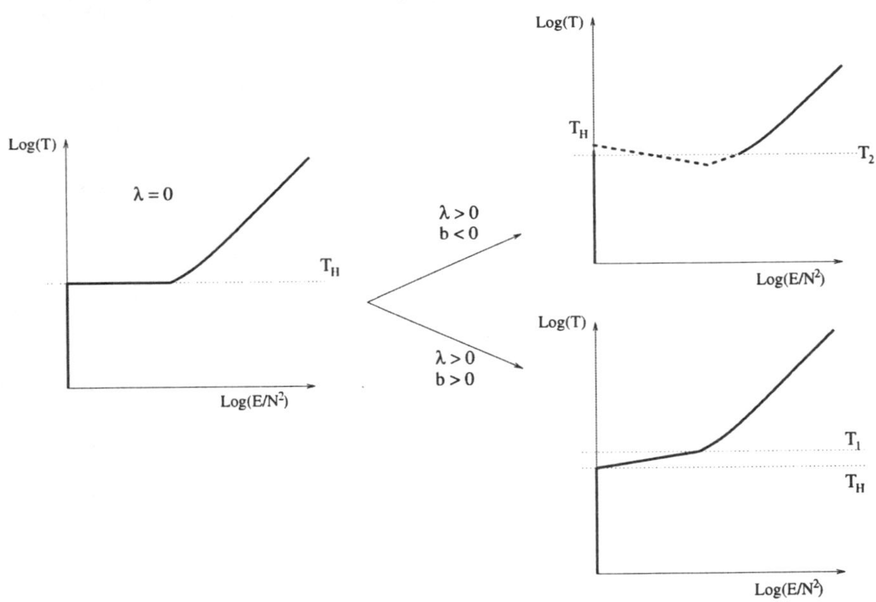

Fig. 4. Plot of $\log(T)$ as a function of $\log(E)$ (for energies of order N^2, in the microcanonical ensemble), for $\lambda = 0$ and for $\lambda > 0$ with $b > 0$ and $b < 0$.

The behavior of the density of states (or $\frac{\partial S}{\partial E}$) as a function of temperature may be deduced directly from the behavior of $F(T)$ using the relation $E = \partial_\beta(\beta F)$. However, this information is incomplete where there are first order phase transitions, since these correspond to temperatures at which the energy jumps discontinuously in going from one phase to an-

[23] We thank R. Gopakumar for suggesting this to us.

other (non-zero latent heat). For the case $\lambda = 0$, we have seen that the first order transition is a Hagedorn transition, so we may conclude that for microcanonical energies intermediate between the two phases, the density of states is Hagedorn, corresponding to a horizontal line in the graph of $\ln(T)$ as a function of $\ln(E)$. Thus, the $F(T)$ behavior depicted in figure 1 leads to the microcanonical behavior shown on the left in figure 4. For small $\lambda > 0$, the density of states should be a small perturbation of this behavior, and we may deduce from the results of §5.3 that the qualitatively different behavior obtained for $b > 0$ and $b < 0$ corresponds to whether the flat Hagedorn region in the $\lambda = 0$ plot tilts upwards ($b > 0$), resulting in two continuous phase transitions at T_H and some higher temperature T_1, or downwards ($b < 0$), preserving the first order transition (dashed lines in figure 4 indicate points not accessible in the canonical ensemble).

The plots in figure 4 correspond to energies of order N^2, and they are drawn in the large N limit where the canonical ensemble exhibits sharp transitions and the boundaries between phases in the microcanonical ensemble are distinct. To understand in more detail the phases available in the microcanonical ensemble at general energies (including energies of order one), it is convenient to stretch out the low energy regions of the three diagrams for large but finite N, as we do in figures 5, 6, and 7 below.

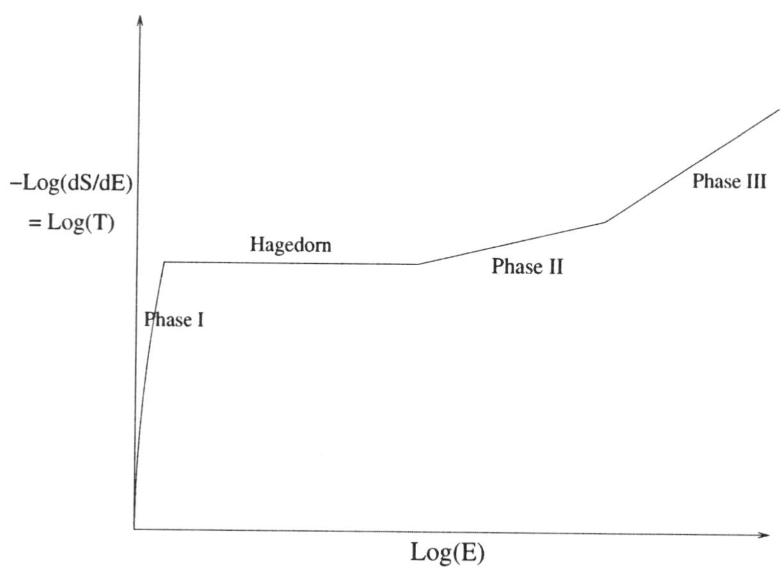

Fig. 5. Plot of $\log(T)$ as a function of $\log(E)$ (in the microcanonical ensemble) when $b > 0$.

In the case $b > 0$ this plot is displayed in figure 5. As is apparent from the figure, at any given temperature the theory has exactly one phase available to it; this is consistent with the behavior described by the series of Landau-Ginzburg diagrams in figure 2 and with the analysis of the toy model of the previous section. For $T < T_H$ the theory is in Phase I. At T_H, the theory undergoes a second order Hagedorn phase transition, emerging into Phase II (the energy after the transition is of order N^2). At a still higher temperature, the theory undergoes yet another continuous phase transition into Phase III.

Note that Phase I corresponds to $u_1 = 0$, Phase II corresponds to the minimum of (for instance) the second plot in figure 2, and Phase III is at the boundary of the configuration space of eigenvalue distributions.

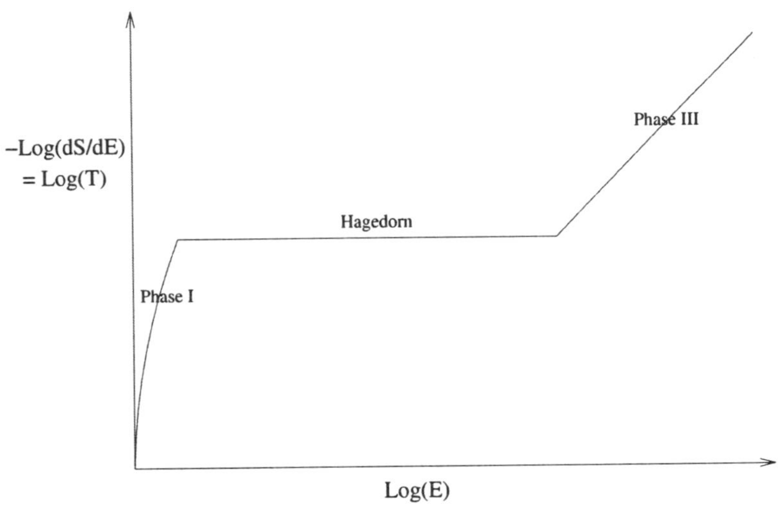

Fig. 6. Plot of $\log(T)$ as a function of $\log(E)$ (in the micro-canonical ensemble) in the free Yang-Mills theory.

The density of states of the free theory may be obtained from the limit $b \to 0$ of the previous discussion; the relevant plot takes the form shown in figure 6. As is apparent from this figure, the free theory makes a phase transition (of first order) directly from Phase I to Phase III at the Hagedorn temperature, in agreement with the analysis of §4.

When $b < 0$, we believe that the density of states is characterized by the plot displayed in figure 7. The thermodynamics induced by the density of states plotted in figure 7 is consistent with the behavior described by the Landau Ginzburg plots in figure 3. In particular, the five graphs shown in figure 3 (from top to bottom) should be taken to represent $S_{eff}(u_1)$ for $T < T_1$, $T = T_1$, $T = T_2$, $T_2 < T < T_H$ and $T > T_H$, respectively. Phase I lies at $u_1 = 0$ in figure 3. Phase II is the unstable maximum in the 3rd and 4th graphs in figure 3, and Phase III is the saddle point at the boundary $u_1 = \frac{1}{2}$. Note that, according to both figures 7 and 3, Phase III exists only for $T > T_1$, and Phase II exists only for $T_1 < T < T_H$.

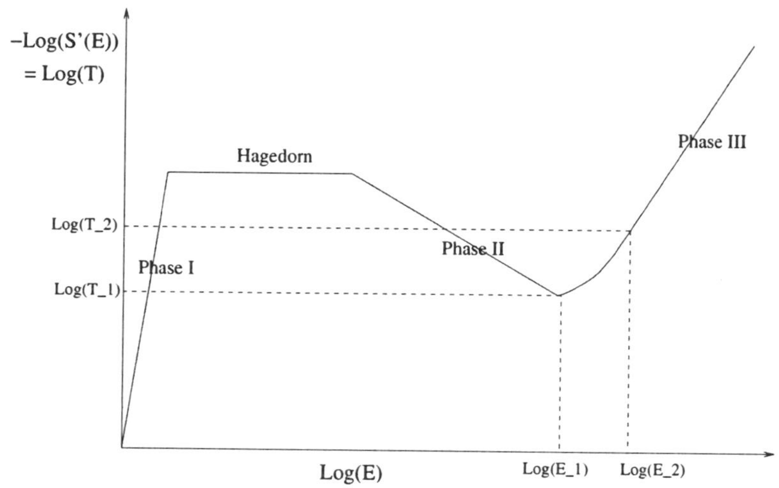

Fig. 7. Plot of $\log(T)$ as a function of $\log(E)$ (in the microcanonical ensemble) when $b < 0$.

In ending this section we would like to emphasize that adding interactions has two qualitatively different effects on the spectrum of weakly coupled Yang-Mills theories. First, single-trace states involving a number of fields which is much smaller than N pick up corrections to their energy, starting at first order in λ (for conformal field theories on S^{d-1} these corrections are equivalent to the anomalous dimensions of the single-trace operators). These corrections result in a renormalization of $m_1^2(T)$, and hence of the Hagedorn temperature. However, the physics of the phase transition (and, in particular, the crucial sign of the coefficient b) is governed by a different effect, which is the deviation from Hagedorn-type behavior in states which involve a number of fields of order N^2. [24] This effect appears even in the free Yang-Mills theory, and it appears to be modified in a qualitatively

[24] When the number of fields is of order $\sqrt{N}$ or larger, single-trace and multi-trace states mix, and one cannot distinguish between them. This was highlighted recently in the study of the effects of non-planar graphs in the Yang-Mills/plane wave duality [21].

important fashion by coupling constant effects at $\mathcal{O}(\lambda^2)$, as shown in figure 4. These modifications determine the sign of b and the nature of the phase transition at finite coupling.

6. Extrapolation to strong coupling and the dual description

6.1. *Dual interpretation of the $\mathcal{N}=4$ SYM thermodynamics at strong coupling*

It should be possible to understand the possible phase diagrams presented earlier in this section in terms of a dual string theory description of the relevant gauge theories. Unfortunately, among the $3+1$ dimensional gauge theories with adjoint fields, the only one whose string dual is known is the $\mathcal{N}=4$ supersymmetric Yang-Mills theory (and theories related to it by renormalization group flows), and this dual is mostly understood only in the limit of strong 't Hooft coupling. In this subsection we will argue that, at least in this one case, the thermodynamics of the string theory dual fits rather nicely with the picture presented in this paper. Previous discussions of the thermodynamics of $\mathcal{N}=4$ SYM as a function of the coupling, which are consistent with ours, appear in [22], and a detailed discussion of the transitions discussed in this subsection appears in [20,23,24].

The string dual to $\mathcal{N}=4$ SYM on $S^3\times\mathbb{R}$ (where the S^3 is taken to be of unit radius) is type IIB string theory on $AdS_5\times S^5$, whose metric may be written as

$$ds^2 = R_0^2\left(-\cosh^2\rho\, d\tau^2 + d\rho^2 + \sinh^2\rho\, d\Omega_3^2 + d\Omega_5^2\right). \qquad (6.1)$$

Under this duality the Hamiltonian of the Yang-Mills theory is identified with the generator of global time translations ∂_τ in the geometry (6.1). The energy E of the gauge theory on a sphere of unit radius is related to the proper energy in string theory at $\rho=0$ by $E_{prop}=E/R_0$. The radius R_0 of the $AdS_5\times S^5$ space is related to the 't Hooft coupling λ and the inverse string tension α' by $R_0\simeq\lambda^{\frac{1}{4}}\sqrt{\alpha'}$ (we will ignore all numbers of order unity in the qualitative discussion of this subsection).

For large $R_0/\sqrt{\alpha'}$ (large 't Hooft coupling) the density of states $\rho(E)$ of type IIB string theory on (6.1) has four distinct regimes (see [25] and references therein). The only states in the spectrum with proper energy[25]

[25] Recall that states of finite energy on (6.1) are all localized about $\rho=0$. AdS_5 space behaves effectively like a four dimensional box of physical radius R_0, as is apparent from the fact that the spectrum of ∂_τ is discrete with discretization step unity.

smaller than the string scale, $E \ll \lambda^{1/4}$, are ten dimensional supergravitons [26] whose entropy scales as $S(E) \propto E^{9/10}$. For $E \gg \lambda^{1/4}$, excited string states are added to the spectrum; the contribution of these states to the entropy is $S \simeq E_{prop}\sqrt{\alpha'} \simeq \lambda^{-1/4}E$ (see §2.1), and this contribution dominates over the graviton gas for $E \gg \lambda^{5/2}$. When the proper energy exceeds the Planck mass (namely, $E \gg m_s R_0/g_s^{1/4} = N^{1/4}$) small ten dimensional Schwarzschild black holes are also added to the spectrum (note that these energies are inaccessible in the strict $N \to \infty$ limit). The entropy of such black holes is proportional to $S \sim (l_P E/R_0)^{8/7} = \left(E/N^{1/4}\right)^{8/7}$; they have negative specific heat and positive free energy. This entropy of black holes dominates over that of the Hagedorn strings for $E \gg N^2/\lambda^{7/4}$. Finally, at an energy $E_1 \sim N^2$ (an energy at which the radius of these Schwarzschild black holes becomes comparable to the *AdS* radius R_0) the black hole horizon covers the whole S^5, and the specific heat of these black holes becomes positive. For $E > E_1$ these black holes are referred to as *AdS*-Schwarzschild or big black holes. At a higher energy E_2 (also of $\mathcal{O}(N^2)$) the free energy of these black holes becomes negative. Finally, for $E \gg N^2$, the entropy of these black holes is $S \propto N^{1/2}E^{3/4}$, so the thermodynamics of these big black holes resembles that of a four dimensional conformal field theory.

For comparison with the canonical ensemble it is most convenient (see §5.4) to display the density of states described in the previous paragraph as a plot (see figure 8) of $\ln(T(E)) = -\ln(S'(E))$ [20]. Quite remarkably, figure 8 is identical in its general form to figure 7, even though figure 7 was derived for a general gauge theory (with $b < 0$) at weak coupling while figure 8 applies to the $\mathcal{N} = 4$ Yang Mills theory at strong coupling, and has been derived using its stringy dual.

To complete this subsection we now use figure 8 to discuss the behavior of the $\mathcal{N} = 4$ SYM theory at strong coupling in the canonical ensemble [20] (see §5.4 for a very similar discussion). For $T < T_1 (= \frac{\sqrt{2}}{\pi})$ the only available saddle point is the thermal gas of gravitons. Over the temperature range $T_1 < T < T_H \simeq \lambda^{\frac{1}{4}}$ the theory has three saddle points to choose from. One of these saddle points (small black holes) is unstable, so it cannot appear in the canonical ensemble. The other two saddle points are stable and compete with each other. The gas of gravitons has lower free energy for $T_1 < T < T_2 (= \frac{3}{2\pi})$, while the big black hole dominates the canonical ensemble for $T_2 < T < T_H$. Consequently, the theory should undergo a

[26] In terms of the gauge theory, the only states in this regime are those created by the chiral primary operators, their products and their descendants.

first order phase transition at $T = T_2$, which (as we mentioned above) is indeed the case [9, 11]. $T = T_H$ is the Hagedorn temperature at which the graviton gas saddle point stops existing. For $T > T_H$ the generic state of the theory continues to be a big black hole, which is now the only available phase.

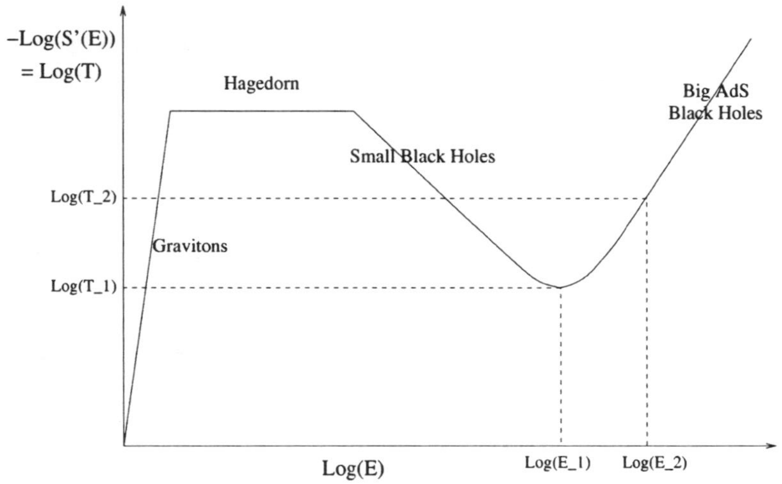

Fig. 8. $\log(T)$ as a function of $\log(E)$ (in the microcanonical ensemble) for all the 'phases' of type IIB string theory on $AdS_5 \times S^5$, when $R_0 \gg \sqrt{\alpha'}$.

6.2. *Deconfinement and black holes*

We do not, as yet, understand the string dual of weakly coupled gauge theories, such as pure Yang-Mills or $\mathcal{N} = 1$ supersymmetric Yang-Mills on a small S^3, or even the $\mathcal{N} = 4$ SYM theory at weak 't Hooft coupling. Nonetheless, each of these theories is continuously related to the $\mathcal{N} = 4$ theory at strong coupling, so the AdS/CFT correspondence provides a demonstration (in the physicist's sense of the term) of the existence of string duals for all these gauge theories. Furthermore, these string duals may be expected to qualitatively resemble type IIB string theory on $AdS^5 \times S^5$, in the sense that they should all include at least one additional "holographic" dimension beyond the dimensions of the Yang-Mills theory.

As we reviewed above, the deconfined phase of the strongly coupled $\mathcal{N} = 4$ theory admits a dual description in terms of black holes.[27] We

[27] The connection between deconfinement transitions and black holes has also been

would now like to argue that this is a general feature (see [4] for related comments): large N deconfined phases should always be associated with black holes.[28] Our simple argument follows directly from ideas presented in [9].

In order to measure the Polyakov loop we need to put an external quark on a trajectory that wraps around the circle of the time direction. Since in the large N gauge/string correspondence quarks appear at the boundaries of the worldsheet, we expect the Polyakov loop in string theory to be realized as the partition function (in Euclidean space) for a string whose boundary wraps around the time-like loop, and this is indeed the case in the AdS/CFT correspondence [31, 32].[29] This partition function clearly vanishes if the Euclidean time circle is non-contractible,[30] but it is generically non-zero (when suitably defined, as in our discussion of §4.5 and the analogous discussions in [9, 35]) if the time circle is contractible. In all the examples that we know of, a static Euclidean time cycle is contractible in general relativity only when the corresponding Minkowskian solution has a horizon. Outside the range of validity of general relativity it would seem reasonable to define black holes by this requirement. Putting all this together, we conclude that suitably defined Polyakov loops are non-zero only in a black hole phase. Thus, we generally expect that the string theory (at finite temperature) in Phases II and III will be in a generalized black hole background, while the theory in Phase I (where the time cycle is non-contractible) should be in a background similar to the naive gauge theory background (in which the time direction is simply compactified).

A similar conclusion follows by analyzing the other order parameter

verified in examples of confining field theories in flat space, such as the duality between a cascading $d = 4$ $SU(N+M) \times SU(N)$ $\mathcal{N} = 1$ supersymmetric gauge theory and type IIB string theory on a manifold which includes the resolved conifold [26, 27]. Similar results for the Maldacena-Nunez background [28] were obtained in [29], and for other theories in (for example) [30].

28 By this we mean that if the background has a geometrical interpretation it would resemble a black hole; we generalize this to other backgrounds by calling such phases "black holes".

29 In this case the boundary of the string is actually not precisely the Polyakov loop, as it contains also couplings to scalar fields. We expect that these additional couplings will not change the qualitative features of the order parameter, and that these "generalized Polyakov loops" will vanish if and only if the usual Polyakov loop vanishes. Computations of such "generalized Polyakov loops" in the AdS/CFT correspondence first appeared in [9, 33, 34].

30 Again, this is true when we have a good geometrical description of the background, and otherwise we take the vanishing of the Polyakov loop to be the definition of a non-contractible circle.

for confinement, $\lim_{N\to\infty} F(T)/N^2$. In string theory this order parameter maps to the sphere partition function, which naively always vanishes. However, this is not true even in the low-energy general relativity approximation for some types of unbounded spaces [9,11], where the naive classical action diverges and one needs a subtraction procedure to define it. This can lead to a non-zero result for the relative classical partition function between two spaces with the same asymptotics but a different interior. This again suggests that deconfined phases must involve different space-times than the confined phase (which are the same asymptotically but differ in the interior), which have different classical actions. The discussion of the previous paragraph suggests that these spaces should be generalized black holes.

References

1. G. 't Hooft, "A planar diagram theory for strong interactions", Nucl. Phys. **B72**, 461 (1974).
2. A.M. Polyakov, "String theory and quark confinement", Nucl. Phys. Proc. Suppl. **68**, 1 (1998) [hep-th/9711002] and references therein.
3. R. Hagedorn, "Statistical thermodynamics of strong interactions at high-energies", Nuovo Cim. Suppl. **3**, 147 (1965).
4. B. Sundborg, "The Hagedorn transition, deconfinement and $N = 4$ SYM theory", Nucl. Phys. **B573**, 349 (2000) [hep-th/9908001].
5. A.M. Polyakov, "Gauge fields and space-time", Int. J. Mod. Phys. **A17S1**, 119 (2002) [hep-th/0110196].
6. J. Hallin and D. Persson, "Thermal phase transition in weakly interacting, large N_c QCD", Phys. Lett. **B429**, 232 (1998) [hep-ph/9803234].
7. D.J. Gross and E. Witten, "Possible third order phase transition in the large N lattice gauge theory", Phys. Rev. **D21**, 446 (1980); S.R. Wadia, "A study of $U(N)$ lattice gauge theory in two-dimensions", preprint EFI-79/44-CHICAGO; S.R. Wadia, "$N = \infty$ phase transition in a class of exactly soluble model lattice gauge theories", Phys. Lett. **B93**, 403 (1980).
8. O. Aharony, J. Marsano, S. Minwalla, K. Papadodimas and M. Van Raamsdonk, work in progress.
9. E. Witten, "Anti-de Sitter space, thermal phase transition, and confinement in gauge theories", Adv. Theor. Math. Phys. **2**, 505 (1998) [hep-th/9803131].
10. J.M. Maldacena, "The large N limit of superconformal field theories and supergravity", Adv. Theor. Math. Phys. **2**, 231 (1998) [Int. J. Theor. Phys. **38**, 1113 (1999)] [hep-th/9711200]; S.S. Gubser, I.R. Klebanov and A.M. Polyakov, "Gauge theory correlators from non-critical string theory", Phys. Lett. **B428**, 105 (1998) [hep-th/9802109]; E. Witten, "Anti-de Sitter space and holography", Adv. Theor. Math. Phys. **2**, 253 (1998) [hep-th/9802150].
11. S.W. Hawking and D.N. Page, "Thermodynamics of black holes in anti-de Sitter space", Commun. Math. Phys. **87**, 577 (1983).
12. O. Aharony, J. Marsano, S. Minwalla, K. Papadodimas and M. Van Raamsdonk, hep-th/0310285.

13. Y.Y. Goldschmidt, "$1/N$ expansion in two-dimensional lattice gauge theory", J. Math. Phys. **21**, 1842 (1980).
14. J. Polchinski, "High temperature limit of the confining phase", Phys. Rev. Lett. **68**, 1267 (1992) [hep-th/9109007].
15. D. Kutasov, "Two-dimensional QCD coupled to adjoint matter and string theory", Nucl. Phys. **B414**, 33 (1994) [hep-th/9306013].
16. C.P. Burgess, N.R. Constable and R.C. Myers, "The free energy of $N = 4$ Super Yang-Mills and the AdS/CFT correspondence", *J. High En. Phys.* 08 (1999) 017 [hep-th/9907188].
17. D. Kutasov and F. Larsen, "Partition sums and entropy bounds in weakly coupled CFT", *J. High En. Phys.* 01 (2001) 001 [hep-th/0009244].
18. K. Landsteiner and E. Lopez, "The thermodynamic potentials of Kerr-AdS black holes and their CFT duals", *J. High En. Phys.* 12 (1999) 020 [hep-th/9911124].
19. R.D. Pisarski and M. Tytgat, "Why the $SU(infinity)$ deconfining phase transition might be of second order", hep-ph/9702340.
20. J.L. Barbon and E. Rabinovici, "Closed-string tachyons and the Hagedorn transition in AdS space", *J. High En. Phys.* 03 (2002) 057 [hep-th/0112173].
21. N.R. Constable, D.Z. Freedman, M. Headrick, S. Minwalla, L. Motl, A. Postnikov and W. Skiba, "PP-wave string interactions from perturbative Yang-Mills theory", *J. High En. Phys.* 07 (2002) 017 [hep-th/0205089]; C. Kristjansen, J. Plefka, G.W. Semenoff and M. Staudacher, "A new double-scaling limit of $N = 4$ super Yang-Mills theory and PP-wave strings", Nucl. Phys. **B643**, 3 (2002) [hep-th/0205033]; N. Beisert, C. Kristjansen, J. Plefka, G.W. Semenoff and M. Staudacher, "BMN correlators and operator mixing in $N = 4$ super Yang-Mills theory", Nucl. Phys. **B650**, 125 (2003) [hep-th/0208178]; N.R. Constable, D.Z. Freedman, M. Headrick and S. Minwalla, "Operator mixing and the BMN correspondence", *J. High En. Phys.* 10 (2002) 068 [hep-th/0209002]; D.J. Gross, A. Mikhailov and R. Roiban, "Operators with large R charge in $N = 4$ Yang-Mills theory", Annals Phys. **301**, 31 (2002) [hep-th/0205066]; R. de Mello Koch, A. Jevicki and J.P. Rodrigues, "Collective string field theory of matrix models in the BMN limit", hep-th/0209155; N. Beisert, C. Kristjansen, J. Plefka and M. Staudacher, "BMN gauge theory as a quantum mechanical system", Phys. Lett. **B558**, 229 (2003) [hep-th/0212269]; J. Gomis, S. Moriyama and J.W. Park, "SYM description of SFT Hamiltonian in a pp-wave background", hep-th/0210153; J. Gomis, S. Moriyama and J.W. Park, "SYM description of pp-wave string interactions: Singlet sector and arbitrary impurities", Nucl. Phys. **B665**, 49 (2003) [hep-th/0301250]; R. de Mello Koch, A. Donos, A. Jevicki and J.P. Rodrigues, "Derivation of string field theory from the large N BMN limit", Phys. Rev. **D68**, 065012 (2003) [hep-th/0305042].
22. J.L. Barbon and E. Rabinovici, "Extensivity versus holography in anti-de Sitter spaces", Nucl. Phys. **B545**, 371 (1999) [hep-th/9805143]; J.L. Barbon, I.I. Kogan and E. Rabinovici, "On stringy thresholds in SYM/AdS thermodynamics", Nucl. Phys. **B544**, 104 (1999) [hep-th/9809033]; S.A. Abel, J.L. Barbon, I.I. Kogan and E. Rabinovici, "String thermodynamics in D-brane backgrounds", *J. High En. Phys.* 04 (1999) 015 [hep-th/9902058]; S.A. Abel, J.L. Barbon, I.I. Kogan and E. Rabinovici, "Some thermodynamical aspects of string theory", hep-th/9911004.
23. D.S. Berman and M.K. Parikh, "Confinement and the AdS/CFT correspon-

dence", Phys. Lett. **B483**, 271 (2000) [hep-th/0002031].
24. J.L. Barbon and E. Rabinovici, "Remarks on black hole instabilities and closed string tachyons", Found. Phys. **33**, 145 (2003) [hep-th/0211212].
25. O. Aharony, S.S. Gubser, J.M. Maldacena, H. Ooguri and Y. Oz, "Large N field theories, string theory and gravity", Phys. Rep. **323**, 183 (2000) [hep-th/9905111].
26. I.R. Klebanov and E. Witten, "Superconformal field theory on three-branes at a Calabi-Yau singularity", Nucl. Phys. **B536**, 199 (1998) [hep-th/9807080]; I.R. Klebanov and N.A. Nekrasov, "Gravity duals of fractional branes and logarithmic RG flow", Nucl. Phys. **B574**, 263 (2000) [hep-th/9911096]; I.R. Klebanov and A.A. Tseytlin, "Gravity duals of supersymmetric $SU(N) \times SU(N+M)$ gauge theories", Nucl. Phys. **B578**, 123 (2000) [hep-th/0002159]; I.R. Klebanov and M.J. Strassler, "Supergravity and a confining gauge theory: Duality cascades and χSB-resolution of naked singularities", *J. High En. Phys.* 08 (2000) 052 [hep-th/0007191].
27. A. Buchel, "Finite temperature resolution of the Klebanov-Tseytlin singularity", Nucl. Phys. **B600**, 219 (2001) [hep-th/0011146]; A. Buchel, C.P. Herzog, I.R. Klebanov, L.A. Pando Zayas and A.A. Tseytlin, "Non-extremal gravity duals for fractional D3-branes on the conifold", *J. High En. Phys.* 04 (2001) 033 [hep-th/0102105]; S.S. Gubser, C.P. Herzog, I.R. Klebanov and A.A. Tseytlin, "Restoration of chiral symmetry: A supergravity perspective", *J. High En. Phys.* 05 (2001) 028 [hep-th/0102172].
28. J.M. Maldacena and C. Nunez, "Towards the large N limit of pure $N = 1$ super Yang Mills", Phys. Rev. Lett. **86**, 588 (2001) [hep-th/0008001].
29. A. Buchel and A.R. Frey, "Comments on supergravity dual of pure $N = 1$ super Yang Mills theory with unbroken chiral symmetry", Phys. Rev. **D64**, 064007 (2001) [hep-th/0103022]; A. Buchel, "On the thermodynamic instability of LST", hep-th/0107102; S.S. Gubser, A.A. Tseytlin and M.S. Volkov, "Non-Abelian 4-d black holes, wrapped 5-branes, and their dual descriptions", *J. High En. Phys.* 09 (2001) 017 [hep-th/0108205].
30. D.Z. Freedman and J.A. Minahan, "Finite temperature effects in the supergravity dual of the $N = 1^*$ gauge theory", *J. High En. Phys.* 01 (2001) 036 [hep-th/0007250]; A. Buchel and J.T. Liu, "Thermodynamics of the $N = 2^*$ flow", hep-th/0305064.
31. J.M. Maldacena, "Wilson loops in large N field theories", Phys. Rev. Lett. **80**, 4859 (1998) [hep-th/9803002].
32. S.-J. Rey and J.-T. Yee, "Macroscopic strings as heavy quarks in large N gauge theories and anti-de Sitter supergravity", Eur. Phys. J. **C22**, 379 (2001) [hep-th/9803001].
33. S.-J. Rey, S. Theisen and J.-T. Yee, "Wilson-Polyakov loop at finite temperature in large N gauge theory and anti-de Sitter supergravity", Nucl. Phys. **B527**, 171 (1998) [hep-th/9803135].
34. A. Brandhuber, N. Itzhaki, J. Sonnenschein and S. Yankielowicz, "Wilson loops in the large N limit at finite temperature", Phys. Lett. **B434**, 36 (1998) [hep-th/9803137]; A. Brandhuber, N. Itzhaki, J. Sonnenschein and S. Yankielowicz, "Wilson loops, confinement, and phase transitions in large N gauge theories from supergravity", *J. High En. Phys.* 06 (1998) 001 [hep-th/9803263].
35. O. Aharony and E. Witten, "Anti-de Sitter space and the center of the gauge group", *J. High En. Phys.* 11 (1998) 018 [hep-th/9807205].

TWO-LOOP COMMUTING CHARGES AND THE STRING/GAUGE DUALITY

GLEB ARUTYUNOV, MATTHIAS STAUDACHER
Max-Planck-Institut für Gravitationsphysik
Albert-Einstein-Institut
Am Mühlenberg 1, D-14476 Potsdam, Germany
agleb@aei.mpg.de, matthias@aei.mpg.de

We briefly review the status quo of the application of integrable systems techniques to the AdS/CFT correspondence in the large charge approximation, a rapidly evolving topic. Intricate string and gauge computations of, respectively, energies and scaling dimensions agree at the one and two-loop level, but disagree starting from three loops. To add to this pattern, we present further computations which demonstrate that for folded and circular spinning strings the full tower of infinitely many hidden commuting charges, responsible for the integrability, also agrees up to *two*, but not three, loops.

1. The status quo

A fresh approach to the AdS/CFT correspondence was initiated in [1]. The proposal involved e.g. the study of composite operators in the planar $\mathcal{N} = 4$ gauge theory containing as constituents complex scalar fields Z,Φ of the form

$$\mathrm{Tr}\, Z^{J_1} \Phi^{J_2} + \dots \,, \tag{1}$$

and to consider the situation where $J = J_1 + J_2$ becomes large, while J_2 remains small. The dots indicate arbitrary orderings of the scalars inside the trace. It was argued that in this limit the conjectured dual string becomes essentially free, and the spectrum therefore explicitly calculable. Furthermore, the string energies turned out to be *analytic* at small BMN coupling

$$\lambda' = \frac{\lambda}{J^2} \,, \tag{2}$$

where λ is the usual large color 't Hooft coupling. The string energy is interpreted on the gauge side as the anomalous dimension of the corresponding operator eq.(1). Expanding in λ' therefore *looks* like a perturbative expansion of the dimension in the dual gauge theory. And indeed the one- and two-loop predictions for the BMN operators eq.(1) were successfully reproduced in [1],[2].

The large J limit of [1] was given in [3] an alternative interpretation as a semi-classical approximation to the string sigma model. This way of thinking allows for a generalization of the BMN limit: One may, more generally, assume several charges to be large [4] (see also [5], and the closely related earlier work [6]). In the case of our operators in eq.(1) one thus assumes both J_1 and J_2 to be large. It was then shown [4] that the semi-classical computations of string energies could be performed exactly, leading to predictions for the anomalous dimensions of these operators which are rather intricate as they generically depend in mathematically complicated ways on both the coupling λ' and the ratio

$$\alpha = \frac{J_2}{J} \,. \tag{3}$$

In subsequent papers it was understood that the reason for the exact solvability of the semi-classical string motions may be traced back to *integrability* [7]. For a string propagating on a five-sphere the evolution equations are those of the O(6) sigma model. In light-cone coordinates (ξ, η), where $\tau = \xi + \eta$ and $\sigma = \xi - \eta$, they read

$$X_{\xi\eta} + (X_\xi \cdot X_\eta) X = 1 \,, \qquad (X \cdot X) = 1. \tag{4}$$

Here X is a six-dimensional vector describing the embedding of the string world-sheet in S^5. The equations are supplemented by the Virasoro constraints. The simplest solutions correspond to rigid strings, i.e. string configurations with a time-independent shape. Rigid strings are naturally classified in terms of the Neumann integrable system [7], which is a special one-dimensional reduction of eqs.(4).

The string equations of motion inherit their integrability from the one of the string sigma model [8]. The underlying reason for the solvability is then an infinite family of mutually commuting Pohlmeyer charges $\{\mathcal{Q}_k\}$, where $\{\mathcal{Q}_2\}$ denotes the string energy.

How to check these predictions for the $\mathcal{N} = 4$ gauge theory? The problem is a priori difficult, even at one-loop, due to the complicated mixing between operators differing by the ordering of the scalars inside the trace. To deal with these problems, operatorial methods were developed in [9], not necessarily restricted to the planar case. However, for the planar case, a very important feature was noticed in [10]: The one-loop anomalous dimensions may be obtained by the diagonalization of an *integrable* quantum spin chain. In the example of the operators of eq.(1) the Z and Φ fields correspond to "up" and "down" spins. This integrable structure leads to the appearance of an infinite set of mutually commuting one-loop gauge charges $\{Q_k^{(1)}\}$, which generate (planar) symmetries which, mysteriously and yet-to-be-understood, are *not* part of the known symmetries of the superconformal $\mathcal{N} = 4$ theory. With one exception: $\{Q_2^{(1)}\}$, which is the one-loop approximation to the model's dilatation operator. Excitingly, investigating two-loop corrections, it was found that the charges continue to commute beyond the one-loop level [11]. This led to the conjecture [11] that the one-loop charges are actually only the leading order approximation of coupling constant dependent commuting charges $\{Q_k(\lambda)\}$, and that the exact dilatation operator $\{Q_2(\lambda)\}$ might thus be integrable.

The integrability of the one-loop planar gauge theory for scalar operators leads to the existence of a Bethe Ansatz for the anomalous dimensions [10]. Incidentally, one can show that this remains true for the set of all planar one-loop anomalous dimensions [12], as the one-loop dilatation operator, not necessarily restricted to the planar limit, is known for the most general $\mathcal{N} = 4$ composite operator [13]. The Bethe Ansatz allows one to quantitatively address the problem of the large J_1,J_2 limit of the operators in eq.(1), and to subsequently compare to the semi-classical string results. And indeed, this programme recently enjoyed considerable success. Investigating the two simplest types of string motions corresponding to the operators eq.(1) (the *folded* and the *circular* string), it was shown in [14],[16] that the one-loop string energies [4],[15],[7] and the gauge theory anomalous dimensions agree: $\mathcal{Q}_2^{(1)} \sim Q_2^{(1)}$. For agreement for other types of motion, see the string [17] and gauge computations [18]; for a review, see also [19].

In [20] it was shown, for folded and circular strings, that *all* one-loop charges $\mathcal{Q}_{2k}^{(1)} \sim Q_{2k}^{(1)}$ agree, establishing a direct connection between the respective integrable structures. This involved applying a Bäcklund transformation to the string motions, and extracting the charges from the Bethe Ansatz for the gauge theory. For a beautiful extension of this approach to the case involving three complex scalars, see [21]. In [22] it was shown that the two-loop energies $\mathcal{Q}_2^{(2)} \sim Q_2^{(2)}$ agree. This required a generalization of the spin chain picture to a long range chain, and an extension of the Bethe

Ansatz to higher loops[a].

Below we will extend these matchings by proving that also the two-loop charges $\mathcal{Q}_{2k}^{(2)} \sim Q_{2k}^{(2)}$ coincide.

While the matching is done for the two specific cases we are convinced that the relations between string and gauge charges are universal. As we will see, they are not only independent of the type of motion, but also of the parameter α in eq.(3). Of course it would be nice to present a general, solution independent proof. A first important step into this direction was undertaken, at one loop, in [24].

Very recently the most general one- and two-loop solution of the large J gauge theory scaling dimensions for the operators eq.(1) was obtained by Kazakov, Marshakov, Minahan and Zarembo [25]. Furthermore, a Bethe equation for the classical string sigma model was derived, and it was argued that the quantum spin chain Bethe equations may be obtained from it, mapping the integrable structures up to two loops for the most general solution. These results based on the monodromy approach are completely consistent with our extension of the Bäcklund approach to two loops described in this work below.

Not all is well, however. Serious problems appear at the *three* loop level. Indeed, the three-loop dilatation operator for the operators eq.(1) had been conjectured in [11], see also [23], and rigorously derived in [26]. Completely independent confirmation comes from a conceptually, but not technically related study of plane wave matrix theory at three loops [27]. Strictly speaking, the three loop dilatation generator is only known up to two unknown constants: The procedure of [26] allows, to the considered order, to add to $Q_2^{(3)}$, with arbitrary coefficients, the second ($Q_2^{(1)}$) and fourth ($Q_4^{(1)}$) one-loop conserved charge. However, both additions would be inconsistent with the perturbative BMN limit, and therefore certainly also with the large J_1,J_2 "spinning limit". However, three-loop integrability is *proven* in [26]. Therefore, the Bethe Ansatz of [22] applies, but the obtained three loop string and gauge energies *disagree*, $\mathcal{Q}_2^{(3)} \nsim Q_2^{(3)}$, and the same is true for all other charges. Below we will show that the difference is nevertheless "small" in the sense that it may always be accounted for

[a]The procedure applied in [22], using a long range chain invented by Inozemtsev, accurately diagonalizes the dilatation operator for the states eq.(1) up to three loops. At four loops, a violation of BMN scaling was detected. This might either indicate that BMN scaling is indeed violated in the gauge theory, or alternatively, that a further integrable long range chain, different from Inozemtsev, exists. Some evidence for the second scenario comes from the results of [23]. However, this issue is beyond the present scope.

by a simple non-local combination of charges, generalizing the "curious observation" of [22]. An earlier three-loop string-gauge disagreement was reported for the easier (on the gauge side) case of $1/J$ corrections to the BMN limit in [28].

Let us assume that the AdS/CFT correspondence is not a near-symmetry, but actually correct. One possible explanation for the disagreement might be the fact that the parameter λ' in eq.(2) is not obtained in the same fashion in string and gauge theory. This possibility was first discussed, to our knowledge, in [22]. For an earlier discussion in a slightly different but related context, see [29]. In the "semiclassical" string theory calculations we take λ and J to be large in a coupled fashion, and form the finite coupling λ'. In *perturbative* gauge theory, we first assume λ to be small, and subsequently take J to infinity. The difference in procedure might be expressible as certain non-perturbative corrections on either side of the correspondence. Taking them into account will hopefully eventually show that indeed IIB string theory on the background $AdS_5 \times S^5$ and $\mathcal{N} = 4$ gauge theory are one and the same thing.

2. Two-loop string/gauge matching of infinitely many higher commuting charges

Let us now show explicitly that the integrable structures on both sides of the correspondence agree, up to two loops, in the two distinct cases of the folded and the circular spinning string. This will be done by combining the results of [20] and [22] in order to show that the full tower of commuting charges matches up to $\mathcal{O}(\lambda'^2)$. In the following we will concentrate on the new, two-loop aspects of this matching, and refer to [20],[22] for further background, technical details, and precise notations. On the string side the generating function of charges (resolvent) was found by applying the Bäcklund transformation to the solutions of the classical string equations of motion, *cf.* eqs.(4). The result for the "nearly improved" string resolvent reads, in both cases,

$$\tilde{\mathcal{E}}(\mu) = \frac{4\mu^3}{\pi} \frac{\sqrt{(1-z)(1-tz)}}{\sqrt{z}} \, \Pi(tz, t). \tag{5}$$

Here μ is the string spectral parameter, t is the string modulus, and the function $\Pi(tz, t)$ is the complete elliptic integral of the third kind. The auxiliary parameter z depends on μ, t, and $\mathcal{J} = J/\sqrt{\lambda}$, i.e. the charge in units of the string tension. In order to expand this expression, it is convenient to use the form

$$\tilde{\mathcal{Q}}(\varphi,\lambda') = \frac{1}{2} - \frac{1}{2}\,\mu^{-2}\,\frac{\tilde{\mathcal{E}}(\mu)}{\mathcal{J}} = \frac{1}{2} - 2\varphi\frac{\sqrt{\mp(1-z)(1-tz)}}{\sqrt{z}}\,\Pi(tz,t)\,, \quad (6)$$

The string resolvent then becomes a function of the rescaled spectral parameter φ and the BMN coupling constant λ',

$$\varphi^2 = \mp\frac{\mu^2}{\pi^2\mathcal{J}^2} \qquad \text{and} \qquad \lambda' = \frac{1}{\mathcal{J}^2}\,. \quad (7)$$

Here and in the following the sign $\mp$ depends on the type of string motion (folded: upper sign, circular: lower sign), which is entirely due to our conventions. The modulus $t = t(\lambda';\alpha)$ depends on the coupling constant λ', and also encodes information on the filling fraction $\alpha = J_2/J$, *cf.* eqs.(19),(20) and eqs.(31),(32). The auxiliary parameter $z = z(\varphi,\lambda';\alpha)$ also slightly differs in the two considered cases. It is determined by eqs. (21),(33) below.

This "nearly improved" string resolvent generates a set of string charges $\{\tilde{\mathcal{Q}}_{2k}(\lambda')\}$

$$\tilde{\mathcal{Q}}(\varphi,\lambda') = \sum_{k=0}^{\infty}\tilde{\mathcal{Q}}_{2k}(\lambda')\,\varphi^{2k}, \quad (8)$$

such that the lowest (zeroth) charge is linearly related to the string energy $\mathcal{Q}_2(\lambda') = \mathcal{E}/\mathcal{J}$ by $\tilde{\mathcal{Q}}_0(\lambda') = \frac{1}{2}(1-\mathcal{Q}_2(\lambda'))$. Of course we can alternatively, or in addition, expand in the coupling constant λ':

$$\tilde{\mathcal{Q}}(\varphi,\lambda') = \sum_{n=0}^{\infty}\tilde{\mathcal{Q}}^{(n+1)}(\varphi)\,\lambda'^n = \sum_{n,k=0}^{\infty}\tilde{\mathcal{Q}}_{2k}^{(n+1)}\,\lambda'^n\,\varphi^{2k}. \quad (9)$$

On the gauge side the resolvent for a set of gauge charges $\{\bar{Q}_{2k}(\lambda')\}$ was also found perturbatively, up to three loops, by using a Bethe Ansatz for a long range spin chain originally invented by Inozemtsev [30],[31]:

$$H(\varphi,\lambda') = \pm\sum_{k=1}^{\infty}\bar{Q}_{2k}(\lambda')\,\varphi^{2k}\,. \quad (10)$$

Here φ is a gauge spectral parameter which is obtained from the scaled rapidity of the Inozemtsev-Bethe Ansatz. Expanding in λ', we have

$$H(\varphi,\lambda') = \pm\sum_{n=0}^{2}\bar{Q}^{(n+1)}(\varphi)\,\lambda'^n = \pm\sum_{n=0}^{2}\sum_{k=1}^{\infty}\bar{Q}_{2k}^{(n+1)}\,\lambda'^n\,\varphi^{2k}. \quad (11)$$

It is important to note that the charges $\bar{Q}_{2k}^{(n)}$ are not the correct "observables" of the spin chain. E.g. the three-loop gauge anomalous dimension is obtained by the following specific linear combination:

$$Q_2(\lambda') = 1 \mp \frac{\lambda'}{4\pi^2}\, \bar{Q}_2(\lambda') - \frac{3\lambda'^2}{64\pi^4}\, \bar{Q}_4(\lambda') \mp \frac{5\lambda'^3}{512\pi^6}\, \bar{Q}_6(\lambda'). \tag{12}$$

Likewise, while in the spin chain any linear combination of local charges leads again to a set of local charges, in the gauge theory these linear superpositions are essentially fixed: At one loop, the charges are required to obey BMN scaling [20], and loop corrections are then determined by quantum field theory. The proper three-loop charges will thus look like ($k > 0$)

$$Q_{2k}(\lambda') = \bar{Q}_{2k}(\lambda') \mp e_{k,1}\lambda'\, \bar{Q}_{2k+2}(\lambda') + e_{k,2}\lambda'^2\, \bar{Q}_{2k+4}(\lambda')\,. \tag{13}$$

While it is certainly possible to find these (universal) numbers $e_{k,1}$, $e_{k,2}$, it is easily seen that we do not require them for the present purposes of two-loop matching: All we need to show[b] is that a linear map from the set of "nearly improved" string charges $\{\tilde{\mathcal{Q}}_{2k}(\lambda')\}$ to the set of gauge charges $\{\bar{Q}_{2k}(\lambda')\}$ exists to leading and next-to-leading order in λ'! By explicit inspection of the first few two-loop string and gauge charges, one finds that ($k \geq 0$, and $\bar{Q}_0^{(2)} \equiv 0$)

$$\tilde{\mathcal{Q}}_{2k}^{(2)} = \pm\bar{Q}_{2k}^{(2)} \pm \frac{1}{8\pi^2}(2k+1)\, \bar{Q}_{2k+2}^{(1)}\,. \tag{14}$$

This leads to the following proposition

$$\tilde{\mathcal{Q}}(\varphi,\lambda') = \bar{Q}^{(1)}(\varphi) \pm \lambda' \left[\bar{Q}^{(2)}(\varphi) + \frac{1}{8\pi^2}\frac{\partial}{\partial\varphi}\frac{1}{\varphi}\, \bar{Q}^{(1)}(\varphi)\right] + \mathcal{O}(\lambda'^2) \tag{15}$$

It can be written in an elegant fashion as follows:

$$\tilde{\mathcal{Q}}(\varphi,\lambda') = \Big(1 \mp \frac{\lambda'}{8\pi^2\varphi^2}\Big) H\Big(\varphi \pm \frac{\lambda'}{8\pi^2\varphi}, \lambda'\Big) + \mathcal{O}(\lambda'^2)\,. \tag{16}$$

This is our main new result, and will be proven in the next sections for the case of the folded and circular string. It proves our assertion of the two-loop matching of string and gauge charges. Although derived starting from particular solutions, we believe that (16) is universal in the sense that

---

[b]This also relieves us from applying the procedure of "full improvement" to the string resolvent eq.(5), as discussed in [20], p. 24. At any rate, the latter would only assure us of the correct leading BMN scaling behavior of the charges; one can show that the obtained charges do *not* agree with the "proper" gauge charges at the two-loop level. Thus a further linear, upper triangle redefinition would be required.

it does not depend on specific solutions[c], and that it expresses the general matching of string/gauge integrable structures at two loops.

Now we discuss the three-loop case. As was already noticed in [22] the gauge/string energies disagree starting from three loops. It is however remarkable that the disagreement between the whole towers of gauge/string commuting charges admits a uniform description.

First, we expand eq.(6) up to λ'^2 and identify the term $\tilde{\mathcal{Q}}^{(3)}(\varphi)$. Then we perform the Gauss-Landen transformation (30) to bring this term to the gauge theory frame. Our two-loop matching formula (14) would seem to suggest that the matching of the three-loop gauge and string resolvents should again be given by a linear relation

$$\tilde{\mathcal{Q}}_{2k}^{(3)} \stackrel{?}{=} \bar{Q}_{2k}^{(3)} + \alpha_{k,1}\, \bar{Q}_{2k+2}^{(2)} + \alpha_{k,2}\, \bar{Q}_{2k+4}^{(1)}\,. \tag{17}$$

Working out explicitly the first few Inozemtsev charges one may verify that this proposal does not work, since the coefficients $\alpha_{k,1}$ and $\alpha_{k,2}$ then appear to be functions of the modular parameter q_0. Thus, linear combinations of Inozemtsev charges with constant coefficients cannot reproduce the string result. This motivates us to extend the set of charges by adding also their products. By trial and error we found the following remarkable formula ($k \geq 0$, and $\bar{Q}_0^{(2)} \equiv \bar{Q}_0^{(3)} \equiv 0$)

$$\begin{aligned} \tilde{\mathcal{Q}}_{2k}^{(3)} &= \bar{Q}_{2k}^{(3)} + \frac{2k+1}{8\pi^2}\, \bar{Q}_{2k+2}^{(2)} + \frac{(2k+1)(2k+3)}{128\pi^4}\, \bar{Q}_{2k+4}^{(1)} \\ &\quad - \frac{1}{8\pi^2}\, \bar{Q}_2^{(1)}\, \bar{Q}_{2k}^{(2)} \end{aligned} \tag{18}$$

which describes the relation between the gauge and string towers of commuting charges up to and including three loops.

2.1. *Folded string*

The classical motion of the folded string leads to the following parametric result for the all-loop string energy $\mathcal{Q}_2$:

$$\lambda' = \frac{\pi^2}{4\mathrm{K}(t_0)^2}\left[\left(\frac{\mathrm{K}(t_0)-\mathrm{E}(t_0)}{\mathrm{K}(t)-\mathrm{E}(t)}\right)^2 - \left(\frac{\mathrm{E}(t_0)}{\mathrm{E}(t)}\right)^2\right], \tag{19}$$

$$\mathcal{Q}_2 = \frac{\mathrm{K}(t)}{\mathrm{K}(t_0)}\sqrt{(1-t)\left(\frac{\mathrm{E}(t_0)}{\mathrm{E}(t)}\right)^2 + t\left(\frac{\mathrm{K}(t_0)-\mathrm{E}(t_0)}{\mathrm{K}(t)-\mathrm{E}(t)}\right)^2}\,.$$

[c]The sign flips are entirely due to a slight difference in the normalization of gauge theory charges for the folded and circular cases, *cf.* [20]. Defining the charges in an identical fashion in both cases removes this difference.

Here the modulus t contains the information about the coupling λ'. In addition, its constant piece t_0 parametrizes the "filling fraction"

$$\alpha = 1 - \frac{\mathrm{E}(t_0)}{\mathrm{K}(t_0)} \,. \tag{20}$$

The charges are found by the Bäcklund transformation; the final result for the string resolvent is given in eq.(6), where the "auxiliary" and true spectral parameters z, φ are related through

$$\varphi^2 + \left(\frac{1}{4\mathrm{K}(t_0)} \frac{\mathrm{E}(t_0)}{\mathrm{E}(t)}\right)^2 \frac{z(1-tz)}{1-z} + \frac{\lambda'}{4\pi^2} tz = 0\,. \tag{21}$$

Expanding this in λ' one finds

$$\tilde{\mathcal{Q}}(\varphi, \lambda') = \tilde{\mathcal{Q}}^{(1)}(\varphi) + \lambda' \tilde{\mathcal{Q}}^{(2)}(\varphi) + \mathcal{O}(\lambda'^2)\,. \tag{22}$$

Here the leading piece is

$$\tilde{\mathcal{Q}}^{(1)}(\varphi) = \frac{1}{2} - 2\varphi \frac{\sqrt{(1-z_0)(t_0 z_0 - 1)}}{\sqrt{z_0}} \,\Pi(t_0 z_0, t_0)\,. \tag{23}$$

where

$$\varphi^2 + \left(\frac{1}{4\mathrm{K}(t_0)}\right)^2 \frac{z_0(1-t_0 z_0)}{1-z_0} = 0\,. \tag{24}$$

The next term $\tilde{\mathcal{Q}}^{(2)}(\varphi)$ is rather complicated and we will not write it out.

The Bethe Ansatz for the spin chain describing two-loop gauge theory in the large J limit quite generally leads to a singular integral equation. In the present case this equation is of elliptic type, and closely related to the one appearing in the so-called $\mathrm{O}(N)$ matrix model, see e.g. [32]. Its solution leads to the following generating functions for the one-loop charges $\bar{Q}^{(1)}_{2k}$

$$\bar{Q}^{(1)}(\varphi) = \frac{1}{4} - \frac{a_0^2}{b_0} \sqrt{\frac{b_0^2 - \varphi^2}{a_0^2 - \varphi^2}} \,\Pi\left(-q_0 \frac{\varphi^2}{a_0^2 - \varphi^2}, q_0\right) \tag{25}$$

and two-loop charges $\bar{Q}^{(2)}_{2k}$

$$\bar{Q}^{(2)}(\varphi) = \frac{1}{32\pi^2 \varphi^2} \left(1 - \frac{b_0}{2a_0}\sqrt{\frac{a_0^2 - \varphi^2}{b_0^2 - \varphi^2}} - \frac{a_0}{2b_0}\sqrt{\frac{b_0^2 - \varphi^2}{a_0^2 - \varphi^2}}\right)\,. \tag{26}$$

Here the natural modulus q_0 parametrizes the filling fraction through

$$\alpha = \frac{1}{2} - \frac{1}{2\sqrt{1-q_0}} \frac{\mathrm{E}(q_0)}{\mathrm{K}(q_0)}\,. \tag{27}$$

while the leading-order rapidity boundaries a_0, b_0 are given by

$$a_0 = \frac{1}{4\mathrm{K}(q_0)}, \qquad b_0 = \frac{1}{4\sqrt{1-q_0}\,\mathrm{K}(q_0)}. \tag{28}$$

It is now straightforward to calculate the expression

$$\bar{Q}^{(2)}(\varphi) + \frac{1}{8\pi^2}\frac{\partial}{\partial\varphi}\frac{1}{\varphi}\,\bar{Q}^{(1)}(\varphi) = \frac{a_0^2 + b_0^2 - 8a_0 b_0^2 \mathrm{E}(q_0)}{64\pi^2 a_0 b_0 \sqrt{(a_0^2-\varphi^2)(b_0^2-\varphi^2)}}. \tag{29}$$

As was shown in [20] after a Gauss-Landen transformation of the moduli

$$t_0 = -\frac{(1-\sqrt{1-q_0})^2}{4\sqrt{1-q_0}}, \tag{30}$$

eq.(23) becomes identical to eq.(25). Now we observe that the same transformation nicely turns the rather complicated expression $\tilde{Q}^{(2)}(\varphi)$ in eq.(22) into the r.h.s. of eq.(29), proving our main assertion eq.(15).

2.2. *Circular string*

The analysis of the circular string is very similar to one for the folded case. The BMN coupling constant λ' and the string energy $\mathcal{Q}_2$ are given by

$$\lambda' = \frac{\pi^2 t}{4t_0^2 \mathrm{K}(t_0)^2}\left[\left(\frac{\mathrm{K}(t_0)-\mathrm{E}(t_0)}{\mathrm{K}(t)-\mathrm{E}(t)}\right)^2 - \left(\frac{\mathrm{E}(t_0)-(1-t_0)\mathrm{K}(t_0)}{\mathrm{E}(t)-(1-t)\mathrm{K}(t)}\right)^2\right], \tag{31}$$

$$\mathcal{Q}_2 = \frac{t\mathrm{K}(t)}{t_0\mathrm{K}(t_0)}\sqrt{\frac{1}{t}\left(\frac{\mathrm{K}(t_0)-\mathrm{E}(t_0)}{\mathrm{K}(t)-\mathrm{E}(t)}\right)^2 - \frac{1-t}{t}\left(\frac{\mathrm{E}(t_0)-(1-t_0)\mathrm{K}(t_0)}{\mathrm{E}(t)-(1-t)\mathrm{K}(t)}\right)^2},$$

where the parameter t_0 is determined via the filling fraction as follows

$$\alpha = 1 - \frac{1}{t_0} + \frac{1}{t_0}\frac{\mathrm{E}(t_0)}{\mathrm{K}(t_0)}. \tag{32}$$

The string spectral parameters z and φ are related through

$$\varphi^2 - \frac{t^2}{t_0^2}\left(\frac{1}{4\mathrm{K}(t_0)}\frac{\mathrm{K}(t_0)-\mathrm{E}(t_0)}{\mathrm{K}(t)-\mathrm{E}(t)}\right)^2 \frac{z(1-z)}{1-tz} - \frac{\lambda'}{4\pi^2}\frac{(1-t)z}{1-tz} = 0. \tag{33}$$

On the gauge theory side the configuration of Bethe roots corresponding to the circular string is described by two cuts on the imaginary axis: $ic < i\varphi < id$ and $-id < i\varphi < -ic$, with a constant condensate in between. For the one-loop problem we denote the endpoints of the cut as c_0 and d_0 and introduce the gauge theory modulus as $r_0 = \frac{c_0^2}{d_0^2}$, where $c_0 = \frac{1}{8\mathrm{K}(r_0)}$. The modulus

r_0 is related to the filling fraction α as follows $\alpha = \frac{1}{2} - \frac{1}{2\sqrt{r_0}} + \frac{1}{2\sqrt{r_0}}\frac{\mathrm{E}(r_0)}{\mathrm{K}(r_0)}$. Then the one-loop resolvent is

$$\bar{Q}^{(1)}(\varphi) = \frac{1}{4} - \frac{2}{d_0}\sqrt{(d_0^2-\varphi^2)(c_0^2-\varphi^2)}\;\Pi\left(\frac{\varphi^2}{d_0^2}, r_0\right). \tag{34}$$

The two-loop correction found from the Inozemtsev-Bethe Ansatz reads

$$\bar{Q}^{(2)}(\varphi) = \frac{1}{32\pi^2\varphi^2}\left(1 - \frac{d_0}{2c_0}\sqrt{\frac{c_0^2-\varphi^2}{d_0^2-\varphi^2}} - \frac{c_0}{2d_0}\sqrt{\frac{d_0^2-\varphi^2}{c_0^2-\varphi^2}}\right). \tag{35}$$

Using the two formulae above one finds

$$\bar{Q}^{(2)}(\varphi) + \frac{1}{8\pi^2}\frac{\partial}{\partial\varphi}\frac{1}{\varphi}\,\bar{Q}^{(1)}(\varphi) = \frac{c_0^2 - d_0^2 + 16c_0d_0^2\mathrm{E}(r_0)}{64\pi^2 c_0 d_0\sqrt{(c_0^2-\varphi^2)(d_0^2-\varphi^2)}} \tag{36}$$

Expanding in λ' the resolvent (6) corresponding to the circular string one identifies the leading and the subleading terms, which are $\tilde{\mathcal{Q}}^{(1)}(\varphi)$ and $\tilde{\mathcal{Q}}^{(2)}(\varphi)$ respectively. Quite remarkably, the Gauss-Landen transformation

$$t_0 = -\frac{4\sqrt{r_0}}{(1-\sqrt{r_0})^2} \tag{37}$$

transforms these expressions into eqs.(34) and (36).

Acknowledgments

We would like to thank Niklas Beisert for important discussions. M.S. thanks Vladimir Dobrev and all other organizers of the *Fifth International Workshop on Lie Theory and its Applications in Physics* in Varna (Bulgaria) in June 2003 for an inspiring conference, and their warm hospitality.

References

1. D. Berenstein, J.M. Maldacena and H. Nastase, *J. High En. Phys.* 04 (2002) 013, [hep-th/0202021].
2. D.J. Gross, A. Mikhailov and R. Roiban, *Annals Phys.* **301**, 31 (2002), [hep-th/0205066].
3. S.S. Gubser, I.R. Klebanov and A.M. Polyakov, *Nucl. Phys.* **B636**, 99 (2002), [hep-th/0204051].
4. S. Frolov and A.A. Tseytlin, *Nucl. Phys.* **B668**, 77 (2003), [hep-th/0304255].
5. S. Frolov and A.A. Tseytlin, *J. High En. Phys.* 06 (2002) 007, [hep-th/0204226].
6. H.J. de Vega, A.L. Larsen and N. Sanchez, *Phys. Rev.* **D51**, 6917 (1995), [hep-th/9410219].

7. G. Arutyunov, S. Frolov, J. Russo and A.A. Tseytlin, *Nucl. Phys.* **B671**, 3 (2003), [hep-th/0307191]; G. Arutyunov, J. Russo and A.A. Tseytlin, hep-th/0311004.
8. K. Pohlmeyer, *Commun. Math. Phys.* **46**, 207 (1976).
9. C. Kristjansen, J. Plefka, G.W. Semenoff and M. Staudacher, *Nucl. Phys.* **B643**, 3 (2002), [hep-th/0205033]; N.R. Constable, D.Z. Freedman, M. Headrick, S. Minwalla, L. Motl, A. Postnikov and W. Skiba, *J. High En. Phys.* 07 (2002) 017, [hep-th/0205089]; N. Beisert, C. Kristjansen, J. Plefka, G.W. Semenoff and M. Staudacher, *Nucl. Phys.* **B650**, 125 (2003), [hep-th/0208178]; N.R. Constable, D.Z. Freedman, M. Headrick and S. Minwalla, *J. High En. Phys.* 10 (2002) 068, [hep-th/0209002]; N. Beisert, C. Kristjansen, J. Plefka and M. Staudacher, *Phys. Lett.* **B558**, 229 (2003), [hep-th/0212269].
10. J.A. Minahan and K. Zarembo, *J. High En. Phys.* 03 (2003) 013, [hep-th/0212208].
11. N. Beisert, C. Kristjansen and M. Staudacher, *Nucl. Phys.* **B664**, 131 (2003), [hep-th/0303060].
12. N. Beisert and M. Staudacher, *Nucl. Phys.* **B670**, 439 (2003), [hep-th/0307042].
13. N. Beisert, *Nucl. Phys.* **B676**, 3 (2004), [hep-th/0307015].
14. N. Beisert, J.A. Minahan, M. Staudacher and K. Zarembo, *J. High En. Phys.* 09 (2003) 010, [hep-th/0306139].
15. S. Frolov and A.A. Tseytlin, *Phys. Lett.* **B570**, 96 (2003), [hep-th/0306143].
16. N. Beisert, S. Frolov, M. Staudacher and A.A. Tseytlin, *J. High En. Phys.* 10 (2003) 037, [hep-th/0308117].
17. J.G. Russo, *J. High En. Phys.* 06 (2002) 038, [hep-th/0205244]; J.A. Minahan, *Nucl. Phys.* **B648**, 203 (2003), [hep-th/0209047]; A.A. Tseytlin, *Int. J. Mod. Phys.* **A18**, 981 (2003), [hep-th/0209116]; S. Frolov and A.A. Tseytlin, *J. High En. Phys.* 07 (2003) 016, [hep-th/0306130]; B.J. Stefanski, hep-th/0312091; B. Chen, X.J. Wang and Y.S. Wu, hep-th/0403004.
18. J. Engquist, J.A. Minahan and K. Zarembo, *J. High En. Phys.* 11 (2003) 063, [hep-th/0310188]; C. Kristjansen, hep-th/0402033.
19. A.A. Tseytlin, hep-th/0311139.
20. G. Arutyunov and M. Staudacher, *J. High En. Phys.* 03 (2004) 004, [hep-th/0310182].
21. J. Engquist, hep-th/0402092.
22. D. Serban and M. Staudacher, hep-th/0401057.
23. N. Beisert, *J. High En. Phys.* 09 (2003) 06, [hep-th/0308074].
24. M. Kruczenski, hep-th/0311203.
25. V.A. Kazakov, A. Marshakov, J.A. Minahan and K. Zarembo, hep-th/0402207.
26. N. Beisert, hep-th/0310252.
27. T. Klose and J. Plefka, *Nucl. Phys.* **B679**, 127 (2004), [hep-th/0310232].
28. C.G. Callan, H.K. Lee, T. McLoughlin, J.H. Schwarz, I. Swanson and X. Wu, *Nucl. Phys.* **B673**, 3 (2003), [hep-th/0307032].

29. I.R. Klebanov, M. Spradlin and A. Volovich, *Phys. Lett.* **B548**, 111 (2002), [hep-th/0206221].
30. V.I. Inozemtsev, *J. Stat. Phys.* **59**, 1143 (1990).
31. V.I. Inozemtsev, Phys. Part. Nucl. **34**, 166 (2003), [hep-th/0201001].
32. I.K. Kostov and M. Staudacher, *Nucl. Phys.* **B384**, 459 (1992), [hep-th/9203030].

BOUNDARY GROUND RING AND DISC CORRELATION FUNCTIONS IN LIOUVILLE QUANTUM GRAVITY

IVAN KOSTOV

Service de Physique Théorique, CNRS – URA 2306, C.E.A. - Saclay,
F-91191 Gif-Sur-Yvette, France;
Associate member of INRNE - BAS, Sofia, Bulgaria
kostov@spht.saclay.cea.fr

We construct the boundary ground ring in $c \leq 1$ open string theories with non-zero boundary cosmological constant (FZZT brane), using the Coulomb gas representation. The ring relations yield an over-determined set of functional recurrence equations for the boundary correlation functions, which involve shifts of the the target space momenta of the boundary fields as well as the boundary parameters on the different segments of the boundary.

1. Introduction

The non-critical string theory, known also as 2D quantum gravity, has two complementary descriptions: the world sheet description in terms of Liouville CFT and the target space description in terms of random matrix models (for reviews see [1, 2, 3, 4, 5]). It has been observed in the early 90's that both descriptions possess integrable structures, the KP and Toda integrable hierarchies for the matrix models and the 'ground ring' for the world sheet CFT. There are some similarities between the integrable structures on both sides, which might be used, if well understood, to build a more direct connection between the two dual descriptions of 2D string theory more direct. Such a connection could be very helpful when studying strong perturbations of the 2D string theory as the 2D Euclidean black hole.

The integrable structure of the world sheet CFT [6, 7, 8] is based on the so called ground ring of operators. It has been shown by Witten [6] that there exists a ring of dimensionless ghost-number zero operators, with respect to which the local observables form a module. The action of ground ring leads to to a set of recurrence equations [9] for the correlation functions of the primary fields (the closed string 'tachyons'). It has been also observed [6, 9] that the ground ring relations resemble the string equations in the corresponding matrix models[a]. The ground ring structure was generalized in [10] to open string theory, described by a boundary CFT. The action of the 'boundary ground ring' yields a set of recurrence relations for the correlation functions of boundary operators, or open string 'tachyons'. These (over determined) relations reproduced elegantly the results obtained previously by Coulomb gas integrals [12, 13, 14, 15].

Originally the ring relations were applied only for the so called 'resonant' amplitudes. These are the correlation functions for vanishing cosmological constant, when the Liouville interaction is not taken into account. In the full theory, the resonant amplitudes give the residues of the 'on-mass-shell' poles of the exact amplitudes. It was conjectured [6] that the perturbations like the cosmological term are described by certain deformations of the ground ring structure, but the exact field-theoretical meaning of this conjecture remained obscure. Only after the impressive developments in Liouville CFT from the last several years [16, 17, 18, 19, 20] it has been realized that the ground ring structure is an *exact* symmetry of the full CFT [21, 11] . The recurrence relations that follow from the ground ring are the string theory version of the relations obtained by the conformal bootstrap in Liouville theory.

In these notes we will focus on boundary 2D quantum gravity, which is characterized by two cosmological constants: the bulk constant μ coupled to the area of the world sheet and the boundary constant μ_B coupled to the length of the boundary. In the CFT description these are the coupling constants for the bulk and boundary Liouville interaction. In Liouville theory the boundary term can be as well introduced as a non-homogeneous Neumann boundary condition [18], which is also referred to as FZZT brane.

A deformation of the boundary ground ring structure by bulk and boundary Liouville interactions was proposed in [22]. It was shown that the ring structure yields a set of functional recurrence equations, which completely determine all boundary correlation functions on a disc. Crucial role in the derivation plays the precise definition of Liouville degenerate boundary primaries given in [18], which we sketch below.

The observables in Liouville CFT with boundary are meromorphic functions of the boundary cosmological constant μ_B with a cut along the semi-infinite interval $[-\infty, -\mu_B^0]$, where $\mu_B^0 \sim \sqrt{\mu}$. The branch point singularity can be resolved by the uniformization map

$$\mu_B = \mu_B^0 \cosh(\pi b s),$$

[a]This analogy explained and developed further in the recent paper [11].

where b is the Liouville coupling constant and s is a dimensionless parameter. It happens that it is more appropriate to label the Liouville boundary conditions by the values of s instead of μ_B. Thus a Liouville boundary operator is specified by three numbers: the Liouville charge β and the two boundary parameters s_{left} and s_{right}. In [18], Fateev, Zamolodchikov and Zamolodchikov (FZZ) observed that the correctly defined lowest degenerate boundary operators introduce shifts of the boundary parameter $s \to s \pm ib$ or $s \to s \pm ib^{-1}$ at the points of the boundary where they are inserted[b]. The fusion rules with such an operator yield a functional equation for the boundary two-point function (see the concluding section of [18]), which can be written as a *difference equation* with respect to one of the boundary parameters. It was subsequently found [23] that all boundary Liouville structure constants satisfy similar difference equations.

In these notes we review and slightly generalize the result of [22] (there the matter field obeys $U(1)$ fusion rules) by adding the two screening charges to the effective action. The ring relations lead to a set of functional recurrence identities for all boundary correlators, from which we derive a pair of difference equations with respect to one of the Liouville boundary parameters. We will see that the screening charges add an extra term to the functional equations of [22], but the difference equations remain the same.

2. Preliminaries: world-sheet CFT for the disc amplitudes

2.1. *The effective action*

We consider the Gaussian field realization [24] of the matter CFT, which is very convenient for actual calculations. Then the Euclidean 2D string theory is defined in terms of a Liouville field ϕ with background charge Q and a Gaussian matter field χ with background charge e_0. The simplest form of the effective action is achieved by mapping the disc to the upper half plane $\{\text{Im } x \geq 0\}$. Then the bulk and boundary curvatures are concentrated at

[b]The general validity of this suggestion, which was proven in [18] only for the two-point function, follows from the formula for the three point function derived subsequently in [19].

infinity and the effective action takes the form [18, 30]

$$\mathcal{A}[\phi,\chi] = \int\limits_{\text{Im } x\geq 0} d^2x \left(\frac{1}{4\pi}[(\nabla\phi)^2 + (\nabla\chi)^2] + \mu e^{2b\phi}\right) + \int\limits_{-\infty}^{\infty} dx\, \mu_B\, e^{b\phi} + \text{ghosts} \tag{2.1}$$

The couplings μ and μ_B are are referred to as bulk and the boundary cosmological constants. The matter and Liouville background charges

$$e_0 = 1/b - b, \quad Q = 1/b + b \tag{2.2}$$

are introduced through the asymptotics of the fields at infinity

$$\phi(x,\bar{x}) \sim -Q \log|x|^2, \quad \chi(x,\bar{x}) \sim -e_0 \log|x|^2. \tag{2.3}$$

The two background charges satisfy $Q^2 - e_0^2 = 4$, which is equivalent to the balance of the central charge $c_{\text{tot}} \equiv c_\phi + c_\chi + c_{\text{ghosts}} = (1+6Q^2) + (1 - 6e_0^2) - 26 = 0$.

The boundary term encodes a inhomogeneous Neumann boundary condition for the Liouville field

$$i(\partial - \bar{\partial})\phi(x,\bar{x}) = 4\pi\mu_B\, e^{b\phi(x,\bar{x})} \qquad (\bar{x} = x), \tag{2.4}$$

the so called FZZT brane. The matter field is assumed to satisfy pure Neumann boundary condition:

$$i(\partial - \bar{\partial})\chi(x,\bar{x}) = 0 \qquad (\bar{x} = x), \tag{2.5}$$

which is of course Dirichlet boundary condition for the T-dual field $\tilde{\chi}(x,\bar{x})$.

The Coulomb gas mapping requires also to add to the action (2.1) the bulk and boundary screening charges

$$Q_+ = \mu_+ \int_{\text{Im } x\geq 0} d^2x\, e^{-2ib\chi}, \quad Q_- = \mu_- \int_{\text{Im } x\geq 0} d^2x e^{2i\chi/b}), \tag{2.6}$$

$$Q_+^B = \mu_+^B \int_{-\infty}^{\infty} dx e^{-ib\chi}, \quad Q_-^B = \mu_-^B \int_{-\infty}^{\infty} dx\, e^{i\chi/b} \tag{2.7}$$

The values of the 'couplings' $\mu_\pm$ and $\mu_\pm^B$ are related to the normalization of the correlation functions. Below we will choose particular values for which the fusion rules acquire simplest form.

2.2. *Bulk and boundary vertex operators*

The bulk primary fields (closed string tachyons) $\mathcal{V}_P^{(\pm)}(x,\bar{x})$ are defined as the bulk vertex operators

$$\begin{aligned} \mathcal{V}_P^{(+)} &= \tfrac{1}{\pi}\gamma(bP)\ e^{i(e_0-P)\chi+(Q-P)\phi} \\ \mathcal{V}_P^{(-)} &= \tfrac{1}{\pi}\gamma\left(-\tfrac{1}{b}P\right)\ e^{i(e_0-P)\chi+(Q+P)\phi} \end{aligned} \tag{2.8}$$

where $\gamma(x) \equiv \Gamma(x)/\Gamma(1-x)$. This normalization removes the external 'leg pole' factors in the correlation functions and is the one to be used when comparing with the microscopic realizations of 2D QG. The boundary primary fields $\mathcal{B}_P^{(\pm)}(x)$, or left/right moving on-mass-shell open string tachyons, are defined as the boundary vertex operators

$$\begin{aligned} \mathcal{B}_P^{(+)} &= \tfrac{1}{\pi}\ \Gamma(2bP)\ e^{i\left(\frac{1}{2}e_0-P\right)\chi+\left(\frac{1}{2}Q-P\right)\phi} \\ \mathcal{B}_P^{(-)} &= \tfrac{1}{\pi}\ \Gamma\left(-\tfrac{2}{b}P\right)\ e^{i\left(\frac{1}{2}e_0-P\right)\chi+\left(\frac{1}{2}Q+P\right)\phi}. \end{aligned} \tag{2.9}$$

As any CFT boundary operator, the open string tachyon is completely defined only after the left and right boundary conditions are specified [25]. The matter component is assumed to satisfy homogeneous Neumann boundary condition, and the Liouville left and right boundary conditions are determined by the values of the uniformization parameter s defined as

$$\mu_B = \mu_B^0 \cosh(\pi b s). \tag{2.10}$$

Following [18], we will denote such an operator as ${}^{s_1}[\mathcal{B}_P^{(\pm)}]^{s_2}$.

The tachyons of opposite chiralities are related by the Liouville bulk and boundary reflection amplitudes

$$\mathcal{V}_P^{(+)} = S_P^{(+-)}\ \mathcal{V}_P^{(-)}, \qquad {}^{s_1}[\mathcal{B}_P^{(+)}]^{s_2} = D_P^{(+-)}(s_1,s_2)\ {}^{s_1}[\mathcal{B}_P^{(-)}]^{s_2}. \tag{2.11}$$

For each value of the momentum there is only one vertex operator that corresponds to a physical state. The physical operators are

$$\mathcal{V}_P = \begin{cases} \mathcal{V}_P^{(+)} & \text{if } P<0, \\ \mathcal{V}_P^{(-)} & \text{if } P>0. \end{cases} \qquad \mathcal{B}_P = \begin{cases} \mathcal{B}_P^{(+)} & \text{if } P<0, \\ \mathcal{B}_P^{(-)} & \text{if } P>0. \end{cases} \tag{2.12}$$

The "wrongly dressed" operators are related to the physical ones by (2.11).

2.3. *Normalization of the bulk and boundary cosmological constants and the screening operators*

It is convenient to redefine the cosmological constants $\mu \to \Lambda$ and $\mu_B \to z$ according to the normalizations 2.8 of the vertex operators :

$$\mu e^{2b\phi} = \Lambda\ \mathcal{V}_{e_0}^{(+)}, \quad \mu_B\ e^{b\phi} = z\ \mathcal{B}_{e_0/2}^{(+)}. \tag{2.13}$$

The new cosmological constants are related to the old ones by

$$\Lambda = \pi\gamma(b^2)\mu, \qquad z = \pi[\Gamma(1-b^2)]^{-1}\mu_B. \tag{2.14}$$

Now the uniformization map 2.10 reads

$$z = M\cosh(\pi b s), \qquad M = \sqrt{\Lambda}. \tag{2.15}$$

Also, it will be convenient to normalize thé screening charges as

$$Q_\pm = \int\limits_{\mathrm{Im}\ x>0} d^2x\ \mathcal{V}^{(\pm)}_{\pm Q}, \quad Q^B_\pm = \int\limits_{-\infty}^{\infty} dx\ \mathcal{V}^{(\pm)}_{\pm Q/2}\ . \tag{2.16}$$

In the normalization (2.13) the self-duality property [18] of Liouville quantum gravity reads

$$b \to \tilde{b} = 1/b, \qquad \Lambda \to \tilde{\Lambda}, \qquad M \to \tilde{M},$$

with

$$\tilde{\Lambda} = \Lambda^{1/b^2} \qquad \text{or} \qquad \tilde{M} = M^{1/b^2}. \tag{2.17}$$

The boundary parameter s is self-dual. The duality transformation for the correlation functions follows from the duality transformation of left and right chiral tachyons

$$\tilde{\mathcal{V}}^{(\pm)}_P = \overline{\mathcal{V}^{(\mp)}_{-P}}, \qquad {}^{s_1}[\tilde{\mathcal{B}}^{(\pm)}_P]^{s_2} = \overline{{}^{s_1}[\mathcal{B}^{(\mp)}_{-P}]^{s_2}}, \tag{2.18}$$

where the bar means complex conjugation.

2.4. *Degenerate fields*

By degenerate fields in Liouville quantum gravity we will understand the gravitationally dressed degenerate matter fields. In our case these are the on-shell vertex operators (2.12) with degenerate matter components

$$\mathcal{V}_{rs} = \mathcal{V}_{r/b-sb}, \qquad \mathcal{B}_{rs} = \mathcal{B}_{(r/b-sb)/2} \qquad (r, s \in N). \tag{2.19}$$

The flat scaling dimensions of these fields are

$$\Delta_{rs} = \frac{(r/b - sb)^2 - e_0^2}{4}.$$

The fusion rules for the degenerate in Liouville quantum gravity are the same as the fusion rules for the degenerate fields in the matter CFT [29]. Note that the Liouville components of these fields are not degenerate Liouville fields.

3. Bulk ground ring

3.1. *The ring relations*

Before considering the boundary ground ring, we will give a short review of the ground ring structure for a theory without boundary. The operators that span the ground ring are obtained by applying a Virasoro raising operator of level $rs-1$ to the product of two degenerate matter and Liouville fields with Kac labels r, s. The resulting operators have conformal weights $\Delta = \bar{\Delta} = 0$. The ring is generated by the first two lowest operators [6]

$$\begin{aligned} a_+ &= -|\mathbf{bc} - b\partial_x(\phi - i\chi)|^2 \, e^{-b^{-1}(\phi + i\chi)} \\ a_- &= -|\mathbf{bc} - b^{-1}\partial_z(\phi + i\chi)|^2 \, e^{-b(\phi - i\chi)} \end{aligned} \tag{3.20}$$

where $\mathbf{b}, \mathbf{c}$ are the reparametrization ghost and anti-ghost fields. The ground ring is spanned on the polynomials $(a_+)^m(a_-)^n$ with $m, n \in Z_+$. In the case of non-rational b^2 the ground ring contains no other relations and has an infinite number of elements labeled by the integers $r, s \geq 1$.

A crucial property of the operators $a_\pm$ is that their derivatives $\partial_x a_\pm$ and $\partial_{\bar{x}} a^\pm$ are BRST exact: $\partial_x a_- = \{Q_{\text{BRST}}, \mathbf{b}_{-1} a_-\}$. Therefore, any amplitude that involves $a_\pm$ and other BRST invariant operators does not depend on the position of $a_\pm$. This property allows to write recurrence equations for the correlation functions from the OPE of $a_\pm$ and the other BRST-invariant fields.

The vertex operators (2.8) form a module under the ground ring:

$$a_+ \mathcal{V}_P^{(+)} = -\mathcal{V}_{P+\frac{1}{b}}^{(+)}, \qquad a_- \mathcal{V}_P^{(-)} = -\mathcal{V}_{P-b}^{(-)}. \tag{3.21}$$

and also

$$a_+ \mathcal{V}_P^{(-)} = a_- \mathcal{V}_P^{(+)} = 0. \tag{3.22}$$

Both relations (3.21) and (3.22) follow from the free field OPE and are true up to commutators with the BRST charge. While the first one survives in the correlation functions, the second one receives non-linear corrections. Due to the contact terms, the last relation should be modified in presence of an integrated vertex operator to

$$a_+ \mathcal{V}_P^{(-)} \int d^2 z \; \mathcal{V}_{P_1}^{(-)} = \mathcal{V}_{P+P_1+b}^{(-)} \tag{3.23}$$

$$a_- \mathcal{V}_P^{(+)} \int d^2 z \; \mathcal{V}_{P_1}^{(+)} = \mathcal{V}_{P+P_1-1/b}^{(+)}. \tag{3.24}$$

These relations imply a set of recurrence equations [9, 10, 26], which determine completely the resonant tachyon amplitudes.

After taking into account the Liouville interaction and the screening charges second relation (3.22) gets deformed. Remarkably, the deformation can be calculated perturbatively by applying (3.23) and (3.24):

$$\begin{aligned} a_-\mathcal{V}_P^{(+)} &= -\Lambda\ \mathcal{V}_{P-b}^{(+)} + \mathcal{V}_{P+b}^{(+)} \\ a_+\mathcal{V}_P^{(-)} &= -\tilde{\Lambda}\ \mathcal{V}_{P+1/b}^{(-)} + \mathcal{V}_{P-1/b}^{(-)}. \end{aligned} \tag{3.25}$$

It is not at all evident that the linear insertions allow to calculate the exact action of $a_\pm$, but this has been justified by various self-consistency checks in Liouville theory [16, 17]. The ground ring structure in presence of screening charges has been considered in [27, 28].

3.2. *Functional equations for the bulk correlators*

Using the ring relations (3.21) and (3.25) one can obtain a set of recurrence equations for the correlation functions of the bulk tachyons

$$\begin{aligned} G(P_1,...,P_n|P_{n+1},...,P_{n+m}) = \Big\langle \left[\textstyle\prod_{k=1}^{n-1}\int \mathcal{V}_{P_k}^{(-)}\right] \mathbf{c}\bar{\mathbf{c}}\mathcal{V}_{P_n}^{(-)}(0)\times \\ \times \mathbf{c}\bar{\mathbf{c}}\mathcal{V}_{P_{n+1}}^{(+)}(1) \left[\textstyle\prod_{j=n+2}^{n+m-1}\int \mathcal{V}_{P_j}^{(+)}\right] \mathbf{c}\bar{\mathbf{c}}\mathcal{V}_{P_{n+m}}^{(+)}(\infty)\Big\rangle. \end{aligned} \tag{3.26}$$

Consider the auxiliary function

$$\begin{aligned} F(x,\bar{x}|P_1,...,P_n|P_{n+1},...,P_{n+m}) = \langle \left[\textstyle\prod_{k=1}^{n-1}\int \mathcal{V}_{P_k}^{(-)}\right] \mathbf{c}\bar{\mathbf{c}}\mathcal{V}_{P_n+b}^{(-)}(0)\times \\ \times a_-(x,\bar{x})\ \mathbf{c}\bar{\mathbf{c}}\mathcal{V}_{P_{n+1}}^{(+)}(1) \left[\textstyle\prod_{j=n+2}^{m+n-1}\int \mathbf{c}\bar{\mathbf{c}}\mathcal{V}_{P_j}^{(+)}\right] \mathbf{c}\bar{\mathbf{c}}\mathcal{V}_{P_{n+m}}^{(+)}(\infty)\rangle \end{aligned} \tag{3.27}$$

The function F does not depend on x and $\bar{x}$. This can be proved by using $\partial_x a_- = \{Q_{\mathrm{BRST}}, \mathbf{b}_{-1}a_-\}$ and deforming the contour, commuting Q_{BRST} with the other operators in (3.27) [10]. Therefore one can evaluate this function at $x = 0$ and $x = 1$ by using the fusion rules (3.21) and (3.25). As a result one obtains the recurrence relation

$$\begin{aligned} G(P_1,...,P_n|P_{n+1},...,P_{n+m}) = \Lambda\ G(P_1,...,P_n+b|P_{n+1}-b,...,P_{n+m}) - \\ +G(P_1,...,P_n-b|P_{n+1}-b,...,P_{n+m}) \\ -\textstyle\sum_{j=1}^{m-1} G(P_1,...,P_n+b|P_{n+2},...,P_{n+j}+P_{n+m}-\frac{1}{b},...,P_{n+m}). \end{aligned} \tag{3.28}$$

Similarly, by inserting a_+ we get the dual recurrence relation

$$\begin{aligned} G(P_1,...,P_n|P_{n+1},...,P_{n+m}) = \tilde{\Lambda}\ G(P_1,...,P_n+\tfrac{1}{b}|P_{n+1}-\tfrac{1}{b},...,P_{n+m}) \\ +\ G(P_1,...,P_n-\tfrac{1}{b}|P_{n+1}-\tfrac{1}{b},...,P_{n+m}) \\ -\textstyle\sum_{k=1}^{n-1} G(P_1,...,P_k+P_n+b,...,P_{n-1}|P_{n+1},...,P_{n+m}). \end{aligned} \tag{3.29}$$

Note that the three-point function $m + n = 3$ coincides with the corresponding Liouville three-point function. The latter has been evaluated by using a very similar argument in [30].

4. The boundary ground ring

4.1. *The ring relations*

Similarly to the bulk case the two generators of the boundary ground ring are defined as

$$\begin{aligned} A_+ &= -[\mathrm{bc} - \tfrac{1}{2}b\partial_x(\phi - i\chi)]\, e^{-\frac{1}{2}b^{-1}(\phi+i\chi)} \\ A_- &= -[\mathrm{bc} - \tfrac{1}{2}b^{-1}\partial_x(\phi + i\chi)]\, e^{-\frac{1}{2}b(\phi-i\chi)}. \end{aligned} \tag{4.1}$$

where b, c are the boundary reparametrization ghosts. The two operators are related by a duality transformation combined with complex conjugation:

$$\tilde{A}_+ = \overline{A}_-, \qquad \tilde{A}_- = \overline{A}_+. \tag{4.2}$$

These operators are BRST closed: $\partial_x A_\pm = \{Q_{\mathrm{BRST}}, \mathrm{b}_{-1}A_\pm\}$ and have $\Delta = 0$. The open string tachyons (2.9) form a module with respect to the ring generated by these two operators.

Let us first consider the simplest case of pure Neumann b.c. for the Liouville field ($\mu_B = 0$), following [10]. Then the action of the ring on the tachyon modules is generated by the relations

$$A_+\mathcal{B}_P^{(+)} = \mathcal{B}_{P+\frac{1}{2b}}^{(+)}, \qquad A_-\mathcal{B}_P^{(-)} = \mathcal{B}_{P-\frac{b}{2}}^{(-)} \tag{4.3}$$

$$A_+\mathcal{B}_P^{(-)} = A_-\mathcal{B}_P^{(+)} = 0. \tag{4.4}$$

Again, the first relation (4.3) is exact and the second relation (4.4) gets deformed in presence of integrated boundary tachyon fields [10] :

$$A_-\mathcal{B}_{P_1}^{(+)} \int dx \mathcal{B}_{P_2}^{(+)} = \frac{1}{\sin 2\pi b P_1}\, \mathcal{B}^{(+)}_{P_1+P_2-\frac{1}{2b}}. \tag{4.5}$$

$$\int dx \mathcal{B}_{P_1}^{(+)} A_-\mathcal{B}_{P_2}^{(+)} = \frac{\sin 2\pi b(P_1+P_2)}{\sin 2\pi b P_1\, \sin 2\pi b P_2}\, \mathcal{B}^{(+)}_{P_2+P_1-\frac{1}{2b}}. \tag{4.6}$$

$$A_+\mathcal{B}_{P_1}^{(-)} \int dx \mathcal{B}_{P_2}^{(-)} = \frac{1}{\sin \frac{2\pi}{b} P_1}\, \mathcal{B}^{(-)}_{P_1+P_2+\frac{b}{2}} \tag{4.7}$$

$$\int dx \mathcal{B}_{P_1}^{(-)} A_+\mathcal{B}_{P_2}^{(-)} = \frac{\sin \frac{2\pi}{b}(P_1+P_2)}{\sin \frac{2\pi}{b} P_2\, \sin \frac{2\pi}{b} P_1}\, \mathcal{B}^{(-)}_{P_1+P_2+\frac{b}{2}}. \tag{4.8}$$

The coefficients on the r.h.s. are obtained as standard Coulomb integrals. As in the bulk case, these relations yield an over-determined set of recurrence equations for the n-point open string amplitudes [10].

4.2. *Deformation of the ring relations by the boundary Liouville term and the screening charges*

With the screening charges (2.16) added to the effective action, the action of the ring on the vertex operators gets deformed according to the general relations (4.3) and (4.5). Namely the relations (4.4) become

$$A_- \mathcal{B}_P^{(+)} = n_P \cdot \mathcal{B}_{P+\frac{b}{2}}^{(+)}, \quad A_+ \mathcal{B}_P^{(-)} = \tilde{n}_P \cdot \mathcal{B}_{P-\frac{1}{2b}}^{(-)}, \tag{4.9}$$

where the coefficients m_P and $\tilde{m}_P$ are given by

$$\begin{aligned} n_P &= \tan(\pi b P) - \tan(\pi b e_0/2) \\ \tilde{n}_P &= \tan(\pi P/b) + \tan(\pi e_0/2b). \end{aligned} \tag{4.10}$$

The evaluation of the deformation of the the ring relations by the boundary Liouville interaction is more subtle. It is based on the observation made in [18] and subsequently confirmed in [19, 23] that a level-n degenerate boundary Liouville field $e^{-nb\phi}$ has vanishing null-vector and therefore a truncated OPE with the other primary fields if either $s_{\text{left}} - s_{\text{right}} = ibk$ or $s_{\text{left}} + s_{\text{right}} = ibk$, with $k = -n/2, -n/2+1, ..., n/2$. By the duality property of Liouville theory, the degenerate boundary fields $e^{-n\phi/b}$ exhibit a similar property with b replaced by $1/b$. No direct proof is supplied for this statement, but it was shown to be consistent with the exact results obtained in boundary Liouville theory. In particular, the above property leads to a pair of functional equations for the Liouville boundary reflection amplitude, whose unique solution coincides with the result of the standard conformal bootstrap [18].

According to [18], the operators (4.1) should be defined by

$$\begin{array}{lcl} A_+ & \to & {}^{s}[A_+]^{s\pm i/b} \\ A_- & \to & {}^{s}[A_-]^{s\pm ib} \end{array} \tag{4.11}$$

and the relations (4.3), (4.5), (4.7) and (4.8) understood as

$${}^{s_1}[\mathcal{B}_P^{(-)}]^{s}[A_-]^{s\pm ib} = -\ {}^{s_1}[\mathcal{B}_{P-\frac{b}{2}}^{(-)}]^{s\pm ib} \tag{4.12}$$

etc. Taking into account the three possible insertions of the boundary Liouville interaction boundary interaction one finds [22] that the relations (4.9) deform further to

$${}^{s_1}[A_-]^{s_1\pm ib}[\mathcal{B}_P^{(+)}]^{s_2} = m_P^{\pm} \cdot {}^{s_1}[\mathcal{B}_{P-\frac{b}{2}}^{(+)}]^{s_2} + n_P \cdot {}^{s_1}[\mathcal{B}_{P+\frac{b}{2}}^{(+)}]^{s_2} \tag{4.13}$$

$${}^{s_1}[A_+]^{s_1\pm i/b}[\mathcal{B}_P^{(-)}]^{s_2} = \tilde{m}_P^{\pm} \cdot {}^{s_1}[\mathcal{B}_{P+\frac{1}{2b}}^{(+)}]^{s_2} + \tilde{n}_P \cdot {}^{s_1}[\mathcal{B}_{P-\frac{1}{2b}}^{(+)}]^{s_2} \tag{4.14}$$

with the coefficients $m_P^{\pm} = m_P^{\pm}(s_1, s_2)$ and $\tilde{m}_P^{\pm} = \tilde{m}_P^{\pm}(s_1, s_2)$ given by

$$m_P^{\pm} = -M \,\frac{\cosh[\pi b(s_1 \pm 2iP)] + \cosh(\pi b s_2)}{\sin 2\pi b P}. \tag{4.15}$$

$$\tilde{m}_P^{\pm} = -M^{1/b^2} \,\frac{\cosh[\pi\,(s_1 \pm 2iP)\,/b] + \cosh(\pi s_2/b)}{\sin 2\frac{\pi}{b} P}. \tag{4.16}$$

The derivation of (4.13) and (4.14) essentially repeats the one presented in [18] for pure Liouville vertex operators and leads to the same fusion coefficients.

4.3. *Functional and difference equations for the correlation functions of boundary primary fields*

Consider a boundary correlation function of the form

$$W_{P_1,P_2,\ldots}(s_1, s, s_2, s_3, \ldots) = \langle\, {}^{s_1}[\mathcal{B}_{P_1}]^{s}[\mathcal{B}_{P_2}]^{s_2}[\mathcal{B}_{P_3}]^{s_3}\ldots\rangle \tag{4.17}$$

The realization of the physical boundary fields depends on the sign of the momenta and is given by (2.12). Our notations do not distinguish between integrated and non-integrated fields; it is assumed that three of the integrations should be replaced by ghost insertions. The amplitude (4.17) is analytic in the boundary parameters $s, s_1, \ldots$ but not in the target space momenta, since the fields with positive and negative momenta are realized by different vertex operators. By the momentum conservation the correlation function is zero unless $\sum_k(\frac{1}{2}e_0 - P_k) = e_0 + m/b - nb$ for some positive integers m and n. Equation (2.18) implies the identity

$$W_{P_1,P_2,\ldots} = \overline{W}_{-P_1,-P_2,\ldots} = W_{-P_1,-P_2,\ldots}. \tag{4.18}$$

Let us assume that $P_1 < 0,\ P_2 > 0$. Then the amplitude (4.17) is realized as

$$W_{P_1,P_2,P_3,\ldots}(s_1, s, s_2, s_3, \ldots) = \langle\, {}^{s_1}[\mathcal{B}^{(-)}_{P_1}]^{s}[\mathcal{B}^{(+)}_{P_2}]^{s_2}[\mathcal{B}^{(+)}_{P_3}]^{s_3}\ldots\rangle \tag{4.19}$$

By the symmetry (4.18) the amplitude can be written also as

$$\overline{W}_{P_1,P_2,P_3,\ldots}(s_1, s, s_2, s_3, \ldots) = \langle \ldots {}^{s_3}[\mathcal{B}^{(-)}_{-P_3}]^{s_2}[\mathcal{B}^{(-)}_{-P_2}]^{s}[\mathcal{B}^{(+)}_{-P_1}]^{s_1}\rangle. \tag{4.20}$$

We will use these two representations to derive two independent functional identities for W.

Consider the auxiliary correlation function F with an operator A_- inserted in the r.h.s. of (4.17):

$$F = \langle\, {}^{s_1}[\mathcal{B}^{(-)}_{P_1+\frac{b}{2}}]^{s}[A_-]^{s\pm ib}[\mathcal{B}^{(+)}_{P_2}]^{s_2}[\mathcal{B}^{(+)}_{P_3}]^{s_3}\ldots\rangle.$$

Let us assume further that $P_1 < -\frac{b}{2}$ and $P_2 > \frac{b}{2}$. Then, evaluating F by (4.12) and by (4.13) and equating the results we get the following functional identities for the correlation functions of three or more fields:

$$\begin{aligned} & W_{P_1,P_2,P_3,\dots}(s_1, s \pm ib, s_2, s_3, \dots) = \\ & + m_{P_2}(s, s_2) \cdot\ W_{P_1+\frac{b}{2},P_2-\frac{b}{2},P_3,\dots}(s_1, s, s_2, s_2, \dots) \\ & +\ n_{P_2} \cdot\ W_{P_1-\frac{b}{2},P_2-\frac{b}{2},P_3,\dots}(s_1, s, s_2, s_2, \dots) \\ & + \sin^{-1}(2\pi b P_2) \cdot\ W_{P_1+\frac{b}{2},P_2+P_3-\frac{1}{2b},\dots}(s_1, s, s_3, \dots). \end{aligned} \tag{4.21}$$

The last, 'contact', term represents a correlation function with one operator less. A dual equation can be obtained in the same way, assuming that $P_1 < -\frac{1}{2b}$ and $P_2 > \frac{1}{2b}$, by evaluating the complex conjugated function $\overline{F}$ using the representation (4.20) and the relations (4.2).

The symmetric part of (4.21) generalize the recurrence equations of [10] derived in the free-field ($\mu_B = 0$) limit. The antisymmetric part equation (4.21) and its dual represents a pair of homogeneous difference equations which do not have analog in the $\mu_B = 0$ limit. We write them in operator form:

$$\left[\sin(b\partial_s) - M \sinh(\pi b s)\ e^{\frac{b}{2}(\partial_{P_1}-\partial_{P_2})}\right] W_{P_1,P_2,P_3,\dots}(s_1, s, s_2, s_3, \dots) = 0 \tag{4.22}$$

for $P_1 < -\frac{b}{2}$ and $P_2 > \frac{b}{2}$, and

$$\left[\sin(\frac{1}{b}\partial_s) - M^{\frac{1}{b^2}} \sinh(\frac{\pi}{b}s)\ e^{\frac{1}{2b}(\partial_{P_1}-\partial_{P_2})}\right] W_{P_1,P_2,P_3,\dots}(s_1, s, s_2, s_3, \dots) = 0 \tag{4.23}$$

for $P_1 < -\frac{1}{2b}$ and $P_2 > \frac{1}{2b}$.

Note that these equations have the same form for matter CFT with or without screening charges. Difference equations for all values of the momenta can be obtained by the following rule [31, 23]. If the sign of one or both momenta changes after the shift, one should apply the reflection property with the amplitude $D_P^{(+-)}(s_1, s_2)$. The details are explained in [23, 22].

5. Discussion

We have constructed the boundary ground ring in the Coulomb gas realization of the matter CFT. Since any matter CFT can be constructed from the Coulomb gas, the validity of the difference equations (4.22) and (4.23) is universal. Unlike underlying functional equations, the form of the difference equations is not altered by adding the screening charges. Remarkably, these equations have their equivalent in the microscopic realization of 2D QG as loop gas on random planar graphs [32]. In this realization the degenerate boundary operators are represented geometrically as sources of open lines with endpoints at the boundary. As a matter of fact, the difference equations (4.22) and (4.23) have been first derived in the microscopic approach [31, 23] by cutting open the sum over triangulated surfaces along one of the lines and then using certain factorization property of the sum over surfaces. Of course, these equations are equivalent to certain word identities in the corresponding matrix model. This confirms the deep relation between the integrable structures in the world sheet and microscopic formulations of 2D QG.

Acknowledgments

I would like to express my gratitude to Valya Petkova for many discussions and suggestions and to all organizers of Lie Theory - 5 for inviting me to this nice workshop. This research is supported in part by the European network EUCLID, HPRN-CT-2002-00325.

References

1. P. Ginsparg and G. Moore, "Lectures on 2D gravity and 2D string theory (TASI 1992)", hep-th/9304011.
2. P. Di Francesco, P. Ginsparg and J. Zinn-Justin,"2-D Gravity and random matrices", *Phys. Rep.* **254**, 1 (1995), hep-th/9306153.
3. J. Polchinski, "What is string theory", Lectures presented at the 1994 Les Houches Summer School "Fluctuating Geometries in Statistical Mechanics and Field Theory", hep-th/9411028.
4. I. Klebanov, Lectures delivered at the ICTP Spring School on String Theory and Quantum Gravity, Trieste, April 1991, hep-th/9108019.
5. A. Jevicki, "Developments in 2-d string theory", hep-th/9309115.
6. E. Witten, "Ground ring of two-dimensional string theory", *Nucl. Phys.* **B373**, 187 (1992), hep-th/9108004.
7. E. Witten and B. Zwiebach, "Algebraic structures and differential geometry in 2D string theory", *Nucl. Phys.* **B377**, 55 (1992), hep-th/9201056.
8. I.R. Klebanov and A.M. Polyakov, "Interaction of discrete states in two-dimensional string theory", *Mod. Phys. Lett.* **A6**, 3273 (1991) hep-th/9109032.
9. D. Kutasov, E. Martinec and N. Seiberg, "Ground rings and their modules in 2-D gravity with $c \leq 1$ matter", *Phys. Lett.* **B276**, 437 (1992), hep-th/9111048.
10. M. Bershadsky and D. Kutasov, "Scattering of open and closed strings in (1+1)-dimensions", *Nucl. Phys.* **382**, 213 (1992), hep-th/9204049.

11. N. Seiberg and D. Shih, "Branes, rings and matrix models on minimal (super)string theory", hep-th/0312170.
12. M. Goulian and B. Li, *Phys. Rev. Lett.* **66**, 2051 (1991).
13. V. Dotsenko, *Mod. Phys. Lett.* **A6**, 3601 (1991).
14. M. Bershadsky and D. Kutasov, *Phys. Lett.* **274B**, 331 (1992), hep-th/9110034.
15. Y. Tanii and S.-I. Yamaguchi, "Two-dimensional quantum gravity on a disc", *Mod. Phys. Lett.* **A7**, 521 (1992), hep-th/9110068; "Disk Amplitudes in Two-Dimensional Open String Theories", hep-th/9203002.
16. H. Dorn and H.J. Otto, "Two and three point functions in Liouville theory", *Nucl. Phys.* **429**, 375 (1994), hep-th/9403141.
17. A.B. Zamolodchikov and Al.B. Zamolodchikov, "Structure constants and conformal bootstrap in Liouville field theory", *Nucl. Phys.* **477**, 577 (1996), hep-th/9506136.
18. V. Fateev, A.B. Zamolodchikov and Al.B. Zamolodchikov, "Boundary Liouville field theory. I: Boundary state and boundary two-point function", hep-th/0001012.
19. B. Ponsot and J. Teschner, "Boundary Liouville Field Theory: Boundary three point function", *Nucl. Phys.* **B622**, 309 (2002), hep-th/0110244.
20. K. Hosomichi, "Bulk-Boundary Propagator in Liouville Theory on a Disc", *J. High En. Phys.* 11 (2001) 044, hep-th/0108093.
21. M.R. Douglas, I.R. Klebanov, D. Kutasov, J. Maldacena, E. Martinec and N. Seiberg, "A new hat for the c = 1 matrix model", hep-th/0307195.
22. I. Kostov, hep-th/0312301.
23. I. Kostov, B. Ponsot and D. Serban, "Boundary Liouville Theory and 2D Quantum Gravity", hep-th/0307189.
24. V.S. Dotsenko and V.A. Fateev, "Four point correlations functions and the operator algebra in the two dimensional conformal invariant theories with the central charge $c < 1$", *Nucl. Phys.* **251**, 691 (1985).
25. J. Cardy, "Conformal invariance and surface critical behavior", *Nucl. Phys.* **240**, 514 (1984); "Boundary conditions, fusion rules and the Verlinde formula", *Nucl. Phys.* **324**, 581 (1989).
26. S. Kachru, "Quantum Rings and Recursion Relations in 2D Quantum Gravity", *Mod. Phys. Lett.* **A7**, 1419 (1992), hep-th/9201072.
27. S. Govindarajan, T. Jayaraman and V. John, "Genus Zero Correlation Functions in $c < 1$ String Theory", *Phys. Rev.* **D48**, 839 (1993), hep-th/9208064.
28. S. Govindarajan, T. Jayaraman and V. John, "Correlation Functions and Multicritical Flows in $c < 1$ String Theory", *Int. J. Mod. Phys.* **A10**, 477 (1995), hep-th/9309040.
29. A.A. Belavin, A.M. Polyakov and A.B. Zamolodchikov, "Infinite conformal symmetry in two-dimensional quantum field theory", *Nucl. Phys.* **B241**, 333-380 (1984).
30. J. Teschner, "On the Liouville Three-Point Function", *Phys. Lett.* **B363**, 65 (1995).
31. I. Kostov, "Boundary Correlators in 2D Quantum Gravity: Liouville versus Discrete Approach", *Nucl. Phys.* **658**, 397 (2003), hep-th/0212194.
32. I. Kostov, "Strings with discrete target space", *Nucl. Phys.* **376**, 539 (1992), hep-th/9112059.

STRINGS, INTEGRABLE SYSTEMS, GEOMETRY AND STATISTICAL MODELS

A. MARSHAKOV

Lebedev Physics Institute and ITEP, Moscow, Russia

The role of integrable systems in string theory is discussed. We remind old examples of the correspondence between stringy partition functions or effective actions and integrable equations, based on effective application of the matrix model technique. Then we turn to a new example, coming from the Nekrasov deformation of the Seiberg-Witten prepotential. In the last case the deformed theory is described by a different statistical model, which becomes equivalent to a partition function of a topological string. The full partition function of string theory arises therefore always as a certain "quantization" of its quasiclassical geometry.

1. Introduction

There is lack of algorithmic direct computational methods in non-perturbative physics. One of the working ideas is to replace the divergent series of perturbation theory by differential equations in some parameter space (say, in the space of couplings) and to study their solutions, which can be in many cases treated non-perturbatively. Another way is based on some discrete (e.g. lattice) regularization of continuous theory, which in practice turn it into some statistical model, where one may hope to reach more progress. Amazingly enough these two ways are closely related, at least in the class of models, where the exact non-perturbative results have been reached during last fifteen years, (see e.g. [1], where these issues were discussed from different angles of view).

Our experience shows, that general scheme of formulation of the non-perturbative results is always the same: the generating function $F = F(t)$ should be identified with a solution to a hierarchy of nonlinear differential equations and, when the system is known, is given by a logarithm of tau-function of this hierarchy. The string theory genus expansion

$$F = \sum_{g=0}^{\infty} \hbar^{2g-2} F_g \tag{1}$$

is then equivalent to the so-called *quasiclassical* expansion of F and therefore we denote the string coupling as a Planck constant $\hbar$.

The leading term of this expansion $F_0 = F_0(t, S)$, (depending usually on the set of extra parameters or moduli S) is already well-known for all examples and can be defined in pure *geometric* terms. The geometric data contain a complex manifold (in practically all studied cases a curve) Σ endowed with pair of functions (x, y), or with the generating meromorphic form $dS = ydx$, then F_0 can be identified with the Krichever quasiclassical tau-function [2], sometimes also called a "prepotential".

The whole sum in (1) can be thought of as a "quantization" of this geometry. Except for few simple examples (the most well-known is Kontsevich model and its generalizations [3, 4, 5]) yet there is no strict formulation of the full function F, but it is widely believed, that

$$Z = \exp(-F) \tag{2}$$

satisfies some (infinite system of) relatively simple linear differential equations, like Virasoro or W-constraints [6] together with nonlinear equations of an integrable hierarchy.

The pair of functions (x, y) entering the geometric data for F_0 may be thought of as "symplectic pair", and different solutions (their Baker-Akhiezer functions) are related by a sort of Fourier transform [7]

$$\Psi_i(y) \propto \int d\mu(x)\Psi_i(x)e^{\int ydx} \tag{3}$$

In particular, this is the origin of existence of integral representations of the solutions to Kontsevich-like models since by (3) they are "dual" to trivial tau-functions. The corresponding partition functions

$$Z\left(t_k = -\frac{1}{k}\sum x_j^{-k}\right) \propto \frac{\det \Psi_i(x_j)}{\Delta(x)} \tag{4}$$

are given by matrix version of (3), since for trivial Ψ-functions in the r.h.s. formula (3) gives rise to the matrix integral representation of Z. However, these exact full answers are known only for the cases when quasiclassical geometry is extremely simple, or, basically, when Σ is sphere. Already for the (p, q) critical points of two-dimensional gravity (with $p, q \neq 1$; $(p, 1)$ points correspond to generalized Kontsevich models [4]) generic curves are of genus $\frac{1}{2}(p-1)(q-1)$ [8] and nothing strictly is known about explicit form of the full F. Recently these problems were re-addressed to in [9] and from a different angle of view in [10].

Hence, the quasiclassical part of the story looks like being totally based on geometry. However, its "quantization" leads to relatively simple statistical models, like matrix integrals. The present hope is that this is a general phenomenon, and as a new example of this phenomenon we will consider the Nekrasov deformation [11] of the Seiberg-Witten theory, which leads to "Lie-algebraic construction" of the full partition function based on summing over the Young diagrams.

2. Matrix models and integrable systems

Matrix models are the most simple example of the "gauge-string correspondence" (2) where $\log Z = F$ obeys (1). A typical matrix model partition function is given by an integral

$$Z = \int d\Phi e^{-\frac{1}{\hbar}\mathrm{Tr}V(\Phi)} = \frac{1}{N!}\int \prod_{i=1}^{N}\left(dx_i e^{-\frac{1}{\hbar}V(x_i)}\right)\Delta^2(x) \tag{5}$$

where Φ is $N \times N$ matrix (in general situation the integral is taken over ensemble of matrices), for example Hermitean, the measure $d\Phi \propto \prod d\Phi_{ij}$ and potential $V(\Phi) = \sum t_k \Phi^k$ in this case is an arbitrary polynomial. The simple matrix integral like (5) can be rewritten as an integral over eigenvalues x_i of the matrix Φ. The full partition function (5) can be considered as a "quantization" of the geometry, corresponding to its limit $N \to \infty$, $\hbar \to 0$, with $N\hbar = t_0 =$ fixed.

In this quasiclassical limit $\hbar \to 0$ the first term F_0 of the expansion (1) for partition function (5) can be determined from a variational problem

$$F_0 = \left[\int V\rho - \int\int \rho_1 \log|x_1 - x_2|\,\rho_2 + \sum_k \Pi_k\left(\int_{D_k}\rho - S_k\right)\right]_{\frac{\delta F_0}{\delta\rho}=0} \tag{6}$$

where S_k are the new quasiclassical parameters – the fractions of eigenvalues on various supports D_k, and Π_k are corresponding Lagrange multipliers. This function can be described in pure geometric terms as a *prepotential*, i.e. the variational problem (6) is solved by $F_0 = \mathfrak{F}(t,S)$, where

$$\frac{\partial\mathfrak{F}}{\partial S_i} = \Pi_i = \oint_{B_i} dS, \qquad \frac{\partial\mathfrak{F}}{\partial t_k} = \mathrm{res}\,(x^k dS) \tag{7}$$

with the matrix model geometry appearing in the following way.

Loop equations [12] in the main order define the complex curve of the matrix model Σ, which is of genus $g_1 = (\deg V - 2)$ for the 1-matrix model [13] (where always hyperelliptic), and of (now always non-hyperelliptic) of genus $g_2 = (\deg V_1 - 1)(\deg V_2 - 1) - 1$ for the 2-matrix model with potential $V = -xy + V_1(x) + V_2(y)$ [14]. The density of eigenvalues ρ endows Σ with a meromorphic 1-form dS, so that the fractions of eigenvalues are given by its periods $S_i = \oint_{A_i} dS$ on Σ.

For the symmetric 2-matrix model the matrix integral (5) and the variational problem (6) become now two-dimensional, so that the inverse to the kernel in (6) is simply a two-dimensional Laplace operator. In this way the problem for 2-matrix model can be reformulated as a potential problem for two-dimensional domains in complex plane, some aspects of which are sometimes referred to as "Laplacian growth" of domains w.r.t. $t_0 = N\hbar$ with fixed coefficients t of the potential V, (see, e.g. [15] and references therein). In the simplest case this problem is solved in terms of the dispersionless Toda lattice tau-function (the corresponding geometric data and prepotential are sometimes referred to as tau-functions of curves [16]), but for many domains the equations are much less trivial, their explicit form in this case is derived in [17].

The simplicity of dispersionless solution comes from the fact that curve Σ for this case is just sphere, which allows to express differentials $d\Omega_k = \frac{\partial dS}{\partial t_k}$ in terms of rational functions of the uniformizing global co-ordinate $w = e^{\Omega_0}$ as $\Omega_k = P_k(w)$. This gives rise to dispersionless Hirota equations on the (second derivatives of) F_0. An alternative way is to write dispersionless W-constraints of the form $\sum \operatorname{res}(x^n y^m dx) = 0$.

In all known examples the dispersionless W-constraints are reminiscent of the W-constraints [6] imposed onto the full partition function (2). It is possible then to define the full partition function (2) as a correlator of some free fields on Σ, bosons or fermions $\tilde{\psi}_k = e^{i\phi_k}$, $\psi_k = e^{-i\phi_k}$ [18], (see also [19]). The matrix integral (5) is an example of such correlator $Z = \left\langle \exp\left(\oint V(x)\partial\phi(x)\right) G\right\rangle$ with a very simple "element of Grassmannian" $G = \exp\left(\int e^{\sqrt{2}\phi}\right)$ commuting with the W-generators, but for less trivial cases, like in (4), the explicit representation of G in terms of bosons or fermions is not yet known.

Note, that the matrix integrals (5) from string theory point of view correspond to the gauge or open string theory, in contrast to the Kontsevich like integrals, which are rather simplest examples of the closed string field theory [20]. For example, for the Gaussian potential $V(\Phi) \propto \Phi^2$ both descriptions are known [21] which results in a nontrivial identity for two integrals over matrices of different sizes.

3. Seiberg-Witten geometry, Young diagrams and topological strings

Another example, now of more realistic theory, where geometry determines F_0 is the famous Seiberg-Witten theory [22], or the exact solution for the low-energy prepotential of the $N = 2$ SUSY four-dimensional Yang-Mills vector multiplet with the gauge field A, complex scalar Φ and adjoint fermions. The geometric data again contain complex curve Σ of finite genus, endowed with a meromorphic differential dS with special properties or a finite-gap integrable system. The Seiberg-Witten prepotential is introduced by analog of (7)

$$\frac{\partial \mathfrak{F}^{SW}}{\partial a_i} = \oint_{B_i} dS \tag{8}$$

which is equivalent to $\frac{\partial^2 \mathfrak{F}}{\partial a_i \partial a_j} = T_{ij}$, where T_{ij} is period matrix of Σ, defined together with dS by equations

$$w + \frac{\Lambda^{2N}}{w} = P_N(\lambda), \qquad dS = ydx \equiv \lambda \frac{dw}{w}, \qquad S_i = a_i \equiv \oint_{A_i} dS \tag{9}$$

i.e. $F_0 = \mathfrak{F}^{SW}(a, \Lambda)$ is again a quasiclassical tau-function [23]. The role of quasiclassical parameters S is now played by the condensates (BPS masses) a (in addition to $t_0 = \tau_0 \propto \log \Lambda$) and generating parameters t can be introduced in a standard way [24] along the general lines of [2].

From the example with matrix models we have learned that the "quantization" of the matrix model geometry is given in terms of a relatively simple statistical model. One may ask, what is the analog of matrix model for the Seiberg-Witten geometry, and what is the physical sense of corresponding "quantization"? An answer to this question was proposed by Nekrasov [11], which leads to different from integrals over matrices statistical model.

One should consider the *deformation* of the Seiberg-Witten theory by putting it into nontrivial supergravity background with twisted boundary conditions on the fields. A natural way is to use the Scherk-Schwarz mechanism in six-dimensional theory on $\mathbf{R}^4 \times \mathbf{T}^2$ with metric $ds^2 = r^2 dz d\bar{z} + (dx^\mu + 2\mathrm{Re}(\Omega^\mu_\nu x^\nu dz))^2$ and an extra Wilson loop $\mathbf{A}(\Omega)$, where holonomy matrix $||\Omega^\nu_\mu|| = \mathrm{diag}(i\epsilon_1, i\epsilon_2) \otimes \sigma_2$.

Another part of deformation is topological twisting of the SUSY algebra, and the *holomorphic* perturbation of the effective action, which comes from the perturbation of the ultraviolet action by the operators $\mathrm{Tr}\Phi^J$ (and their products) provided the conjugated "prepotential" is kept classical $\bar{\tau}_0 \mathrm{Tr}\bar{\Phi}^2$.

This leads to the deformation of the Seiberg-Witten prepotential by holomorphic operators $\mathfrak{O}_{\vec{n}}(a) = \left\langle \prod_{J=1}^{\infty} \frac{1}{n_J!} \left(\frac{1}{J}\mathrm{Tr}\Phi^J\right)^{n_J} \right\rangle_{\langle\Phi\rangle=a}$

$$\mathfrak{F}(a,\tau_{\vec{n}}) = \mathfrak{F}^{SW}(a;\tau_0) + \sum_{\vec{n}} \tau_{\vec{n}}\mathfrak{O}_{\vec{n}}(a) + \sum_{\vec{n},\vec{m}} \tau_{\vec{n}}\tau_{\vec{m}}\mathfrak{O}_{\vec{n}\vec{m}}(a) + \ldots \tag{10}$$

where $\mathfrak{O}_{\vec{n}\vec{m}}(a)$ are the "contact terms".

It was shown in [11, 25], that computation of so deformed deformed partition function leads to statistical model of "random growth" of the Young diagrams. The standard argument with Q_{BRST}-exact terms allows computation of the chiral observables at $\bar{\tau}_0 \to \infty$ so that the term $\bar{\tau}_0 ||F_A^+||^2$ leads to the self-duality constraint $F_A^+ = 0$ and the partition function

$$Z(\tau_{\vec{n}}; a, \Omega) = \int_{\phi(\infty)=a} D\Phi DA \ldots e^{-S(\Omega)} \tag{11}$$

localizes onto instanton solutions, described by the ADHM "matrix model". As a result, the localization on the fixed points in the moduli space $\mathfrak{M} = \{F_A^+ = 0\}/\mathfrak{G}$ gives rise to a statistical model with summing taken over partitions $\vec{\mathbf{k}}$

$$Z = \int_{\mathfrak{M}} d\mu \exp\left(-\sum_{\vec{n}} \tau_{\vec{n}}\mathfrak{O}_{\vec{n}}\right) = \sum_{\vec{\mathbf{k}}} \mu_{\vec{\mathbf{k}}} \exp\left(-\sum_J t_J \mathfrak{O}_J[\vec{\mathbf{k}}]\right) \tag{12}$$

with the chiral operators replaced by quantities

$$\mathfrak{O}_{\vec{n}}(\Phi) \to \mathfrak{O}_J[\vec{\mathbf{k}}] =$$

$$= \frac{1}{J}\sum_{l,i} \left[\left((a_l + \hbar(k_{li}+1-i)\right)^J - \left(a_l + \hbar(k_{li}-i)\right)^J \right]$$

In the $U(1)$ case [25] one gets the sum over the Young diagrams, where, due to the "hook formulas"

$$\mu_{\mathbf{k}} = (-1)^k \left[\prod_{i<j}^{n} \left(\hbar\left(k_i - k_j + j - i\right)\right) \prod_{i=1}^{n} \frac{1}{\hbar^{k_i+n-i}(k_i+n-i)!} \right]^2 =$$

$$= (-1)^k \left[\frac{\mathrm{dim}R_{\mathbf{k}}}{\hbar^k\ k!}\right]^2$$

and summing using Burnside's theorem we get (with $t_K = 0$ for $K > 1$)

$$Z = e^{-t_1\frac{a^2}{2\hbar^2}} \sum_{\mathbf{k}} \mu_{\mathbf{k}} e^{t_1|\mathbf{k}|} = \exp\left[-\frac{1}{\hbar^2}\left(t_1\frac{a^2}{2} + e^{t_1}\right)\right] = \exp\left(-\frac{F}{\hbar^2}\right) \tag{13}$$

where $F = F_0$ (in this case exactly!) is the partition function of the topological string on $\mathbf{CP}^1$!

To sum effectively over partitions one may apply fermionic formalism of integrable systems, see for example [26]. For the generating function of all chiral operators in $U(1)$ theory we get [25]

$$Z = \left\langle \frac{a}{\hbar} \left| e^{\frac{J_1}{\hbar}} \exp\left[\sum_{p=1}^{\infty} \hat{t}_p W_{p+1}\right] e^{-\frac{J_{-1}}{\hbar}} \right| \frac{a}{\hbar} \right\rangle \tag{14}$$

where W-background is introduced by the formulas

$$W_{p+1} = \frac{1}{\hbar} \oint \ : \ \widetilde{\psi}\, (\hbar D)^p\, \psi, \qquad D = z\partial_z$$

$$\sum_{p=1}^{\infty} \hat{t}_p\ x^p = \sum_{p=1}^{\infty} \frac{t_p}{(p+1)!} \frac{(x+\frac{\hbar}{2})^{p+1} - (x-\frac{\hbar}{2})^{p+1}}{\hbar}$$

As in matrix model case, this is the tau-function of the Toda lattice hierarchy, which can be checked [25] by direct comparison with [27], after applying a Bogolyubov transformation.

For the $U(N)$ theory it was shown in [28] that growth of random partitions or Young diagrams (which replaces here the Laplacian growth of the domains of eigenvalues) leads to appearance of the Seiberg-Witten curves, which generalize the Vershik-Kerov asymptotic curves for the Young diagrams.

4. Discussion

In this notes we have concentrated on the geometric properties of the quasiclassical exact string theory solutions and their quantization which leads to representation of stringy partition functions in terms of simple statistical models.

The quasiclassical part is relatively well-known, and it always has a geometric formulation. The situation with its full or "quantum" analog is rather more poor. Already the first correction to F_0 in (1) or F_1 is much less studied (see, however the computations of [29, 30, 31] and discussion of this issue in [19, 32, 33]), though there is a common belief that it is determined by determinant of the Dirac $\bar{\partial}$-operator on Σ, i.e. $F_1 = -\frac{1}{24}\log\det\bar{\partial}_\Sigma$. As for the full function F it is not known (and even hardly believed to be possible) how to define it in terms of geometry on Σ.

It is necessary to point out that already the sense of S-variables, naturally appearing in (6) due to condensates of eigenvalues and corresponding to possible nontrivial form of Σ, is unclear for the full "quantum" theory (5). In particular cases, when the full solution is known, the dependence on these parameters is almost trivial [34]. There is a clear understanding that dispersionless geometry of sphere is "quantized" into solutions of the ordinary KP or Toda hierarchies, but it is not at all clear what is the hypothetical "dispersional" analog of generic quasiclassical integrable systems of [2].

The example of sect. 3 shows also that the nature of quasiclassical expansion can be rather nontrivial for interesting physical theories, and for the supersymmetric gauge theory it is not naive and corresponds to some dual topological string. This is a new interesting example of gauge/string duality, which nevertheless fits in the general scheme we discussed above. The "quantization" of the Seiberg-Witten geometry leads in this way to appearance of different from known before in this context statistical model (other examples of statistical models arising in a similar way were recently considered in [35, 36]). However, all them concern only the topological string models. One may hope, nevertheless, that analogous phenomena occur even for the "realistic" four-dimensional gauge theories, and the first sign of this is the appearance of the algebro-geometric solutions to spin chains [37] in study of renormalization of operators in $N = 4$ SUSY Yang-Mills. These solutions correspond to the classical approximation in dual string theory, and one may try to understand the "quantization" of corresponding geometry, discussed above, in terms of usual physical quantization of corresponding string theory.

Acknowledgments

I am indebted to I. Krichever, A. Losev, A. Mironov, A. Morozov and N. Nekrasov for very important discussions, and I am grateful to Vlado Dobrev and other organizers of the workshop *Lie Theory and Its Applications in Physics - 5* for the warm hospitality in Varna. The work was partially supported by RFBR grant 02-02-16496, the grant for support of scientific schools 1578.2003.2 and the Russian Science Support Foundation.

References

1. V.G. Knizhnik, *Sov. Phys. Usp.* **32**, 945 (1989);
   A. Morozov, *Sov. Phys. Usp.* **35**, 671 (1992);
   R. Dijkgraaf, hep-th/9201003;
   M. Bershadsky, S. Cecotti, H. Ooguri and C. Vafa, *Commun. Math. Phys.* **165**, 311 (1994) [hep-th/9309140];

J. Polchinski, hep-th/9411028;
A. Marshakov, in: *Integrable models and strings*, Lectures at 3rd Baltic Student Seminar, Helsinki, 1993, [hep-th/9404126]; *Seiberg-Witten Theory and Integrable Systems*, (World Scientific, 1999); *Phys. Usp.* **45**, 915 (2002) [hep-th/0212114].
2. I. Krichever, *Commun. Pure. Appl. Math.* **47**, 437 (1994), hep-th/9205110.
3. M. Kontsevich, *Funct. Anal. & Appl.* **25**, 50 (1991); *Commun. Math. Phys.* **147**, 1 (1992).
4. S. Kharchev, A. Marshakov, A. Mironov, A. Morozov and A. Zabrodin, *Phys. Lett.* **B275**, 311 (1992) [hep-th/9111037]; *Nucl. Phys.* **B380**, 181 (1992) [hep-th/9201013].
5. M. Adler and P. van Moerbeke, *Commun. Math. Phys.* **147**, 25 (1992);
C. Itzykson and J.-B. Zuber, *Int. J. Mod. Phys.* **A7**, 5661 (1992) [hep-th/9201001].
6. M. Fukuma, H. Kawai and R. Nakayama, *Int. J. Mod. Phys.* **A6**, 1385-1406 (1991);
R. Dijkgraaf, H. Verlinde and E. Verlinde, *Nucl. Phys.* **B348**, 435 (1991);
A. Gerasimov, A. Marshakov, A. Mironov, A. Morozov and A. Orlov, *Nucl. Phys.* **B357**, 565-618 (1991);
E. Witten, IASSNS-HEP-91-24;
A. Marshakov, A. Mironov and A. Morozov, *Phys. Lett.* **B274**, 280 (1992) [hep-th/9201011];
T. Eguchi, K. Hori and C.S. Xiong, *Int. J. Mod. Phys.* **A12**, 1743 (1997) [hep-th/9605225].
7. S. Kharchev and A. Marshakov, in: *String Theory, Quantum Gravity and the Unification of the Fundamental Interactions*, Proceedings of International Workshop on String Theory, Quantum Gravity and the Unification of Fundamental Interactions, Rome, Italy, 1993, World Scientific; hep-th/9210072; *Int. J. Mod. Phys.* **A10**, 1219 (1995) [hep-th/9303100].
8. I. Krichever, "On Heisenberg relations for the ordinary linear differential operators", preprint ETH (1990);
S. Novikov, *Funct. Anal. & Appl.* **24**, 43-53 (1990);
R. Schimmrigk, *Phys. Rev. Lett.* **65**, 2483-2486 (1990);
G. Moore, *Commun. Math. Phys.* **133**, 261-304 (1990);
A. Marshakov, *Mod. Phys. Lett.* **A11**, 1169 (1996) [hep-th/9602005].
9. M. Aganagic, R. Dijkgraaf, A. Klemm, M. Marino and C. Vafa, hep-th/0312085;
N. Seiberg and D. Shih, hep-th/0312170.
10. A. Alexandrov, A. Mironov and A. Morozov, hep-th/0310113.
11. N. Nekrasov, hep-th/0206161.
12. A.A. Migdal, *Phys. Rep.* **102**, 199 (1983).
13. F. David, *Phys. Lett.* **B302**, 403-410 (1993), hep-th/9212106;
G. Bonnet, F. David and B. Eynard, *J. Phys.* **A33**, 6739-6768 (2000), cond-mat/0003324.
14. V. Kazakov and A. Marshakov, *J. Phys.* **A36**, 3107 (2003) hep-th/0211236.
15. M. Mineev-Weinstein, P. Wiegmann and A. Zabrodin, *Phys. Rev. Lett.* **84**,

5106-5109 (2000), nlin.SI/0001007;
I. Krichever, M. Mineev-Weinstein, P. Wiegmann and A. Zabrodin, nlin.SI/0311005.

16. P. Wiegmann and A. Zabrodin, *Commun. Math. Phys.* **213**, 523 (2000) [hep-th/9909147];
I.K. Kostov, I. Krichever, M. Mineev-Weinstein, P.B. Wiegmann and A. Zabrodin, hep-th/0005259;
A. Marshakov, P. Wiegmann and A. Zabrodin, *Commun. Math. Phys.* **227**, 131 (2002) [hep-th/0109048].
17. I. Krichever, A. Marshakov and A. Zabrodin, hep-th/0309010;
A. Marshakov and A. Zabrodin, hep-th/0305259.
18. A. Marshakov, A. Mironov and A. Morozov, *Phys. Lett.* **B265**, 99 (1991);
S. Kharchev, A. Marshakov, A. Mironov, A. Morozov and S. Pakuliak, *Nucl. Phys.* **B404**, 717 (1993), hep-th/9208044.
19. I.K. Kostov, hep-th/9907060.
20. A. Marshakov, in: Proc. of 16th Johns Hopkins Workshop on Current Problems in Particle Theory: Pathways to Fundamental Theories, hep-th/9208022.
21. S. Kharchev, A. Marshakov, A. Mironov and A. Morozov, *Nucl. Phys.* **B397**, 339 (1993) [hep-th/9203043].
22. N. Seiberg and E. Witten, *Nucl. Phys.* **B426**, 19-52 (1994); **B431**, 484-550 (1994).
23. A. Gorsky, I. Krichever, A. Marshakov, A. Mironov and A. Morozov, *Phys. Lett.* **B355**, 466-477 (1995), hep-th/9505035.
24. A. Gorsky, A. Marshakov, A. Mironov and A. Morozov, *Nucl. Phys.* **B527**, 690-716 (1998), hep-th/9802004.
25. A.S. Losev, A. Marshakov and N. Nekrasov, hep-th/0302191.
26. S. Kharchev, hep-th/9810091.
27. A. Okounkov and R. Pandharipande, math.AG/0207233, math.AG/0204305.
28. N. Nekrasov and A. Okounkov, hep-th/0306238.
29. G. Akemann, *Nucl. Phys.* **B482**, 403 (1996) [hep-th/9606004]
30. B. Eynard, *J. High En. Phys.* 01 (2003) 051 [hep-th/0210047].
31. L. Chekhov, hep-th/0401089.
32. A. Klemm, M. Marino and S. Theisen, *J. High En. Phys.* 03 (2003) 051 [hep-th/0211216];
R. Dijkgraaf, A. Sinkovics and M. Temurhan, hep-th/0211241.
33. B. Eynard, A. Kokotov and D. Korotkin, hep-th/0401166.
34. S. Kharchev, A. Marshakov, A. Mironov and A. Morozov, *Mod. Phys. Lett.* **A8**, 1047 (1993) [hep-th/9208046].
35. A. Okounkov, N. Reshetikhin and C. Vafa, hep-th/0309208.
36. A. Iqbal, N. Nekrasov, A. Okounkov and C. Vafa, hep-th/0312022.
37. J.A. Minahan and K. Zarembo, *J. High En. Phys.* 03 (2003) 013 [hep-th/0212208];
N. Beisert, J.A. Minahan, M. Staudacher and K. Zarembo, *J. High En. Phys.* 09 (2003) 010 [hep-th/0306139].

IMPACT OF DYNAMICAL TENSIONS IN MODIFIED STRING AND BRANE THEORIES

E.I. GUENDELMAN, A. KAGANOVICH

Department of Physics,
Ben-Gurion University, Beer-Sheva, Israel
guendel@bgumail.bgu.ac.il, alexk@bgumail.bgu.ac.il

E. NISSIMOV, S. PACHEVA

Institute for Nuclear Research and Nuclear Energy,
Bulgarian Academy of Sciences, Sofia, Bulgaria
nissimov@inrne.bas.bg, svetlana@inrne.bas.bg

First we briefly outline the general construction of Dp-brane models with *dynamical* tensions. We then proceed to a more detailed discussion of a modified string model where the string tension is related to the potential of (an external) world-sheet electric current. We show that cancellation of the pertinent conformal anomaly on the quantum level requires the dynamical string tension to be a square of a free massless world-sheet scalar field.

1. Main Motivation

Dirichlet p-branes (Dp-branes) [1] are $p+1$-dimensional extended objects in space-time which carry the end points of fundamental open strings. Their crucial relevance in modern string theory is due to several basic properties of theirs such as providing explicit realization of non-perturbative string dualities, microscopic description of black-hole physics, gauge theory/gravity correspondence, large-radius compactifications of extra dimensions, brane-world scenarios in particle phenomenology, *etc.* . For a background on string and brane theories, see refs.[2].

In an independent recent development two of us have proposed a broad class of new models involving Gravity called *Two-Measure Gravitational Models* [3], whose actions are typically of the form:

$$S = \int d^D x\, \Phi(\varphi)\, L_1 + \int d^D x\, \sqrt{-g}\, L_2 \, , \tag{1}$$

$$L_{1,2} = e^{\frac{\alpha\phi}{M_P}} \Big[-\frac{1}{\kappa} R(g,\Gamma) - \frac{1}{2} g^{\mu\nu} \partial_\mu \phi \partial_\nu \phi + (\text{Higgs}) + (\text{fermions}) \Big] \, . \tag{2}$$

with the standard notations: $R(g,\Gamma)$ is the scalar curvature in the first-order formalism (*i.e.*, the connection Γ is independent of the metric $g_{\mu\nu}$), ϕ is the dilaton field, M_P is the Planck mass, *etc.*. The main new ingredient appears in the first term of (1) – it is an alternative non-Riemannian (i.e. independent of the metric $g_{\mu\nu}$) generally-covariant integration measure density $\Phi(\varphi)$ built up in terms of additional auxiliary scalar fields φ^i ($i = 1,\ldots,D$ where D is the space-time dimension) :

$$\Phi(\varphi) \equiv \frac{1}{D!}\varepsilon^{\mu_1\ldots\mu_D}\varepsilon_{i_1\ldots i_D}\partial_{\mu_1}\varphi^{i_1}\ldots\partial_{\mu_D}\varphi^{i_D} \ . \tag{3}$$

Although naively the additional "measure-density" scalars φ^i appear in (1) as pure-gauge degrees of freedom (due to the invariance under arbitrary diffeomorphisms in the φ^i-target space), there is still a remnant – the so called "geometric" field $\chi(x) \equiv \frac{\Phi(\varphi)}{\sqrt{-g}}$, which remains an additional dynamical degree of freedom beyond the standard physical degrees of freedom characteristic to the ordinary gravity models with the standard Riemannian-metric integration measure. The most important property of the "geometric" field $\chi(x)$ is that its dynamics is determined solely through the matter fields locally (*i.e.*, without gravitational interaction). The latter turns out to have a significant impact on the physical properties of the two-measure gravity models which allows them to address various basic problems in cosmology and particle physics phenomenology and provide physically plausible solutions, for instance: (i) the issue of scale invariance and its dynamical breakdown, *i.e.*, spontaneous generation of dimensionfull fundamental scales; (ii) cosmological constant problem; (iii) geometric origin of fermionic families. In the very recent papers [4] it has been demonstrated that two-measure gravity theories are of significant interest in the context of modern brane-world scenarios, namely, a new conformally invariant brane-world model in $D=6$ without (bulk) cosmological constant fine tuning has been constructed there.

Subsequently, the idea of employing alternative non-Riemannian reparametrization-covariant integration measures was applied in the context of strings and branes theories [5, 6]. A common basic property of these modified string/brane models is that the ratio of both integration measure densities (the alternative versus the standard Riemannian) becomes a dynamical string/brane tension – an additional dynamical degree of freedom beyond the original string/brane degrees of freedom. In particular, in ref.[6] we systematically constructed a new class of modified Dp-brane models with dynamical brane tension – this construction is briefly reviewed in Section 2. We have shown in [5, 6], that the dynamical nature of the string/brane tension leads to some new interesting physical effects such as simple mechanisms of confinement of "color" point-like charges (in the string case) and of charged lower-dimensional sub-branes (in the Dp-brane case).

In Section 3 below we study in some detail the quantum properties of a modified string model where the dynamical tension becomes related to the potential of an external electric charge current on the string world-sheet. We explicitly show that quantum consistency, *i.e.*, cancellation of the pertinent conformal anomaly, apart from the well-known restriction on D ($D=26$ in the simplest bosonic case) implies that the dynamical string tension must be a square of a free massless world-sheet scalar field.

2. Dp-Branes with Dynamical Tension

First, let us recall the standard formulation of Dp-branes given in terms of the Dirac-Born-Infeld (DBI) action (see e.g. third ref.[2]) :

$$S_{DBI} = -T \int d^{p+1}\sigma \left[e^{-\alpha U} \sqrt{-\det ||G_{ab} - \mathcal{F}_{ab}||} + \ldots \right] , \tag{4}$$

with the following short-hand notations:

$$G_{ab} \equiv \partial_a X^\mu \partial_b X^\nu G_{\mu\nu}(X) \quad , \quad \mathcal{F}_{ab} \equiv \partial_a X^\mu \partial_b X^\nu B_{\mu\nu}(X) - F_{ab}(A) \, . \tag{5}$$

Here $G_{\mu\nu}(X)$, $U(X)$, and $B_{\mu\nu}(X)$ and are the background metric, the dilaton, and the 2-form Neveu-Schwarz, respectively, whereas $F_{ab}(A) = \partial_a A_b - \partial_b A_b$ is the field-strength of the Abelian world-volume gauge field A_a. The dots in (4) indicate coupling to the $(p+1)$-form Ramond-Ramond background gauge field which is omitted for simplicity. All world-volume indices take values $a, b = 0, 1, \ldots, p$ and $\varepsilon^{a_1 \ldots a_{p+1}}$ is the $(p+1)$-dimensional totally antisymmetric tensor ($\varepsilon^{01\ldots p} = 1$).

Similarly to the gravity case (1)–(3) we now introduce a modified world-volume integration measure density in terms of $p+1$ auxiliary scalar fields φ^i ($i = 1, \ldots, p+1$) :

$$\Phi(\varphi) \equiv \frac{1}{(p+1)!} \varepsilon_{i_1 \ldots i_{p+1}} \varepsilon^{a_1 \ldots a_{p+1}} \partial_{a_1} \varphi^{i_1} \ldots \partial_{a_{p+1}} \varphi^{i_{p+1}} , \tag{6}$$

and use it to construct the following new p-brane-type action :

$$S = - \int d^{p+1}\sigma \, \Phi(\varphi) \left[e^{-\beta U} \frac{1}{2} \zeta^{ab} \left(G_{ba} - \mathcal{F}_{ba} \right) + \frac{1}{\sqrt{-\zeta}} \Omega(\mathcal{A}) \right] + \int d^{p+1}\sigma \, \mathcal{L}(\mathcal{A}) \, . \tag{7}$$

Here apart from (5) the following new notations are used. The $(p+1) \times (p+1)$ matrix ζ_{ab} of auxiliary variables is an arbitrary world-volume 2-tensor, ζ^{ab} denotes the corresponding inverse matrix ($\zeta^{ac}\zeta_{cb} = \delta^a_b$) and

$\zeta \equiv \det ||\zeta_{ab}||$. The term $\Omega(\mathcal{A})$ indicates a topological density given in terms of some additional auxiliary gauge fields $\mathcal{A}^I$ living on the world-volume:

$$\frac{\partial\Omega}{\partial\mathcal{A}^I} - \partial_a\left(\frac{\partial\Omega}{\partial\partial_a\mathcal{A}^I}\right) = 0 \text{ identically} \quad , \quad \text{i.e. } \delta\Omega(\mathcal{A}) = \partial_a\left(\frac{\partial\Omega}{\partial\partial_a\mathcal{A}^I}\delta\mathcal{A}^I\right) . \tag{8}$$

Finally, $\mathcal{L}(\mathcal{A})$ describes possible coupling of the auxiliary fields A^I to external "currents" on the brane world-volume.

The requirement for $\Omega(\mathcal{A})$ to be a topological density is dictated by the requirement that the new brane action (7) (in the absence of the last gauge-coupling term $\int d^{p+1}\sigma\,\mathcal{L}(\mathcal{A})$) reproduces the standard *Dp*-brane equations of motion resulting from the DBI action (4) apart from the fact that the *Dp*-brane tension $T \equiv \Phi(\varphi)/\sqrt{-\zeta}$ becomes now an *additional dynamical degree of freedom* (note that no *ad hoc* dimensionfull tension factor T has been introduced in (7)). Let us particularly stress that the modified-measure brane model (7) naturally requires (through the necessity to introduce topological density $\Omega(\mathcal{A})$) the existence on the world-volume of an additional (higher-rank tensor) gauge field $\mathcal{A}$ apart from the standard world-volume Abelian vector gauge field A_a.

Splitting the auxiliary tensor variable $\zeta^{ab} = \gamma^{ab} + \zeta^{[ab]}$ into symmetric and anti-symmetric parts and setting $\zeta^{[ab]} = 0$, the action (7) reduces to the action of the modified-measure model of ordinary *p*-branes [5] with Neveu-Schwarz field $B_{\mu\nu}$ and world-volume gauge field A_a disappearing and γ_{ab} assuming the role of world-volume Riemannian metric.

The most obvious example of a topological density $\Omega(\mathcal{A})$ for the additional auxiliary world-volume gauge fields in (7) is:

$$\Omega(\mathcal{A}) = -\frac{\varepsilon^{a_1\ldots a_{p+1}}}{p+1}F_{a_1\ldots a_{p+1}}(\mathcal{A}) \quad , \quad F_{a_1\ldots a_{p+1}}(\mathcal{A}) = (p+1)\partial_{[a_1}\mathcal{A}_{a_2\ldots a_{p+1}]} \, , \tag{9}$$

where $\mathcal{A}_{a_1\ldots a_p}$ denotes rank p antisymmetric tensor (Abelian) gauge field on the world-volume. More generally we can have (for $p+1=rs$) :

$$\Omega(\mathcal{A}) = \frac{1}{rs}\varepsilon^{a_{11}\ldots a_{1r}\ldots a_{s1}\ldots a_{sr}}F_{a_{11}\ldots a_{1r}}(\mathcal{A})\ldots F_{a_{s1}\ldots a_{sr}}(\mathcal{A}) \tag{10}$$

with rank $r-1$ (smaller than p) auxiliary world-volume gauge fields.

We may also employ *non-Abelian* auxiliary world-volume gauge fields $\mathcal{A}_a$. For instance, when $p+1=2q$:

$$\Omega(\mathcal{A}) = \frac{1}{2q}\varepsilon^{a_1b_1\ldots a_qb_q}\,\mathrm{Tr}\left(F_{a_1b_1}(\mathcal{A})\ldots F_{a_qb_q}(\mathcal{A})\right) \, , \tag{11}$$

where $F_{ab}(\mathcal{A}) = \partial_a\mathcal{A}_b - \partial_b\mathcal{A}_a + i\left[\mathcal{A}_a, \mathcal{A}_b\right]$.

The equations of motion w.r.t. φ^i and ζ^{ab} corresponding to the modified-measure brane action (7) read:

$$e^{-\beta U}\frac{1}{2}\zeta^{ab}\left(G_{ba}-\mathcal{F}_{ba}\right)+\frac{1}{\sqrt{-\zeta}}\Omega(\mathcal{A})=M\equiv\text{const}\;, \tag{12}$$

$$e^{-\beta U}\left(G_{ab}-\mathcal{F}_{ab}\right)+\zeta_{ab}\frac{1}{\sqrt{-\zeta}}\Omega(\mathcal{A})=0\;. \tag{13}$$

Both Eqs.(12)–(13) imply:

$$\zeta^{ab}\left(G_{ba}-\mathcal{F}_{ba}\right)=2M\frac{p+1}{p-1}e^{\beta U}\quad,\quad\frac{1}{\sqrt{-\zeta}}\Omega(\mathcal{A})=-\frac{2M}{p-1} \tag{14}$$

which when substituted in (13) give:

$$G_{ab}-\mathcal{F}_{ab}=\frac{2M}{p-1}e^{\beta U}\,\zeta_{ab} \tag{15}$$

Next we consider the equations of motion w.r.t. auxiliary (gauge) fields $\mathcal{A}^I$:

$$\partial_a\left(\frac{\Phi(\varphi)}{\sqrt{-\zeta}}\right)\frac{\partial\Omega}{\partial\partial_a\mathcal{A}^I}+j_I=0\;, \tag{16}$$

where $j_I\equiv\frac{\partial\mathcal{L}}{\partial\mathcal{A}^I}-\partial_a\left(\frac{\partial\mathcal{L}}{\partial\partial_a\mathcal{A}^I}\right)$ is the corresponding "current" coupled to $\mathcal{A}^I$. These are the equations determining the dynamical brane tension $T\equiv\Phi(\varphi)/\sqrt{-\zeta}$. In deriving Eq.(16) crucial use was made of the identity (8) satisfied by the topological density $\Omega(\mathcal{A})$.

In particular, in the absence of coupling of external world-volume currents to the auxiliary (gauge) fields $\mathcal{A}^I$ Eq.(16) imply:

$$T\equiv\Phi(\varphi)/\sqrt{-\zeta}=C\equiv\text{const} \tag{17}$$

Now, using Eqs.(12) and (15) it is straightforward to show that the modified brane action (7) with $\mathcal{L}(\mathcal{A})=0$ classically reduces to the standard *Dp*-brane DBI-action (4) :

$$S'_{DBI}=-T'\int d^{p+1}\sigma\,e^{-\beta' U}\sqrt{-\det\|G_{ab}-\mathcal{F}_{ab}\|}\;, \tag{18}$$

$$T'\equiv\frac{1}{2}C(2M)^{-\frac{p-1}{2}}(p-1)^{\frac{p+1}{2}}\quad,\quad\beta'\equiv\frac{p+1}{2}\beta\;, \tag{19}$$

where, however, the *Dp*-brane tension T' is dynamically generated according to (17) and (19).

For a more detailed analysis of the properties of the modified-measure *Dp*-brane models we refer to [6].

3. Conformal Anomaly and Its Impact on the Dynamical String Tension

Now we turn our attention to a special case of (7) for $p=1$, *i.e.*, a modified string model with dynamical tension:

$$S = -\int d^2\sigma\, \Phi(\varphi)\Big[\frac{1}{2}\gamma^{ab}\partial_a X^\mu \partial_b X_\mu - \frac{\varepsilon^{ab}}{2\sqrt{-\gamma}}F_{ab}(A)\Big] + \epsilon\int d^2\sigma\, A_a \varepsilon^{ab}\partial_b u\,. \tag{20}$$

Notice that the last term in Eq.(20) can be rewritten in the reparametrization-invariant form:

$$\int d^2\sigma\,\sqrt{-\gamma}\,A_a J^a \quad , \quad J^a \equiv \epsilon\frac{\varepsilon^{ab}}{\sqrt{-\gamma}}\partial_b u\,, \tag{21}$$

where J^a indicates the general expression for a covariantly conserved world-sheet electric current. Let us stress that such coupling of string degrees of freedom to an external world-sheet electric current is natural only in the present context of modified-measure string models due to the inevitable appearance of the auxiliary world-sheet gauge field A_a. Let us also note that in (21) the ordinary Riemannian world-sheet integration measure density $\sqrt{-\gamma}$ is used unlike the modified one:

$$\Phi(\varphi) = \frac{1}{2}\varepsilon_{ij}\varepsilon^{ab}\partial_a\varphi^i\partial_b\varphi^j \quad , \quad a,b=0,1 \;\; , \;\; i,j=1,2 \tag{22}$$

in the main action term in (20), much in the spirit of the previously proposed two-measure gravity theories [3] (cf. Eq.(1)).

Quantization of the modified string model (20) within the functional integral framework can be performed in the standard way based on the canonical Hamiltonian formalism for constrained systems a'la Dirac. As already shown in the third ref.[5], the total canonical Hamiltonian $H_T \equiv \sum_A \Lambda_A \Phi_A$ is a linear combination of the following first-class constraints Φ_A (the letters π and $\mathcal{P}$ indicating the pertinent canonical momenta) :

$$\pi_{\gamma^{ab}} = 0 \quad , \quad \mathcal{T}_\pm \equiv \frac{1}{4}\Big(\frac{\mathcal{P}}{E} \pm \partial_\sigma X\Big)^2 = 0\,, \tag{23}$$

which are of the same form as in the ordinary bosonic string case modulo the fact that now the string tension $T \equiv E$ is a dynamical degree of freedom (see last Eq.(25) below), plus the new constraints:

$$\partial_\sigma\varphi^i\pi^\varphi_i = 0 \quad , \quad \frac{\pi^\varphi_2}{\partial_\sigma\varphi^1} = 0\,; \tag{24}$$

$$\pi_{A_0} = 0 \quad , \quad \partial_\sigma(E+\epsilon u) = 0 \quad , \quad \text{where } E \equiv \pi_{A_1} = \frac{\Phi(\varphi)}{\sqrt{-\gamma}}\,. \tag{25}$$

Relations (24) imply that the auxiliary measure-fields φ^i are pure-gauge degrees of freedom which decouple from the string dynamics. The only remnant of the latter appears through E – the canonical momentum of A_1 according to the last expression (25), which together with the second relation (25) tells us that E has the physical meaning of a world-sheet electric field-strength obeying the $D = 2$ Gauss law.

To quantize the modified-measure string model (20) one starts with the standard Faddeev's functional integral:

$$Z = \int \mathcal{D}X\, \mathcal{D}\mathcal{P}\, \mathcal{D}E\, \mathcal{D}A_1 \mathcal{D}\Lambda_A\, \Delta_{\Phi\Pi}\, \delta(\text{conf. gauge}) \times \exp\Big\{ i \int d^2\sigma \Big[\mathcal{P}_\mu \partial_\tau X^\mu + E\partial_\tau A_1 - \sum_A \Lambda_A \Phi_A \Big] \Big\} \tag{26}$$

where Φ_A are the first-class constraints listed above (23)–(25) and $\Delta_{\Phi\Pi}$ indicates the Faddeev-Popov ghost determinant associated with the conformal gauge-fixing condition. In (26) and in what follows we shall skip the insertion of vertex operators for brevity. Now performing the Gaussian integrations over the canonical momenta we arrive at the following reparametrization-invariant expression:

$$Z = \int \mathcal{D}X\, \mathcal{D}\gamma_{ab}\, \mathcal{D}E\, \mathcal{D}A_a\, \Delta_{\Phi\Pi}\, \delta(\text{conf. gauge}) \times \exp\Big\{ i \int d^2\sigma \Big[-E\frac{1}{2}\sqrt{-\gamma}\gamma^{ab}\partial_a X^\mu \partial_b X_\mu + \frac{1}{2}(E + \epsilon u)\varepsilon^{ab} F_{ab}(A) \Big] \Big\} \tag{27}$$

Integration over the auxiliary gauge field A_a yields functional delta-function $\delta\left(\varepsilon^{ab}\partial_b(E + \epsilon u)\right)$ which in turn reduces the functional integration over E to an ordinary integration over the overall world-sheet constant C :

$$Z = \int dC\, \mathcal{D}X\, \mathcal{D}\gamma_{ab}\, \Delta_{\Phi\Pi}\, \delta(\text{conf. gauge}) \times \exp\Big\{ -i\frac{1}{2} \int d^2\sigma\, (C - \epsilon u)\sqrt{-\gamma}\gamma^{ab}\partial_a X^\mu \partial_b X_\mu \Big\} \tag{28}$$

Note that $T \equiv C - \epsilon u$ is the dynamical string tension. Thus, integration over string coordinates X^μ amounts to computation of the determinant of the modified d'Alembertian (or Laplace-Beltrami upon Euclidean rotation) operator:

$$-\frac{1}{\sqrt{-\gamma}}\partial_a \left(v^2 \sqrt{-\gamma}\gamma^{ab}\partial_b \right) \tag{29}$$

where for later convenience we have introduced the notation v for the square-root of the dynamical tension: $v^2 \equiv C - \epsilon u$. Due to the presence of the latter there is an additional contribution to the well-known conformal anomaly (see *e.g.* [7] and references therein where the "dilaton"-like notation $v = e^{-\phi}$ is employed) :

$$Z = \int dC\,\mathcal{D}\gamma_{ab}\,\delta(\text{conf. gauge})\exp\Big\{i\int d^2\sigma\,\sqrt{-\gamma}\Big[\frac{D-26}{96\pi}\,R\Box^{-1}R + \frac{D}{8\pi}\,\frac{\Box v}{v}\,\Box^{-1}R - b\frac{D}{4\pi}\gamma^{ab}(\nabla_a \ln v)(\nabla_b \ln v)\Big]\Big\} \tag{30}$$

Here R denotes the usual scalar curvature for the intrinsic world-sheet metric γ_{ab} and $\Box \equiv \frac{1}{\sqrt{-\gamma}}\partial_a\big(\sqrt{-\gamma}\gamma^{ab}\partial_b\big)$ is the ordinary covariant d'Alembertian. The last non-anomalous term in (30) is determined up to a regularization-dependent constant b. Therefore, absence of conformal anomaly implies in the present case, apart from the usual critical value for the space-time dimension, an additional condition on the square-root of the dynamical string tension:

$$\Box v = 0 \quad , \quad v^2 \equiv T = C - \epsilon u \; , \tag{31}$$

i.e., the dynamical string tension must be a square of a free massless world-sheet scalar field. Note that this condition is a dynamical constraint on the external world-sheet electric current (21) coupled to the modified-measure string.

Going back to the string action in (28) and taking into account relations (31) we can rewrite it in the following simple form:

$$-\frac{1}{2}\int d^2\sigma\,\sqrt{-\gamma}\gamma^{ab}\partial_a\big(vX^\mu\big)\partial_b\big(vX_\mu\big) \tag{32}$$

where we used the following identity for the modified d'Alembertian operator (29) :

$$\frac{1}{\sqrt{-\gamma}}\partial_a\big(v^2\sqrt{-\gamma}\gamma^{ab}\partial_b\big)(\cdot) = v\,\Box(v\cdot) - (\Box v)\,v = v\,\Box(v\cdot) \tag{33}$$

due to the free masslessness of the square-root tension v. Therefore, in the quantized modified-measure string model (20) it is the dynamically rescaled fields $\widetilde{X}^\mu = v\,X^\mu$ which describe free wave mode propagation along the string rather than the usual string coordinates X^μ.

4. Conclusions

Replacing the standard Riemannian world-sheet/world-volume integration measure density with an metric-independent reparametrization-invariant one (6) in the Lagrangian formulation of string and brane models has significant impact on the string/brane dynamics. Consistency of dynamics requires the introduction of additional auxiliary (higher-rank) gauge fields on the world-sheet/world-volume which are absent in the standard string/brane theories. The main new property of the modified-measure string/brane models is that the string/brane tension appears as an additional dynamical degree of freedom which is canonically conjugated to the auxiliary world-sheet/world-volume gauge fields. It acquires the physical meaning of world-sheet electric field-strength (in the string case) or field-strength of higher-rank world-volume gauge fields (in the brane case) obeying the Maxwell (or Yang-Mills) equations of motion or their higher-rank generalizations. As a simple consequence of the latter, modified-measure string/brane models provide (already on the classical level) simple mechanisms for "color" charge confinement. Furthermore, in the quantized string context the interplay between the intrinsic conformal anomaly and the dynamical nature of the string tension imply an important constraint on the form of the dynamical string tension forcing it to be a square of a free massless world-sheet scalar. It is curious to note that the last property resembles the property found in the context of the string-inspired low-energy effective field theory in $D = 10$ [8] where the square of the target-space dilaton field plays the role of a covariant integration measure density

More detailed study of the effects resulting from the new physical properties of the modified-measure string models with dynamical tension reported above will be done in a separate work.

Acknowledgments

Two of us (E.N. and S.P.) are sincerely grateful to the organizers of the Fifth Workshop on *Lie Theory and Applications in Physics* (Varna, Bulgaria) for the kind invitation to present there the above results. E.N. and S.P. also acknowledge cordial hospitality and illuminating discussions with Prof. Yannis Bakas at the University of Patras (Greece). The work of E.N. and S.P. is partially supported by NATO collaborative linkage grant *PST.CLG.978785* and Bulgarian NSF grant *F-904/99*. Finally, all of us acknowledge support of our collaboration through the exchange agreement between the Ben-Gurion University of the Negev and the Bulgarian Academy of Sciences.

References

1. J. Polchinski, "TASI Lectures on D-Branes", hep-th/9611050 (1996).

2. Y. Ne'eman and E. Eizenberg, *Membranes and Other Extendons*, (World Scientific, 1995);
M. Green, J. Schwarz and E. Witten, *Superstring Theory*, Vol. 1 & 2 (Cambridge Univ. Press, 1987);
J. Polchinski, *String Theory*, Vol. 1 & 2 (Cambridge Univ. Press, 1998);
C. Johnson, *D-Branes* (Cambridge Univ. Press, 2002), and references therein (see also hep-th/0007170).
3. E.I. Guendelman and A. Kaganovich, *Phys. Rev.* **D53**, 7020 (1996);
E.I. Guendelman and A. Kaganovich, *Phys. Rev.* **D60**, 065004 (1999);
E.I. Guendelman, *Mod. Phys. Lett.* **A14**, 1043 (1999);
E.I. Guendelman, *Class. Quant. Grav.* **17**, 261 (2000);
A. Kaganovich, *Phys. Rev.* **D63**, 025022 (2001);
E.I. Guendelman and A. Kaganovich, *Int. J. Mod. Phys.* **A17**, 417 (2002) (hep-th/0106152);
E.I. Guendelman and A. Kaganovich, *Mod. Phys. Lett.* **A17** (2002) 1227 (hep-th/0110221);
E.I. Guendelman and A. Kaganovich, gr-qc/0312006 (2003).
4. E.I. Guendelman, *Phys. Lett.* **412B**, 42 (1997);
E.I. Guendelman, gr-qc/0303048 (2003);
E.I. Guendelman and E. Spallucci, hep-th/0311102 (2003).
5. E.I. Guendelman, *Class. Quantum Grav.* **17**, 3673 (2000) (hep-th/0005041);
E.I. Guendelman, *Phys. Rev.* **D63**, 046006 (2001) (hep-th/0006079);
E.I. Guendelman, A. Kaganovich, E. Nissimov and S. Pacheva, *Phys. Rev.* **D66**, 046003 (2002) (hep-th/0203024);
E.I. Guendelman, A. Kaganovich, E. Nissimov and S. Pacheva, in: *"First Workshop on Gravity, Astrophysics and Strings"*, eds. P. Fiziev et al (Sofia Univ. Press, 2003) (hep-th/0210062).
6. E.I. Guendelman, A. Kaganovich, E. Nissimov and S. Pacheva, in: Proc. of the *"Second Internat. School on Modern Math. Physics"*, Kopaonik (Serbia and Montenegro), eds. B. Dragovich et al (2003) (hep-th/0304269).
7. D. Grumiller, W. Kummer and D. Vassilevich, *Phys. Reports*, **369** 327 (2002) (hep-th/024253);
D. Vassilevich, hep-th/0306138 (2003).
8. W. Siegel, *Phys. Rev.* **D47**, 5453 (1993) (hep-th/9302036);
W. Siegel, *Phys. Rev.* **D48**, 2826 (1993) (hep-th/9305073).

IV. Integrable Systems

QUANTUM GROUP IN ROOTS OF UNITY AND THE RESTRICTION OF XXZ MODEL

A. BELAVIN

Landau Institute for Theoretical Physics

We consider the restrictions of XXZ models in roots of unity.

1. Center of the quantum group and the restriction in roots of unity

In the first part of the talk, I describe the center of the algebra of monodromy matrices connected with the R-matrix of the eight-vertex model in the case where the crossing parameter has rational values [1]. In the second part, I discuss the connection between the extended center of the quantum group in roots of unity and the restriction of the XXZ model. I also give explicit expressions for operators that respect the restriction and act on the state space of the restricted models.

As usual, let the R-matrix $R(u)$ act in the tensor product of two n-dimensional spaces $V \otimes V$ and satisfy the Yang–Baxter equation

$$R_{12}(u_1 - u_2)R_{13}(u_1 - u_3)R_{23}(u_2 - u_3) = \\ = R_{23}(u_2 - u_3)R_{13}(u_1 - u_3)R_{12}(u_1 - u_2), \qquad (1)$$

where $R_{12}(u)$, $R_{13}(u)$, and $R_{23}(u)$ are matrices acting in $V_1 \otimes V_2 \otimes V_3$ and $R_{12}(u) = R(u) \otimes I$, $R_{23} = I \otimes R(u)$, etc. Then we call the algebra whose generators $L_i^j(u)$, considered as entries of the $n \times n$ matrix $L(u)$, satisfy

$$R_{12}(u - v)L_1(u)L_2(v) = L_2(v)L_1(u)R_{12}(u - v), \qquad (2)$$

where $L_1(u) = L(u) \otimes I$ and $L_2(u) = I \otimes L(u)$, the *Yang–Baxter algebra* A.

The simplest and best-known solutions of (1) arise for $n = 2$ and correspond to six- and eight-vertex models [2]. In the last case, the elements of the R-matrix can be written in terms of Jacobian elliptic theta functions with the half-periods K and K' as

$$\begin{aligned}
R^{\alpha\alpha}_{\alpha\alpha}(u) &= \rho\Theta(\lambda)\Theta(\lambda u)H(\lambda(u+1)),\\
R^{\alpha\beta}_{\alpha\beta}(u) &= \rho\Theta(\lambda)H(\lambda u)\Theta(\lambda(u+1)),\\
R^{\alpha\alpha}_{\beta\beta}(u) &= \rho H(\lambda)H(\lambda u)H(\lambda(u+1)),\\
R^{\alpha\beta}_{\beta\alpha}(u) &= \rho H(\lambda)\Theta(\lambda u)\Theta(\lambda(u+1)),
\end{aligned} \tag{3}$$

where $\alpha, \beta = 1, 2$, $\alpha \neq \beta$. The crossing parameter λ plays an important role in the structure of A, its representations, and the integrable models connected with A. As can be seen from (2), the parameter ρ is inessential.

We will concentrate on the case of the Yang–Baxter algebra for the eight-vertex model and investigate the question of the center of A, i.e., the set of its elements that commute with all elements of A. For the ice-model case corresponding to trigonometric degeneration of the Baxter model (i.e., for $K' = 0$), the answer was given by V.Tarasov [3]. It was shown in [3] that although the center of A is generated by only the so-called $q\det L(u)$ for a general value of λ, the situation is very different in the "roots of unity" case, i.e., where $\lambda = 2mK/N$, N is a positive integer, and m is coprime to N. In this case, the center of A is much wider and (if m is even) is generated by four new generators

$$\langle L^j_i(u)\rangle = L^j_i(u+N-1)\cdot L^j_i(u+N-2)\cdots L^j_i(u), \tag{4}$$

where summing with respect to i and j is absent.

Our goal is to show [1] that in the elliptic case with $\lambda = 2mK/N$ and even m, the center of A is also extended and is generated by four new elements:

$$\Lambda_{ab}(u) = \sum_{i,j} \overline{\Psi}^{i_1\ldots i_N}_a L^{j_1}_{i_1}(u+N-1)\cdots L^{j_N}_{i_N}\cdot\Psi_{a,j_1\ldots j_N},$$

where $a, b = \pm 1$ and the Ψ_a are some special vectors of the tensor product of N two-dimensional spaces $V_1 \otimes \cdots \otimes V_N$. The vectors Ψ_a and the dual vectors $\overline{\Psi}_a$ are defined below.

The idea of the construction of the central elements of A in a root of unity is as follows. We define $L_{1\ldots N}(u)$ and $R_{1\ldots N}(u)$ by

$$\begin{aligned}
L_{1\ldots N}(u) &= L_1(u+N-1)\cdots L_N(u),\\
R_{1\ldots N}(u) &= R_{10}(u+N-1)\cdots R_{N0}(u).
\end{aligned} \tag{5}$$

We can consider $L_{1\ldots N}(u)$ and $R_{1\ldots N}(u)$ operators in $V_1 \otimes \cdots \otimes V_N$ if the entries of the operator $L_{1\ldots N}(u)$ belong to A and the entries of the operator

$R_{1\ldots N}(u)$ act in an additional two-dimensional space V_0. Using explicit expressions (3) for the R-matrix and Eqs. (1) and (2), we show that for any value of the parameter λ, the operators $R_{1\ldots N}(u)$ and $L_{1\ldots N}(u)$ have an invariant $(2^N - N - 1)$-dimensional subspace $W_0 \subset V_1 \otimes \cdots \otimes V_N$.

Moreover, if λ is a "root of unity," i.e., $\lambda = 2mK/N$, then the invariant subspace of the operators is extended up to the space $W \supset W_0$ and $\dim W = \dim W_0 + 2$.

We give an explicit construction for the vectors Ψ_a, $a = \pm 1$, completing W_0 up to W.

Restricting the operators $L_{1\ldots N}(u)$ and $R_{1\ldots N}(u)$ on the subfactor W/W_0 and letting $\Lambda(u)$ and $\hat{R}(u)$ denote the resulting 2×2 operators, we obtain the equation

$$\hat{R}(u-v)\Lambda(u)L(v) = L(v)\Lambda(u)\hat{R}(u-v). \tag{6}$$

We can use the results in [4] to show that

$$\hat{R}(u) = f(u) I \otimes I, \tag{7}$$

where $f(u)$ is some scalar factor. From (6), we therefore have

$$[\Lambda(u), L(v)] = 0 \tag{8}$$

or, in components,

$$[\Lambda_{ab}(u), L_i^j(v)] = 0,$$

where $a, b = \pm 1$ and $i, j = 1, 2$. The entries of $\Lambda(u)$ are therefore central elements of A.

We now explain the construction in more detail. We introduce the additional operators $R_{i,i+1}$ and S also acting in $V_1 \otimes \cdots \otimes V_N$. We define $R_{i,i+1}$ as above and set

$$S = \prod_{i<j} R_{ij}(j-i), \tag{9}$$

where the factors are ordered with respect to both the indices i and j. For example, if $N = 3$, then

$$S = R_{12}(1)R_{13}(2)R_{23}(1).$$

Using (1), we can easily show that the factors of S can be permuted such that $R_{i,i+1}(1)$ for a given i is in the rightmost position. Thus, for $N = 3$ and $i = 1$,

$$S = R_{23}(1)R_{13}(2)R_{12}(1).$$

From (1) and (2), we have

$$\begin{aligned} R_{i,i+1}(1)R_{1\dots N}(u) &= R^{i,i+1}_{1\dots N}(u)R_{i,i+1}(1), \\ R_{i,i+1}(1)L_{1\dots N}(u) &= L^{i,i+1}_{1\dots N}(u)R_{i,i+1}(1), \end{aligned} \tag{10}$$

where $R^{i,i+1}_{1\dots N}(u)$ and $L^{i,i+1}_{1\dots N}(u)$ differ from $R_{1\dots N}(u)$ and $L_{1\dots N}(u)$ by permutation of the ith and $(i+1)$th factors. Moreover, it follows from (1) and (2) that

$$\begin{aligned} SR_{1\dots N}(u) &= R_{N\dots 1}(u)S, \\ SL_{1\dots N}(u) &= L_{N\dots 1}(u)S, \end{aligned} \tag{11}$$

where $R_{N\dots 1}(u)$ and $L_{N\dots 1}(u)$ differ from $R_{1\dots N}(u)$ and $L_{1\dots N}(u)$ by permutation of all factors to the opposite order.

Taking $u = 1$ in (3), we can easily see that $R_{i,i+1}(1)$ has a nontrivial kernel consisting of the vectors

$$e_{\alpha_1} \otimes \cdots \otimes e_{\alpha_{i-1}} \otimes [e_+ \otimes e_- - e_- \otimes e_+] \otimes e_{\alpha_{i+1}} \otimes \cdots \otimes e_{\alpha_N},$$

where $\alpha_k = \pm 1$ and e_+ and e_- form the canonical basis of V.

We set $W_0 \subset V_1 \otimes \cdots \otimes V_N$

$$W_0 = \sum_{i=1}^{N-1} \ker R_{i,i+1}(1). \tag{12}$$

It follows from Eqs. (10) that W_0 is an invariant subspace of $R_{1\dots N}(u)$ and $L_{1\dots N}(u)$.

Following Baxter, we now introduce the vectors $\phi \in V$

$$\phi_{s,s+\epsilon}(u) = H(\lambda(s - \epsilon u))e_+ + \Theta(\lambda(s - \epsilon u))e_-, \tag{13}$$

where $\epsilon = \pm 1$ and s is a positive integer. If $|a - b| \neq 1$, then $\phi_{ab}(u) = 0$. Here we use the notation in [4].

Baxter showed that

$$R(u-v)\phi_{s,s\pm 1}(u)\otimes\phi_{s\pm 1,s\pm 2}(v) = \frac{[u - v + 1]}{[u]}\phi_{s\pm 1,s\pm 2}(u)\otimes\phi_{s,s\pm 1}(v), \tag{14}$$

where $[u] = H(\lambda u)\Theta(\lambda u)$. Our notation in the last definition slightly differs from that in [4].

We now introduce the vectors $\Psi_a(u) \in V_1 \otimes \cdots \otimes V_N$, $a = \pm 1$

$$\Psi^s_a(u) = \phi_{s+aN,s+a(N-1)}(u + N - 1)\cdots\phi_{s+a,s}(u). \tag{15}$$

It is easy to verify that

$$S\Psi^s_a(u) = 0 \tag{16}$$

if $\lambda = 2mK/N$ and that

$$\Psi_a^s(u) \notin W_0.$$

For even m,

$$\Psi_a^{s+1}(u) = \Psi_a^s(u) \mod W_0$$

as we will argue in the appendix.

We set

$$W = c_1\Psi_+^s(v) + c_2\Psi_-^s(v) + W_0. \tag{17}$$

We can show that $\dim W = \dim W_0 + 2$.

It follows from Eqs. (11) that $W \supset W_0$ is an invariant subspace for $R_{1\ldots N}(u)$ and $L_{1\ldots N}(u)$. Let $\hat{R}(u)$ and $\Lambda(u)$ be restrictions of $R_{1\ldots N}(u)$ and $L_{1\ldots N}(u)$ on W/W_0, explicitly,

$$\begin{aligned} R_{1\ldots N}(u)\Psi_a^s(v) &= \hat{R}_{ab}(u)\Psi_b^s(v) \mod W_0, \\ L_{1\ldots N}(u)\Psi_a^s(v) &= \Lambda_{ab}(u)\Psi_b^s(v) \mod W_0. \end{aligned} \tag{18}$$

From (2), we can obtain

$$R_{1\ldots N}(u-v)L_{1\ldots N}(u)L_0(v) = L_0(v)L_{1\ldots N}(u)R_{1\ldots N}(u-v). \tag{19}$$

Restricting (19) on W/W_0, we obtain

$$\hat{R}(u-v)\Lambda(u)L_0(v) = L_0(v)\Lambda(u)\hat{R}(u-v). \tag{20}$$

Using the results in [4], we can now prove Eq. (7),

$$\hat{R}(u) = f(u)I \otimes I.$$

Our main statement (8) follows from this.

2. Restriction of the XXZ model in roots of unity

We now consider the R-matrix $R(u)$ corresponding to the six-vertex model; the matrix elements can be written as

$$\begin{aligned} R_{\alpha\alpha}^{\alpha\alpha}(u) &= \rho\sin(u+\eta), \\ R_{\alpha\beta}^{\alpha\beta}(u) &= \rho\sin u, \\ R_{\beta\alpha}^{\alpha\beta}(u) &= \rho\sin\eta, \end{aligned} \tag{21}$$

where $\alpha, \beta = 1, 2$, $\alpha \neq \beta$, and η is the so-called anisotropy parameter.

The quantum group $\mathbf{A}$ connected with $R(u)$ is generated by $A(u)$, $B(u)$, $C(u)$, and $D(u)$, entries of the monodromy matrix $L(u)$.

The center of $\mathbf{A}$ in roots of unity, i.e., $\eta = \pi m/N$, is generated by the following elements of $\mathbf{A}$:

$$\begin{aligned}\langle A(u)\rangle &= A(u)A(u+\eta)\cdots A(u+(N-1)\eta),\\ \langle B(u)\rangle &= B(u)B(u+\eta)\cdots B(u+(N-1)\eta),\\ \langle C(u)\rangle &= C(u)C(u+\eta)\cdots C(u+(N-1)\eta),\\ \langle D(u)\rangle &= D(u)D(u+\eta)\cdots D(u+(N-1)\eta).\end{aligned}$$

For convenience, we let $\langle B(u)\rangle$ denote the central element,

$$\langle B(u)\rangle = B(u)B_1(u),$$

where

$$B_1(u) = B(u+\eta)\cdots B(u+(N-1)\eta).$$

We now fix $V = C^2 \otimes \cdots \otimes C^2$ as the representation space of our quantum group. It is easy to see that for arbitrary v,

$$\langle B(v)\rangle = B(v)B_1(v) = 0$$

on this space. We can then define the state space of a restricted model as

$$W(v) = \operatorname{Ker} B(v)/\operatorname{Im} B_1(v).$$

In the limit $v \to \infty$, $B(v)$ coincides up to a scalar factor with X, one of the generators of $U_q(sl(2))$. As a result, $W(\infty)$ coincides with the space of the "good" highest vectors of Pasqier–Saleur, which also shows that the Hamiltonian of the XXZ chain with OBC of special type,

$$H_{XXZ} = \sum_{n=1}^{L-1}\left[\sigma_n^+\sigma_{n+1}^- + \sigma_n^-\sigma_{n+1}^+ + \frac{\cos\eta}{2}\sigma_n^z\sigma_{n+1}^z + i\frac{\sin\eta}{2}(\sigma_n^z - \sigma_{n+1}^z)\right],$$

is invariant under the quantum algebra $U_q(sl(2))$. Here, $q = e^\eta$, and H_{XXZ} is therefore properly defined on $W(\infty) = \operatorname{Ker} X/\operatorname{Im} X^{(N-1)}$.

In the thermodynamic limit, where $L \to \infty$, the spectrum of low-lying states coincides (in Cardy's sense) with $M(N-1/N)$, one of the Minimal Models of CFT. Below, we show that this construction can be generalized to arbitrary values of the parameter v. The corresponding Hamiltonian is [5], [6]

$$\begin{aligned}H_{XXZ} = &\sum_{n=1}^{L-1}\left[\sigma_n^+\sigma_{n+1}^- + \sigma_n^-\sigma_{n+1}^+ + \frac{\cos\eta}{2}\sigma_n^z\sigma_{n+1}^z\right]\\ &+ i\frac{\sin\eta}{2}(\cot v\sigma_1^z - \cot(v+\eta)\sigma_L^z).\end{aligned} \quad (22)$$

We now find the conditions that the operators Q must satisfy in order to be projected on $W(v)$. It is easy to see that sufficient conditions are

$$B(v)Q = \widehat{Q}B(v) \tag{23}$$

and

$$QB_1(v) = B_1(v)\widehat{Q}_1, \tag{24}$$

where $\widehat{Q}$ and $\widehat{Q}_1$ are some properly chosen operators. Indeed, Eq. (23) guarantees that if a vector $\psi \in \operatorname{Ker} B(v)$, then the vector $Q\psi \in \operatorname{Ker} B(v)$. It follows from Eq. (24) that if the difference of two vectors ψ_1 and ψ_2 belong to $\operatorname{Im} B_1(v)$, i.e., if $\psi_1 - \psi_2 = B_1(v)\chi$, then this also holds for $Q\psi_1$ and $Q\psi_2$.

In [5], E. Sklyanin explained the integrability of models with OBC by constructing a special monodromy matrix and using it to diagonalize Hamiltonian (22) and the corresponding transfer matrix via the algebraic Bethe ansatz.

Let $K_+(u) = K(u + \eta/2, \xi_+)$ and $K_-(u) = K(u - \eta/2, \xi_-)$, where

$$K(u, \xi) = \begin{bmatrix} \sin(u+\xi) & 0 \\ 0 & -\sin(u-\xi) \end{bmatrix}.$$

Then $K_\pm$ satisfies the boundary Yang–Baxter equations [5].

The Sklyanin monodromy matrix is defined [5] as

$$\begin{aligned} \Lambda(u) &= \sigma^2 L^t(-u)\sigma^2 K_+(u)L(u) \\ &= \begin{bmatrix} \Lambda_1^1 & \Lambda_1^2 \\ \Lambda_2^1 & \Lambda_2^2 \end{bmatrix}. \end{aligned}$$

Using (2) and the boundary Yang–Baxter equation, we can prove that $\Lambda(u)$ satisfies the same equation as $K_+(u)$. It gives the commutation relations between Λ_j^i. The Sklyanin transfer matrix is defined as

$$\begin{aligned} T_{\mathrm{S}}(u) &= \operatorname{tr} \Lambda(u)K_-(u) \\ &= \sin(u - \eta/2 + \xi_-)\Lambda_1^1 - \sin(u - \eta/2 - \xi_-)\Lambda_2^2. \end{aligned}$$

Explicit expressions for the elements of the Sklyanin monodromy matrix are

$$\begin{aligned} \Lambda_1^1(u) &= \sin(u+\eta/2+\xi_+)A(u)D(-u) + \sin(u+\eta/2-\xi_+)C(u)B(-u), \\ \Lambda_2^2(u) &= -\sin(u+\eta/2+\xi_+)B(u)C(-u) - \sin(u+\eta/2-\xi_+)D(u)A(-u), \\ \Lambda_1^2(u) &= \sin(u+\eta/2+\xi_+)B(u)D(-u) + \sin(u+\eta/2-\xi_+)D(u)B(-u), \\ \Lambda_2^1(u) &= -\sin(u+\eta/2+\xi_+)A(u)C(-u) - \sin(u+\eta/2-\xi_+)C(u)A(-u). \end{aligned}$$

Let $\xi_+ = v - \eta/2$ and $\xi_- = -v - \eta/2$. Then the relations

$$B(v)T_S(u) = \widehat{T}_S(u)B(v) \tag{25}$$

are satisfied, where the explicit expressions for the Sklyanin transfer matrix for these values of ξ_{pm} are

$$\begin{aligned} T_S(u) = &\sin(u-v-\eta)\sin(u+v)A(u)D(-u) \\ &+ \sin(u-v-\eta)\sin(u-v+\eta)C(u)B(-u) \\ &+ \sin(u+v)\sin(u+v)B(u)C(-u) \\ &+ \sin(u+v)\sin(u-v+\eta)D(u)A(-u) \end{aligned} \tag{26}$$

and

$$\begin{aligned} \widehat{T}_S(u) = &\sin(u-v)\sin(u+v+\eta)A(u)D(-u) \\ &+ \sin(u-v)\sin(u-v)C(u)B(-u) \\ &+ \sin(u+v+\eta)\sin(u+v-\eta)B(u)C(-u) \\ &+ \sin(u+v+\eta)\sin(u-v)D(u)A(-u). \end{aligned}$$

We also have

$$B(v)\Lambda_1^2(u) = \hat{\Lambda}_1^2(u)B(v), \tag{27}$$

where explicitly

$$\begin{aligned} \Lambda_1^2(u) &= \sin(u+v)B(u)D(-u) + \sin(u-v+\eta)D(u)B(-u), \\ \hat{\Lambda}_1^2(u) &= \sin(u+v+\eta)B(u)D(-u) + \sin(u-v)D(u)B(-u). \end{aligned}$$

Equations (25) and (27) can be verified directly.

The operators $T_S(u)$ and $\Lambda_1^2(u)$ thus satisfy the first condition, Eq. (23), for the restriction. We conjecture that they also satisfy the second one, Eq. (24), if q is a root of unity. This conjecture was explicitly verified for $\eta = \pi/3$ and $\eta = 2\pi/3$ for the case $T_S(u)$.

The operators $T_S(u)$ and $\Lambda_1^2(u)$ depend on one parameter. There also exists a two-parameter family of operators satisfying (23) and (24).

By definition, let

$$\begin{aligned} T(x_{ij}; u_1, u_2) = x_{11}A(u_1)D(u_2) + x_{22}D(u_1)A(u_2) \\ + x_{12}B(u_1)C(u_2) + X_{21}C(u_1)B(u_2). \end{aligned}$$

Then

$$B(v)T(x_{ij}(v); u_1, u_2) = T(\hat{x}_{ij}(v); u_1, u_2)B(v), \tag{28}$$

where
$T(x_{ij}(v);u_1,u_2)$ and $T(\hat{x}_{ij}(v);u_1,u_2)$ are obtained from $T(x_{ij};u_1,u_2)$ by suitably substituting $x_{ij}(v)$ and $\hat{x}_{ij}(v)$ for x_{ij} and

$$\begin{aligned}
x_{11}(v) &= \sin(u_1-\eta-v)\sin(u_2-v),\\
x_{22}(v) &= \sin(u_1+\eta-v)\sin(u_2-v),\\
x_{12}(v) &= -\sin(u_2-v)\sin(u_2-v),\\
x_{21}(v) &= -\sin(u_1+\eta-v)\sin(u_1-\eta-v),\\
\hat{x}_{11}(v) &= \sin(u_1-v)\sin(u_2-\eta-v),\\
\hat{x}_{22}(v) &= \sin(u_1-v)\sin(u_2+\eta-v),\\
\hat{x}_{12}(v) &= -\sin(u_2-\eta-v)\sin(u_2+\eta-v),\\
\hat{x}_{21}(v) &= -\sin(u_1-v)\sin(u_1-v).
\end{aligned}$$

We can verify that

$$\begin{aligned}
T(x'_{ij}(v);u_1,u_2)&B(v+\eta)B(v+2\eta)\\
&= B(v+\eta)B(v+2\eta)T(\hat{x}'_{ij}(v);u_1,u_2) \qquad (29)
\end{aligned}$$

with some x'_{ij} if $\eta=\pi/3$ or $\eta=2\pi/3$.

As discussed above, it follows that the two-parameter family of operators $T(x_{ij}(v);u_1,u_2)$ can be restricted on $W(v)$.

We conjecture that $T(x_{ij}(v);u_1,u_2)$ satisfies restriction conditions (23) and (24) if $\eta=m\pi/p$, where m and p are coprime integers.

It is easy to verify that the relation

$$T_S(u) = T(x_{ij}(v);u,-u)$$

holds, where $T_S(u)$ is the Sklyanin transfer matrix in (26).

These formulas give some of the operators surviving under the restriction. It would be interesting to find local operators with the same property that is needed for constructing the correlation functions following the approach in [7].

Acknowledgments

I thank the organizers of the conference and especially Vlado Dobrev for their warm hospitality during this interesting meeting.

References

1. A. Belavin and M. Jimbo, "Central elements of the elliptic Yang–Baxter algebra at roots of unity", hep-th/0208224.

2. R.J. Baxter, *Exactly Solved Models in Statistical Mechanics* (Academic Press, 1982).
3. V. Tarasov, *Int. J. Mod. Phys.* **A7**, Suppl. 1B, 963 (1992).
4. E. Date, M. Jimbo, A. Kuniba, T. Miwa and M. Okado, *Adv. Stud. Pure Math.* **16**, 17 (1988).
5. E.K. Sklyanin, "Boundary conditions for integrable quantum systems", *J. Phys. A: Math. Gen.* **21**, 2375-2389 (1988).
6. F.C. Alcaraz, M. Baake, U. Grimm and V. Rittenberg, *J. Phys. A. Math. Gen* **22**, L5-L11 (1989).
7. N. Kitanine, J.M. Maillet and V. Terras, *Nucl. Phys.* **B554**, 647 (1999).

SPACES OF BOUNDARY VALUES RELATED TO A MULTIPOINT VERSION OF THE KP-HIERARCHY

G.F. HELMINCK

Faculty of Applied Mathematics,
University of Twente,
P.O.Box 217, 7500 AE Enschede,
The Netherlands
helminckgf@cs.utwente.nl

In this paper one considers a finite number of points in the complex plane and various spaces of boundary values on circles surrounding these points. To this geometric configuration one associates a Grassmann manifolds that is shown to yield solutions of a multipoint version of the linearization of the KP-hierarchy. These Grassmann manifolds are built in such a way that the determinant line bundle and its dual over them still make sense. The same holds for the so-called τ-functions, determinants of certain Fredholm operators that measure the failure of equivariance at lifting the commuting flows of the hierarchy to these bundles. Solutions of the linearization are described by wave functions of a certain type. They are perturbations of the trivial solution with the leading term of the perturbation determining the type. One concludes with showing that, if a plane in the Grassmann manifold yields wave functions of different types, they are connected by a differential operator in the coordinates of the flows.

1. Introduction

As Sato has indicated, see e.g. [10], spaces of boundary values on the circle yield solutions of the KP-hierarchy, which are the compatibility conditions for an infinite linear system, called its linearization, see e.g. [1].

For the case of the square integrable functions on the unit circle this connection was treated extensively in [11]. There they considered a Grassmann manifold consisting of planes close to that of the square integrable functions on the unit circle that have an analytic continuation to infinity. The flows lying at the basis of the linearization correspond to the group of holomorphic functions on the unit disc around infinity and vanishing there. To each subspace in this manifold they associated a solution of this linearization, which carries the name wavefunction or Baker-Akhiezer function. These wavefunctions can be expressed in so-called τ-functions,

Fredholm-determinants, which have an interpretation in terms of the determinant bundle over this Grassmann manifold and its dual. A Lie algebraic interpretation of these functions was given by V.Kac, see e.g. [9].

In [2] it was suggested to consider a multipoint version of the KP-hierarchy corresponding to n points $\{a_1, \ldots, a_n\}$ in $\mathbb{C}$ and the point infinity. In this paper the multipoint analogue of the linearization of the KP-hierarchy is taken under consideration. The group of flows that are a proper translation of the ones for the KP-hierarchy consists of the functions that are holomorphic on the complement of circles around each of the points a_i and that vanish at infinity. These continuous flows can be complemented with a natural lattice group that commutes with them. The next step will be to choose a number of disjoint circles around the points $\{a_1, \ldots, a_n\}$ and on each circle an appropriate space of boundary values. The total space of boundary values possesses a decomposition to which one associates a suitable Grassmannian. For each plane in this manifold one can construct solutions of the linearization of a specific nature depending of the lattice points. Solutions corresponding to the same plane and different lattice points are shown to be related by a differential operator.

A specification of the content of the various sections is as follows: the first is devoted to an algebraic set-up for the linear evolution equations corresponding to the flows. In the next section the properties of the spaces of boundary values are discussed and the basic decomposition is presented. It is different of the one suggested in [2]. The subsequent section starts with a description of the Grassmannian, it discusses furthermore its structure, various realizations and how the flows fit in. Note that already in the square integrable case, as considered e.g. in [11], the Grassmannian introduced here is significantly more general. Despite this general character one can still define the determinant bundle and its dual over this manifold and that is done in the following section. The final section contains the construction for each plane in the connected component of the Grassmann manifold of multipoint wavefunctions of a certain type and a discussion of the dependence of the lattice.

2. An algebraic set-up for the linearization

The independent variables $\{t_i \mid i \geq 1\}$ occurring in the KP-hierarchy are the coordinates of the commuting flows of this integrable system,

$$\gamma(t, z) = \exp(\sum_{i=1}^{\infty} t_i z^{-i}), \tag{1}$$

where t is a short-hand notation for the $\{t_i\}$. The nonlinear equations of the KP-hierarchy are obtained by cross-differentiation of the so-called linearization. In this system the relevant differential operators of the KP-hierarchy act on objects like

$$\{\sum_{i \geq N} a_i(t) z^i\} \gamma(t, z), \tag{2}$$

which are formal products of a power series in z, depending of t, that has at the utmost a pole in $z = 0$ and an exponential factor corresponding to the commuting flows that describes the essential singularity around zero. Often the role of the points zero and infinity is interchanged, but for the present generalization one sticks to the decomposition (2). Consider thereto, instead of the point zero, n different points $a_1, \ldots, a_n$ in $\mathbb{P}^1(\mathbb{C}) \setminus \{\infty\}$. The exponential factor describing the essential singularity around each of the points $a_1, \ldots, a_n$ has the form

$$\gamma(t, z) := \exp \sum_{k=1}^{n} \sum_{l=1}^{\infty} t_l^{(k)} (z - a_k)^{-l}. \tag{3}$$

Here t is again the short-hand notation for the collection of coordinates $\{t_l^{(k)}\}$. It is convenient to use a notation related to this group of transformations. One will write

$$\xi_k(z, t) = \sum_{l=1}^{\infty} t_l^{(k)} (z - a_k)^{-l} \quad \text{and} \quad \xi(z, t) = \sum_{k=1}^{n} \xi_k(z, t).$$

Let $\mathcal{R}$ be a complex algebra of functions in the $\{t_l^{(k)}\}$ that is stable under all the partial derivatives $\partial_k^{(i)} := \frac{\partial}{\partial t_k^{(i)}}$. For each $k, 1 \leq k \leq n$, let $\mathcal{H}_k(\mathcal{R})$ be the space of all power series in $z - a_k$ with coefficients from $\mathcal{R}$ that only have a pole in $z = a_k$, i.e.

$$\mathcal{H}_k(\mathcal{R}) = \{\sum_{s \geq N} a_s (z - a_k)^s \mid a_s \in \mathcal{R}, \ N \in \mathbb{Z}\}. \tag{4}$$

The analogue of the other factor in decomposition (2) is taken from the space

$$\mathcal{H}(\mathcal{R}) := \oplus_{k=1}^{n} \mathcal{H}_k(\mathcal{R}). \tag{5}$$

The linearization in the multipoint setting is now realised on the space $\mathcal{M}(\mathcal{R})$ of formal products of elements of $\mathcal{H}(\mathcal{R})$ and the exponential factor $\gamma(t, z)$. Note that we have for each $s \in \mathbb{Z}$ a natural action of $(z - a_k)^s$ on the whole space $\mathcal{H}(\mathcal{R})$. On the k-th factor this action consists of multiplication with $(z-a_k)^s$ and for $i \neq k$ it acts on the i-th component by multiplying with the power series $((z-a_i)+a_i-a_k)^s$. This enables you to define an $\mathcal{R}[\partial_k^{(i)}]$-module structure on $\mathcal{M}(\mathcal{R})$ by extending the following actions: for each element

$$\psi = (\ldots, \sum_l h_l^{(k)}(z-a_k)^l, \ldots)\gamma(t,z)$$

in the space $\mathcal{M}(\mathcal{R})$ and each $r \in \mathcal{R}$ one defines

$$\begin{aligned} r.\psi :=& (\ldots, \sum_l r h_l^{(k)}(z-a_k)^l, \ldots)\gamma(t,z) \qquad (6) \\ \partial_s^{(i)}(\psi) :=& (\ldots, \sum_l \partial_s^{(i)}(h_l^{(k)})(z-a_k)^l, \ldots)\gamma(t,z) \\ &+ (\ldots, \sum_l h_l^{(k)}(z-a_k)^l, \ldots)(z-a_i)^{-s}\gamma(t,z). \end{aligned}$$

Now that one can consider differential equations w.r.t. the $\{\partial_k^{(i)}\}$ inside $\mathcal{M}(\mathcal{R})$, one still wants to fix a gauge of the eigenfunctions, since there is a natural group of transformations on $\mathcal{M}(\mathcal{R})$ that commutes with these derivatives. Recall that the group Σ given by

$$\Sigma = \{\sigma(\underline{m}) := \prod_{k=1}^{n}(z-a_k)^{m_k} \mid \underline{m} = \{m_k\} \in \mathbb{Z}^n\} \qquad (7)$$

acts on the space $\mathcal{H}(\mathcal{R})$ from the right and this action is extended to $\mathcal{M}(\mathcal{R})$ by

$$h\gamma(t,z) \mapsto h\sigma(\underline{m})\gamma(t,z). \qquad (8)$$

An element $\psi \in \mathcal{M}(\mathcal{R})$ is called *of type $(i, \sigma(\underline{m}))$* if it can be written as

$$\psi = \hat{\psi}\sigma(\underline{m})\gamma(t,z)$$

, where $\hat{\psi} \in \mathcal{M}(\mathcal{R})$ has the form

$$\hat{\psi} = (\ldots, \sum_{l \geq -1} h_l^{(k)}(z-a_k)^l, \ldots), \text{ with } h_{-1}^{(i)} = 1 \text{ and } h_{-1}^{(k)} = 0 \text{ for } k \neq i. \qquad (9)$$

Now one can introduce the linearization of the multipoint KP-hierarchy. For an element $\psi_i^{(\sigma(\underline{m}))} \in \mathcal{M}(\mathcal{R})$ of type $(i, \sigma(\underline{m}))$, it consists of all the equations such that

$$\partial_k^{(j)}(\psi_i^{(\sigma(\underline{m}))}) = B_k^{(j,i)}(\psi_i^{(\sigma(\underline{m}))}), k \geq 1, 1 \leq j \leq n, \qquad (10)$$

where $B_k^{(j,i)}$ is a differential operator of order k in $\partial_j := \partial_1^{(j)}$ with coefficients from $\mathcal{R}$. Following the terminology used in the case of the KP-hierarchy, one calls $\psi_i^{(\sigma(\underline{m}))}$ a *wavefunction of the multipoint KP-hierarchy of type $(i,\sigma(\underline{m}))$.*

The equations of the *multipoint KP-hierarchy* are then the compatibility conditions for this linearization, which amount to nonlinear differential equations for the coefficients of $\hat{\psi}_i^{(\sigma(\underline{m}))}$ that can in a compact way be written as

$$\{\partial_m^{(s)}(B_k^{(j,i)}) - \partial_k^{(j)}(B_m^{(s,i)}) + [B_k^{(j,i)}, B_m^{(s,i)}]\}(\psi_i^{(\sigma(\underline{m}))}) = 0. \tag{11}$$

Remark 2.1. Note that if one considers a wavefunction ψ of the multipoint KP-hierarchy of type $(i,\sigma(\underline{m}))$ merely for the variables $\{t_l^{(k)} \mid l \geq 1,\ k \text{ fixed }\}$, then it is a solution of the KP-hierarchy. Hence the function ψ is an intermingling of n solutions of the KP-hierarchy.

To produce concrete solutions of the multipoint KP-hierarchy one constructs an analytic setting that generates elements of $\mathcal{M}(\mathcal{R})$ of type $(i,\sigma(\underline{m}))$ for which the formal product is real and that satisfy all the equations (10). The next section intends to illustrate the freedom of choice one has in carrying out this construction.

3. Spaces of boundary values

Let the $\{D_i \mid 1 \leq i \leq n\}$ be disjoint open discs in $\mathbb{C}$ with center a_i, boundary C_i and radius r_i. On each circle C_i one assumes given a Banach space H_i with norm $\|\cdot\|_i$. All these boundary values are joined together inside $H = \oplus_{i=1}^n H_i$, which is again a Banach space w.r.t. the norm

$$\| (h_i) \|_H := \sum_{i=1}^{n} \| h_i \|_i \tag{12}$$

The spaces H_i should satisfy a number of conditions that are listed here

Properties 3.1.

(1) First of all, each space H_i should consist of power series in $z - a_i$, i.e.

$$H_i = \left\{ f(z) = \sum_{n\in\mathbb{Z}} a_n (z - a_i)^n \right\}$$

and all the $\{(z - a_i)^n \mid n \in \mathbb{Z}\}$ should belong to H_i.

(2) Inside H_i, one has the subspace

$$H_i^+ = \left\{ f \in H_i \,\middle|\, f(z) = \sum_{n \geq 0} a_n (z - a_i)^n \right\}$$

of all series that possibly converge around a_i. It is the closure w.r.t. the norm $\|\cdot\|_i$ of the span of the $\{(z-a_i)^n | \, n \geq 0\}$. Moreover to each $f_i \in H_i^+$ corresponds a function F_i holomorphic on D_i with boundary value f_i.

(3) The second subspace of H_i is

$$H_i^- = \left\{ f \in H_i \,\middle|\, f(z) = \sum_{n < 0} a_n (z - a_i)^n \right\}$$

the space of all series in H_i that have a potential zero at infinity. Likewise, the space H_i^- is the closure w.r.t. the same norm of the space spanned by the $\{(z-a_i)^n | \, n < 0\}$. For each $f_i \in H_i^-$ corresponds a function F_i holomorphic on the complement of $\overline{D_i}$ with boundary value $F_i(i) := f_i$. For all $k \neq i$ write $F_i(k)$ for the boundary value of F_i at C_k, then each $F_i(k)$ should belong to H_k^+ and the map $f_i \mapsto F_i(k)$ should be continuous.

(4) The spaces H_i^+ and H_i^- generate H_i: $H_i = H_i^+ \oplus H_i^-$ and let for each $f_i \in H_i$ the corresponding decomposition be $f_i = f_i^+ + f_i^-$. Moreover, the projections $p_i^+ : H_i \mapsto H_i^+$ and $p_i^- : H_i \mapsto H_i^-$ are continuous.

(5) Finally, on each space H_i multiplication with $z - a_j, 1 \leq j \leq n$, is a bounded invertible operator $M_j(i)^{-1}$. Consequently the same holds for multiplying with any of the $\{(z-a_i)^n | \, n \in \mathbb{Z}\}$.

Examples of such spaces are the L^p-functions on the boundary C_i and variations on this theme. For the KP-hierarchy, Sato, see e.g. [10], realized that there is a great diversity in spaces of boundary values that lead to solutions of the KP-hierarchy. The analytic context for the L^2-boundary values was given in [11]. A class of Banach spaces yielding solutions to the KP-hierarchy occurs in [4].

Inside the space H one considers the boundary values that are holomorphic inside all the circles, i.e. $H^+ = \oplus_{i=1}^n H_i$, and those that are holomorphic outside all the discs D_i and disappear at infinity,

$$H^* = \overline{\mathrm{Span}(((z-a_i)^m, \ldots, (z-a_i)^m) \,|\, 1 \leq i \leq n, \; m < 0)}. \qquad (13)$$

By making use of the properties 3.1(2) and 3.1(3) and by invoking the edge of the wedge theorem as formulated e.g. in [3], one proves that the subspaces H^+ and H^* satisfy: $H^+ \cap H^* = \{0\}$. From property 3.1(3) one sees that the map $p^* : H \mapsto H^*$ given by

$$p^*((f_i)) = (\sum_{j=1}^{n} F_j^-(1), \ldots, \sum_{j=1}^{n} F_j^-(n)), \tag{14}$$

projects H onto H^* and that its restriction to H^- is a continuous bijection. Thus one arrives at

Proposition 3.1. *The subspaces H^+ and H^* satisfy: $H = H^* \oplus H^+$.*

This is the central decomposition leading to solutions of the linearization of the multipoint KP-hierarchy, since one can construct for any subspace of H 'close' to H^* a wavefunction of a certain type. The precise specification of what is meant by 'close" is done in the next section.

4. The Grassmannian related to $H = H^* \oplus H^+$

Before describing the infinite dimensional Grassmann manifold, we recall some definitions and notations. For any closed subspace W of H let $i_W : W \mapsto H$ the natural continuous embedding. If X and Y are Banach spaces, then the space of Fredholm operators from X to Y consists of the continuous linear operators with a closed image and a finite dimensional kernel and cokernel. It is denoted by $\Phi(X,Y)$. The connected components of $\Phi(X,X)$ are described by the index

$$i(A) = \dim(\mathrm{Ker}(A)) - \dim(\mathrm{Coker}(A)).$$

Definition 4.1. The *Grassmann manifold* $\mathrm{Gr}(H)$ corresponding to the decomposition $H = H^* \oplus H^+$ consists of all closed subspaces W such that $p^* \circ i_W$ belongs to $\Phi(W, H^*)$.

This is a Banach manifold based on the space $L(H^*, H^+)$ of bounded linear operators from H^* to H^+ and its connected components are given by

$$\mathrm{Gr}_k(H) := \{W \in \mathrm{Gr}(H) \mid i(p^* \circ i_W) = k\},\ k \in \mathbb{Z}.$$

This manifold is a homogeneous space for various groups. The group that is presented here enables you to define the determinant bundle Det over $\mathrm{Gr}_0(H)$ and its dual Det^* and needs the notion of nuclear operator. One recalls some of their features.

Let E and F be Banach spaces. Equip the topological dual F^* with the operator norm. There is a natural embedding v of $E \otimes F^*$ into the bounded operators $L(F,E)$, namely

$$v(\sum_j e_j \otimes f_j^*)(x) = \sum_j < f_j^*, x > e_j, \tag{15}$$

where the e_j belong to E, the f_j^* to F^* and x to F. The image of v are the finite dimensional operators in $L(F,E)$. On the tensor product $E \otimes F^*$ one defines

$$\| u \|_{\mathrm{n}} = \inf \sum_j \| e_j \|_E \| f_j^* \|_{F^*}, \tag{16}$$

where the infimum is taken over all $\sum_j e_j \otimes f_j^*$ equal to u. It is shown in [5] that $\| \cdot \|_{\mathrm{n}}$ defines a norm on $E \otimes F^*$, the Schatten norm. The completion of $E \otimes F^*$ w.r..t the Schatten norm is denoted by $E \hat{\otimes} F^*$. According to [6], every element of $E \hat{\otimes} F^*$ can be written, though not uniquely, in the form

$$\sum_{i=0}^{\infty} \lambda_i (e_i \otimes f_i^*), \text{ with } \lambda_i \geq 0, \sum_{i=0}^{\infty} \lambda_i < \infty, \| e_i \|_E \leq 1, \| f_i^* \|_{F^*} \leq 1. \tag{17}$$

The map v has a continuous extension $v: E \hat{\otimes} F^* \mapsto L(F,E)$ and the image of this last map is the space of nuclear operators $\mathcal{N}(F,E)$ from F to E. Equipped with the quotient norm, also denoted by $\| \cdot \|_{\mathrm{n}}$, it is a Banach space. In the case that E and F both equal a Hilbert space K, the nuclear operators are the trace class operators on K and the norm $\| \cdot \|_{\mathrm{n}}$ is the trace-norm.

Nuclear operators are stable w.r.t. composition from the left and right with bounded operators. More concretely, let D be yet another Banach space and consider operators $X \in \mathcal{N}(F,E)$, $Y \in L(F,D)$ and $Z \in L(D,E)$. Then one has $Y \circ X \in \mathcal{N}(E,F)$ resp. $X \circ Z \in \mathcal{N}(D,F)$ and the following inequalities hold

$$\| Y \circ X \|_{\mathrm{n}} \leq \| Y \| \| X \|_{\mathrm{n}} \text{ and } \| X \circ Z \|_{\mathrm{n}} \leq \| X \|_{\mathrm{n}} \| Z \| . \tag{18}$$

Here $\| \cdot \|$ denotes again the operator norm. Using these properties of nuclear operators one shows that $Gl_{\mathrm{n}}(H)$, consisting of

$$Gl_{\mathrm{n}}(H) = \left\{ g = \begin{pmatrix} a & b \\ c & d \end{pmatrix} \in Gl(H) \middle| \begin{matrix} a \in \Phi(H^*, H^*) & d \in \Phi(H^+, H^+) \\ c \in L(H^*, H^+) & b \in \mathcal{N}(H^+, H^*) \end{matrix} \right\},$$

is a group that acts transitively on $\mathrm{Gr}(H)$. In particular, the component $\mathrm{Gr}_0(H)$ is a homogeneous space for the subgroup

$$Gl_{\mathrm{n}}^{(0)}(H) = \left\{ g = \begin{pmatrix} a & b \\ c & d \end{pmatrix} \in Gl_{\mathrm{n}}(H) \middle| i(a) = i(d) = 0 \right\},$$

In $\mathrm{Gr}_0(H)$ we have the open dense subset of all planes W such that $p^* : W \to H^*$ is a bijection. Like in the finite dimensional situation, this is called the big cell of $\mathrm{Gr}(H)$ and its elements are called transverse to H^*.

For each $j, 1 \leq j \leq n$, consider the operator $M_j = \oplus_{i=1}^{n} M_j(i) : H \mapsto H$ and let $M > 0$ satisfy $M >\| M_j \|$, where $\| \cdot \|$ denotes the operator norm. For each such M one considers the group of power series

$$\tilde{\Gamma}(M) = \left\{ \exp \sum_{k=1}^{n} \sum_{l=1}^{\infty} t_l^{(k)} (z - a_k)^{-l} \middle| \sum_{k=1}^{n} \sum_{l=1}^{\infty} |t_l^{(k)}| \, M^l < \infty \right\}.$$

Its elements act continuously on H by multiplication and if $\begin{pmatrix} a & b \\ c & d \end{pmatrix}$ is the decomposition of

$$\sum_{k=1}^{n} \sum_{l=1}^{\infty} t_l^{(k)} M_k^l \tag{19}$$

w.r.t. $H = H^* \oplus H^+$ then a direct computation of the operator b shows that it can be written in the form (17). Hence it is nuclear, so that the group of flows $\tilde{\Gamma}(M)$ embeds in this way into $Gl_n^{(0)}(H)$. Likewise one can consider the action of the group Σ on H by multiplying on each factor with the element $\sigma(\underline{m})$ of Σ and this gives you also an operator in $Gl_n(H)$. The subgroup

$$\Sigma^{(0)} = \{\sigma(\underline{m} \mid \sum_{k=1}^{m} m_k = 0\}$$

corresponds to the elements that yield an operator in $Gl_n^{(0)}(H)$. This last group is generated by the elements $\sigma_{ij} := \frac{(z-a_i)}{(z-a_j)}$.

5. The Line bundles Det and Det* over $\mathrm{Gr}_0(H)$

Let W belong to $\mathrm{Gr}_0(H)$. Since a Fredholm operator of index zero can be written as the sum of a finite dimensional one and an invertible operator, one can write W as the image of an embedding $w : H^* \to H$ that decomposes w.r.t. $H = H^* \oplus H_+$ as

$$w = \begin{pmatrix} w^* \\ w^+ \end{pmatrix}, \text{ with } w^* - \mathrm{Id} \text{ finite dimensional and } w^+ \text{ bounded.} \tag{20}$$

This class of embeddings is somewhat too restricted for the action of the group $Gl_n^{(0)}(H)$. Henceforth one chooses to vary $w^* - \mathrm{Id}$ inside a wider class of compact operators, the nuclear ones. The collection of all embeddings of this form is denoted by $\mathcal{P}^*$. The component $\mathrm{Gr}_0(H)$ can be described then as the quotient $\mathcal{P}^*/\mathcal{T}^*$, where $\mathcal{T}^*$ is the group defined by

$$\mathcal{T}^* = \{t \in Gl(H^*) \mid t - \mathrm{Id} \text{ is nuclear } \}, \tag{21}$$

which acts on $\mathcal{P}^*$ by composition on the right. For operators from H^* to H^* of the form "identity + nuclear" one has a well-defined notion of determinant. If E is a Banach space and $u = \sum_j e_j \otimes e_j^* \in E \otimes E^*$, then one defines the trace by

$$\mathrm{Tr}(v(\sum_j e_j \otimes e_j^*)) = \sum_j e_j^*(e_j) \tag{22}$$

and for the determinant one has the formula

$$\det(\mathrm{Id} + v(u)) = 1 + \sum_{k=1}^{\mathrm{rank}(v(u))} \mathrm{Tr}(\bigwedge^k v(u)) := 1 + \sum_{k=1}^{\mathrm{rank}(v(u))} \alpha_k(u), \tag{23}$$

where $\bigwedge^k v(u)$ is the k-th exterior power of $v(u)$ and each α_k is a k-homogeneous polynomial from $E \otimes E^*$ to $\mathbb{C}$. Now Grothendieck has shown in [6] that all the α_k extend to k-homogeneous polynomials from $E\hat{\otimes}E^*$ to $\mathbb{C}$ and that $\sum_k \alpha_k(u)$ converges absolutely.

Recall that the map $v : E\hat{\otimes}E^* \mapsto L(E, E)$ is injective if and only if the space E has the approximation property, see [5]. Since this is the case for $E = H^*$, one can define for $v(u) \in v(H^*\hat{\otimes}(H^*)^*)$, the analytic function

$$\det(\mathrm{Id} + v(u)) = 1 + \sum_{k=1}^{\infty} \alpha_k(u) \tag{24}$$

It has the usual properties for a determinant: for two nuclear operators u_1 and u_2 on H^*

$$\det((\mathrm{Id} + u_1)(\mathrm{Id} + u_2)) = \det(\mathrm{Id} + u_1)\det(\mathrm{Id} + u_2) \tag{25}$$

$$, \mathrm{Id} + u_1 \text{ is invertible } \Leftrightarrow \det(\mathrm{Id} + u_1) \neq 0. \tag{26}$$

Now one can define holomorphic line bundles Det and Det^* over $\mathrm{Gr}_0(H)$, by dividing out on the product space $\mathcal{P} \times \mathbb{C}$ respectively the following two actions of $\mathcal{T}^*$,

$$(w, \lambda) \mapsto (w \circ t^{-1}, \lambda det(t)^{-1}), \tag{27}$$

$$(w, \lambda) \mapsto (w \circ t^{-1}, \lambda det(t)). \tag{28}$$

The class of (w,λ) in Det is denoted by $[w,\lambda]_1$ and that in Det* by $[w,\lambda]_2$. As in the finite dimensional case, the line bundle Det* has nontrivial sections, e.g. $\sigma : \mathrm{Gr}_0(H) \mapsto \mathrm{Det}^*$ given by

$$\sigma(\text{ Image of } w) = [w, det(w^*)]_2. \tag{29}$$

The connected component $Gl_{\mathrm{n}}^{(0)}(H)$ of $Gl_{\mathrm{n}}(H)$ acts transitively on $\mathrm{Gr}_0(H)$ and to lift this action to the line bundles Det and Det* one has to pass to an extension $\mathcal{E}$ of $Gl_{\mathrm{n}}^{(0)}(H)$. It is defined by

$$\mathcal{E} = \left\{ (g,q) = (\begin{pmatrix} a & b \\ c & d \end{pmatrix}, q) \middle| \begin{array}{c} \text{with } g \in Gl_{\mathrm{n}}^{(0)}(H), q \in Gl(H^*) \\ \text{and } aq^{-1} - \mathrm{Id} \text{ nuclear} \end{array} \right\}.$$

The group $\mathcal{E}$ acts on Det and Det* by

$$(g,q).[w,\lambda]_i = [g \circ w \circ q^{-1}, \lambda]_i. \tag{30}$$

This induces an action of $\mathcal{E}$ on the holomorphic sections of Det*, in particular for σ we get

$$(g,q).\sigma(Im(w)) = [w, \det((g^{-1} \circ w \circ q)^*)]_2. \tag{31}$$

Hence, if we define the function τ_w on $\mathcal{E}$ by

$$\tau_w((g,q)) = \det((g^{-1} \circ w \circ q)^*) = \det((aw^*q + bw^+q)^*), \text{ for } g^{-1} = \begin{pmatrix} a & b \\ c & d \end{pmatrix},$$

then this function measures the failure of the $\mathcal{E}$-equivariance of the section σ. Note that choosing another embedding from $\mathcal{P}^*$ with image W only results in multiplying this function with a non zero constant.

The next step will be the construction of the wave functions of type $(i, \sigma(\underline{m}))$ of the multipoint KP-hierarchy.

6. The construction of the multipoint wavefunctions

For any plane W belonging to $\mathrm{Gr}_0(H)$ one considers in the group $\Sigma^{(0)}$ the subset

$$\Sigma_W^{(0)} = \left\{ \sigma(\underline{m}) \middle| \begin{array}{l} \sigma(\underline{m}) \in \Sigma^{(0)}, \text{ there is a } \gamma \in \tilde{\Gamma}(M) \text{ such that} \\ p^* : \gamma^{-1}\sigma(\underline{m})^{-1} W \to H^* \text{ is a bijection} \end{array} \right\} \tag{32}$$

Note that each $\sigma(\underline{m})$ belongs to $\Sigma_W^{(0)}$ for each W in the inverse image under $\sigma(\underline{m})$ of the big cell. Since the translates of the big cell under the group $\Sigma^{(0)}$ cover $\mathrm{Gr}_0(H)$, one knows that the set $\Sigma_W^{(0)}$ is non-empty.

Starting with a $\sigma(\underline{m})$ in $\Sigma_W^{(0)}$ one considers for a sufficiently large M the group of continuous flows $\tilde{\Gamma}(M)$ and in it the open subset $\Gamma(\sigma(\underline{m}), W)$ given by

$$\Gamma(\sigma(\underline{m}), W) = \{\gamma \mid \gamma \in \tilde{\Gamma}(M), \gamma^{-1}\sigma(\underline{m})^{-1}W \text{ is transverse to } H^*\}.$$

Let $\mathcal{R}$ be the ring of analytic functions on $\Gamma(\sigma(\underline{m}), W)$. Since for each γ in $\Gamma(\sigma(\underline{m}), W)$, the plane $\gamma^{-1}\sigma(\underline{m})^{-1}W$ is transverse to H^*, this enables you to define for each $k, n \leq k \leq 1$,

$$\hat{\psi}_{W,k}^{(\sigma(\underline{m}))}(\gamma) = (p^*|_{\gamma^{-1}\sigma(\underline{m})^{-1}W})^{-1}(((z-a_k)^{-1}, \cdots, (z-a_k)^{-1})).$$

By multiplying with $\gamma\sigma(\underline{m})$ one translates the element $\hat{\psi}_{W,k}^{(\sigma(\underline{m}))}(\gamma)$ in $\gamma^{-1}\sigma(\underline{m})^{-1}W$ back to W and thus we obtain a $\psi_{W,k}^{(\sigma(\underline{m}))} : \Gamma(\sigma(\underline{m}), W) \to W$ that can be written as

$$\begin{aligned}\psi_{W,k}^{(\sigma(\underline{m}))}(\gamma) &= \hat{\psi}_{W,k}^{(\sigma(\underline{m}))}(\gamma)\sigma(\underline{m})e^{\xi} \\ &= \{(\ldots, \frac{1}{(z-a_k)^1} + \sum_{r\geq 0} \alpha_{r\ell}^{(k,j)}(\sigma(\underline{m}), \gamma)(z-a_j)^r, \ldots)\}\sigma(\underline{m})e^{\xi}.\end{aligned}$$

Hence each function $\psi_{W,k}^{(\sigma(\underline{m}))} \in \mathcal{M}(\mathcal{R})$ is of type $(k, \sigma(\underline{m}))$ and they are the candidate wavefunctions. To loosen up the notations one leaves out in the sequel the W- resp. $\sigma(\underline{m})$-dependence of the $\{\psi_{W,k}^{(\sigma(\underline{m}))}\}$ and writes $\hat{\psi}_k$ resp. ψ_k instead of $\hat{\psi}_{W,k}^{(\sigma(\underline{m}))}$ resp. $\psi_{W,k}^{(\sigma(\underline{m}))}$. The j-th component of $\hat{\psi}_k$ resp. ψ_k is denoted by $\hat{\psi}_{kj}$ resp. ψ_{kj} and the last one is given by

$$\psi_{kj}(\gamma) = \{\frac{1}{(z-a_k)^1} + \sum_{r\geq 0} \alpha_{r\ell}^{(i,j)}(\sigma(\underline{m}), \gamma)(z-a_j)^r\}\sigma(\underline{m})e^{\xi}. \tag{33}$$

In a number of steps it will be shown that the ψ_i satisfy the equations (10) and one starts with the $t_k^{(i)}$-derivatives of ψ_i.

Proposition 6.1. *For all $k \geq 1$ and all $i, 1 \leq i \leq n$, there is a unique differential operator $P_k^{(i)}$ of the form $P_k^{(i)} = \sum_{s=0}^{k} p_{ks}^{(i)}\partial_i^s$, with $p_{kk}^{(i)} = 1$, such that*

$$\partial_k^{(i)}(\psi_i) = P_k^{(i)}(\psi_i).$$

The coefficients of the differential operator $P_k^{(i)}$ are polynomial expressions in the coefficients $\{\alpha_{r1}^{(i,i)} \mid 0 \leq r \leq k-1\}$ and their ∂_i-derivatives.

Proof. For each s in $\mathbb{N}$, one has

$$\partial_i^s(\psi_i) = \sum_{r=0}^{s} \binom{s}{r} \partial_i^{s-r}(\hat{\psi}_i)\frac{1}{(z-a_i)^r}\sigma(\underline{m})e^{\xi}.$$

This formula implies that, for $j \neq i$, the j-th component of $\partial_i^s(\psi_i)$ contains only positive powers of $(z-a_j)$ and that the i-th component of $\partial_i^s(\psi_i)$ looks like

$$\left\{\frac{r_i}{(z-a_i)^{s+1}} + \text{ “higher order in } z-a_i\text{”}\right\}\sigma(\underline{m})e^{\xi}.$$

On the other hand one has

$$\partial_k^{(i)}(\psi_i) = \{\partial_k^{(i)}(\hat{\psi}_i) + \hat{\psi}_i(z-a_i)^{-k}\}\sigma(\underline{m})e^{\xi},$$

and from both relations one sees that $\{\partial_k^{(i)}(\psi_i) - \partial_i^k(\psi_i)\}\sigma(\underline{m})^{-1}e^{-\xi}$ has no singularities around the points $\{a_j, j \neq i\}$ and has a lower order singularity than $k+1$ around a_i. Therefore, the j-th component of $\{\partial_k^{(i)}(\psi_i) - \partial_i^k(\psi_i)\}\sigma(\underline{m})^{-1}e^{-\xi}$ for $j \neq i$ belongs to H_j^+ and the i-th component $\{\partial_k^{(i)}(\psi_i) - \partial_i^k(\psi_i)\}$ looks like

$$\{c_t(z-a_i)^t + \text{ “higher order in } z-a_i\text{” }\}\sigma(\underline{m})e^{\xi}$$

with $t > -k-1$ and c_t a polynomial expression in the $\alpha_{r1}^{(i,i)}, 0 \leq r \leq k-1$, and their ∂_i-derivatives. If $t < 0$, then one continues to reduce the order of the singularity around a_i and one considers

$$\partial_k^{(i)}(\psi_i) - \partial_i^k(\psi_i) - c_t\partial_i^{-t-1}(\psi_i).$$

Continuing in this fashion, one ends up with a differential operator $P_k^{(i)}$ of the required form such that for all $\gamma \in \Gamma(\sigma(\underline{m}), W)$

$$\partial_k^{(i)}(\psi_i)(\gamma) - P_k^{(i)}(\psi_i)(\gamma) \in H^+\sigma(\underline{m})\gamma.$$

These vectors, however, also belong to W and, by construction, the space $W \cap H^+\sigma(\underline{m})\gamma = \{0\}$. Hence one gets

$$\partial_k^{(i)}(\psi_i)(\gamma) = P_k^{(i)}(\psi_i)(\gamma).$$

The differential operator $P_k^{(i)}$ is unique, because one easily shows, by comparing coefficients of the power series involved, that each differential operator $P = \sum \alpha_s \partial_i^s$ such that $P(\psi_i) = 0$ has to be zero. This proves the proposition. □

Next one considers the action of $\partial_k^{(j)}$, with $j \neq i$, and thus obtain the “mixed” equations.

Proposition 6.2. *For all $k \geq 1$ and all $j \neq i$, there exists a unique differential operator $R_k^{(j,i)}$ of the form*

$$R_k^{(j,i)} = \sum_{s=0}^{k-1} r_s^{(j,i)} \partial_j^s$$

satisfying

$$\partial_k^{(j)}(\psi_i) = \frac{1}{(a_i - a_j)^k}\psi_i + R_k^{(j,i)}(\psi_j).$$

The coefficients of the differential operator $R_k^{j,i}$ are polynomial expressions in the functions $\{\alpha_{r1}^{(i,j)} \mid 0 \leq r \leq k-1\}$, the $\{\alpha_{r1}^{(j,j)} \mid 0 \leq r \leq k-1\}$ and their ∂_j-derivatives.

Proof. Consider again the singularities of $\{\partial_k^j(\psi_i)\}\sigma(\underline{m})^{-1}e^{-\xi}$ around the points $\{a_1, \ldots, a_n\}$. Since

$$\partial_k^{(j)}(\psi_i) = \{\partial_k^{(j)}(\hat{\psi}_i) + \hat{\psi}_i(z - a_j)^{-k}\}\sigma(\underline{m})e^{\xi},$$

there is a k-th order singularity around a_j and a first order around a_i. The first can be eliminated as in the proof of proposition 6.1 with a suitable $(k-1)$-th order differential operator $R_k^{j,i} = \sum_{0 \leq s \leq k-1} r_s^{(j,i)} \partial_j^s$. The second with a proper multiple of ψ_i. Then one has again for all $\gamma \in \Gamma(\sigma(\underline{m}), W)$

$$\partial_k^{(j)}(\psi_i)(\gamma) - \frac{1}{(a_i - a_j)^k}\psi_i(\gamma) - R_k^{(j,i)}(\psi_j)(\gamma) \in W \cap H^+\sigma(\underline{m})e^{\xi} = \{0\}.$$

The first term in the right hand side of the expression for $\partial_k^{(j)}(\psi_i)$ is fixed and since the action of the differential operators in ∂_j on ψ_j has no torsion, this equation also implies the uniqueness of $R_k^{(j,i)}$. $\square$

The lowest order operator $R_1^{(j,i)}$ has a simple form. The equation from proposition 6.2 becomes then

$$\begin{aligned}\partial_j(\psi_i) &= \tfrac{1}{(a_i - a_j)}\psi_i + \left\{\alpha_{01}^{(i,j)} - \tfrac{1}{a_i - a_j}\right\}\psi_j \\ &:= \tfrac{1}{(a_i - a_j)}\psi_i + c_W^{(i,j)}\psi_j\end{aligned}$$

Note that for all $i \neq j, c_{H^*}^{(i,j)} = \frac{-1}{a_i - a_j}$ is non-zero. This holds more generally

Lemma 6.1. *For each W in $\mathrm{Gr}_0(H)$ and all $i \neq j$, the function $c_W^{(i,j)}$ is not identical zero.*

From the construction of the functions $\{\psi_k\}$, one deduces directly that it suffices to prove this for all W in the big cell corresponding to H^*. Then one uses that $c_{H^*}^{(i,j)} \neq 0$, the actual form of the flows in $\tilde{\Gamma}(M)$ and the analyticity of $c_W^{(i,j)}(\gamma)$ in γ to prove the lemma.

To derive a relation between the candidate wavefunctions one restricts them now to the dense open part $\tilde{\Gamma}(\sigma(\underline{m}), W)$ of $\Gamma(\sigma(\underline{m}), W)$, given by

$$\tilde{\Gamma}(\sigma(\underline{m}), W) = \{\gamma \mid \gamma \in \Gamma(\sigma(\underline{m}), W), c_W^{(i,j)}(\gamma) \neq 0 \text{ for all } i \neq j\}.$$

Clearly this implies that the ring $\mathcal{R}$ has to be adapted to that of the analytic functions on $\tilde{\Gamma}(\sigma(\underline{m}), W)$ and one introduces the differential operators T_{ji} by

$$T_{ji} = \frac{1}{c_W^{(i,j)}}(\partial_j - \frac{1}{a_i - a_j}).$$

The interrelations between the wavefunctions $\{\psi_1, \ldots, \psi_n\}$ are resumed in the following proposition

Proposition 6.3. *For $i \neq j$ and $\gamma \in \tilde{\Gamma}(\sigma(\underline{m}), W)$ one has*

$$T_{ji}(\psi_i)(\gamma) = \psi_j(\gamma).$$

With the help of the operators $\{T_{ji}\}$ one can put the equations from the propositions 6.1 and 6.2 in the same form: for each $k \geq 1$ and all i and j, there is differential operator $B_k^{(j,i)}$ of order k in ∂_j such that

$$\partial_k^{(j)}(\psi_i) = B_k^{(j,i)}(\psi_i). \tag{34}$$

This shows that each ψ_i is a wavefunction of type $(i, \sigma(\underline{m}))$ and it suffices to have one of them, since the others can be derived by differentiating this one.

Until now one has restricted oneself to the equations coming from the continuous flows from $\tilde{\Gamma}(M)$. One concludes with taking the action of $\Sigma^{(0)}$ into account. Assume that σ_1 and $\sigma_1\sigma_2$ belong to $\Sigma_W^{(0)}$. One has then wavefunctions $\psi_{W,k}^{(\sigma_1)}$ and $\psi_{W,k}^{(\sigma_1\sigma_2)}$ and as in [8] these solutions corresponding to different sites in the lattice $\Sigma^{(0)}$ are connected by a differential operator

Proposition 6.4. *There is a differential operator $R_{\sigma_1}^{\sigma_1\sigma_2} \in \mathcal{R}[\partial_j, 1 \leq j \leq n]$ with $\mathcal{R}$ the ring of analytic functions on $\tilde{\Gamma}(\sigma_1, W) \cap \tilde{\Gamma}(\sigma_1\sigma_2, W)$ such that*

$$\psi_{W,k}^{(\sigma_1\sigma_2)} = R_{\sigma_1}^{\sigma_1\sigma_2}(\psi_{W,k}^{(\sigma_1)})$$

Proof. Since σ_2 belongs to $\Sigma^{(0)}$, it is the product of the elements

$$\sigma_{ij} := \frac{(z-a_i)}{(z-a_j)} = 1 - \frac{a_i - a_j}{z - a_j}.$$

Hence σ_2 is a polynomial in the $\{\frac{1}{z-a_j} \mid 1 \leq j \leq n\}$ and let its highest order term be given by

$$c\prod_{j=1}^{n}(\frac{1}{z-a_j})^{s_j}, \text{ with } s_j \geq 0 \text{ and } c \in \mathbb{C}, c \neq 0.$$

Consider then the function

$$(\psi_{W,k}^{(\sigma_1\sigma_2)} - c\prod_{j=1}^{n}\partial_j^{s_j}(\psi_{W,k}^{(\sigma_1)}))\sigma_1^{-1}e^{-\xi}.$$

It has on each component singularities in the points $\{a_1, \ldots, a_n\}$ of order less then $\sum_j s_j$. Hence by subtracting from $\psi_{W,k}^{(\sigma_1\sigma_2)}$ a finite number of terms of the form $\prod_{j=1}^{n}\partial_j^{n_j}(\psi_{W,k}^{(\sigma_1)})$ one can remove by a step by step procedure the singularities in all components and finally one ends up with a differential operator $R_{\sigma_1}^{\sigma_1\sigma_2} \in \mathcal{R}[\partial_j, 1 \leq j \leq n]$ such that

$$\psi_{W,k}^{(\sigma_1\sigma_2)} - R_{\sigma_1}^{\sigma_1\sigma_2}(\psi_{W,k}^{(\sigma_1)}) \in W \cap H^+\sigma_1 e^{\xi} = \{0\}.$$

This proves the claim in the proposition. □

Remark 6.1. Note in particular that proposition 6.3 implies that each wavefunction ψ_i is annihilated by each of the second order differential operators $T_{ij}T_{ji} - 1$, with $j \neq i$. Hence one can not, contrary to the case of the KP-hierarchy, scratch the wavefunction in the equations (11).

Remark 6.2. We have associated to each plane W in $\mathrm{Gr}_0(H)$ a set of wavefunctions $\{\psi_{W,k}^{(\sigma(\underline{m}))} \mid 1 \leq i \leq n\}$ of which each member determines the others. There holds even something stronger: the whole space W can be recovered from any of these wavefunctions by taking arbitrary derivatives w.r.t. the variables $\{t_1^{(i)}\}$.

Remark 6.3. Since the computations in [7] for the relation between the wavefunction and the τ-function in the L^2-case are based on the formula (24), they carry over to the present more general setting with τ_w defined as in section 5.

References

1. E. Date, M. Jimbo, M.Kashiwara and T. Miwa, "Transformation groups for soliton equations", in: *Nonlinear integrable systems – Classical theory and quantum theory*, eds. M. Jimbo and T. Miwa (World Scientific, 1983) pp. 39-120.
2. L.A. Dickey, "Why the general Zakharov-Shabat equations form a hierarchy?", *Commun. Math. Phys.* **163**, 509-521 (1994).
3. J.W. de Roever, *Complex Fourier Transformation and analytic functionals with unbounded carriers*, Mathematical Centre Tracts 89 (1978).
4. J. Dorfmeister, E. Neher and J. Szmigielski, "Automorphisms of Banach manifolds associated with the KP-equation", *Quart. J. Math.* **40** (158) 161-195 (1989).
5. A. Grothendieck, "Produits tensoriels topologiques et espaces nucléaires", *Mem. Amer. Math. Soc.* **16** (1955).
6. A. Grothendieck, "La Théorie de Fredholm", *Bull. Soc. Math. de France*, **84**,319-384 (1956).
7. G.F. Helminck, "τ-functions for a two-point version of the KP-hierarchy", to appear in *Indag. Math.*
8. G.F. Helminck and G.F. Post, "The geometry of differential difference equations", *Indag. Math., N.S.* **5** (4) 411-438 (1994).
9. V.G. Kac., *Infinite dimensional Lie algebras*, third edition, *Progress in Mathematics*, Vol. 44 (Birkhäuser, 1988).
10. Mikio Sato and Yasuko Sato, "Soliton equations as dynamical systems on infinite-dimensional Grassmann manifold", Nonlinear partial differential equations in applied science (Tokyo, 1982), North-Holland Math. Stud., vol. 81 (North-Holland, Amsterdam, 1983) pp. 259-271.
11. G. Segal and G. Wilson, "Loop groups and equations of KdV type", *Publ. Math. IHES* **63**, 1-64 (1985).

INTEGRABLE SYSTEMS WITH IMPURITY

E. RAGOUCY
Laboratoire d'Annecy-le-Vieux de Physique Théorique LAPTH,
UMR 5108 du CNRS associée à l'Université de Savoie,
9 Chemin de Bellevue, BP 110, F-74941 Annecy-le-Vieux Cedex, France,
ragoucy@lapp.in2p3.fr

We review some basic properties of RT algebras, which appear to be the natural framework to deal with integrable systems in presence of an impurity.
Then, we show how any integrable system (including these possessing translation invariance) can be promoted to an integrable system with an impurity which can reflect and transmit particles. The technics allows bulk translation invariant S-matrices while avoiding the no-go theorem stated recently about these laters.

1. Introduction

In the realm of factorized scattering theories in integrable models on the line, the Zamolodchikov-Faddeev (ZF) algebra [1, 2] plays a fundamental role. It encodes (quadratically) the two-body scattering, which is the only particle interaction present in this case. There have been various generalizations, preserving integrability, to the case with a purely reflecting boundary [3, 4, 5, 6] or to the case of point-like impurities [7, 8]. In the later context, the relevant algebraic structure emerging recently [9, 10, 11], is the so-called reflection-transmission (RT) algebra. It captures both the two-body scattering in the bulk, and the particle interaction with the impurity.

After reviewing some basic properties of RT algebras, we present a procedure which allows, starting from any integrable model on the line, to built an integrable model with a point-like impurity. In particular, despite the property stated in [8], we show that the present technics produces theories possessing a (non trivial) translation invariant scattering matrix in the bulk together with an impurity which reflects and transmits particles.

2. An example of reflection–transmission algebra

To sustain the introduction of RT algebra in the study of quantum impurity problems, we start with the simple example of (free) non-relativistic bosonic particles on $\mathbb{R}$, which interact with a δ-type impurity localized in the origin. This model is well-known to be exactly solvable (see e.g. [12]). It has been presented from the RT algebra point of view in [10], so that we just sketch the results, referring to the original paper for more details. In particular, we will consider here only the case $\eta \geq 0$.

It is sufficient to investigate the spectral problem associated with the one-particle Hamiltonian

$$H^{(1)} = \sum_{i=1}^{n} -\frac{1}{2}\partial_{x_i}^2 + \eta\,\delta(x_i)\,, \qquad \eta \in \mathbb{R}\,, \tag{2.1}$$

defined on a suitable domain $\mathcal{D}_\eta$ of continuous functions on $\mathbb{R}$, which are twice differentiable in $\mathbb{R} \setminus \{0\}$ and satisfy

$$\lim_{x\to 0^+} [(\partial_x \psi)(x) - (\partial_x \psi)(-x)] = 2\eta\,\psi(0)\,. \tag{2.2}$$

$H^{(1)}$ is self-adjoint on $\mathcal{D}_\eta$. A set $\Phi = \{\psi_k^\pm(x)\,;\, k \in \mathbb{R}\}$ of orthogonal (generalized) eigenstates, verifying (2.2), is provided by:

$$\psi_k^\pm(x) = \theta(\mp k)\left\{\theta(\mp x)T(\mp k)e^{ikx} + \theta(\pm x)\left[e^{ikx} + R(\mp k)e^{-ikx}\right]\right\}\,, \tag{2.3}$$

where θ denotes the standard Heaviside function and

$$T(k) = \frac{k}{k+i\eta}\,, \qquad R(k) = \frac{-i\eta}{k+i\eta}\,. \tag{2.4}$$

The family $\overline{\Phi} = \{\overline{\psi}_{-k}^{\pm}(x)\,;\, k \in \mathbb{R}\}$, where the bar stands for complex conjugation, is also orthonormal. The systems Φ and $\overline{\Phi}$ represent physically scattering states and are separately complete in $L^2(\mathbb{R}, dx)$.

The reflection and transmission from the impurity give rise to a non-trivial scattering operator, which preserves the particle number. One can interpret $\psi_k^\pm(x)$ asymptotically as incoming waves, traveling in $\mathbb{R}_\pm$ with momentum $k \neq 0$ towards the impurity. Accordingly, the vectors

$$|k\rangle^{\text{in}} = \psi_k^+(x) + \psi_k^-(x) \text{ and } |k\rangle^{\text{out}} = \overline{\psi}_{-k}^{+}(x) + \overline{\psi}_{-k}^{-}(x) \tag{2.5}$$

form a basis of one-particle "in" states and "out" states resp. The one-particle scattering operator is defined by $\mathbf{S}^{(1)}\,|k\rangle^{\text{out}} = |k\rangle^{\text{in}}$. By construction, $\mathbf{S}^{(1)}$ is a unitary operator on $L^2(\mathbb{R}, dx)$. The one-particle transition amplitude reads

$$\begin{aligned} {}^{\text{out}}\langle p|k\rangle^{\text{in}} = {}^{\text{out}}\langle p|\mathbf{S}^{(1)}|k\rangle^{\text{out}} &= [\theta(p)T(p) + \theta(-p)T(-p)]2\pi\delta(p-k) \\ &+ [\theta(p)R(p) + \theta(-p)R(-p)]2\pi\delta(p+k). \end{aligned} \tag{2.6}$$

Thus, R and T indeed represent the transmission and reflection amplitudes.

The n-particle amplitude, with initial and final configurations satisfying $k_1 < ... < k_n$ and $p_1 > ... > p_n$ respect., can be expressed in terms of (2.6):

$$ {}^{\text{out}}\langle p_1, ..., p_n | k_1, ..., k_n \rangle^{\text{in}} = \prod_{i=1}^{n} {}^{\text{out}}\langle p_i | k_i \rangle^{\text{in}} . \tag{2.7} $$

The basic feature of RT algebra is to provide an algebraic framework for dealing with the above system when $\eta > 0$, similar to the familiar canonical commutation approach, which works in the case $\eta = 0$. We introduce the associative algebra $\mathcal{C}$ with identity element $\mathbf{1}$, generated by $\{a^{\dagger\xi}(k),\, a_\xi(k) : \xi = \pm,\, k \in \mathbb{R}\}$ obeying the commutation relations:

$$ a_{\xi_1}(k_1)\, a_{\xi_2}(k_2) - a_{\xi_2}(k_2)\, a_{\xi_1}(k_1) = 0 \,, \tag{2.8} $$

$$ a^{\dagger\xi_1}(k_1)\, a^{\dagger\xi_2}(k_2) - a^{\dagger\xi_2}(k_2)\, a^{\dagger\xi_1}(k_1) = 0 \,, \tag{2.9} $$

$$ \begin{aligned} a_{\xi_1}(k_1)\, a^{\dagger\xi_2}(k_2) - a^{\dagger\xi_2}(k_2)\, a_{\xi_1}(k_1) = {} & \left[\delta_{\xi_1}^{\xi_2} + \mathcal{T}_{\xi_1}^{\xi_2}(k_1)\right] 2\pi\delta(k_1 - k_2)\, \mathbf{1} \\ & + \mathcal{R}_{\xi_1}^{\xi_2}(k_1) 2\pi\delta(k_1 + k_2)\, \mathbf{1} \,, \end{aligned} \tag{2.10} $$

where

$$ \mathcal{T}(k) = \begin{pmatrix} 0 & T(k) \\ \overline{T}(k) & 0 \end{pmatrix} , \qquad \mathcal{R}(k) = \begin{pmatrix} R(k) & 0 \\ 0 & \overline{R}(k) \end{pmatrix} . \tag{2.11} $$

The right-hand side of eq. (2.10) captures the presence of the impurity. The term proportional to $\delta(k_1 + k_2)$ reflects in particular the breaking of translation invariance due to the impurity. We will see that $\mathcal{C}$ is a particular RT algebra. For the moment we focus on the Fock representation $\mathcal{F}$ of $\mathcal{C}$, referring for the explicit construction to sect. 3.1. An essential feature of $\mathcal{F}$ is that the operators $a^\dagger(k) = (a^{\dagger +}(k), a^{\dagger -}(k))$ and $a(k) = \begin{pmatrix} a_+(k) \\ a_-(k) \end{pmatrix}$ in this representation satisfy

$$ a(k) = \mathcal{T}(k) a(k) + \mathcal{R}(k) a(-k) \,, \tag{2.12} $$

$$ a^\dagger(k) = a^\dagger(k) \mathcal{T}(k) + a^\dagger(-k) \mathcal{R}(-k) \,. \tag{2.13} $$

The relations (2.12), (2.13) encode the interaction with the impurity.

The vacuum state $\Omega \in \mathcal{F}$ obeys as usual $a_\xi(k)\Omega = 0$. We denote by $(\cdot\,,\,\cdot)$ the scalar product in $\mathcal{F}$ and consider the vacuum expectation value

$$ (a^{\dagger\epsilon_1}(p_1)...a^{\dagger\epsilon_n}(p_n)\Omega\, ,\, a^{\dagger\xi_1}(k_1)...a^{\dagger\xi_n}(k_n)\Omega) \,, \tag{2.14} $$

$$ k_1 < \cdots < k_n \,, \quad \xi_i = -\epsilon(k_i) \,, \qquad p_1 > \cdots > p_n \,, \quad \epsilon_i = \epsilon(p_i) \tag{2.15} $$

ϵ being the sign function. By means of eqs. (2.8)–(2.10) it is easily verified that (2.14) reproduces the amplitudes (2.7) for any n. Therefore, $\mathcal{C}$ provides a purely algebraic framework for constructing the scattering operator. In this framework, the Hamiltonian takes the form

$$H = \frac{1}{2}\int_{-\infty}^{+\infty} \frac{dk}{2\pi} k^2 a^{\dagger\xi}(k) a_\xi(k)\,. \tag{2.16}$$

The restriction of H to the n-particle subspace is the algebraic counterpart of the starting Hamiltonian.

Thus, $\mathcal{C}$ is a universal and powerful tool for handling δ-type impurities. One can view $\mathcal{C}$ as a central extension of the algebra of canonical commutation relations. A direct generalization is to substitute $\mathcal{T}(k)\mathbf{1}$ and $\mathcal{R}(k)\mathbf{1}$ in eq. (2.10) with new generators $t(k)$ and $r(k)$ which are no longer central. Moreover, in the spirit of the ZF algebra, it is possible to replace the bosonic exchange factor between $\{a^\dagger(k),\, a(k)\}$ with a more general one.

3. The RT algebra

Adopting throughout the paper the compact auxiliary spaces notation (see e.g. [13]), we define RT algebras [10] by:

Definition 3.1. (RT algebra $\mathcal{C}_S$)
A RT algebra is generated by $\mathbf{1}$ and the generators $A(k)$, $A^\dagger(k)$, (which are column and line vector valued resp.) $t(k)$ and $r(k)$ (which are matrix valued) obeying:

$$A_1(k_1)A_2(k_2) = S_{21}(k_2,k_1)A_2(k_2)A_1(k_1) \tag{3.1}$$

$$A_1^\dagger(k_1)A_2^\dagger(k_2) = A_2^\dagger(k_2)A_1^\dagger(k_1)S_{21}(k_2,k_1) \tag{3.2}$$

$$\begin{aligned} A_1(k_1)A_2^\dagger(k_2) = {} & A_2^\dagger(k_2)S_{12}(k_1,k_2)A_1(k_1) + \delta(k_1-k_2)\Big(\delta_{12}\mathbf{1} + t_{12}(k_1)\Big) \\ & + r_{12}(k_1)\delta(k_1+k_2) \end{aligned} \tag{3.3}$$

$$A_1(k_1)t_2(k_2) = S_{21}(k_2,k_1)t_2(k_2)S_{12}(k_1,k_2)A_1(k_1) \tag{3.4}$$

$$A_1(k_1)r_2(k_2) = S_{21}(k_2,k_1)r_2(k_2)S_{12}(k_1,-k_2)A_1(k_1) \tag{3.5}$$

$$t_1(k_1)A_2^\dagger(k_2) = A_2^\dagger(k_2)S_{21}(k_2,k_1)t_1(k_1)S_{21}(k_2,k_1) \tag{3.6}$$

$$r_1(k_1)A_2^\dagger(k_2) = A_2^\dagger(k_2)S_{21}(k_2,k_1)r_1(k_1)S_{21}(k_2,-k_1) \tag{3.7}$$

$$\begin{aligned} & S_{12}(k_1,k_2)t_1(k_1)S_{21}(k_2,k_1)t_2(k_2) = \\ & \qquad t_2(k_2)S_{12}(k_1,k_2)t_1(k_1)S_{21}(k_2,k_1) \end{aligned} \tag{3.8}$$

$$\begin{aligned} & S_{12}(k_1,k_2)t_1(k_1)S_{21}(k_2,k_1)r_2(k_2) = \\ & \qquad r_2(k_2)S_{12}(k_1,-k_2)t_1(k_1)S_{21}(-k_2,k_1) \end{aligned} \tag{3.9}$$

$$\begin{aligned} & S_{12}(k_1,k_2)r_1(k_1)S_{21}(k_2,-k_1)r_2(k_2) = \\ & \qquad r_2(k_2)S_{12}(k_1,-k_2)r_1(k_1)S_{21}(-k_2,-k_1) \end{aligned} \tag{3.10}$$

where $r(k)$ and $t(k)$ satisfy:

$$t(k)t(k) + r(k)r(-k) = \mathbb{I} \quad \text{and} \quad t(k)r(k) + r(k)t(-k) = 0 \tag{3.11}$$

We refer to $r(k)$ and $t(k)$ as reflection and transmission generators. Here and below the S-matrix obeys the following well-known Yang-Baxter equation and unitarity relation:

$$\begin{aligned} &S_{12}(k_1,k_2)\, S_{13}(k_1,k_3)\, S_{23}(k_2,k_3) = S_{23}(k_2,k_3)\, S_{13}(k_1,k_3)\, S_{12}(k_1,k_2) \\ &S_{12}(k_1,k_2)\, S_{21}(k_2,k_1) = \mathbb{I} \otimes \mathbb{I} \end{aligned} \tag{3.12}$$

A special case of RT algebra is the boundary algebra $\mathcal{B}_S$ [14] defined as the coset of $\mathcal{C}_S$ by the relation $t(k) = 0$. In the same way, $r(k) = 0$ leads to an algebra with a purely transmitting impurity [15].

The generators $r(k)$ and $t(k)$ form a subalgebra $\mathcal{K}_S \subset \mathcal{C}_S$. This subalgebra contains itself two subalgebras: $\mathcal{R}_S \subset \mathcal{K}_S$, generated by $r(k)$, appears already in [4], called reflection algebra and $\mathcal{T}_S \subset \mathcal{K}_S$, generated by $t(k)$, called transmission algebra.

RT algebras possesses two important automorphisms: the (antilinear) adjunction I given by

$$I : \begin{cases} a^\dagger(\chi) \mapsto a(\chi) \quad \text{and} \quad a(\chi) \mapsto a^\dagger(\chi) \\ r(\chi) \mapsto r(-\chi) \quad \text{and} \quad t(\chi) \mapsto t(\chi) \end{cases} \tag{3.13}$$

and the reflection-transmission automorphism:

$$\varrho : \begin{cases} a^\dagger(\chi) \mapsto a^\dagger(\chi)\, t(\chi) + a^\dagger(-\chi)\, r(\chi) \\ a(\chi) \mapsto t(\chi)\, a(\chi) + r(\chi)\, a(-\chi) \\ r(\chi) \mapsto r(\chi) \quad \text{and} \quad t(\chi) \mapsto t(\chi) \end{cases} \tag{3.14}$$

They both square to identity. We use them for the construction of Fock spaces for RT algebras.

3.1. *Fock spaces*

A Fock representation of the RT algebra with adjunction $\{\mathcal{C}_S, I\}$ is specified by the conditions:

(i) the representation space is a complex Hilbert space $\mathcal{H}$ with scalar product $(\cdot\,,\,\cdot)$;

(ii) the generators $\{a(\chi),\ a^\dagger(\chi),\ r(\chi),\ t(\chi)\}$ are operator-valued distributions with common and invariant dense domain $\mathcal{D} \subset \mathcal{H}$, where eqs. (3.1)–(3.11) hold;

(iii) the adjunction I is realized as a conjugation with respect to $(\cdot\,,\,\cdot)$;

(iv) there exists a vacuum state $\Omega \in \mathcal{D}$, which is annihilated by $a(\chi)$. The vector Ω is cyclic with respect to $\{a^\dagger(\chi)\}$ and $(\Omega\,,\,\Omega) = 1$.

There is a number of simple, but quite important consequences from the assumptions (i)–(iv). They are proved in [10].

Prop 3.1. The reflection–transmission automorphism ϱ is realized in any Fock representation by the identity operator.

Any RT algebra $\mathcal{C}_S$ admits in general a whole family $\mathcal{F}(\mathcal{C}_S)$ of Fock representations, parametrized by the vacuum expectation values

$$\mathcal{R}(\chi) = (\Omega\,,\, r(\chi)\Omega)\,, \qquad \mathcal{T}(\chi) = (\Omega\,,\, t(\chi)\Omega)\,, \tag{3.15}$$

called transmission and reflection matrices. They obey:

Prop 3.2. In each Fock representation of $\{\mathcal{C}_S, I\}$, $\mathcal{T}(\chi)$ and $\mathcal{R}(\chi)$ satisfy the Hermitian analyticity conditions

$$\mathcal{R}^\dagger(\chi) = \mathcal{R}(-\chi)\,, \qquad \mathcal{T}^\dagger(\chi) = \mathcal{T}(\chi)\,, \tag{3.16}$$

the consistency relations

$$\mathcal{S}_{12}(\chi_1,\chi_2)\,\mathcal{R}_1(\chi_1)\,\mathcal{S}_{21}(\chi_2,-\chi_1)\,\mathcal{R}_2(\chi_2) = \\ \mathcal{R}_2(\chi_2)\,\mathcal{S}_{12}(\chi_1,-\chi_2)\,\mathcal{R}_1(\chi_1)\,\mathcal{S}_{21}(-\chi_2,-\chi_1)\,, \tag{3.17}$$

$$\mathcal{S}_{12}(\chi_1,\chi_2)\,\mathcal{T}_1(\chi_1)\,\mathcal{S}_{21}(\chi_2,\chi_1)\,\mathcal{T}_2(\chi_2) = \\ \mathcal{T}_2(\chi_2)\,\mathcal{S}_{12}(\chi_1,\chi_2)\,\mathcal{T}_1(\chi_1)\,\mathcal{S}_{21}(\chi_2,\chi_1)\,, \tag{3.18}$$

$$\mathcal{S}_{12}(\chi_1,\chi_2)\,\mathcal{R}_1(\chi_1)\,\mathcal{S}_{21}(\chi_2,-\chi_1)\,\mathcal{T}_2(\chi_2) = \\ \mathcal{T}_2(\chi_2)\,\mathcal{S}_{12}(\chi_1,\chi_2)\,\mathcal{R}_1(\chi_1)\,\mathcal{S}_{21}(\chi_2,-\chi_1) \tag{3.19}$$

and unitarity conditions

$$\mathcal{T}(\chi)\mathcal{T}(\chi) + \mathcal{R}(\chi)\mathcal{R}(-\chi) = \mathbb{I}\,, \qquad \mathcal{T}(\chi)\mathcal{R}(\chi) + \mathcal{R}(\chi)\mathcal{T}(-\chi) = 0\,. \tag{3.20}$$

Moreover, the vacuum state Ω is unique (up to a phase factor) and satisfies

$$r(\chi)\Omega = \mathcal{R}(\chi)\Omega\,, \qquad t(\chi)\Omega = \mathcal{T}(\chi)\Omega\,. \tag{3.21}$$

We thus recover at the level of Fock representation the well-known boundary Yang–Baxter equation (3.17), together with a transmission (3.18) and a transmission–reflection (3.19) Yang–Baxter equation.

4. Kinematics and RT algebras

The goal of this section is to present a diagrammatic framework for integrable models with impurity, which sustains the RT algebra presentation. We parameterize the asymptotic particles by spectral parameter $k \in \mathbb{R}$ and an "isotopic" index $i = 1, ..., N$, corresponding to the auxiliary space. We will use as fundamental building blocks the two-body bulk scattering matrix $S_{12}(k_1, k_2) = S_{i_1 i_2}^{j_1 j_2}(k_1, k_2)\, E_{i_1 j_1} \otimes E_{i_2 j_2}$; the reflection matrix $R(k)$, describing the reflection of a particle from the impurity and the transmission matrix $T(k)$, describing the transmission of a particle by the impurity.

It is worth stressing that S is allowed to depend on k_1 and k_2 separately [6, 9, 10]. Allowing S to depend on k_1 and k_2 separately, leads to a natural generalization of the inverse scattering method, which avoids the no-go theorem of [8] and describes a large set of integrable systems, not covered there. We will come back on this point in sect. 6.

It is convenient to define at this stage the matrices

$$R^{\pm}(k) \equiv \theta(\pm k) R(k)\,, \qquad T^{\pm}(k) \equiv \theta(\pm k) T(k)\,, \tag{4.1}$$

which have a simple physical interpretation: $R^{\pm}$ describe the reflection of a particle propagating $\mathbb{R}_{\pm}$, whereas $T^{\pm}$ correspond to the transmission of a particle from $\mathbb{R}_-$ to $\mathbb{R}_+$ and vice versa, see fig. 1. Time is flowing along the vertical direction and single lines represent particles. The double line corresponds to the impurity world line.

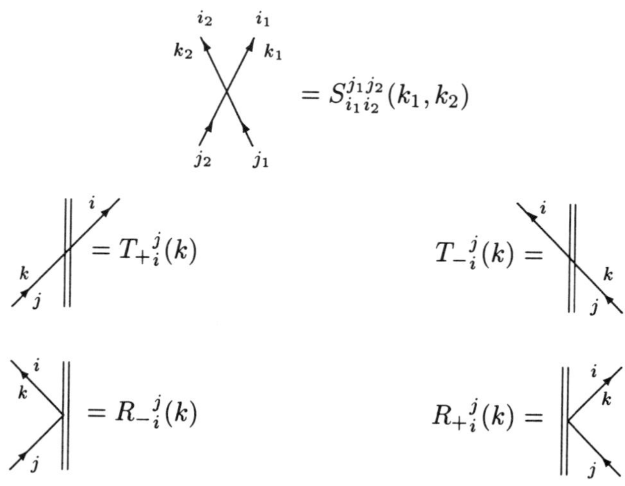

Figure 1. The two-body processes.

The construction of the possible three-body processes in terms of S, $R^{\pm}$ and $T^{\pm}$ leads to a series of relations [7, 8]. Considering the scattering of three particles, we get the well-known quantum Yang-Baxter equation (in its braid form). The consistency conditions implied by the scattering of two particles among each other and with the impurity are organized in the following three groups:

(a) pure reflection (fig. 2):

$$S_{12}(k_1,k_2)R_2^+(k_1)S_{12}(k_2,-k_1)R_2^+(k_2) = \\ R_2^+(k_2)S_{12}(k_1,-k_2)R_2^+(k_1)S_{12}(-k_2,-k_1)\,, \qquad (4.2)$$

$$S_{12}(k_1,k_2)R_1^-(k_2)S_{12}(-k_2,k_1)R_1^-(k_1) = \\ R_1^-(k_1)S_{12}(-k_1,k_2)R_1^-(k_2)S_{12}(-k_2,-k_1)\,. \qquad (4.3)$$

Eqs. (4.2) and (4.3) concern the reflection on $\mathbb{R}_+$ and $\mathbb{R}_-$ respectively. Using the rules of fig. 1, one gets from eq. (4.2) the fig. 2. To any figure, one associates a new one, deduced by reflection with respect to the impurity world line: eq. (4.3) gives the reflection of fig. 2.

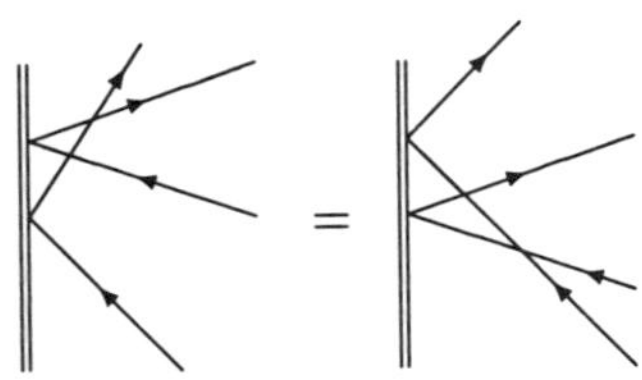

Figure 2. Pure reflection.

(b) pure transmission (fig. 3):

$$T_1^+(k_1)S_{12}(k_1,k_2)T_1^-(k_2) = T_2^-(k_2)S_{12}(k_1,k_2)T_2^+(k_1)\,, \qquad (4.4)$$

$$S_{12}(\chi_1,\chi_2)T_1^-(\chi_2)T_2^-(\chi_1) = T_1^-(k_1)T_2^-(k_2)S_{12}(k_1,k_2)\,, \qquad (4.5)$$

$$S_{12}(\chi_1,\chi_2)T_1^+(\chi_2)T_2^+(\chi_1) = T_1^+(k_1)T_2^+(k_2)S_{12}(k_1,k_2)\,, \qquad (4.6)$$

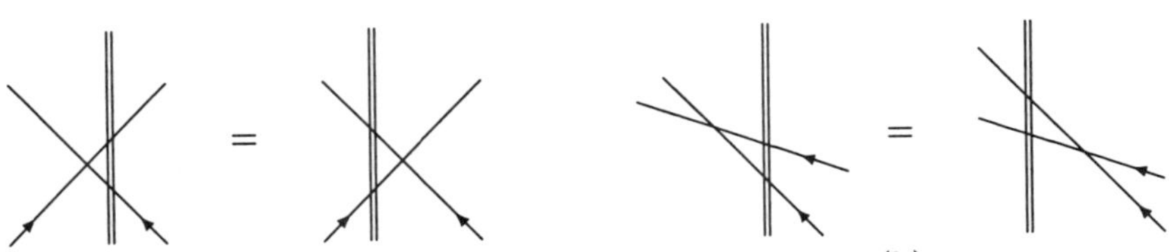

(a) (b)

Figure 3. Pure transmission.

(c) mixed relations (fig. 4):

$$R_1^+(k_1)T_2^-(k_2) = T_2^-(k_2)S_{12}(k_1,k_2)R_2^+(k_1)S_{12}(k_2,-k_1)\,, \qquad (4.7)$$
$$T_1^+(k_1)R_2^-(k_2) = T_1^+(k_1)S_{12}(k_1,k_2)R_1^-(k_2)S_{12}(-k_2,k_1)\,, \qquad (4.8)$$
$$R_1^+(k_1)T_2^+(k_2) = S_{12}(k_1,k_2)R_2^+(k_1)S_{12}(k_2,-k_1)T_2^+(k_2)\,, \qquad (4.9)$$
$$T_1^-(k_1)R_2^-(k_2) = S_{12}(k_1,k_2)R_1^-(k_2)S_{12}(-k_2,k_1)T_1^-(k_1)\,, \qquad (4.10)$$

$$R_1^+(k_1)T_2^-(k_2)S_{12}(-k_1,k_2) = T_2^-(k_2)S_{12}(k_1,k_2)R_2^+(k_1)\,, \qquad (4.11)$$
$$T_1^+(k_1)R_2^-(k_2)S_{12}(k_1,-k_2) = T_1^+(k_1)S_{12}(k_1,k_2)R_1^-(k_2)\,, \qquad (4.12)$$
$$R_2^+(k_1)S_{12}(k_2,-k_1)T_2^+(k_2) = S_{12}(k_2,k_1)R_1^+(k_1)T_2^+(k_2)\,, \qquad (4.13)$$
$$R_1^-(k_2)S_{12}(-k_2,k_1)T_1^-(k_1) = S_{12}(k_2,k_1)T_1^-(k_1)R_2^-(k_2)\,. \qquad (4.14)$$

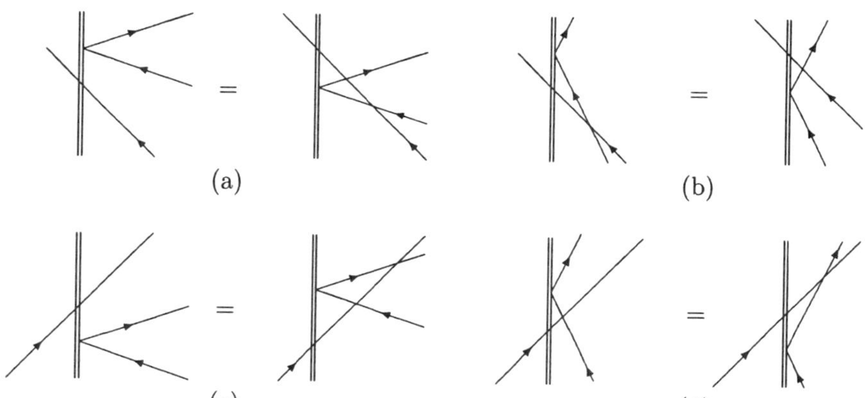

Figure 4. Mixed relations.

Finally, the requirements of unitarity and Hermitian analyticity lead for S to the familiar [3, 4, 5] conditions (3.12). Concerning R and T, these conditions read

$$T(k)T(k) + R(k)R(-k) = 1\,, \qquad T(k)R(k) + R(k)T(-k) = 0\,, \qquad (4.15)$$
$$[T]^\dagger(k) = T(k)\,, \qquad [R]^\dagger(k) = R(-k)\,, \qquad (4.16)$$

The two sets of eqs. (4.7)-(4.10) and (4.11)-(4.14) being equivalent, one is left with the study eqs. (4.2-4.10) and (4.15)-(4.16).

The impurity at $x = 0$ breaking translation invariance, we equip the generators of our algebra with a double index $\alpha = (\xi, i)$, which amounts to extend the auxiliary space from $End(\mathbb{C}^N)$ to $End(\mathbb{C}^{2N})$. The index $\xi = \pm$ indicates the half line $\mathbb{R}_\pm$ where the particle is created or annihilated. With this notation, we consider an associative algebra $\mathcal{C}$ with identity element $\mathbf{1}$, generated by $\{a(k),\, a^\dagger(k)\}$, which are subject to the constraints:

$$a_1(k_1)\, a_2(k_2) \;-\; \mathcal{S}_{21}(k_2,k_1)\, a_2(k_2)\, a_1(k_1) = 0\,, \tag{4.17}$$

$$a_1^\dagger(k_1)\, a_2^\dagger(k_2) - a_2^\dagger(k_2)\, a_1^\dagger(k_1)\, \mathcal{S}_{21}(k_2,k_1) = 0\,, \tag{4.18}$$

$$a_1(k_1)\, a_2^\dagger(k_2) \;-\; a_2^\dagger(k_2)\, \mathcal{S}_{12}(k_1,k_2)\, a_1(k_1) = \delta(k_1-k_2)\,[\delta_{12} + \mathcal{T}_{12}(k_1)]\; \mathbf{1} + \delta(k_1+k_2)\, \mathcal{R}_{12}(k_1)\, \mathbf{1}\,. \tag{4.19}$$

The constraints on the matrices $\mathcal{R}(k)$ and $\mathcal{T}(k)$ are given by (3.9)-(3.11).

Then, the set $\{\mathcal{S},\mathcal{R},\mathcal{T}\}$ is related to the starting data $\{S,R,T\}$ of an RT algebra in the following way:

$$\mathcal{S}^{(\eta_1,j_1)(\eta_2,j_2)}_{(\xi_1,i_1)\,(\xi_2,i_2)}(k_1,k_2) \equiv \delta^{\eta_2}_{\xi_1}\delta^{\eta_1}_{\xi_2}\; S^{j_2 j_1}_{i_1 i_2}(k_1,k_2)\,, \tag{4.20}$$

$$\mathcal{R}^{(\eta,j)}_{(\xi,i)}(k) \equiv \delta^{\eta}_{\xi}\; R^{j}_{i}(k)\,, \qquad \mathcal{T}^{(\eta,j)}_{(\xi,i)}(k) \equiv \delta^{\eta}_{-\xi}\; T^{j}_{i}(k)\,, \tag{4.21}$$

where the auxiliary space indices have been explicated.

5. From ZF to RT algebras

We present here a connection between the ZF algebras and the RT algebras, in the same way boundary algebras (defined for integrable systems on the half-line [6, 14]) can be constructed from ZF algebras [16]. We remind that a ZF algebras $\mathcal{A}_S$ is the polynomial algebra generated by a unit $\mathbf{1}$ and the generators $a(k)$ and $a^\dagger(k)$, subject to:

$$a_1(k_1)a_2^\dagger(k_2) = a_2^\dagger(k_2)S_{12}(k_1,k_2)a_1(k_1) + \delta_{12}\delta(k_1-k_2)\mathbf{1} \tag{5.1}$$

$$a_1(k_1)a_2(k_2) = S_{21}(k_2,k_1)a_2(k_2)a_1(k_1) \tag{5.2}$$

$$a_1^\dagger(k_1)a_2^\dagger(k_2) = a_2^\dagger(k_2)a_1^\dagger(k_1)S_{21}(k_2,k_1) \tag{5.3}$$

We employ below an extension $\overline{\mathcal{A}}_S$ of $\mathcal{A}_S$, involving power series in $a(k)$ and $a^\dagger(k)$ whose individual terms preserve the particle number. The concept of extended ZF (EZF) algebra $\overline{\mathcal{A}}_S$ is relevant for proving [13]:

Prop 5.1. (Well-bred operators)
Each $\overline{\mathcal{A}}_S$ contains an invertible element $L(k)$ satisfying

$$L_1(k_1)a_2(k_2) = S_{21}(k_2,k_1)a_2(k_2)L_1(k_1) \tag{5.4}$$

$$L_1(k_1)a_2^\dagger(k_2) = a_2^\dagger(k_2)S_{12}(k_1,k_2)L_1(k_1) \tag{5.5}$$

$$S_{12}(k_1,k_2)L_1(k_1)L_2(k_2) = L_2(k_2)L_1(k_1)S_{12}(k_1,k_2) \tag{5.6}$$

Hence, $L(k)$ generates a quantum group $\mathcal{U}_S \subset \overline{\mathcal{A}}_S$.

$L(k)$, called well-bred operator, is explicitly constructed in [13] and admits a series representation in terms of $a(k)$ and $a^\dagger(k)$.

We now turn to the construction of RT algebras from ZF ones. Let $R(k)$ and $T(k)$ be any two matrix functions satisfying

$$R^\dagger(k) = R(-k), \qquad T^\dagger(k) = T(k), \tag{5.7}$$

the unitary conditions (3.11) and the boundary Yang-Baxter equation (3.10). It has been shown in [10] that $R(k)$ and $T(k)$ obey automatically the transmission and reflection-transmission Yang-Baxter eqs. (3.8)-(3.9). Then, starting from the EZF algebra $\overline{\mathcal{A}}_S$, one proves [11]:

Prop 5.2. Let $t(k) = L(k)T(k)L^{-1}(k)$ and $r(k) = L(k)R(k)L^{-1}(-k)$, where $L(k)$ is the well-bred operator of $\overline{\mathcal{A}}_S$ and $R(k)$ and $T(k)$ are defined as above. Then, $a(k)$, $a^\dagger(k)$, $t(k)$ and $r(k)$ obey the relations (3.4)–(3.11).

Prop 5.3. The map

$$\sigma \begin{cases} a(k) \;\; \to \alpha(k) \;\; = t(k)a(k) + r(k)a(-k) \\ a^\dagger(k) \to \alpha^\dagger(k) = a^\dagger(k)t(k) + a^\dagger(-k)r(-k) \end{cases} \tag{5.8}$$

extends to a homomorphism on $\overline{\mathcal{A}}_S$.

The homomorphism σ is essential for constructing an RT algebra from EZF. Indeed, we have:

Theorem 5.1. *Let*

$$A(k) = \frac{1}{2}\Big(a(k) + \alpha(k)\Big) = \frac{1}{2}\Big([1 + t(k)]a(k) + r(k)a(-k)\Big) \tag{5.9}$$

$$A^\dagger(k) = \frac{1}{2}\Big(a^\dagger(k) + \alpha^\dagger(k)\Big) = \frac{1}{2}\Big(a^\dagger(k)[1 + t(k)] + a^\dagger(-k)r(-k)\Big) \tag{5.10}$$

Then $A(k)$, $A^\dagger(k)$, $t(k)$ and $r(k)$ form a RT algebra.

In this construction, $\mathcal{K}_S$ becomes a Hopf coideal in $\mathcal{U}_S$.

Note that if one imposes on $R(k)$ the additional relation $R(k)R(-k) = \mathbb{I}$, one recovers the construction [16] of the boundary algebra $\mathcal{B}_S$.

5.1. *Hamiltonians*

One can associate to any ZF algebra a hierarchy of Hamiltonians

$$H^{(n)}_{ZF} = \int_{\mathbb{R}} dk\, k^n\, a^\dagger(k)a(k), \; n \in \mathbb{Z}_+ \tag{5.11}$$

They form an Abelian set and admit as symmetry algebra the quantum group generated by the well-bred operators [13]:

$$[H_{ZF}^{(n)}, L(k)] = 0, \ \forall n \tag{5.12}$$

In a similar way, for any RT algebra, one introduces [11]:

$$H_{RT}^{(n)} = \int_{\mathbb{R}} dk\, k^n \, A^{\dagger}(k)A(k) \tag{5.13}$$

which obey to the following relations:

$$[H_{RT}^{(m)}, H_{RT}^{(n)}] = [(-1)^m - (-1)^n] \int_{\mathbb{R}} dk\, k^{m+n} \, A^{\dagger}(k)r(k)A(-k), \tag{5.14}$$

Thus, the Hamiltonians of the same parity define a hierarchy corresponding to integrable systems with impurity.

Prop 5.4. The subalgebra $\mathcal{K}_S$ is a symmetry algebra of the hierarchy $H_{RT}^{(n)}$: $[H_{RT}^{(n)}, t(k)] = 0$ and $[H_{RT}^{(n)}, r(k)] = 0$.

This result provides an universal model-independent description of the symmetry content of the hierarchy $H_{RT}^{(n)}$. Remark that $r(k)$ and $t(k)$ encode both the particle-impurity interactions and the quantum integrals of motion. This property of $\mathcal{K}_S$ has been applied in [18] for studying the symmetries of the nonlinear Schrödinger equation on the half line.

Let us observe in conclusion that there is a simple relation [11] between the hierarchies with and without impurity:

$$H_{RT}^{(n)} = H_{ZF}^{(n)} + \int_{\mathbb{R}} dk\, k^n \, a^{\dagger}(k)\Big(r(k)a(-k) + t(k)a(k)\Big) \tag{5.15}$$

It extends the results given in [17] in the case of purely reflecting boundary.

6. Including an impurity on the line

It is natural to expect the existence of an integrable system with impurity associated to (almost) any integrable system defined on the whole line. However, the class of integrable systems possessing translation invariant S-matrix seems to be excluded [8]. The aim of this section is to show how RT algebras can circumvent this no-go theorem. Starting from an integrable system defined on a line and possessing a translation invariant S-matrix, we present here a general construction that allows to obtain an integrable system, associated to the original one, but with an impurity located at the origin which reflects and transmits particles.

Let $s_{12}(k_1-k_2) \in End(\mathbb{C}^N) \otimes End(\mathbb{C}^N)$ be the translation invariant S-matrix of the initial system. From this S-matrix, we define a new S-matrix:

$$S(k_1,k_2) = \begin{pmatrix} s_{12}(k_1-k_2) & 0 & 0 & 0 \\ 0 & s_{12}(k_1+k_2) & 0 & 0 \\ 0 & 0 & s_{12}(-k_1-k_2) & 0 \\ 0 & 0 & 0 & s_{12}(k_2-k_1) \end{pmatrix} \tag{6.1}$$

which belongs to $End(\mathbb{C}^{2N}) \otimes End(\mathbb{C}^{2N})$. It is a simple matter of calculation to show that this S-matrix also obey the Yang-Baxter equation and unitary condition (3.12). To this S-matrix we associate a RT-algebra with generators $t(k)$, $r(k)$ and

$$A(k) = \begin{pmatrix} a_+(k) \\ a_-(k) \end{pmatrix} \quad , \quad A^\dagger(k) = \left(a_+^\dagger(k), a_-^\dagger(k)\right) \tag{6.2}$$

where $a_\pm(k)$ and $a_\pm^\dagger(k)$ are N-vectors (column and line respect.), so that $A_\pm(k)$ and $A_\pm^\dagger(k)$ are $2N$-vectors. The defining relations for the RT-algebra are those given in (3.1)-(3.10), with the S-matrix (6.1).

The physical interpretation of such a system is clear: the generators $a_\pm(k)$ and $a_\pm^\dagger(k)$ correspond to annihilation and creation operators in $\mathbb{R}_\pm$. They interact among themselves with the translation invariant $s_{12}(k_1-k_2)$ matrix, in accordance with the naive picture that "far from the impurity", the translation invariance should not be broken. However, the exchange of "+" with "-" operators (which must cross the impurity) is done with the matrix $s_{12}(k_1+k_2)$ which is <u>not</u> translation invariant. Consequently, the complete matrix $S(k_1,k_2)$ is not translation invariant either, so that the property stated in [8] does not apply here, as it was pointed out in [9]. Note also that despite the translation invariance, the generators $\{a_\xi(k),\, a_\xi^\dagger(k)\}$ ($\xi = +$ or $-$) does not obey ZF algebra relations, but rather boundary algebra ones. This is not surprising, this latter algebra being introduced for systems defined on the half-line [6].

In view of the construction, one can look for generators of the form

$$t(k) = \begin{pmatrix} 0 & \tau(k) \\ \tau(-k) & 0 \end{pmatrix} \quad \text{and} \quad r(k) = \begin{pmatrix} \rho(k) & 0 \\ 0 & \rho(-k) \end{pmatrix} \tag{6.3}$$

The defining relations (3.8)-(3.10), expressed for $r(k)$ and $t(k)$ with $S(k_1,k_2)$, then become:

$$s_{12}(k_1-k_2)\tau_1(k_1)\, s_{21}(k_2-k_1)\tau_2(k_2) = \tau_2(k_2)s_{12}(k_1-k_2)\tau_1(k_1)s_{21}(k_2-k_1)$$
$$s_{12}(k_1-k_2)\tau_1(k_1)s_{21}(k_2-k_1)\rho_2(k_2) = \rho_2(k_2)s_{12}(k_1+k_2)\tau_1(k_1)s_{21}(-k_2-k_1)$$
$$s_{12}(k_1-k_2)\rho_1(k_1)s_{21}(k_2+k_1)\rho_2(k_2) = \rho_2(k_2)s_{12}(k_1+k_2)\rho_1(k_1)s_{21}(k_1-k_2)$$

while the unitary relations (3.11) rewrite

$$\tau(k)\tau(-k) + \rho(k)\rho(-k) = \mathbb{I}\,, \qquad \tau(k)^\dagger = \tau(-k) \tag{6.4}$$

$$\tau(k)\rho(-k) + \rho(k)\tau(-k) = 0\,, \qquad \rho(k)^\dagger = \rho(-k) \tag{6.5}$$

Note that this presentation has symmetrized the role of $\tau(k)$ and $\rho(k)$.

Finally, let us remark that this construction can obviously be applied to any S-matrix, translation invariant or not.

References

1. A. Zamolodchikov and A. Zamolodchikov, *Ann. Phys. (N.Y.)* **120**, 253 (1979).
2. L.D. Faddeev, *Sov. Sci. Rev.* **C1**, 107 (1980).
3. I. Cherednik, *Theor. Math. Phys.* **61**, 977 (1984) [Teor. Mat. Fiz. **61** (84) 35].
4. E.K. Sklyanin, *J. Phys.* **A21**, 2375 (1988).
5. S. Ghoshal and A.B. Zamolodchikov, *Int. J. Mod. Phys.* **A9**, 3841 (1994) [Erratum, ibid. **A9**, 4353 (1994)], hep-th/9306002.
6. A. Liguori, M. Mintchev and M. Rossi, *J. Math. Phys.* **38**, 2888 (1997).
7. G. Delfino, G. Mussardo and P. Simonetti, *Nucl. Phys.* **B432**, 518 (1994), hep-th/9409076;
   R. Konik and A. LeClair, *Nucl. Phys.* **B538**, 587 (1999), hep-th/9703085.
8. O.A. Castro-Alvaredo, A. Fring and F. Gohmann, hep-th/0201142.
9. M. Mintchev, E. Ragoucy and P. Sorba, *Phys. Lett.* **B547**, 313 (2002), hep-th/0209052.
10. M. Mintchev, E. Ragoucy and P. Sorba, *J. Phys.* **A36**, 10407 (2003), hep-th/0303187.
11. M. Mintchev and E. Ragoucy, *J. Phys.* **A37**, 425 (2004), math.QA/0306084 .
12. S. Albeverio, F. Gesztesy, R. Hoegh-Krohn and H. Holden, *Solvable models in quantum mechanics* (Springer-Verlag, New York, 1988).
13. E. Ragoucy, *J. Phys.* **A35**, 7929 (2002), math.QA/0108207.
14. A. Liguori, M. Mintchev and L. Zhao, *Commun. Math. Phys.* **194**, 569 (1998), hep-th/9607085.
15. P. Bowcock, E. Corrigan and C. Zambon, hep-th/0305022.
16. E. Ragoucy, *Lett. Math. Phys.* **58**, 249 (2001), math.QA/0108221.
17. E. Ragoucy, *Int. J. Mod. Phys.* **A17**, 3649 (2002), math.QA/0202095.
18. M. Mintchev, E. Ragoucy and P. Sorba, *J. Phys.* **A34**, 8345 (2001), hep-th/0104079.

TWO EXACTLY-SOLVABLE PROBLEMS IN ONE-DIMENSIONAL HYPERBOLIC SPACE

Č. BURDÍK[1,2], G.S. POGOSYAN[1,3]

[1] *Laboratory of Theoretical Physics, Joint Institute for Nuclear Research, Dubna, 141980, Moscow Region, Russia*

[2] *Department of Mathematics, Czech Technical University, Trojanova 13, 120 00, Prague, Czech Republic*

[3] *Centro de Ciencias Físicas UNAM, Apartado Postal 48-3 62251 Cuernavaca, Morelos, México*

In this paper we establish a relation between two exactly-solvable problems on one-dimensional hyperbolic space, namely singular Coulomb and singular oscillator systems.

1. Introduction

In the last few years the relation between Coulomb and oscillator problems on the N-dimensional spaces with constant curvature (both positive on the sphere S_n and negative on hyperboloid H_n) and complex projective spaces CP^N have been discussed in many papers [1]-[5]. The first investigation for one spatial dimension was done in the two recent papers [6, 7]. In particular in the article [6] have constructed the complex mapping $S_{1C} \to S_1$, which extend in spherical geometry well-known for Euclidean space Hurwitz transformation. It was also shown that this transformation establish the correspondence between Coulomb and singular oscillator systems in one dimensions.

In the present note we find the relation between two exactly-solvable potentials on one-dimensional hyperbolic space H_1: $s_0^2 - s_1^2 = R^2$

$$V^{so}(\vec{s}) = \frac{\omega^2 R^2}{2}\frac{s_1^2}{s_0^2} + \frac{1}{2}\frac{k^2 - \frac{1}{4}}{s_1^2}, \qquad V^{sc}(\vec{s}) = -\frac{\mu}{R}\frac{s_0}{|s_1|} + \frac{1}{2}\frac{p^2 - \frac{1}{4}}{s_1^2}, \quad (1)$$

where s_0, s_1 are Cartesian coordinates in the ambient pseudo-euclidean space $E_{1,1}$ and k, and p are positive.

The potentials in (1) belong to the well-known class of superintegrable potentials in the constant curvature spaces [8, 9] restricted to one spatial dimension. The ordinary 1D oscillator system ($k = 1/2$) is singular at the point $s_0 = 0$, which separate the motion on the upper and lower sheets of hyperbola. The ordinary 1D Coulomb system ($p = 1/2$) is attractive on upper and repulsive on lower sheets of hyperbola. Below we will consider only the motion on the upper sheet of hyperbola ($s_0 > 0$).

2. Singular Oscillator

The Schrödinger equation describing the nonrelativistic quantum motion on one-dimensional hyperbolic space H_1 in pseudospherical coordinates $\tau \in (-\infty, \infty)$

$$s_0 = R\cosh\tau\,, \qquad s_1 = R\sinh\tau\,, \tag{2}$$

has the following form ($\hbar = \mu = 1$)

$$\frac{d^2\Psi}{d\tau^2} + 2R^2\left[E - V(\tau)\right]\Psi = 0\,. \tag{3}$$

Let us consider first the eq. (3) for potential $V(\tau) = V^{so}(\tau)$. Taking into account that in pseudospherical coordinates (2)

$$V^{so}(\tau) = \frac{\omega^2 R^2}{2}\frac{s_1^2}{s_0^2} + \frac{1}{2}\frac{k^2-\frac{1}{4}}{s_1^2} = \frac{1}{2R^2}\left(-\frac{k_0^2-\frac{1}{4}}{\cosh^2\tau} + \frac{k^2-\frac{1}{4}}{\sinh^2\tau}\right) + \frac{\omega^2R^2}{2} \tag{4}$$

where $k_0^2 = \omega^2 R^4 + \frac{1}{4}$ we obtain from eq.(3)

$$\frac{d^2\Psi}{d\tau^2} + \left[\epsilon + \frac{k_0^2-\frac{1}{4}}{\cosh^2\tau} - \frac{k^2-\frac{1}{4}}{\sinh^2\tau}\right]\Psi = 0\,, \tag{5}$$

with $\epsilon = 2R^2E - \omega^2R^4$. Thus the singular oscillator problem on the hyperbola H_1 is described by the well-known *modified Pöschl-Teller* equation and, unlike the Schrödinger equation for 1D singular oscillator on the flat space E_1 and the sphere S_1, the eq. (5) possesses both discrete ($\epsilon < 0$) and continuous ($\epsilon \geq 0$) spectrum.

The bound state wave functions is given by [8, 9]

$$\Psi_n^{(\pm k, k_0)}(\tau) = N_n^{(\pm k, k_0)}\left(\sinh\tau\right)^{\frac{1}{2}\pm k}\left(\cosh\tau\right)^{2n-k_0+\frac{1}{2}} \times$$
$$\times\ {}_2F_1(-n, k_0 - n;\ 1\pm k;\ \tanh^2\tau)\,, \tag{6}$$

where $N_n^{(\pm k, k_0)}$ is the normalization constant. The ϵ is quantized as

$$\epsilon = -(2n + 1 \pm k - k_0)^2\,, \tag{7}$$

with $n = 0, 1, ...[\frac{1}{2}(k_0 \mp k - 1)]$ and a bound state solution is possible only for $k_0 \mp k > 1$. Then the discrete energy spectrum takes the value

$$E_n^{\pm k}(R) = -\frac{1}{2R^2}\left[(2n+1\pm k)^2 - 2k_0(2n+1\pm k) + \frac{1}{4}\right]. \tag{8}$$

The wave functions $\Psi_n^{(\pm k, k_0)}(\tau)$ satisfies the orthogonality relation

$$R\int_0^\infty \Psi_n^{(\pm k,k_0)}(\tau)\Psi_{n'}^{(\pm k,k_0)}(\tau)\,d\tau = \frac{1}{2}\delta_{nn'} \tag{9}$$

and the normalization factor has the form

$$N_n^{(\pm k,k_0)}(R) = \frac{1}{\Gamma(1\pm k)}\sqrt{\frac{(-2n+k_0\mp k-1)\Gamma(-n+k_0)\Gamma(n+1\pm k)}{R\Gamma(-n+k_0\mp k)n!}} \tag{10}$$

The positive sign at the k should be taken whenever $k > \frac{1}{2}$, i.e., the additional term to oscillator potential is repulsive at the origin and the motion take place only in one of the domains $\tau \in [0, \infty)$ or $\tau \in [0, -\infty)$. If $0 < k \leq \frac{1}{2}$, i.e., the additional term is attractive at the origin, the motion take place in $\tau \in [-\infty, \infty]$ and both the positive and negative sign must be taken into account in the solution. Thus there exists two family of solutions go over into each other under the change $k \to -k$. This is indicated by the notion $\pm k$, in the formulas.

In the contraction limit $R \to \infty$, $\tau \to 0$, $R\tau \sim x$ - fixed and $k_0 \sim \omega R^2$ we see that the continuous spectrum is vanishing while the discrete spectrum is infinite. It is easy to reproduce the energy spectrum

$$\lim_{R\to\infty} E_n^{\pm k}(R) = E_n^{\pm k} = \omega(2n+1\pm k)\,, \tag{11}$$

and the wave function

$$\lim_{\substack{R\to\infty\\ \tau\to 0}} \Psi_n^{(\pm k,k_0)}(\tau) = \sqrt{\frac{\sqrt{\omega}\Gamma(n+1\pm k)}{n![\Gamma(1\pm k)]^2}} \times$$

$$\times\,(\sqrt{\omega}x)^{\frac{1}{2}\pm k}\, e^{-\frac{\omega x^2}{2}}\, {}_1F_1(-n;\, 1\pm k;\, \omega x^2)\,, \tag{12}$$

for the flat 1D singular oscillator system [10].

3. Singular Coulomb Problem

The 1D singular Coulomb potential in the pseudospherical coordinates (2) has the form

$$V^{sc}(\tau) = -\frac{\mu}{R}\left(\frac{s_0}{|s_1|} - 1\right) + \frac{1}{2}\frac{p^2 - \frac{1}{4}}{s_1^2} =$$

$$= -\frac{\mu}{R}(\coth|\tau| - 1) + \frac{1}{2R^2}\frac{p^2 - \frac{1}{4}}{\sinh^2\tau} . \tag{13}$$

Let us first consider the region $\tau \geq 0$. Substituting the potential (13) in eq.(3), we arrive to equation

$$\frac{d^2\Psi}{d\tau^2} + \left[(2R^2E - 2\mu R) + 2\mu R\coth\tau - \frac{p^2 - \frac{1}{4}}{\sinh^2\tau}\right]\Psi = 0\,, \tag{14}$$

which is known as the *Manning-Rosen* potential problem [11]. Making now the transformation from variable τ to the new variable $\alpha \in [0, \infty)$

$$e^{\tau} = \cosh\alpha\,, \tag{15}$$

and setting $\Psi(\alpha) = W(\alpha)/\sqrt{\coth\alpha}$. As result we obtain the equation

$$\frac{d^2W}{d\alpha^2} + \left[2R^2E + \frac{(-2R^2E + 4\mu R) - \frac{1}{4}}{\cosh^2\alpha} - \frac{4p^2 - \frac{1}{4}}{\sinh^2\alpha}\right]W = 0\,. \tag{16}$$

It is easy to see that, up to the substitution

$$\epsilon = 2R^2E, \qquad k_0^2 = -2R^2E + 4\mu R, \qquad k^2 = 4p^2\,. \tag{17}$$

the equation (16) for the 1D singular Coulomb problem coincides with the 1D singular oscillator equation (5). The regular for $\alpha \in [0, \infty)$ solution of this equation according to (6) is

$$W_n(\alpha) = A_n\,(\sinh\alpha)^{\frac{1}{2}\pm k}\,(\cosh\alpha)^{2n-k_0+\frac{1}{2}}\ \times$$

$$\times\,{}_2F_1\left(-n, -n + k_0; 1 \pm k; \tanh^2\alpha\right)\,, \tag{18}$$

where A_n is a normalization constant. The constant A_n is computed from the requirement that the wave function (18) satisfies the normalized condition

$$R\int_0^{\infty}|\Psi_n(\tau)|^2\,d\tau = R\int_0^{\infty}|W_n(\alpha)|^2\,\tanh^2\alpha\,d\alpha = \frac{1}{2}\,, \tag{19}$$

and has the following form

$$A_n = \sqrt{\frac{k_0(k_0 - 2n \mp k - 1)\Gamma(k_0 - n)\Gamma(n + 1 \pm k)}{R(2n + 1 \pm k)(n)![\Gamma(1 \pm k)]^2\Gamma(k_0 - n \mp k)}}\,. \tag{20}$$

Let us now consider the two most interesting cases.

3.1. *Relation between Coulomb problem and singular oscillator.*

In the case when $p = \frac{1}{2}$ the centrifugal term in V^{sc} disappears and the motion take place in the region $\tau \in (-\infty, \infty)$. Then the transformation (15) establishes the connection between ordinary 1D Coulomb problem and 1D singular oscillator with $k = 1$. Comparing now eq.(17) with (7), we get

$$k_0 = (n + 1 + \sigma), \qquad \sigma = \frac{\mu R}{n+1}\ . \tag{21}$$

and for the discrete energy spectrum

$$E_n(R) = -\frac{(n+1)^2}{2R^2} - \frac{\mu^2}{2(n+1)^2} + \frac{\mu}{R}, \qquad n = 0, 1, 2, ...[\sigma - 1]\ . \tag{22}$$

Returning to the variable τ, we find for bound state wave function ($\tau > 0$)

$$\Psi_{n\sigma}(\tau) = A_n(\sigma)\ \sinh\tau\, e^{(n-\sigma)\tau}\ {}_2F_1\left(-n, 1+\sigma;\ 2;\ 1 - e^{-2\tau}\right), \tag{23}$$

where the normalization constant $A_n(\sigma)$ is

$$A_n(\sigma) = \sqrt{\frac{2\sigma[\sigma^2 - (n+1)^2]}{R}}\ . \tag{24}$$

In the domain $-\infty < \tau < 0$ 1D Coulomb wave function are determined from eq. (23) after reflection $\tau \to -\tau$. Finally the general solution of the Schrödinger equation for $-\infty < \tau < \infty$ then can be presented in form of even and odd functions

$$\Psi^{(+)}_{n\sigma}(\tau) = A_n(\sigma)\ \sinh|\tau|\, e^{(n-\sigma)|\tau|}\ {}_2F_1\left(-n, 1+\sigma;\ 2;\ 1 - e^{-2|\tau|}\right), \tag{25}$$

$$\Psi^{(-)}_{n\sigma}(\tau) = A_n(\sigma)\ \sinh\tau\, e^{(n-\sigma)|\tau|}\ {}_2F_1\left(-n, 1+\sigma;\ 2;\ 1 - e^{-2|\tau|}\right)\ . \tag{26}$$

Thus, we have seen that using the transformation (15) we can completely solve 1D Coulomb system on hyperbola H_1, including eigenfunctions and discrete energy spectrum.

Now let us consider the flat space contraction. In the contraction limit $R \to \infty$ the energy spectrum for finite n goes to the energy spectrum of the 1D hydrogen atom [12]

$$E_n = \lim_{R\to\infty} E^{\nu}_n(R) = -\frac{\mu^2}{2(n+1)^2}, \qquad n = 0, 1, 2, ... \tag{27}$$

In the limit $R \to \infty$, putting $\tau \sim x/R$ in eqs. (25) and (26), we obtain (up to the factor $\sqrt{2}$) the well-known 1D Coulomb wave function [12]

$$\lim_{\substack{R\to\infty \\ \tau\to 0}} \Psi^{(+)}_{n\sigma}(\varphi) = \sqrt{\frac{2\mu^3}{(n+1)^3}}\, |x|\, e^{-\frac{\mu x}{n}}\, {}_1F_1\left(-n;\, 2;\, \frac{2\mu|x|}{n}\right),$$

$$\lim_{\substack{R\to\infty \\ \tau\to 0}} \Psi^{(-)}_{n\sigma}(\varphi) = \sqrt{\frac{2\mu^3}{(n+1)^3}}\, x\, e^{-\frac{\mu x}{n}}\, {}_1F_1\left(-n;\, 2;\, \frac{2\mu|x|}{n}\right).$$

3.2. *Relation between ordinary oscillator and singular Coulomb problem.*

Let us now choose $k = \frac{1}{2}^{-}$. Then $p^2 - \frac{1}{4} = -3/16$ and the centrifugal term is attractive at the origin for potential V^{sc}. As in the previous case the motion take place in the domain $-\infty < \tau < \infty$. For potential V^{so} this term vanish and therefore transformation (15) connect the ordinary oscillator with 1D singular Coulomb system.

Let us now introduce the quantity $\nu = \frac{1}{2}(1 \pm k)$, which takes two values $\nu = \frac{1}{4}$ and $\nu = \frac{3}{4}$. Making the all calculation by analogy to previous case, it easy to get the discrete energy spectrum

$$E^{\nu}_n(R) = -\frac{(n+\nu)^2}{2R^2} - \frac{\mu^2}{2(n+\nu)^2} + \frac{\mu}{R}, \qquad n = 0, 1, 2, ... \tag{28}$$

and normalized bound state wave function

$$\Psi^{\nu}_{n\sigma_\nu}(\tau) = 2^{\nu} \sqrt{\frac{[\sigma^2_\nu - (n+\nu)^2]\Gamma(\sigma_\nu + \nu)\Gamma(n+2\nu)}{2R(n+\nu)n!\Gamma(\sigma_\nu + 1 - \nu)[\Gamma(2\nu)]^2}} \times$$

$$\times (\sinh\tau)^{\nu}\, e^{\tau(n-\sigma_\nu)}\, {}_2F_1(-n,\, \nu+\sigma_\nu;\, 2\nu;\, 1 - e^{-2\tau}), \tag{29}$$

where $\sigma_\nu = \mu R/(n+\nu)$.

In the flat space limit $R \to \infty$, it is easy to see

$$E^{\nu}_n = \lim_{R\to\infty} E^{\nu}_n(R) = -\frac{\mu^2}{2(n+\nu)^2}, \qquad n = 0, 1, 2, ... \tag{30}$$

and

$$\Psi^{\nu}_n(y) = \lim_{\substack{R\to\infty \\ \tau\to 0}} \Psi^{\nu}_{n\sigma}(\varphi) = \frac{\sqrt{\mu}}{(n+\nu)} \sqrt{\frac{\Gamma(n+2\nu)}{2n![\Gamma(2\nu)]^2}}\, y^{\nu}\, e^{-\frac{y}{2}}\, {}_1F_1(-n; 2\nu; y) \tag{31}$$

where $y = 2\mu x/(n+\nu)$. Formulas (30) and (31) coincides with the formulas for energy levels E^{ν}_n and up to the factor $\sqrt{2}$ for wave functions $\Psi^{\nu}_n(y)$ for the two type of one-dimensional Coulomb anyons with $\nu = \frac{1}{4}$ and $\nu = \frac{3}{4}$ respectively [13].

References

1. E.G. Kalnins, W. Miller Jr. and G.S. Pogosyan, *J.Math.Phys.* **41**, 2629-2657 (2000).
2. A. Nersessian and G.S. Pogosyan, *Phys.Rev* **A63**, 20103 (R) (2000) [quant-ph/0006118].
3. A. Nersessian, *Physics of Atomic Nuclei*, **65**, 1103-1108 (2002).
4. E.G. Kalnins, W. Miller Jr. and G.S. Pogosyan, *Physics of Atomic Nuclei*, **65** (6) 1086-1094 (2002).
5. S. Bellucci and A. Nersessian, *Phys. Rev.* **D67**, 065013 (2003) [hep-th/0211070].
6. L.G. Mardoyan, G.S. Pogosyan and A.N. Sissakian, *Theoretical and Mathematical Physics*, **135** (3) 808-813 (2003).
7. L.G. Mardoyan, G.S. Pogosyan and A.N. Sissakian, [quant-ph/0303016].
8. C. Grosche, G.S. Pogosyan and A.N. Sissakian, *Fortschritte der Physik*, **43** (6) 523-563 (1995); *Phys. Part. Nucl.* **27** (3) 244-278 (1996), **28**, 486-519 (1997).
9. E.G. Kalnins, W. Miller Jr. and G.S. Pogosyan, *J. Math. Phys.* **38**, 5416-5433 (1997).
10. Ye.M. Hakobyan, M. Kibler, G.S. Pogosyan and A.N. Sissakian, *Physics of Atomic Nuclei*, **61** (10) 1782-1788 (1998).
11. M.F. Manning and N. Rosen. *Phys. Rev.* **44**, 953 (1933).
12. I.V. Lutsenko, L.G. Mardoyan, G.S. Pogosyan, A.N. Sissakian and V.M. Ter-Antonyan, *J. Phys.* **A22**, 2739 (1989).
13. V.M. Ter-Antonyan. Dyon-Oscillator Duality, quant-ph/0003106 (2000).

THE DIAGONAL AFFINE COSET CONSTRUCTION OF THE $\mathbb{Z}_K$ PARAFERMION HALL STATES

L. GEORGIEV

Institute for Nuclear Research and Nuclear Energy
72 Tsarigradsko Chaussee, 1784 Sofia, Bulgaria
lgeorg@inrne.bas.bg

We construct the $\mathbb{Z}_k$ parafermions as diagonal affine cosets and apply them to the quantum Hall effect. This realization is particularly convenient for the analysis of the $\mathbb{Z}_k$ pairing rules, the modular S-matrices, the W_k symmetry and quantum group symmetry. The results are used for the computation of the mesoscopic chiral persistent currents in presence of Aharonov–Bohm flux.

1. Introduction

Read and Rezayi [1] have described a new hierarchy of incompressible Hall states in the second Landau level with filling factors

$$\nu_k = 2 + \frac{k}{k+2}, \quad k = 2, 3, 4 \dots \tag{1}$$

which have been observed later in an experiment [2] with thermal activation of quasiparticle–quasihole pairs in extremely high mobility samples. The exact quantization of the Hall conductance $\sigma_H = \nu(e^2/h)$ for filling factors $\nu = 5/2$ and $8/3$ ($k = 2$ and 4) was obvious while at $\nu = 13/5$ and $19/7$, ($k = 3$ and 5) there were only well-expressed nonzero minimums of the resistance.

The two-dimensional conformal field theory (CFT) has been successfully applied for the description of the dynamics of the edge excitations in the fractional quantum Hall (FQH) states. Due to the bulk incompressibility in the FQH sample the effective large-scale field theory turns out to be a (2+1) dimensional Chern–Simons theory [3] (say, on a dense cylinder) which is known to be equivalent to a $(1 + 1)$ dimensional CFT (on a cylinder). Thus, the universality classes of the incompressible FQH states can be ultimately labeled by unitary rational conformal field theories (RCFT).

The RCFT for the parafermion FQH states with filling factors (1) has been originally formulated [1] in terms of a $u(1)$ boson describing the electric properties and the $\mathbb{Z}_k$ parafermionic CFT describing the neutral degrees of freedom. The ground state wave function is realized as [1]

$$\Psi_{\mathrm{GS}}(z_1,\ldots,z_N) = \prod_{1\leq a<b\leq N} (z_a - z_b)^{1+\frac{2}{k}} \quad \langle \Psi_1(z_1)\cdots\Psi_1(z_N)\rangle \qquad (2)$$

where Ψ_1 is the first among the $\mathbb{Z}_k$-parafermion currents Ψ_i for $i = 1,\ldots,k-1$ with CFT dimensions $\Delta_i = i(k-i)/k$, and the following operator product expansion [4]

$$\Psi_i(z)\Psi_j(w) \sim (z-w)^{-\frac{2ij}{k}}\,\Psi_{i+j}(w), \quad i+j<k,$$
$$\Psi_i(z)\Psi_j^*(w) \sim (z-w)^{-\frac{2j(k-i)}{k}}\,\Psi_{i-j}(w). \quad i<j$$

In this paper we focus on an alternative realization of the $\mathbb{Z}_k$ parafermions in terms of a diagonal affine coset [5] which turns out to be more natural for the parafermion FQH states.

2. The RCFT for the $\mathbb{Z}_k$ parafermion FQHE

The RCFT for the $\mathbb{Z}_k$ parafermion FQH states has the same structure like the original one as formulated in Ref. [1]

$$\left(\widehat{u(1)} \oplus \mathrm{PF}_k\right)^{\mathbb{Z}_k}, \quad \mathrm{PF}_k = \frac{\widehat{su(k)_1} \oplus \widehat{su(k)_1}}{\widehat{su(k)_2}}. \qquad (3)$$

However, now the $\mathbb{Z}_k$ parafermions are realized as a diagonal affine coset denoted as PF_k and we have used the superscript $\mathbb{Z}_k$ to denote a $\mathbb{Z}_k$ selection rule coupling the electric and neutral sectors of the theory. This rule can be formulated as follows: the edge excitations for the parafermion FQH system are represented by certain primary conformal fields which are labeled by pairs of the $u(1)$ charge λ and the parafermion primary fields Φ

$$(\lambda,\Phi) \;\Rightarrow\; :\mathrm{e}^{2\pi i\frac{\lambda}{\sqrt{k(k+2)}}\phi(z)}: \otimes\, \Phi(z), \quad \Phi(z) \in \mathrm{PF}_k^*.$$

Then the $\mathbb{Z}_k$ paring rule (PR) states that an excitation labeled by (λ,Φ) is admissible only if

$$P[\Phi] = \lambda \mod k, \qquad (4)$$

where P is the parafermion $\mathbb{Z}_k$-charge [4, 5]. The topological order of the parafermion states, which is an important characteristics of any FQH state and gives basically the number of the topologically inequivalent excitations, in this case is [5]

$$\mathrm{TO} = \frac{(k+1)(k+2)}{2}.$$

It is instructive to note that the complete parafermion CFT (3) can be obtained by projecting neutral degrees of freedom in a *maximally symmetric chiral quantum Hall lattice* (in the terminology of Ref. [3]) denoted by the symbol $(3|{}^1A_{k-1}\,{}^1A_{k-1})$ that reproduces [5] the same filling factor (1). The structure of this K-matrix theory corresponds to the left hand side of

$$\left(\widehat{u(1)} \oplus \widehat{su(k)_1} \oplus \widehat{su(k)_1}\right)^{\mathbb{Z}_k} \simeq \left(\widehat{u(1)} \oplus \frac{\widehat{su(2)_k}}{\widehat{u(1)}} \oplus \widehat{su(k)_2}\right)^{\mathbb{Z}_k}$$

while the $\widehat{su(k)_2}$ current algebra in the right hand side expresses the layer (or flavor) symmetry of k pairs of Dirac–Weyl fermions that could be gauged away without changing the filling factor since the subalgebra $\widehat{su(k)_1} \oplus \widehat{su(k)_1}$ is completely decoupled from the electric charge. This projection provides a constructive way to obtain a non-abelian theory from an abelian parent (the latter being identified within the ADE classification of abelian FQH states [3]). It is worth-noting that the PR appears naturally [5] in the abelian parent CFT and is preserved by the coset projection so that the projected theory inherits it.

3. The diagonal coset PF_k

In this section we shall analyse in more detail the affine coset

$$\mathrm{PF}_k = \frac{\widehat{su(k)_1} \oplus \widehat{su(k)_1}}{\widehat{su(k)_2}}. \tag{5}$$

The currents in the $\widehat{su(k)_2}$ algebra in the denominator of the coset are defined as

$$J^a = J_1^a + J_2^a \in \widehat{su(k)_2}, \quad \text{where} \quad J_1^a \in \widehat{su(k)_1}^{(1)}, \quad J_1^a \in \widehat{su(k)_1}^{(2)} \tag{6}$$

and the superscript (1) or (2) refers to the first or the second copy of $\widehat{su(k)_1}$ in the numerator of PF_k respectively. That is why this type of cosets are called the "diagonal" cosets.

The stress tensor of the coset CFT (5) is constructed according to the standard Goddard–Kent–Olive procedure [7]

$$T(z) = T^{\widehat{su(k)_1} \oplus \widehat{su(k)_1}}(z) - T^{\widehat{su(k)_2}}(z) \tag{7}$$

and its modes satisfy the Virasoro algebra with central charge

$$c_k = c^{(1)} + c^{(1)} - c^{(2)} = \frac{2(k-1)}{k+2}, \tag{8}$$

which is exactly the central charge of the $\mathbb{Z}_k$ parafermion CFT [4]. Note that the stress tensor (7) commutes with all currents in the $\widehat{su(k)}_2$ subalgebra which allows us to interpret the latter as a gauge symmetry.

The primary fields of the coset PF_k can be labeled generically by triples $(\underline{\Lambda}, \underline{\Lambda}'; \underline{\Lambda}'')$ of SU(k) weights: two level-1 weights $\underline{\Lambda}$ and $\underline{\Lambda}'$ for the two copies of $\widehat{su(k)}_1$ in the numerator and one level-2 weight $\underline{\Lambda}''$ for the $\widehat{su(k)}_2$ in the denominator. Not all combinations of weights are allowed [6]—they must satisfy the following selection rule (conservation of k-ality)

$$[\underline{\Lambda}] + [\underline{\Lambda}'] = [\underline{\Lambda}''], \quad \text{where} \quad [\underline{\Lambda}] = \sum_{i=1}^{k-1} i\lambda^i \quad \text{for} \quad \underline{\Lambda} = \sum_{i=1}^{k-1} \lambda^i \underline{\Lambda}_i,$$

$\underline{\Lambda}_i$ are the fundamental weights for $SU(k)$ and λ^i the Dynkin labels corresponding to the general weight $\underline{\Lambda}$. Since the admissible $\widehat{su(k)}_1$ weights are the fundamental weights $\underline{\Lambda}_\mu$ $(0 \leq \mu \leq k-1;\ \underline{\Lambda}_0 = 0)$ and any admissible $\widehat{su(k)}_2$ weight can be written as a sum of two fundamental weights, $\underline{\Lambda}_\mu + \underline{\Lambda}_\rho$, $(0 \leq \mu \leq \rho \leq k-1)$ the admissible triples should have the form

$$(\underline{\Lambda}_\mu,\ \underline{\Lambda}_\rho;\ \underline{\Lambda}_{\mu+\sigma} + \underline{\Lambda}_{\rho-\sigma}), \quad \mu, \rho, \sigma = 0, \ldots, k-1. \tag{9}$$

Next, not all admissible triples give inequivalent primary fields of the coset. The action of the simple currents [6]

$$J(\underline{\Lambda}_\mu) = \underline{\Lambda}_{(\mu+1 \mod k)} \tag{10}$$

which are outer automorphisms of the Dynkin diagrams, permuting the Dynkin labels as illustrated on Fig. 1, commutes with the coset stress tensor while not commuting with those of the numerator and denominator separately. As a result the triples which can be connected by the action of the simple currents are equivalent and represent the same irreducible representation (IR) of the coset. This is known as the *field identification* [6]

$$\left(J(\underline{\Lambda}),\ J(\underline{\Lambda}');\ J(\underline{\Lambda}'')\right) \simeq \left(\underline{\Lambda},\ \underline{\Lambda}'\ ;\ \underline{\Lambda}''\right).$$

Acting with the simple current (10) on the admissible triples (9) we get the following triple identification

$$(\underline{\Lambda}_{\mu+\gamma},\ \underline{\Lambda}_{\rho+\gamma};\ \underline{\Lambda}_{\mu+\sigma+\gamma} + \underline{\Lambda}_{\rho-\sigma+\gamma}) \simeq (\underline{\Lambda}_\mu,\ \underline{\Lambda}_\rho;\ \underline{\Lambda}_{\mu+\sigma} + \underline{\Lambda}_{\rho-\sigma}). \tag{11}$$

Choosing γ in the above relation such that $\rho+\gamma = 0 \mod k$ one can always eliminate the second weight, which is our canonical choice of representative for the IR's of the coset. In this case the $\widehat{su(k)}_2$-weight $\underline{\Lambda}''$ completely

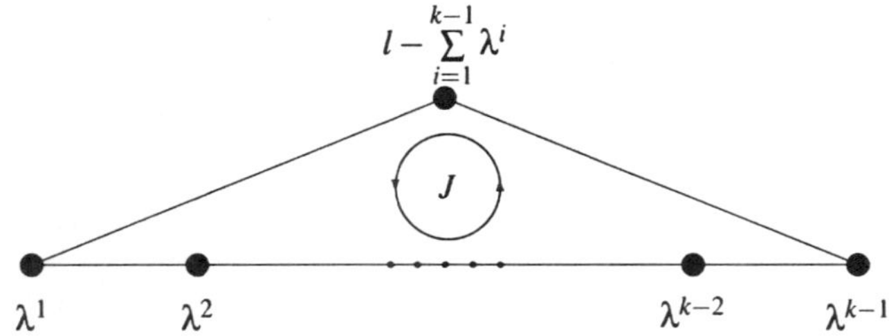

Figure 1. Under the action of the simple current J the Dynkin labels are rotated counter-clockwise, i.e., $\lambda^1 \to \lambda^2, \ldots, \lambda^{k-2} \to \lambda^{k-1}$, $\lambda^{k-1} \to \lambda^0$ and $\lambda^0 \to \lambda^1$.

determines the triple since the first $\widehat{su(k)_1}$ weight is uniquely determined by its k-ality, which should match the k-ality of the level-2 weight

$$\textbf{IR's}: \ \underline{\Lambda}_\mu + \underline{\Lambda}_\rho \quad \Longleftrightarrow \quad (\underline{\Lambda}_{\mu+\rho}, \ \underline{\Lambda}_0; \ \underline{\Lambda}_\mu + \underline{\Lambda}_\rho), \quad 0 \le \mu \le \rho \le k-1. \tag{12}$$

Now the parafermion $\mathbb{Z}_k$ charge is simply

$$P[\underline{\Lambda}_\mu + \underline{\Lambda}_\rho] = \mu + \rho \mod k. \tag{13}$$

The definition of the coset stress tensor (7) as the difference between two other stress tensors gives the following formula for the conformal dimensions

$$\Delta(\underline{\Lambda}, \underline{\Lambda}'; \underline{\Lambda}'') = \Delta^{(1)}(\underline{\Lambda}) + \Delta^{(1)}(\underline{\Lambda}') - \Delta^{(2)}(\underline{\Lambda}'') \mod \mathbb{Z} \tag{14}$$

which reproduces only the fractional part of the CFT dimension. Note that for some triples the level-2 weight $\underline{\Lambda}''$ may have bigger positive CFT dimension than the sum of the CFT dimensions of $\underline{\Lambda}$ and $\underline{\Lambda}'$ in which case the total CFT dimension would appear to be negative, however, this is not the case. That is why " mod $\mathbb{Z}$" in Eq. (14) is necessary and important. Thus obtaining the correct CFT dimensions of the coset primary fields is usually indirect and non-trivial. However, for this particular diagonal coset, using the field identification (11) and choosing appropriate representatives (different from the canonical ones) one can prove [5] the following exact formula for the CFT dimensions of the coset IR's

$$\Delta^{\mathrm{PF}}(\underline{\Lambda}_\mu + \underline{\Lambda}_\rho) = \frac{\mu(k-\rho)}{k} + \frac{(\rho-\mu)(k-\rho+\mu)}{2k(k+2)}, \quad 0 \le \mu \le \rho \le k-1. \tag{15}$$

The characters of the diagonal coset

$$\mathrm{ch}_{\underline{\Lambda}_\mu + \underline{\Lambda}_\rho}(\tau) = \qquad \operatorname*{tr}_{\mathcal{H}_{\Lambda_\mu + \Lambda_\rho}} q^{L_0 - c/24}, \quad q = \mathrm{e}^{2\pi i \tau} \quad (\mathrm{Im}\, \tau > 0)$$

where L_0 is the zero modes of the stress tensor (7) and $\mathcal{H}_{\underline{\Lambda}_\mu+\underline{\Lambda}_\rho}$ is the IR with label $\underline{\Lambda}_\mu + \underline{\Lambda}_\rho$, can be written in the form of the so called *universal chiral partition functions* with statistical matrix equal to the inverse Cartan matrix C^{-1} for $SU(k)$

$$\mathrm{ch}^Q_\sigma(\tau; \mathrm{PF}_k) = q^{\Delta_\sigma - c_k/24} \sum_{\substack{m_1, m_2, \ldots, m_{k-1}=0 \\ \sum_{i=1}^{k-1} i\, m_i \equiv Q \mod k}}^{\infty} \frac{q^{\underline{m}.C^{-1}.(\underline{m}-\underline{\Lambda}_\sigma)}}{(q)_{m_1} \cdots (q)_{m_{k-1}}}, \tag{16}$$

where $(q)_n = \prod_{j=1}^{n}(1-q^j)$, c_k is given by Eq. (8) and $\Delta_\sigma = \Delta^{\mathrm{PF}}(\underline{\Lambda}_0 + \underline{\Lambda}_\sigma)$ is computed from Eq. (15) for $\sigma = 0, \ldots, k-1$. The independent characters are labeled by (σ, Q) with $0 \le \sigma \le Q \le k-1$ and the relation between these new labels and the old ones is $Q = \rho$, $\sigma = \rho - \mu$.

4. The modular S-matrix for the diagonal coset

The S transformation of the modular parameter $\tau' = -1/\tau$ (like in any RCFT) leads to the following linear transformation of the characters

$$\mathrm{ch}_{\underline{\Lambda}_\mu+\underline{\Lambda}_\nu}(\tau') = \sum_{0\le\rho\le\sigma\le k-1} \overset{\circ}{S}_{\underline{\Lambda}_\mu+\underline{\Lambda}_\nu;\ \underline{\Lambda}_\rho+\underline{\Lambda}_\sigma} \quad \mathrm{ch}_{\underline{\Lambda}_\rho+\underline{\Lambda}_\sigma}(\tau)$$

where $\overset{\circ}{S}$ is the coset S matrix. Using the standard Goddard-Kent-Olive procedure for finding the S matrix of the coset we obtain $\overset{\circ}{S}$ as a product of the S matrix of the numerator of the coset (5) and the inverse transposed S matrix of the denominator [7]. Since the numerator is a direct sum of algebras the S matrices for both factors have to be multiplied, however, if we use only canonical triples (12) the second weight $\underline{\Lambda}' = 0$ corresponds to the vacuum sector so that the second S matrix becomes a simple normalization factor $1/\sqrt{k}$. The first S matrix, which is of a rational $u(1)^{k-1}$ torus type, is completely determined from the k-alities of the labels of $\widehat{su(k)}_2$ in the denominator. Finally, since the S matrix for $\widehat{su(k)}_2$ is unitary we get the following expression for the coset S matrix:

$$\overset{\circ}{S}_{\underline{\Lambda}_\mu+\underline{\Lambda}_\nu;\ \underline{\Lambda}_\rho+\underline{\Lambda}_\sigma} = \exp\left(2\pi i \frac{(\mu+\nu)(\rho+\sigma)}{k}\right) \left(S^{(2)}_{\underline{\Lambda}_\mu+\underline{\Lambda}_\nu;\ \underline{\Lambda}_\rho+\underline{\Lambda}_\sigma}\right)^* \tag{17}$$

where the label's parameters satisfy $0 \le \mu \le \nu \le k-1$, $0 \le \rho \le \sigma \le k-1$. Recall that the S matrix for $\widehat{su(k)}_2$ can be computed using Kac formula [7]

$$S^{(2)}_{\underline{\Lambda};\ \underline{\Lambda}'} = \frac{i^{\,k(k-1)/2}}{\sqrt{k(k+2)^{k-1}}} \sum_{w\in\mathcal{W}} \epsilon(w)\ \exp\left(-2\pi i \frac{(\underline{\Lambda}+\underline{\rho}\,|w(\underline{\Lambda}'+\underline{\rho}))}{k+2}\right),$$

where $\mathcal{W}$ is the Weyl group of $SU(k)$, $\epsilon(w)$ is the determinant of the element $w \in \mathcal{W}$ and $\underline{\rho}$ is the Weyl vector [7].

Given the explicit coset S matrix (17), the fusion rules can be obtained by the Verlinde formula [7], however, as can be seen from Eq. (17) the coset S matrix differs from the (conjugated) S matrix for $\widehat{su(k)}_2$ only by a phase. This phase can be interpreted as the action of simple currents on $S^{(2)}$. Since the charge conjugation and the simple currents are automorphisms they preserve the fusion rules. Thus we conclude that the fusion rules for the diagonal coset are the same as those for $\widehat{su(k)}_2$. In particular there are primary fields which obey non-abelian statistics [5].

5. W_k symmetry

The $\widehat{su(k)}_2$ currents which are gauged away in the diagonal coset (5) commute with all independent polynomial Casimir operators of $SU(k)$ which generate a W_k algebra. Therefore, after removing all dimension-one currents, the coset (5) can be completely characterized by its W_k symmetry [5]. The central charge (8) of the diagonal coset coincides with that for the W_k unitary minimal models with the lowest possible parameter $p = k+1$. To make this relation more transparent we shall identify the Fateev–Lukyanov weight β within the Coulomb gas approach to the $\mathbb{Z}_k$ parafermions [4] with the weights of the diagonal coset [5]

$$\beta = \alpha_+ \underline{\Lambda}^{(1)} + \alpha_- \underline{\Lambda}^{(2)} = -\frac{k+1}{\sqrt{(k+1)(k+2)}} \left(\underline{\Lambda}_\mu + \underline{\Lambda}_\rho \right),$$

where, using canonical triples (12) only, we have set $\underline{\Lambda}^{(1)} = 0$ and the Coulomb gas parameters are determined by

$$\alpha_+ + \alpha_- = \alpha_0, \quad \alpha_+ \alpha_- = -1, \quad (\alpha_0)^2 = \frac{1}{(k+1)(k+2)}.$$

The explicit form of the W_k generators can be given by using the quantum Miura transformation [5].

6. Quantum group structure of PF_k

The triples equivalence (11) and the canonical choice (12) demonstrate that there is a 1–1 correspondence between the IRs of PF_k and $\widehat{su(k)}_2$: both labeled by $\underline{\Lambda}_\mu + \underline{\Lambda}_\rho$ (this is not obvious in the $\widehat{su(2)}_k/\widehat{u(1)}$ realization of the $\mathbb{Z}_k$ parafermions). Furthermore, Eq. (17) proves that both S-matrices are equivalent since they are related by charge conjugation and the action of simple currents. These two facts almost prove that the quantum group for the coset (5) is the same as that for $\widehat{su(k)}_2$, i.e., [8]

$$U_q\left(sl(k)\right) \quad \text{with} \quad q = \exp\left(-i\frac{\pi}{k+2}\right), \quad \lfloor m \rfloor_q = \frac{q^m - q^{-m}}{q^1 - q^{-1}}.$$

7. Application: persistent currents in the $\mathbb{Z}_k$ parafermion quantum Hall states

When mesoscopic FQH disk samples are threaded by Aharonov–Bohm flux, the derivative of the free energy $F(T,\phi) = -k_B T \ln Z(T,\phi)$, with respect to the magnetic flux $I(T,\phi) = -\partial_\phi F(T,\phi)$, gives rise to equilibrium persistent currents, which are observable

$$I(T,\phi) = \frac{e}{h} k_B T \frac{\partial}{\partial \phi} \ln Z(\tau, \phi\tau), \quad Z(\tau,\zeta) = \operatorname*{tr}_{\mathcal{H}} q^{L_0 - c/24}\, \mathrm{e}^{2\pi i \zeta J_0},$$

where J_0 is the electric charge operator. The precise thermodynamic identification of the modular parameters is given [10] in terms of the absolute temperature T and the (dimensionless) Aharonov–Bohm flux $\phi = \Phi/(h/e)$

$$q = \mathrm{e}^{2\pi i \tau}, \quad \tau = i\pi\frac{T_0}{T}, \quad \zeta = \phi\tau, \quad T_0 = \frac{\hbar v_F}{\pi k_B L}, \quad \phi \in \mathbb{R}.$$

The complete chiral CFT partition function for the $\mathbb{Z}_k$ parafermion FQH states is constructed as the sum of the characters of all inequivalent IRs [10]

$$Z_k(\tau,\zeta) = \mathrm{e}^{-\pi\nu_H \frac{(\mathrm{Im}\,\zeta)^2}{\mathrm{Im}\,\tau}} \sum_{l \mod k+2} \; \sum_{\rho \geq l-\rho \mod k} \chi_{l,\rho}(\tau,\zeta),$$

where, following Ref. [9], a non-holomorphic exponential factor in front of the sum has been included in order to restore the invariance of the FQH system under the Laughlin spectral flow. The complete IR's characters [5]

$$\chi_{l,\rho}(\tau,\zeta) = \sum_{s=0}^{k-1} K_{l+s(k+2)}(\tau, k\zeta; k(k+2)) \mathrm{ch}\left(\underline{\Lambda}_{l-\rho+s} + \underline{\Lambda}_{\rho+s}\right)(\tau)$$

are special combinations (satisfying the $\mathbb{Z}_k$ pairing rule (4)) of the characters of the $u(1)$ sector describing the electric/magnetic properties of the FQH system

$$K_l(\tau,\zeta;m) = \frac{1}{\eta(\tau)} \sum_{n\in\mathbb{Z}} q^{\frac{m}{2}(n+\frac{l}{m})^2}\, \mathrm{e}^{2\pi i \zeta (n+\frac{l}{m})}, \quad \eta(\tau) = q^{\frac{1}{24}} \prod_{n=1}^{\infty} (1-q^n)$$

and the neutral characters (16) of the diagonal coset PF_k. Using these explicit formulas we have computed numerically [10] the chiral persistent currents for the $\mathbb{Z}_k$ parafermion FQH states with $k = 2, 3$ and 4 for temperatures in the range $0.03 \leq T/T_0 \leq 14$. As illustrated on Fig. 2 these

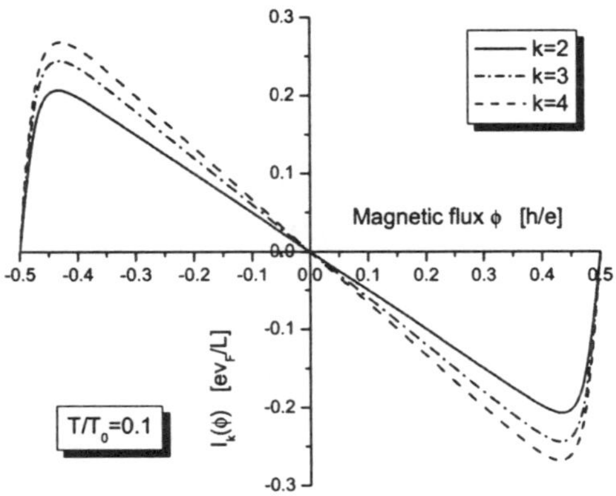

Figure 2. Persistent currents in the $k = 2, 3$ and 4 parafermion FQH states, as functions of the magnetic flux within one period, computed numerically for $T/T_0 = 0.1$.

currents are periodic functions of the AB flux with period exactly one quantum h/e of flux. This is an important result since it forbids any opportunity for a spontaneous breaking of continuous symmetries [10]. The amplitudes of these currents are universal and decay exponentially with increasing the temperature [10] as shown on Fig. 3.

Acknowledgments

Most of the talk followed the work [5]. The author thanks Christoph Schweigert for fruitful discussions, INFN–Firenze for hospitality and acknowledges partial support from the Bulgarian National Council for Scientific Research under Contract F-828 as well as from the FP5-EUCLID Network Program of the European Commission under Contract HPRN-CT-2002-00325.

References

1. N. Read and E. Rezayi, *Phys. Rev.* **B59**, 8084 (1998).

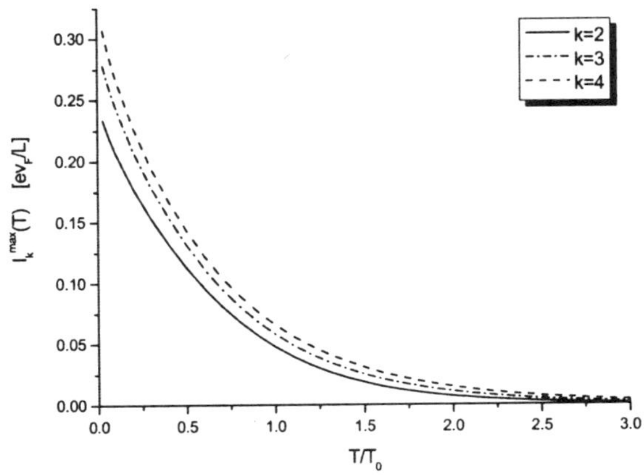

Figure 3. Temperature decay of the persistent current's amplitudes in the $k = 2, 3$ and 4 parafermion FQH states. The zero temperature amplitudes in these units are $\nu_k/2$.

2. W. Pan *et al*, *Phys. Rev. Lett.* **83**, 3530 (1999).
3. J. Fröhlich, U.M. Studer and E. Thiran, *J. Stat. Phys.* **86**, 821 (1997).
4. A.B. Zamolodchikov and V.A. Fateev, *Sov. Phys. JETP*, **62** (1985) 215; *Nucl. Phys.* **B280**, 644 (1987) .
5. A. Cappelli, L. Georgiev and I. Todorov, *Nucl. Phys.* **B599**, 499 (2001).
6. J. Fuchs, A.N. Schellekens and C. Schweigert, *Nucl. Phys.* **B461**, 39 (1996).
7. P. Di Francesco, P. Mathieu and D. Sénéchal, *Conformal Field Theory* (Springer–Verlag, New York, 1997).
8. J.K. Slingerland and F.A. Bais, *Nucl. Phys.* **B612**, 229 (2001).
9. A. Cappelli and G.R. Zemba, *Nucl. Phys.* **B490**, 595 (1997).
10. L. Georgiev, *Phys. Rev.* **B69**, 0453XX (2004).

V. Quantum Mechanical and Probabilistic Systems

MEASURES ON SPACES OF INFINITE-DIMENSIONAL CONFIGURATIONS, GROUP REPRESENTATIONS, AND STATISTICAL PHYSICS

GERALD A. GOLDIN
Departments of Mathematics and Physics, Rutgers University,
118 Frelinghuysen Road, SERC 239, Busch Campus,
Piscataway, NJ 08854 USA
gagoldin@dimacs.rutgers.edu

UGO MOSCHELLA
Dipartimento di Scienze Chimiche, Fisiche e Matematiche,
Università dell'Insubria,
22100 Como, Italy
ugo.moschella@uninsubria.it

TAKAO SAKURABA
Department of Mathematics, Rutgers University,
Piscataway, NJ 08854 USA
tsakuraba@optonline.net

We summarize some of the mathematical results and physical intuitions behind the construction of certain new unitary representations of the group of compactly supported diffeomorphisms of $\mathbb{R}^d$. We include a short discussion of generalized configuration spaces, and measures obtained from self-similar random processes that are quasiinvariant under diffeomorphisms, for $d = 1$ and $d > 1$. This work is motivated by quantum statistical physics and the quantum theory of extended objects, and is descriptive of infinite point configurations with accumulation points.

1. Introduction

Unitary representations of the group of diffeomorphisms of physical space are of fundamental importance in statistical and quantum physics [1, 2, 3, 4, 5, 6, 7, 8]. This article describes the construction of some new unitary representations of the group $Diff^{\,c}(\mathbb{R}^d)$ of compactly-supported diffeomorphisms of d-dimensional Euclidean space. These representations are obtained from self-similar random processes, and can model quantum-mechanical systems having infinitely many degrees of freedom. The key ingredient in the recipe for each such representation is a cylinder set measure on the space of infinite sequences of points in $\mathbb{R}^d$, having the property of quasiinvariance under the action of diffeomorphisms on configurations. Such a measure projects to a measure on the space $\Sigma_{\mathbb{R}^d}$ of finite or countably infinite subsets of $\mathbb{R}^d$.

One physical motivation for the present direction of work is the general goal of constructing quasiinvariant measures for spatially-extended systems, which is in general an unsolved problem. Since $\mathbb{R}^d$ is separable, any closed set in $\mathbb{R}^d$ can be obtained as the closure of an element of $\Sigma_{\mathbb{R}^d}$. Thus such an approach may permit point-like approximations to embedded manifold configurations. Apart from this general consideration, the measures constructed here appear to have a direct interpretation as descriptive of idealized quantum or statistical configurations forming "particle clouds." These allow for a kind of "phase transition" from a rarefied to a condensed phase, as the self-similarity parameter passes through a critical value.

For $d = 1$, some of the results described have been reported earlier [9, 10, 11, 12, 13, 14, 15]. Here we sketch how to generalize from diffeomorphisms of the line or the circle to the higher-dimensional case, using a construction of Sakuraba [16]. Ismagilov earlier proposed a different class of measures, quasiinvariant under diffeomorphisms, that are also concentrated on spaces of configurations having an accumulation point [17, 18, 19, 20]. The measures discussed here and those of Ismagilov are mutually singular.

2. Unitary Representations of Diffeomorphism Groups

In this section we review some background. To see easily how the group of diffeomorphisms of physical space enters into ordinary nonrelativistic quantum mechanics, consider second-quantized canonical nonrelativistic fields $\psi(\mathbf{x})$ and $\psi^*(\mathbf{x})$ at fixed time in a Fock representation. Then the mass density $\rho(\mathbf{x})$ and the momentum density $\mathbf{J}(\mathbf{x})$ are defined as operator-valued distributions over the spatial coordinate $\mathbf{x}$ by

$$\rho(\mathbf{x}) = m\psi^*(\mathbf{x})\psi(\mathbf{x})\,, \quad \mathbf{J}(\mathbf{x}) = (\hbar/2i)\,[\psi^*(\mathbf{x})\,\nabla\psi(\mathbf{x}) - (\nabla\psi^*(\mathbf{x}))\psi(\mathbf{x})]\,. \quad (1)$$

Let $f \in \mathcal{D}(\mathbb{R}^d)$ be a C^∞ function on physical space with compact support, and let $\mathbf{g} \in vect^c(\mathbb{R}^d)$ be a C^∞ vector field with compact support. Then with $\rho(f) = \int \rho(\mathbf{x})f(\mathbf{x})d\mathbf{x}$ and $J(\mathbf{g}) = \int \mathbf{J}(\mathbf{x}) \cdot \mathbf{g}(\mathbf{x})d\mathbf{x}$, there follows from Eq. (1) the Lie algebra bracket

$$[\rho(f_1), \rho(f_2)] = 0\,, \quad [\rho(f), J(\mathbf{g})] = i\hbar\rho(L_{\mathbf{g}}f)\,,$$
$$[\,J(\mathbf{g}_1), J(\mathbf{g}_2)\,] = -i\hbar J([\mathbf{g}_1, \mathbf{g}_2])\,. \quad (2)$$

Here $L_{\mathbf{g}}f = \mathbf{g} \cdot \nabla f$ is the Lie derivative of f (the directional derivative in the $\mathbf{g}$-direction), while $[\mathbf{g}_1, \mathbf{g}_2] = \mathbf{g}_1 \cdot \nabla \mathbf{g}_2 - \mathbf{g}_2 \cdot \nabla \mathbf{g}_1$ is the usual Lie bracket of vector fields. Eqs. (2) are understood as a representation by self-adjoint operators of the natural semidirect sum of the Abelian Lie algebra of scalar functions $\mathcal{D}(\mathbb{R}^d)$, with the algebra of compactly-supported vector fields $vect^c(\mathbb{R}^d)$. It is also possible to work with the Schwartz-space functions $\mathcal{S}(\mathbb{R}^d)$ of rapid decrease (uniformly in all derivatives) at infinity, and with vector fields having components in Schwartz space.

The same *commutator* algebra (2) is obtained regardless of whether ψ and ψ^* are taken to satisfy canonical commutation or anticommutation relations. In both the Bose and Fermi Fock representations of the fields, $\rho(f)$ and $J(\mathbf{g})$ exist as self-adjoint operators, and obey Eqs. (2) on an appropriate common, dense, invariant domain of the Hilbert space. The N-particle symmetric (respectively, antisymmetric) Fock subspaces are invariant under ρ and J. These N-particle spaces describe different numbers of particles in $\mathbb{R}^d$ with the specified exchange statistics. Letting $\Psi_N^{(s)}$ (respectively, $\Psi_N^{(\mathrm{a})}$) be a totally symmetric (respectively, antisymmetric) square-integrable wave function of the N particle positions $(\mathbf{x}_1, \ldots \mathbf{x}_N)$, we have

$$\rho_N(f)\Psi^{(s,\mathrm{a})}(\mathbf{x}_1, \ldots \mathbf{x}_N) = m \sum_{j=1}^{N} f(\mathbf{x}_j)\Psi^{(s,\mathrm{a})}(\mathbf{x}_1, \ldots \mathbf{x}_N)\,,$$

$$J_N(\mathbf{g})\Psi^{(s,\mathrm{a})}(\mathbf{x}_1, \ldots \mathbf{x}_N) = \frac{\hbar}{2i} \sum_{j=1}^{N} [\mathbf{g}(\mathbf{x}_j) \cdot \nabla_j + \nabla_j \cdot \mathbf{g}(\mathbf{x}_j)]\Psi^{(s,\mathrm{a})}(\mathbf{x}_1, \ldots \mathbf{x}_N), \tag{3}$$

where ∇_j acts with respect to $\mathbf{x}_j$. Note that ρ_N and J_N preserve both the particle number and the wave function symmetry. The N-particle representations are mutually unitarily inequivalent, irreducible representations of (2) for distinct values of N. For $d > 1$, the representations respecting different exchange symmetries are also unitarily inequivalent.

Exponentiating Eqs. (3), one obtains continuous unitary representations (CURs) of an infinite-dimensional group: the natural semidirect product of $\mathcal{D}(\mathbb{R}^d)$ under pointwise addition with the group of compactly supported diffeomorphisms $Diff^c(\mathbb{R}^d)$ under composition. The relation between vector fields on $\mathbb{R}^d$ and diffeomorphisms of $\mathbb{R}^d$ is the usual one: given a vector field $\mathbf{g}$ with component in $\mathcal{D}(\mathbb{R}^d)$, one obtains a *flow* $\phi_a^{\mathbf{g}}$, $a \in \mathbb{R}$ from the ordinary differential equation $(\partial/\partial a)\phi_a^{\mathbf{g}}(\mathbf{x}) = \mathbf{g}(\phi_a^{\mathbf{g}}(\mathbf{x}))$, with initial condition $\phi_{a=0}^{\mathbf{g}}(\mathbf{x}) = \mathbf{x}$. We thus have for each $\mathbf{g}$ a one-parameter group of compactly supported C^∞ diffeomorphisms. When $Diff^c(\mathbb{R}^d)$ is endowed with the topology of uniform convergence in all derivatives on compact sets, the flows thus obtained are continuous one-parameter subgroups.

Now with $U(f) = \exp[(i/m)\rho(f)]$ and $V(\phi_a^{\mathbf{g}}) = \exp[(ia/\hbar)J(\mathbf{g})]$, U and V together are a CUR of the semidirect product group, whose group law is

$$U(f_1)V(\phi_1)U(f_2)V(\phi_2) = U(f_1 + f_2 \circ \phi_1)V(\phi_1\phi_2)\,; \tag{4}$$

where $\phi_1\phi_2 = \phi_2 \circ \phi_1$ denotes the composition of diffeomorphisms. The corresponding N-particle representations U_N and V_N are

$$U_N(f)\Psi^{(s,a)}(\mathbf{x}_1, \ldots \mathbf{x}_N) = \prod_{j=1}^{N} \exp[if(\mathbf{x}_j)]\,\Psi^{(s,a)}(\mathbf{x}_1, \ldots \mathbf{x}_N)\,,$$

$$V_N(\phi)\Psi^{(s,a)}(\mathbf{x}_1, \ldots \mathbf{x}_N) = \Psi^{(s,a)}(\phi(\mathbf{x}_1), \ldots \phi(\mathbf{x}_N))\sqrt{\prod_{j=1}^{N} \mathcal{J}_\phi(\mathbf{x}_j)}\,, \tag{5}$$

where $\mathcal{J}_\phi(\mathbf{x})$ is the Jacobian of ϕ at $\mathbf{x}$.

If one begins with a given CUR satisfying Eqs. (4), the self-adjoint generators are recovered in the representation from appropriate one-parameter unitary subgroups, permitting each such representation to be interpreted physically; for example, Eqs. (3) follow from Eqs. (5). Though the group is of course infinite-dimensional, interesting representations exist whose configuration-spaces, like those of Eqs. (5), are finite-dimensional. A configuration describing N identical particles can be taken, in this case, to be an *unordered* N-point subset of $\mathbb{R}^d$ (see below).

Various other inequivalent CURs of this semidirect product group have been obtained, and quantum-mechanical interpretations given. For $d = 2$ the classification of representations led to a mathematically rigorous prediction of "anyon" particle statistics by Goldin, Menikoff, and Sharp [21], as had been earlier conjectured by Leinaas and Myrheim [22] and later studied by Wilczek [23]. Another class of representations was interpreted by Doebner and Goldin as leading to a new family of nonlinear Schrödinger equations, possibly capable of describing dissipative processes in quantum mechanics [24, 25, 26, 27]. Other interesting CURs describe infinite quantum systems—for example, Bose or Fermi gases in the thermodynamic limit (see below).

In a very general framework, the Hilbert space for a CUR of the semidirect product group can be realized as $\mathcal{H} = L^2_\mu(\Delta, \mathcal{W})$; where Δ is a configuration space on which the diffeomorphism group acts (with a right action),

μ is a measure on Δ, $\mathcal{W}$ is an inner product space, and $\mathcal{H}$ is the space of μ-measurable and square-integrable functions $\Psi(\gamma)$ on Δ taking values in $\mathcal{W}$. Typically the operators $U(f), f \in \mathcal{D}(\mathbb{R}^d)$, act as multiplication operators on $\mathcal{H}$,

$$[U(f)\Psi](\gamma) = \exp[i\langle\gamma, f\rangle]\Psi(\gamma)\,; \tag{6}$$

where $\langle\gamma, f\rangle$ denotes a way that $\gamma \in \Delta$ acts on $f \in \mathcal{D}(\mathbb{R}^d)$ as a continuous linear functional. That is, the configuration γ is identified with a *distribution,* so that Δ is identified with a subset of the dual space $\mathcal{D}'(\mathbb{R}^d)$. The operators $V(\phi)$ are given by

$$[V(\phi)\Psi](\gamma) \;=\; \chi_\phi(\gamma)\Psi(\phi\gamma)\sqrt{\frac{d\mu_\phi}{d\mu}(\gamma)}\,, \tag{7}$$

where $\phi\gamma$ refers to the action of the diffeomorphism group on Δ. Let us discuss this formula briefly.

The measure μ is required to have the key property of *quasiinvariance* for the action of diffeomorphisms on Δ. In general, let G be a group of transformations of a measurable space $X, \mathcal{M}$, where $\mathcal{M}$ is a G-invariant σ-algebra for X. A measure μ on $\mathcal{M}$ is said to be quasiinvariant under G if and only if for all $E \in \mathcal{M}$ such that $\mu(E) > 0$, and for all $\phi \in G$, $\mu(\phi(E)) > 0$. For $\phi \in G$, define $\mu_\phi(E) = \mu(\phi(E))$ for any $E \in \mathcal{M}$. Thanks to the group structure and the G-invariance of $\mathcal{M}$, the quasiinvariance of μ under G is equivalent to the absolute continuity of μ_{ϕ_1} with respect to μ_{ϕ_2} for any $\phi_1, \phi_2 \in G$. In particular, the quasiinvariance of μ is necessary and sufficient for the existence of the Radon-Nikodym (RN) derivative $(d\mu_\phi/d\mu)(\gamma)$ in Eq. (7), for all elements $\phi \in \mathit{Diff}^{\,c}(\mathbb{R}^d)$.

In Eq. (7), $\chi_\phi : \mathcal{W} \to \mathcal{W}$ is a family of unitary operators in $\mathcal{W}$ satisfying the following *cocycle equation:* for all ϕ_1, ϕ_2,

$$\chi_{\phi_1}(\gamma)\chi_{\phi_2}(\phi_1\gamma) \;=\; \chi_{\phi_1\phi_2}(\gamma) \tag{8}$$

almost everywhere (a.e.) in Δ; i.e., outside a measure zero set that in general may depend on ϕ_1 and ϕ_2. This equation follows directly from the condition that V respect the group law. The choice $\chi_\phi(\gamma) \equiv I$ is always permitted, while alternate choices of χ_ϕ (noncohomologous cocycles) are associated with nontrivial phase effects and particle statistics (including exotic statistics) [21, 28, 29] as well as with the nonlinear variations of quantum mechanics introduced by Doebner and Goldin [24].

Note that the square root of the RN derivative occurring as a factor in Eq. (7), together with the unitarity of χ_ϕ, are precisely the properties needed to make the representation unitary. In the simplest case $\mathcal{W}$ is just the 1-dimensional space of complex numbers $\mathbb{C}$, so that we have complex-valued wave functions on Δ. Then the χ_ϕ act through multiplication by complex numbers of modulus 1.

The quasiinvariance under diffeomorphisms of Lebesgue measure $d\mathbf{x}$ in $\mathbb{R}^d$ is just the statement that the Jacobian $\mathcal{J}_\phi(\mathbf{x})$ of any diffeomorphism $\phi : \mathbb{R}^d \to \mathbb{R}^d$ is positive and finite. The N-particle representations given by Eqs. (5) can be realized naturally on the configuration space $\Gamma^{(N)}_{\mathbb{R}^d}$ whose elements are (unordered) N-point subsets of $\mathbb{R}^d$. In $\Gamma^{(N)}_{\mathbb{R}^d}$, μ can be chosen to be (locally) a product of Lebesgue measures, $d\mu = d\mathbf{x}_1 \cdots d\mathbf{x}_N$, and is thus quasiinvariant under diffeomorphisms; with $\gamma = \{\mathbf{x}_1, \ldots, \mathbf{x}_N\} \subset \mathbb{R}^d$ and $\phi\gamma = \{\phi(\mathbf{x}_1), \ldots, \phi(\mathbf{x}_N)\}$, we have

$$\frac{d\mu_\phi}{d\mu}(\gamma) = \prod_{j=1}^{N} \mathcal{J}_\phi(\mathbf{x}_j)\,. \tag{9}$$

The representations with $\Delta = \Gamma^{(N)}_{\mathbb{R}^d}$, $\mathcal{W} = \mathbb{C}$, and $\chi_\phi \equiv 1$, are descriptive of N bosons. For $d > 1$, the N-particle Fermi representations correspond to inequivalent choices of χ_ϕ. The reason that for $d = 1$, we cannot describe fermionic statistics through χ_ϕ, is that in $\mathbb{R}^1$ we cannot implement exchanges of points by means of compactly-supported diffeomorphisms.

The existence of quasiinvariant measures on configuration spaces that are finite dimensional is straightforward. Their construction in the infinite-dimensional case is not a minor, technical point, but is essential for unitarity of the representations. It is the quasiinvariant measure that encodes the information about how outcomes of physical measurements distribute in quantum mechanics.

3. Configuration Spaces Containing Countable Subsets

The standard configuration space for statistical physics is the space $\Gamma^{(\infty)}_{\mathbb{R}^d}$ of countably infinite but locally finite subsets of $\mathbb{R}^d$, or sometimes its disjoint union with the spaces of N-point subsets: $\Gamma_{\mathbb{R}^d} = \bigsqcup_{N=1}^{\infty} \Gamma^{(N)}_{\mathbb{R}^d} \bigsqcup \Gamma^{(\infty)}_{\mathbb{R}^d}$. The idea pursued in the balance of this article is the construction of quasiinvariant measures on the space $\Sigma^{(\infty)}_{\mathbb{R}^d}$ of countably infinite subsets of the physical space $\mathbb{R}^d$ that are *not* necessarily locally finite; or alternatively, the space $\Sigma_{\mathbb{R}^d} = \{\gamma \subset \mathbb{R}^d \mid \gamma \text{ is finite or countably infinite}\}$, with

$$\Sigma_{\mathbb{R}^d} = \bigsqcup_{N=1}^{\infty} \Gamma^{(N)}_{\mathbb{R}^d} \bigsqcup \Sigma^{(\infty)}_{\mathbb{R}^d}\,. \tag{10}$$

We call this the space of *generalized configurations*. Our main mathematical motivation for working with this space is that measures on it can be constructed by means of random point processes on spaces of *infinite sequences* of points in $\mathbb{R}^d$. We shall write $\omega = (\mathbf{x}_j) \in [\mathbb{R}^d]^\infty$ for such an infinite sequence, with $j = 1, 2, 3, \ldots$.

Now generalized configurations, like infinite sequences, can have accumulation points. A point $\mathbf{x} \in \mathbb{R}^d$ is an accumulation point of a set $\gamma \subset \mathbb{R}^d$ (respectively, of an infinite sequence $\omega = (\mathbf{x}_j) \in [\mathbb{R}^d]^{(\infty)}$) if for any neighborhood $\mathcal{U}$ of $\mathbf{x}$, the set $\mathcal{U} - \{\mathbf{x}\}$ contains infinitely many points of γ (respectively, ω). An accumulation point of γ may or may not be an element of γ. Evidently diffeomorphisms of $\mathbb{R}^d$ act naturally on generalized configurations, respecting accumulation points: if $\mathbf{x} \in \mathbb{R}^d$ is an accumulation point of $\gamma \in \Sigma_{\mathbb{R}^d}^{(\infty)}$, then $\phi(\mathbf{x})$ is an accumulation point of $\phi\gamma$. The points belonging to configurations in $\Sigma_{\mathbb{R}^d}^{(\infty)}$ can cluster in such a manner as to yield fractals or even more complicated objects. The set of sequences containing coincident points is called the "diagonal" D. Typically D is of measure zero for the point processes of interest, and for technical reasons it is often convenient to exclude it. We have the natural projection from the sequence space to the configuration space, $p : [\mathbb{R}^d]^\infty \to \Sigma_{\mathbb{R}^d}$, given by $p\,[(\mathbf{x}_j)] = \{\mathbf{x}_j\}$. The image of $[\mathbb{R}^d]^\infty$ under p is *all* of $\Sigma_{\mathbb{R}^d}$, since the possibility of repeated entries in elements of $[\mathbb{R}^d]^\infty$ permits the corresponding configurations to be finite as well as infinite.

For any diffeomorphism ϕ of $\mathbb{R}^d$, we have $\phi p\,[(\mathbf{x}_j)] = \{\phi(\mathbf{x}_j)\} = p\,[(\phi(\mathbf{x}_j))]$. Thus we can project a probability measure on the sequence-space $[\mathbb{R}^d]^\infty$ or $[\mathbb{R}^d]^\infty - D$, constructed as is usual from an infinite sequence of conditional probability densities on $\mathbb{R}^d$, to a probability measure on the configuration space $\Sigma_{\mathbb{R}^d}^{(\infty)}$, consistent with the action of $\mathit{Diff}^{\,c}(\mathbb{R}^d)$.

In earlier work, it was shown how for one-dimensional manifolds $\mathbb{R}^1$ or S^1, *self-similar* point processes in the manifold lead quite generally (through such a construction) to *quasiinvariant* measures on the configuration space of countably infinite subsets [9, 10, 11, 12, 15]. Here we shall indicate how the generalization to $d > 1$ can be carried out.

Here let us endow the space $\Sigma_{\mathbb{R}^d}$ with the largest σ-algebra for which the projection p is measurable with respect to the Borel sets of the weak product topology in $[\mathbb{R}^d]^\infty$. A more detailed discussion of topology and measurable structure on $\Sigma_{\mathbb{R}^d}$ is presently in preparation [30].

For measures on the σ-algebra $\mathcal{B}(\,[\mathbb{R}^d]^\infty)$ as we construct them, the RN derivatives under transformations by diffeomorphisms can be expressed (if they exist) as infinite products,

$$\frac{d\mu_\phi}{d\mu}(\omega) = \prod_{j=1}^{\infty} u_{j,\phi}(\omega)\,, \tag{11}$$

where $\omega \in [\mathbb{R}^d]^\infty$ and the $u_{j,\phi}(\omega)$ are measurable functions that depend only on the first j entries of ω. The property of quasiinvariance then requires the convergence of (11) to a non-zero, non-infinite limit almost everywhere in μ, for each ϕ. Under conditions that hold for the measures discussed here, these convergence properties have also been proven sufficient to ensure the quasiinvariance of μ [16].

A general method for constructing a probability measure on $[\mathbb{R}^d]^\infty$ is outlined as follows. For $j = 1, 2, 3 \ldots$, write a conditional (Borel) probability measure μ_j on $\mathbb{R}^d$ depending measurably on the $j-1$ real parameters $\mathbf{x}_1, \ldots, \mathbf{x}_{j-1}$ (the positions of the first $j-1$ particle coordinates), and absolutely continuous with respect to the Lebesgue measure $d\mathbf{x}_j$; i.e., $d\mu_j(\mathbf{x}_j) = f(\mathbf{x}_j|\mathbf{x}_1, \ldots, \mathbf{x}_{j-1})d\mathbf{x}_j$. The joint probability measure for the positions of the first k particles is then specified by $d\mu^{(k)} = \prod_{j=1}^{k} d\mu_j$, and the sequence $(\mu^{(k)})$, $k = 1, 2, 3, \ldots$, is a compatible family of probability measures. By Kolmogorov's theorem, there is a unique measure μ on $[\mathbb{R}^d]^\infty$ determined by the sequence $(\mu^{(k)})$. Under transformation by a diffeomorphism of $\mathbb{R}^d$, the RN derivative for $\mu^{(k)}$ (when it exists) is given by the finite product

$$\frac{d\mu_\phi^{(k)}}{d\mu^{(k)}}(\omega) = \prod_{j=1}^{k} \frac{d\mu_{j,\phi}}{d\mu_j}(\omega)\,, \tag{12}$$

where

$$\frac{d\mu_{j,\phi}}{d\mu_j}(\omega) = u_{j,\phi}(\omega) = \frac{f(\phi(\mathbf{x}_j)|\phi(\mathbf{x}_1), \ldots, \phi(\mathbf{x}_{j-1}))}{f(\mathbf{x}_j|\mathbf{x}_1, \ldots, \mathbf{x}_{j-1})}\mathcal{J}_\phi(\mathbf{x}_j). \tag{13}$$

The quasiinvariance of $\mu^{(k)}$ is assured as long as the RN derivative in Eq. (12) is almost everywhere positive and finite. Now, as anticipated, in the infinite-dimensional case quasiinvariance of the measure μ under diffeomorphisms depends on the behavior of the infinite product in Eq. (11).

Of course, not every measure so constructed will be quasiinvariant. The idea that leads to an interesting class of quasiinvariant measures is to *scale* the probability distribution of the jth particle's position according to the *outcomes* for the previously chosen particle positions. This establishes a self-similar random process. The resulting physical systems behave like an interacting gas of particles with one or more loci of condensation. However, our approach differs from the usual one in that our probability measures are constructed directly, rather than by means of an interaction Hamiltonian.

In general, if the positions of the particle coordinates $\mathbf{x}_j(\omega)$, or the successive difference coordinates $\mathbf{y}_{j+1}(\omega) = \mathbf{x}_{j+1}(\omega) - \mathbf{x}_j(\omega)$, distribute independently but non-identically—so that points can accumulate with non-zero probability—the resulting measure will not be quasiinvariant. However, Ismagilov demonstrated quasiinvariance under diffeomorphisms of the measures resulting from a particular class of processes of this type, in one space dimension [17]. Sakuraba [16] has shown that the quasiinvariant measures constructed by Goldin and Moschella from self-similar random processes, and the quasiinvariant measures of Ismagilov, are mutually singular.

Consider first a "toy model" in one space dimension; with $d = 1$, we write $x_j \in \mathbb{R}$. Choose (x_1, x_2) from a nonvanishing probability density $f_1(x) = f_2(x)$. Let us conventionally write $[a, b]$ to denote the closed interval in $\mathbb{R}$ bounded by the points a and b, regardless of which is the smaller. Then construct the *uniform* probability density $\chi_{[x_1,x_2]}(x)/|x_2 - x_1|$ on the interval $[x_1, x_2]$, where $\chi_{[a,b]}$ denotes the characteristic function (indicator function) of the interval $[a, b]$; and select the pair of points (x_3, x_4) independently from this density. Note that at each stage, the probability is zero that the selected points coincide. Iterating this process, we choose (x_{2m+1}, x_{2m+2}) from the uniform density on $[x_{2m-1}, x_{2m}]$; and we thus have a Markov process. It is easy to show that the resulting measure μ on $\mathbb{R}^\infty$ is concentrated on infinite sequences with one accumulation point, and thus projects to a measure $\hat{\mu}$ on $\Sigma_{\mathbb{R}}$ concentrated on generalized configurations with one accumulation point. This model may be interpreted physically as describing strictly confined clusters formed from infinitely many component particles, and thus having infinitely many internal degrees of freedom.

The expression for $u_{j,\phi}(\omega)$ is in this case, with $j = 2m + 1$,

$$u_{2m+1,\phi}(\omega) = \frac{|x_{2m} - x_{2m-1}|}{|\phi(x_{2m}) - \phi(x_{2m-1})|} \mathcal{J}_\phi(x_{2m+1}), \tag{14}$$

and similarly for $u_{2m+2,\phi}$. Here we can see how the conditional probability (the manner in which the width of the distribution for each pair of points depends on the outcome for the previous pair) affects the convergence: as the x_j become closer and closer together, the first factor approaches the *reciprocal* of the Jacobian. In the limit as $j \to \infty$, each term approaches 1. Thus scale invariance or self-similarity is a property of the configurations, and this is the property that leads naturally to quasiinvariance.

Now, the condition that, with probability one, $\lim_{j\to\infty} u_{j,\phi}(\omega) = 1$ is necessary but not sufficient for convergence of the infinite product in Eq. (11) to a non-zero, finite limit. The result follows because our construction also ensures that the *rate* of convergence is sufficiently rapid. With probability one, we have also $\sum_{j=1}^{\infty} |x_{j+1} - x_j| < \infty$.

The Poisson measure on $\mathbb{R}$ with parameter $\bar{\rho}$, describing the infinite free Bose gas [3], may be obtained from a "reciprocal" Markov process [11]. Choose the first two particles from exponential distributions on the left and right half-lines respectively, and then choose the positions of successive particles to be *outside* the intervals established by preceding choices, with densities decaying exponentially in both directions. The exponential parameter $\bar{\rho}$ is taken to be independent of j in the exponential distributions.

The toy model above generalizes to describe configurations bounded by n intervals, infinitely many intervals, or intervals obtained from random Cantor sets [12, 15]. The latter generalization leads to a family of measures constructed independently by Neretin [31].

So far the examples hinge on the fact that orientation-preserving diffeomorphisms of $\mathbb{R}$ cannot map points *within* an interval to points *outside* the interval; thus ϕ does not take regions of positive measure to regions of zero measure. More realistic examples involve *nowhere vanishing* probability densities. Working still with $d = 1$, let us choose an initial point x_0 from a nowhere vanishing probability density f_0, and the deviation y_1 from a unit normal distribution g_1, with mean 0. Given the points $y_1, \ldots, y_j$, choose y_{j+1} from a normal distribution with mean 0, and standard deviation $\sigma_j = \kappa\,|y_j|$, where $\kappa > 0$ is a fixed *correlation parameter* independent of j. Small values of κ correspond to more tightly bound systems. Thus we have the conditional probability densities for the y_j,

$$g_{j+1}^{\kappa}(y_{j+1}\,|\,y_j) = \frac{1}{(2\pi)^{\frac{1}{2}}\kappa|y_j|}\ \exp\left[-\frac{1}{2\kappa^2}\left(\frac{y_{j+1}}{y_j}\right)^2\right]. \tag{15}$$

For sufficiently small values of κ, the sequence (y_j) converges to 0 (with probability one), and $\sum_{j=1}^{\infty} |y_j| < \infty$. The corresponding measure on the space of sequences (y_j) is quasiinvariant under the stability subgroup $Diff_0^c(\mathbb{R})$, containing the compactly supported diffeomorphisms of $\mathbb{R}$ that leave the origin fixed. We then treat the y_j as successive deviations, letting $x_j = x_{j-1} + y_j$ for $j = 1, 2, 3, \ldots$; so that $x_k = x_0 + \sum_{j=1}^{k} y_j$. Defining the terms $u_{j,\phi}$ in (11) accordingly, for $\phi \in Diff^c(\mathbb{R})$, we obtain $u_j \to 1$ sufficiently rapidly to ensure convergence of the infinite product. More precisely, there exists a critical value κ_0 such that if $0 < \kappa < \kappa_0$, sequences (x_j) converge to an accumulation point with probability one, while if $\kappa_0 < \kappa$, sequences diverge geometrically with probability one. In both cases, the associated measures on $\Sigma_{\mathbb{R}}$ are quasiinvariant under compactly supported diffeomorphism of $\mathbb{R}$ [9, 10, 11, 16]. The proofs make use of the strong law of large numbers.

The above is not tied essentially to the use of normal distributions. All that is really necessary for quasiinvariance under diffeomorphisms is the scaling property. Thus, for a whole class of models, there exists a critical value κ_0 of the scaling parameter κ. For $0 < \kappa < \kappa_0$, the generalized configuration $\{x_j\}$ has an accumulation point with probability one; we call this the *condensed phase*. For $\kappa_0 < \kappa$, $\{x_j\}$ has zero average density; we call this the *rarefied phase*. For each value of κ (except for the critical value itself), we have a bona fide unitary representation of $\mathit{Diff}^c(\mathbb{R})$, describing the associated quantum system.

4. Generalization to Dimension $d > 1$

Some time back it was conjectured that a procedure similar to that encoded in Eq. (15) would work in d space dimensions, $d > 1$, to yield measures on the space of generalized configurations quasiinvariant under $\mathit{Diff}^c(\mathbb{R}^d)$; with the conditional probability density for $\mathbf{y}_{j+1}$ dependent on the preceding d outcomes $(\mathbf{y}_{j-d+1}, \ldots, \mathbf{y}_j)$ through the covariance matrix of a multivariate normal distribution [10, 11]. In fact, the generalization that was obtained by Sakuraba [16] involves some new aspects, which we conclude this article by sketching briefly.

Envision a random point process where at each stage, d vectors in $\mathbb{R}^d$ are to be selected. We are thus choosing a $d \times d$ random matrix V, and thinking of $\omega \in [\mathbb{R}^d]^\infty$ as the sequence of square matrices $([\mathbf{x}_1, \ldots \mathbf{x}_d], [\mathbf{x}_{d+1}, \ldots \mathbf{x}_{2d}], \ldots)$.

For the square matrix $Y = [y_{ij}]$, let $||Y|| = \sqrt{\sum_{i,j=1}^{d} y_{ij}^2}$, noting that $||Y||$ is here a vector norm and *not* the usual operator norm. For $Y \in GL(d, \mathbb{R})$, define the *condition number* $k(Y) := ||Y|| \cdot ||Y^{-1}||$. We may write $Y = P|Y|$ where P is an orthogonal matrix and $|Y| = \sqrt{Y^tY}$ is positive. Let $\tau_1, \ldots, \tau_d$ be eigenvalues of $|Y|$. Then $||Y|| = ||\,|Y|\,|| = \sqrt{\Sigma_{i=1}^{d} \tau_i^2}$, and $||Y^{-1}|| = \sqrt{\Sigma_{j=1}^{d} \tau_j^{-2}}$, so that $k(Y) = \sqrt{\Sigma_{i,j=1}^{d} (\tau_i/\tau_j)^2}$. Evidently $k(Y)$ characterizes the amount of deformation under linear transformation by Y. If Y is not invertible, then $k(Y)$ is undefined (or infinite); but such matrices belong to measure zero sets in the constructions that follow.

We next construct a measure on $[GL(d, \mathbb{R})]^\infty$ and thus on $[\mathbb{R}^d]^\infty$, quasiinvariant under $\mathit{Diff}_0^c(\mathbb{R}^d)$. Define the probability density function

$$f(Y) = C\exp(-\frac{1}{2\kappa^2}[\,||Y||\,k(Y)\,]^2) \tag{16}$$

on the set of $d \times d$ matrices, where C is a constant of normalization so that $\int f(Y)\,dY = 1$; here $dY = d\mathbf{y}_1 \ldots d\mathbf{y}_d$. Let $\mu^{(k)}$ be the probability measure defined by

$$d\mu^{(k)}(Y_1, \ldots, Y_k) = f(Y_1)\frac{f(Y_1^{-1}Y_2)}{|\det Y_1|^d} \cdots \frac{f(Y_{k-1}^{-1}Y_k)}{|\det Y_{k-1}|^d}\,dY_1 \cdots dY_k\,, \tag{17}$$

which is concentrated on $[\,GL(d,\mathbb{R})]^k$. Then we again have a critical value κ_0; for $\kappa < \kappa_0$ the sequence (Y_j) of matrices—and thus the sequence of component vectors $(\mathbf{y}_i)$—converges to 0 with probability one, while for $\kappa_0 < \kappa$ it diverges with probability one. Furthermore the projective limit measure μ on $[\,GL(d,\mathbb{R})]^\infty$ has the desired property of quasiinvariance under $\mathit{Diff}_0^c(\mathbb{R}^d)$. The presence of $k(Y)$ in Eq. (16) is essential for the estimates required in demonstrating convergence of the infinite product, in the resulting expression for the RN derivative. The proof again uses the strong law of large numbers.

Eq. (16) can also be generalized, replacing $k(Y)$ by $k(Y)^\alpha$ ($\alpha \geq 1$), and replacing the Gaussian density by a more general probability density function with appropriate properties.

Finally, we may begin with a matrix of positions $X_0 = [\mathbf{x}_1, \ldots \mathbf{x}_d]$, chosen from a nowhere vanishing probability density. Let $\bar{\mathbf{x}}_0$ be the center of position of the d vectors comprising X_0. Now we may treat each new matrix Y_j as a set of deviations from the center of position of the preceding set of vectors X_{j-1}; so that with obvious notation, $X_j = \bar{\mathbf{x}}_{j-1} + Y_j$. In this manner, we obtain a measure on $[\mathbb{R}^d]^\infty$ quasiinvariant under $\mathit{Diff}^c(\mathbb{R}^d)$, that projects to a quasiinvariant measure on the space $\Sigma_{\mathbb{R}^d}$ of generalized configurations.

More details about the preceding results may be found in the thesis of Sakuraba [16], and in forthcoming publications.

Acknowledgments

G. Goldin thanks Università dell'Insubria for hospitality during his recent visits, and acknowledges support from the Alexander von Humboldt Foundation for related research.

References

1. G.A. Goldin and D.H. Sharp, "Lie algebras of local currents and their representations", in: *1969 Battelle Rencontres: Group Representations*, Lecture Notes in Phys. Vol. 6, ed. V. Bargmann (Springer, Berlin, 1970) p. 300.
2. G.A. Goldin, *J. Math. Phys.* **12**, 462 (1971).
3. G.A. Goldin, J. Grodnik, R.T. Powers and D.H. Sharp, *J. Math. Phys.* **15**, 88 (1974).
4. R. Menikoff, *J. Math. Phys.* **15**, 1138 and 1394 (1974).
5. A.M. Vershik, I.M. Gelfand and M.I. Graev, *Dokl. Akad. Nauk. SSSR*, **232**, 745 (1977).
6. S. Albeverio, Y.G. Kondratiev, and M. Röckner, *J. Funct. Anal.* **154**, 444 (1998); and *J. Funct. Anal.* **157**, 242 (1998).
7. S. Albeverio, Y.G. Kondratiev and M. Röckner, *Rev. Math. Phys.* **11**, 1 (1999).
8. Y.G. Kondratiev and T. Kuna, *Infinite Dimensional Analysis*, **5**, 201 (2002).
9. G.A. Goldin and U. Moschella, "Diffeomorphism groups, quasiinvariant measures, and infinite quantum systems", in: *Symmetries in Science VIII*, ed. B. Gruber (Plenum, New York, 1995) p. 159.
10. G.A. Goldin and U. Moschella, *J. Phys. A: Math. Gen.* **28**, L475 (1995).
11. G.A. Goldin and U. Moschella, "Diffeomorphism group representations and quantum phase transitions in one dimension", in: *XI International Congress on Mathematical Physics,* ed. D. Iagolnitzer (International Press, Cambridge, MA, 1995) p. 445.
12. G.A. Goldin, "Unitary representation of diffeomorphism groups from random fractal configurations", in: *Lie Theory and its Applications,* eds. H.-D. Doebner, V.K. Dobrev and J. Hilgert, (World Scientific, Singapore, 1996) p. 82.
13. G.A. Goldin, "Configuration spaces for self-similar random processes, and measures quasi-invariant under diffeomorphism groups", in: *Mathematical Physics and Stochastic Analysis: Essays in Honour of Ludwig Streit*, eds. S. Albeverio, Ph. Blanchard, L. Ferreira, T. Hida, Y. Kondratiev and R. Vilela Mendes, (World Scientific, Singapore, 2000) p. 228.
14. G.A. Goldin and U. Moschella, "Generalized configuration spaces for quantum systems", in: *Stochastic Processes, Physics and Geometry: New Interplays. II: A Volume in Honor of Sergio Albeverio, Canadian Math. Soc. Conf. Proceedings* **29**, eds. F. Gesztesy, H. Holden, J. Jost, S. Paycha, M. Röckner and S. Scarlatti, (American Math. Soc., Providence, RI, 2000) p. 243.
15. G.A. Goldin and U. Moschella, "Random Cantor sets and measures quasi-invariant for diffeomorphism groups", in: *Coherent States, Quantization and Gravity,* Proceedings of the XVII Workshop on Geometric Methods in Physics, (Bialowieza, July 3-9, 1998) ed. M. Schlichenmaier, A. Strasburger S.T. Ali and A. Odzijewicz, (Wydawnictwa Uniwersytetu Warszawskiego, Warsaw, 2001) p. 263.
16. T. Sakuraba, *Current Algebras, Measures Quasi-Invariant under Diffeomorphism Groups, and Infinite Quantum Systems with Accumulation Points.*

Ph.D. thesis, Rutgers Univ. Dept. of Mathematics, New Brunswick, NJ (2002).
17. R.S. Ismagilov, *Funkts. Anal. i Prilozh.* **5**, 45 (1971); English transl. *Funct. Anal. Appl.* **5** (1971).
18. R.S. Ismagilov, *Izv. Akad. Nauk. SSSR Ser. Mat.* **36**, 180 (1972); English transl. *Math. USSR - Izv.* **6**, 181 (1972).
19. R.S. Ismagilov, *Mat. Sb.* **98**, 55 (1975); English transl. *Math. USSR - Sb.* **27** (1975).
20. R.S. Ismagilov, *Representations of Infinite Dimensional Groups*, AMS Translations of Math. Monographs, **152**, (American Math. Society, Providence, RI, 1996).
21. G.A. Goldin, R. Menikoff and D.H. Sharp, *J. Math. Phys.* **21**, 650 (1980) and *J. Math. Phys.* **22**, 1664 (1981).
22. J.M. Leinaas and J. Myrheim, *Nuovo Cimento*, **37B**, 1 (1977).
23. F. Wilczek, *Phys. Rev. Lett.* **48**, 1144 (1982) and *Phys. Rev. Lett.* **49**, 957 (1982).
24. H.-D. Doebner and G.A. Goldin, *Phys. Lett.* **A162**, 397 (1992).
25. G.A. Goldin, *Int. J. Mod. Phys.* **B6**, 1905 (1992).
26. H.-D. Doebner and G.A. Goldin, *Phys. Rev.* **A54**, 3764 (1996).
27. H.-D. Doebner, G.A. Goldin and P. Nattermann, *J. Math. Phys.* **40**, 49 (1999).
28. G.A. Goldin, R. Menikoff and D.H. Sharp, *Phys. Rev. Lett.* **51**, 2246 (1983).
29. G.A. Goldin, "Parastatistics, θ-statistics, and topological quantum mechanics from unitary representations of diffeomorphism groups", in: *Proceedings of the XV. International Conference on Differential Geometric Methods in Theoretical Physics*, eds. H.-D. Doebner and J.D. Hennig (World Scientific, Singapore, 1987) p. 197.
30. G.A. Goldin, U. Moschella and T. Sakuraba, "Self-similar random processes and infinite-dimensional configuration spaces", presented at the SYMPHYS-10 Conference, August 2003, Yerevan, Armenia (2004, in preparation).
31. Y.A. Neretin, *Sbornik: Mathematics*, **187**, 857 (1996).

ON THE N-PARTICLE WIGNER QUANTUM OSCILLATOR: NON-COMMUTATIVE COORDINATES AND PARTICLE LOCALISATION

R.C. KING

School of Mathematics, University of Southampton, Southampton SO17 1BJ, England. e-mail: R.C.King@maths.soton.ac.uk

T.D. PALEV

INRNE, Boul. Tsarigradsko Chaussee 72, 1784 Sofia, Bulgaria. e-mail: tpalev@inrne.bas.bg

N.I. STOILOVA

Department of Applied Mathematics, Ghent University, Krijgslaan 281-S9, B-9000 Gent, Belgium. e-mail: Neli.Stoilova@UGent.be

J. VAN DER JEUGT

Department of Applied Mathematics, Ghent University, Krijgslaan 281-S9, B-9000 Gent, Belgium. e-mail: Joris.VanderJeugt@UGent.be

After a general introduction to Wigner Quantum Systems, we define the three-dimensional n-particle Wigner Quantum Oscillator, and its relation to the Lie superalgebra $\mathfrak{sl}(1|3n)$. In this framework, coordinates, momentum and the angular momentum of the particles are defined and investigated. This investigation is done in state spaces, using representation theory of $\mathfrak{sl}(1|3n)$. For the case $n = 1$, the main properties are listed, with an emphasis on the non-commutative coordinates of the particle and its unconventional consequences. For general n, we study energy spectra, angular momentum, and in particular the particle configuration and its interpretation. Throughout, a comparison with the canonical oscillator solution is given.

1. Wigner Quantum Systems

In one of his famous papers entitled *Do the equations of motion determine the quantum mechanical commutation relations?* Wigner [1] introduced in 1950 the concept now referred to as a Wigner Quantum System. Wigner himself considered the example of a one-dimensional harmonic oscillator, with Hamiltonian $\hat{H} = \frac{1}{2}(\hat{p}^2 + \hat{q}^2)$ (using here $m = \omega = \hbar = 1$). Working in the Heisenberg picture, he abandoned the requirement $[\hat{q}, \hat{p}] = i$ (the canonical commutation relation - CCR), but instead searched for all operators $\hat{q}$ and $\hat{p}$ such that Hamilton's equations $\dot{\hat{q}} = \hat{p}$ and $\dot{\hat{p}} = -\hat{q}$ are identical with the Heisenberg equations $\dot{\hat{q}} = i[\hat{H}, \hat{q}]$ and $\dot{\hat{p}} = i[\hat{H}, \hat{p}]$. The conclusion of Wigner was that this system now has infinitely many solutions for the "unknown operators" $\hat{q}$ and $\hat{p}$, only one of which satisfies the CCR $[\hat{q}, \hat{p}] = i$.

Nowadays, it is known [2] that this problem posed by Wigner also has a group-theoretical meaning. In fact, what Wigner did is equivalent to the classification of the unitary representations of the Lie superalgebra $\mathfrak{osp}(1|2)$.

Observe that at the time of Wigner it was already known that Hamilton's equations together with the CCRs imply the Heisenberg equations, and that the Heisenberg equations together with the CCRs imply Hamilton's equations (Ehrenfest theorem [3]). Wigner showed, on the example of the one-dimensional harmonic oscillator, that the compatibility of the Heisenberg equations and Hamilton's equations do not imply the validity of the CCRs.

The ideas of Wigner can be further generalized in those cases when the equations of motion admit more than one Hamiltonian [4]. For a recent development along this line we refer to [5] and the references therein. Here we are not concerned with this aspect.

For us, a Wigner Quantum System (WQS) [2, 6] is defined as an n-particle system in three dimensions with Hamiltonian

$$\hat{H} = \sum_{\alpha=1}^{n} \frac{\hat{\mathbf{P}}_\alpha^2}{2m_\alpha} + V(\hat{\mathbf{R}}_1, \hat{\mathbf{R}}_2, \ldots, \hat{\mathbf{R}}_n), \tag{1}$$

depending on $6n$ variables $\hat{\mathbf{R}}_\alpha = (\hat{R}_{\alpha 1}, \hat{R}_{\alpha 2}, \hat{R}_{\alpha 3})$ and $\hat{\mathbf{P}}_\alpha = (\hat{P}_{\alpha 1}, \hat{P}_{\alpha 2}, \hat{P}_{\alpha 3})$, with $\alpha = 1, 2, \ldots, n$, to be interpreted as (Cartesian) coordinates and momenta of the particles, respectively. All postulates of ordinary quantum mechanics should hold, i.e.

- **P1** The state space W is a Hilbert space. To every physical observable O there corresponds a Hermitian (self-adjoint) operator $\hat{O}$ acting in W.
- **P2** The observable O can take on only those values which are eigenvalues of $\hat{O}$. The expectation value of the observable O in a state ψ is given by $\langle \hat{O} \rangle_\psi = (\psi, \hat{O}\psi)/(\psi, \psi)$.

But the postulate on CCRs is replaced by:

P3 Hamilton's equations and the Heisenberg equations hold and are identical (as operator equations) in W.

In the following sections, we shall consider such systems for a simple Hamiltonian, i.e. the harmonic oscillator Hamiltonian.

2. The three-dimensional n-particle Wigner Quantum Oscillator (WQO)

As a special case of (1), we consider the harmonic oscillator Hamiltonian

$$\hat{H} = \sum_{\alpha=1}^{n} \left(\frac{\hat{\mathbf{P}}_\alpha^2}{2m} + \frac{m\omega^2}{2} \hat{\mathbf{R}}_\alpha^2 \right), \tag{2}$$

in terms of the $6n$ variables (operators) $\hat{R}_{\alpha i}$ and $\hat{P}_{\alpha i}$ ($i = 1,2,3$, $\alpha = 1,2\ldots,n$). By postulate **P3**: the 3D vector operators $\hat{\mathbf{R}}_1, \ldots, \hat{\mathbf{R}}_n$ and $\hat{\mathbf{P}}_1, \ldots, \hat{\mathbf{P}}_n$ must satisfy Hamilton's equations

$$\dot{\hat{\mathbf{P}}}_\alpha = -m\omega^2 \hat{\mathbf{R}}_\alpha, \quad \dot{\hat{\mathbf{R}}}_\alpha = \frac{1}{m} \hat{\mathbf{P}}_\alpha, \tag{3}$$

where, as usual, the dot refers to time derivation, and the Heisenberg equations

$$\dot{\hat{\mathbf{P}}}_\alpha = \frac{i}{\hbar} [\hat{H}, \hat{\mathbf{P}}_\alpha], \quad \dot{\hat{\mathbf{R}}}_\alpha = \frac{i}{\hbar} [\hat{H}, \hat{\mathbf{R}}_\alpha], \tag{4}$$

for $\alpha = 1, 2, \ldots, n$. Moreover (3) and (4) should be identical as operator equations. This leads to the following compatibility conditions (CCs):

$$[\hat{H}, \hat{\mathbf{P}}_\alpha] = i\hbar m\omega^2 \hat{\mathbf{R}}_\alpha, \quad [\hat{H}, \hat{\mathbf{R}}_\alpha] = -\frac{i\hbar}{m} \hat{\mathbf{P}}_\alpha. \tag{5}$$

The task is now to find operator solutions of (5). This turns out to be a difficult problem, for which not all solutions are known. On the other hand, one particular solution is well known, namely the canonical bose solution where $[\hat{R}_{\alpha j}, \hat{P}_{\beta k}] = i\hbar \delta_{\alpha\beta} \delta_{jk}$ and $[\hat{R}_{\alpha j}, \hat{R}_{\beta k}] = [\hat{P}_{\alpha j}, \hat{P}_{\beta k}] = 0$.

More than 20 years ago, other classes of solutions for (5) were discovered. One class of solutions can be formulated by means of the Lie superalgebra $\mathfrak{osp}(1|6n)$ [2], and is related to parabose operators [7]. In this paper we shall consider a second class of solutions, first observed by Palev [2], and related to the Lie superalgebra $\mathfrak{sl}(1|3n)$.

For this purpose, consider new (unknown) operators as linear combination of the old ones,

$$A_{\alpha k}^{\pm} = \sqrt{\frac{(3n-1)m\omega}{4\hbar}} \hat{R}_{\alpha k} \pm i \sqrt{\frac{(3n-1)}{4m\omega\hbar}} \hat{P}_{\alpha k}. \tag{6}$$

In terms of these operators, (2) becomes:

$$\hat{H} = \sum_{\alpha=1}^{n} \hat{H}_\alpha \quad \text{with} \quad \hat{H}_\alpha = \frac{\omega\hbar}{3n-1}\sum_{i=1}^{3}\{A_{\alpha i}^{+}, A_{\alpha i}^{-}\}, \tag{7}$$

and the compatibility conditions (5) read:

$$\sum_{\beta=1}^{n}\sum_{j=1}^{3}[\{A_{\beta j}^{+}, A_{\beta j}^{-}\}, A_{\alpha i}^{\pm}] = \mp(3n-1)A_{\alpha i}^{\pm}. \tag{8}$$

All solutions of (8) are not known, but the following yields an important family of solutions:

$$\begin{aligned}
&[\{A_{\alpha i}^{+}, A_{\beta j}^{-}\}, A_{\gamma k}^{+}] = \delta_{jk}\delta_{\beta\gamma}A_{\alpha i}^{+} - \delta_{ij}\delta_{\alpha\beta}A_{\gamma k}^{+},\\
&[\{A_{\alpha i}^{+}, A_{\beta j}^{-}\}, A_{\gamma k}^{-}] = -\delta_{ik}\delta_{\alpha\gamma}A_{\beta j}^{-} + \delta_{ij}\delta_{\alpha\beta}A_{\gamma k}^{-},\\
&\{A_{\alpha i}^{+}, A_{\beta j}^{+}\} = \{A_{\alpha i}^{-}, A_{\beta j}^{-}\} = 0.
\end{aligned} \tag{9}$$

It is easy to verify that (9) implies indeed (8). The main observation of [2] is that the operators $A_{\alpha i}^{\pm}$ ($i = 1,2,3$ and $\alpha = 1,2,\ldots,n$), subject to the above relations (9), are odd elements generating the Lie superalgebra $\mathfrak{sl}(1|3n)$. They are sometimes referred to as creation and annihilation operators (CAOs) of $\mathfrak{sl}(1|3n)$.

Observe that the operators $\hat{R}_{\alpha k}$ and $\hat{P}_{\alpha k}$ depend on time, and so do the CAOs. By Hamilton's equations

$$\dot{A}_{\alpha k}^{\pm}(t) = \mp i\omega A_{\alpha k}^{\pm}(t)$$

one has as solution $A_{\alpha k}^{\pm}(t) = A_{\alpha k}^{\pm}(0)\ e^{\mp i\omega t}$. Therefore, it is sufficient that the defining relations (9) hold at time $t = 0$ and we write $A_{\alpha k}^{\pm} \equiv A_{\alpha k}^{\pm}(0)$. Explicitly, one has:

$$\begin{aligned}
\hat{R}_{\alpha k}(t) &= \sqrt{\frac{\hbar}{(3n-1)m\omega}}(A_{\alpha k}^{+}e^{-i\omega t} + A_{\alpha k}^{-}e^{i\omega t}),\\
\hat{P}_{\alpha k}(t) &= -i\sqrt{\frac{m\omega\hbar}{(3n-1)}}(A_{\alpha k}^{+}e^{-i\omega t} - A_{\alpha k}^{-}e^{i\omega t}).
\end{aligned} \tag{10}$$

For later comparison, we recall here also the canonical solution for (2):

$$\begin{aligned}
\overline{R}_{\alpha k}(t) &= \sqrt{\frac{\hbar}{(3n-1)m\omega}}(B_{\alpha k}^{+}e^{i\omega t} + B_{\alpha k}^{-}e^{-i\omega t}),\\
\overline{P}_{\alpha k}(t) &= i\sqrt{\frac{m\omega\hbar}{(3n-1)}}(B_{\alpha k}^{+}e^{i\omega t} - B_{\alpha k}^{-}e^{-i\omega t}),
\end{aligned} \tag{11}$$

where $B^{\pm}_{\alpha k}$ are ordinary Bose creation and annihilation operators, satisfying $[B^-_{\alpha j}, B^+_{\beta k}] = \delta_{jk}\delta_{\alpha\beta}$ and $[B^{\pm}_{\alpha j}, B^{\pm}_{\beta k}] = 0$. It is known that the elements $B^{\pm}_{\alpha k}$ (as odd generators) and their anti-commutators $\{B^+_{\alpha j}, B^-_{\beta k}\}$ generate the Lie superalgebra $\mathfrak{osp}(1|6n)$ [8].

3. The angular momentum operators

The operators $\hat{\mathbf{R}}_\alpha = (\hat{R}_{\alpha 1}, \hat{R}_{\alpha 2}, \hat{R}_{\alpha 3})$ and $\hat{\mathbf{P}}_\alpha = (\hat{P}_{\alpha 1}, \hat{P}_{\alpha 2}, \hat{P}_{\alpha 3})$ of the WQO are interpreted and referred to as the position and momentum operators of the particle α. We introduce one more physical notion, namely the angular momentum of this particle. Since the operator $\hat{\mathbf{R}}_\alpha \times \hat{\mathbf{P}}_\alpha$ no longer satisfies the usual requirements, we follow an alternative definition for the angular momentum. For the particle α, its angular momentum operator $\hat{\mathbf{M}}_\alpha = (\hat{M}_{\alpha 1}, \hat{M}_{\alpha 2}, \hat{M}_{\alpha 3})$ should be such that

- its components are in the enveloping algebra of $\hat{\mathbf{R}}_\alpha = (\hat{R}_{\alpha 1}, \hat{R}_{\alpha 2}, \hat{R}_{\alpha 3})$ and $\hat{\mathbf{P}}_\alpha = (\hat{P}_{\alpha 1}, \hat{P}_{\alpha 2}, \hat{P}_{\alpha 3})$ (and linear in these components);
- $\hat{M}_{\alpha 1}$, $\hat{M}_{\alpha 2}$ and $\hat{M}_{\alpha 3}$ commute with the Hamiltonian;
- they span a basis of the Lie algebra $\mathfrak{so}(3)$.

This leads to a unique solution [6], given by:

$$\hat{M}_{\alpha j} = -\frac{3n-1}{2\hbar} \sum_{k,l=1}^{3} \epsilon_{jkl}\{\hat{R}_{\alpha k}, \hat{P}_{\alpha l}\} = -i \sum_{k,l=1}^{3} \epsilon_{jkl}\{A^+_{\alpha k}, A^-_{\alpha l}\}, \qquad (12)$$

where ϵ_{jkl} is the familiar antisymmetric tensor. Using (12) and (9), one can verify that the components of $\hat{\mathbf{M}}_\alpha$ and $\hat{\mathbf{M}}_\beta$ commute for $\alpha \neq \beta$, and that indeed the following relations hold:

$$[\hat{M}_{\alpha j}, \hat{M}_{\alpha k}] = i\epsilon_{jkl}\hat{M}_{\alpha l}, \qquad [\hat{M}_{\alpha j}, \hat{H}] = 0. \qquad (13)$$

Apart from the single particle angular momentum, we define as components of the total angular momentum of the n-particle system:

$$\hat{M}_j = \sum_{\alpha=1}^{n} \hat{M}_{\alpha j}, \qquad j = 1, 2, 3. \qquad (14)$$

It is easy to verify that these operators generate an $\mathfrak{so}(3)$ subalgebra of $\mathfrak{sl}(1|3n)$.

4. The state spaces (representations)

In order to investigate physical properties of the WQO, we need to consider the state spaces in which the operators act. According to the postulates, the state space W is a Hilbert space, and $(A^{\pm}_{\alpha j})^{\dagger} = A^{\mp}_{\alpha j}$. The last condition implies that the operators $A^{\pm}_{\alpha j}$ act in "unitary" (i.e. star) representations of $\mathfrak{sl}(1|3n)$.

All the irreducible star representations (irreps) of $\mathfrak{sl}(M|N)$ have been classified [9]: essentially they are the typical, the covariant and the contravariant representations. Here, we make one further choice: instead of considering all star representations, we shall study only one class of them, namely the so-called Fock type representations [10] of $\mathfrak{sl}(1|3n)$. The advantage of these representations is the fact that all operators have a simple action in a particular basis.

The Fock representations are characterized by a positive integer p: $W = W(p)$ is a finite dimensional covariant irrep with highest weight $(p, 0, \ldots, 0)$. It is determined by the relations:

$$A^{-}_{\alpha j}\,|\,0\rangle = 0, \quad A^{-}_{\alpha j}A^{+}_{\beta k}\,|\,0\rangle = p\delta_{\alpha\beta}\delta_{jk}\,|\,0\rangle, \tag{15}$$

a formulation reminiscent of the definition of "parastatistics of order p". Observe that $W(p)$ is a *typical* irrep when $p \geq 3n$ and *atypical* when $p < 3n$.

It is easy to construct a basis for $W(p)$. Let Θ be a string of 0's and 1's:

$$\Theta \equiv (\theta_{11}, \theta_{12}, \theta_{13}, \ldots \theta_{n1}, \theta_{n2}, \theta_{n3}), \qquad \theta_{\alpha i} \in \{0, 1\}. \tag{16}$$

Then, an orthonormal basis of $W(p)$ is given by the vectors

$$\begin{aligned} |\,p;\Theta\rangle &\equiv |p;\theta_{11},\theta_{12},\theta_{13},\ldots,\theta_{n1},\theta_{n2},\theta_{n3}\rangle \\ &= \sqrt{\frac{(p-q)!}{p!}}(A^{+}_{11})^{\theta_{11}}(A^{+}_{12})^{\theta_{12}}(A^{+}_{13})^{\theta_{13}}\cdots \\ &\qquad \cdots(A^{+}_{n1})^{\theta_{n1}}(A^{+}_{n2})^{\theta_{n2}}(A^{+}_{n3})^{\theta_{n3}}\,|\,0\rangle, \end{aligned} \tag{17}$$

where

$$q \equiv \sum_{\alpha=1}^{n}\sum_{i=1}^{3}\theta_{\alpha i} \tag{18}$$

must satisfy:

$$0 \leq q \leq \min(p, 3n). \tag{19}$$

The action of the CAOs reads:

$$A_{\alpha i}^{-}\,|\,p;\Theta\rangle = \theta_{\alpha i}(-1)^{\psi_{\alpha i}}\sqrt{p-q+1}\,|\,p;\Theta\rangle_{\overline{\alpha i}}, \tag{20}$$

$$A_{\alpha i}^{+}\,|\,p;\Theta\rangle = (1-\theta_{\alpha i})(-1)^{\psi_{\alpha i}}\sqrt{p-q}\,|\,p;\Theta\rangle_{\overline{\alpha i}}, \tag{21}$$

where

$$\psi_{\alpha i} = \sum_{(\beta j)<(\alpha i)} \theta_{\beta j},$$

with lexicographical ordering on the pairs (αi). In (20)-(21), $|\,p;\Theta\rangle_{\overline{\alpha i}}$ stands for the state obtained from $|\,p;\Theta\rangle$ after the replacement of $\theta_{\alpha i}$ by $\overline{\theta}_{\alpha i} = (1-\theta_{\alpha i})$.

The explicit Fock space representation of the WQO defined by (16)-(21) is reminiscent of the familiar Fock space representation of the canonical quantum oscillator which has as basis vectors

$$\begin{aligned} |\,\Phi\rangle &\equiv |\phi_{11},\phi_{12},\phi_{13},\ldots,\phi_{n1},\phi_{n2},\phi_{n3}\rangle \\ &= \prod_{\alpha=1}^{n}\prod_{i=1}^{3}\frac{1}{\sqrt{\phi_{\alpha i}!}}(B_{11}^{+})^{\phi_{11}}(B_{12}^{+})^{\phi_{12}}(B_{13}^{+})^{\phi_{13}}\cdots \\ &\quad\cdots(B_{n1}^{+})^{\phi_{n1}}(B_{n2}^{+})^{\phi_{n2}}(B_{n3}^{+})^{\phi_{n3}}\,|\,0\rangle \end{aligned} \tag{22}$$

with $\phi_{\alpha i} \in \{0,1,2,\ldots\}$. Clearly, it is infinite-dimensional, and the action of the Bose creation and annihilation operators is as usual:

$$B_{\alpha i}^{-}\,|\,\Phi\rangle = \sqrt{\phi_{\alpha i}}\,|\,\Phi\rangle_{-\alpha i}, \qquad B_{\alpha i}^{+}\,|\,\Phi\rangle = \sqrt{\phi_{\alpha i}+1}\,|\,\Phi\rangle_{+\alpha i}, \tag{23}$$

where $|\,\Phi\rangle_{\pm\alpha i}$ is obtained from $|\,\Phi\rangle$ through replacing $\phi_{\alpha i}$ by $\phi_{\alpha i}\pm 1$. It is known that this Fock space is a particular (infinite-dimensional) irreducible representation of the Lie superalgebra $\mathfrak{osp}(1|6n)$ [8]

5. Properties of the single particle 3D WQO [11]

Let us first consider the case $n=1$; since the subscript α is now superfluous, it will be dropped from the notation in this section.

The Hamiltonian $\hat{H}$ is diagonal in the basis $|p;\Theta\rangle$, and one finds:

$$\hat{H}\,|\,p;\Theta\rangle = E_q\,|\,p;\Theta\rangle \quad \text{with} \quad E_q = \frac{\hbar\omega}{2}\,(3p-2q)\,, \tag{24}$$

so the energy levels are equidistant in steps of $\hbar\omega/2$. Also the operator $\hat{\mathbf{M}}^2$ is diagonal, with

$$\hat{\mathbf{M}}^2|p;\Theta\rangle = \begin{cases} 0 & \text{if } \theta_1=\theta_2=\theta_3; \\ 2|p;\Theta\rangle & \text{otherwise.} \end{cases} \tag{25}$$

So the angular momentum of the single particle is 0 or 1. This also follows from a decomposition of $W(p)$ with respect to the subalgebra $\mathfrak{so}(3)$.

The most striking properties of the single particle WQO are related to the fact that its position or momentum operators are non-commutative:

$$[\hat{R}_i, \hat{R}_j] \neq 0 \quad \text{and} \quad [\hat{P}_i, \hat{P}_j] \neq 0 \text{ for} \quad i \neq j = 1,2,3. \tag{26}$$

On the other hand, the square operators $\hat{\mathbf{R}}^2$ and $\hat{\mathbf{P}}^2$ are integrals of motion; in fact,

$$\hat{\epsilon} \equiv \frac{2}{\omega\hbar}\hat{H} = \frac{2m\omega}{\hbar}\hat{\mathbf{R}}^2 = \frac{2}{m\omega\hbar}\hat{\mathbf{P}}^2 = \sum_{i=1}^{3}\{A_i^+, A_i^-\}. \tag{27}$$

This implies that in "stationary states" (a superposition of states with the same q) the particle is at a fixed distance $\varrho_q = \sqrt{\frac{\hbar}{2m\omega}(3p-2q)}$ from the origin.

However, one can even be more specific about possible outcomes of a position measurement. Indeed, it is not difficult to verify that the set of operators

$$\hat{H},\ \hat{\mathbf{R}}^2, \hat{\mathbf{P}}^2,\ \hat{R}_1^2,\ \hat{R}_2^2,\ \hat{R}_3^2,\ \hat{P}_1^2,\ \hat{P}_2^2,\ \hat{P}_3^2$$

mutually commute, and are all diagonal in the $|p;\Theta\rangle$-basis. Before deducing physical consequences from this, let us for convenience introduce dimensionless notation for energy, position and momentum:

$$\hat{\varepsilon} = \frac{2}{\omega\hbar}\hat{H}, \quad \hat{r}_i(t) = \sqrt{\frac{2m\omega}{\hbar}}\hat{R}_i(t), \quad \hat{p}_i(t) = \sqrt{\frac{2}{m\omega\hbar}}\hat{P}_i(t). \tag{28}$$

From (20)-(21) it now follows that ($k = 1,2,3$):

$$\hat{r}_k^2|p;\Theta\rangle = \hat{p}_k^2|p;\Theta\rangle = (p - q + \theta_k)|p;\Theta\rangle, \tag{29}$$

$$\hat{r}^2|p;\Theta\rangle = (3p-2q)|p;\Theta\rangle, \qquad (q = \theta_1 + \theta_2 + \theta_3). \tag{30}$$

Furthermore, since $\hat{r}_1^2$, $\hat{r}_2^2$ and $\hat{r}_3^2$ mutually commute, their eigenvalues can be measured simultaneously. So, if the system is in a fixed state $|p;\Theta\rangle$, measurements of position imply that the particle will be detected in one of the eight "nests" with coordinates

$$r_1 = \pm\sqrt{p-q+\theta_1}, \quad r_2 = \pm\sqrt{p-q+\theta_2}, \quad r_3 = \pm\sqrt{p-q+\theta_3}, \tag{31}$$

on a sphere with radius $\rho_q = \sqrt{3p-2q}$.

Consider, as an example, the state $|p;0,0,1\rangle$ (with $p > 2$). The possible outcomes of measurements of the position operators imply that the particle is in one of eight nests indicated in Figure 1(a), even though its precise position cannot be determined due to the non-commutativity of the operators $\hat{r}_k$ $(k = 1, 2, 3)$. For each of the eight basis states $|p; \theta\rangle$, such a configuration holds, except when $p \leq 2$. For $p \leq 2$, the representation is atypical and of smaller dimension; then also the coordinate configurations collapse.

Figure 1. (a) Possible outcomes of the measurements of the coordinates $\hat{r}_k$ of the particle in the stationary state $|p; 0, 0, 1\rangle$ are illustrated by means of eight nests. (b) Possible outcomes of the measurements of rotated coordinates $\hat{s}_k$ of the particle in this eigenstate.

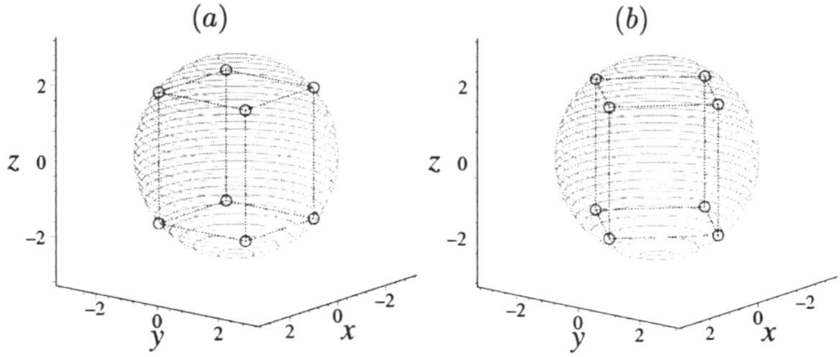

By examining the eigenvectors of $\hat{r}_k$, one can draw conclusions about the occupation probabilities of the 8 nests, even though the 8 probabilities themselves cannot be determined.

Note that similar conclusions can be drawn with respect to the momentum of the particle.

It should be emphasized that due to the non-commutativity, the position of the particle is not absolute but relative to which coordinates are being measured. For example, one can consider measurements of the coordinates $s_k(t)$ with respect to an alternative frame of reference. In case of an orientation obtained by rotating the frame of reference through an angle ϕ about the third axis, one has

$$\begin{aligned} \hat{s}_1(t) &= \cos\phi\, \hat{r}_1(t) + \sin\phi\, \hat{r}_2(t); \\ \hat{s}_2(t) &= -\sin\phi\, \hat{r}_1(t) + \cos\phi\, \hat{r}_2(t); \qquad (32) \\ \hat{s}_3(t) &= \hat{r}_3(t). \end{aligned}$$

Once again, the squares of these operators $\hat{s}_k^2$ mutually commute (but not with $\hat{r}_k^2$), and their eigenvalues are of the same type $p - q + \theta$. The eigenvectors of $\hat{s}_k^2$ are in general linear combination of the states $|p;\theta\rangle$, although our example state $|p;0,0,1\rangle$ happens to be directly an eigenvector of the $\hat{s}_k^2$. The actions indicate that the sites corresponding to possible values of measurements of the coordinates $\hat{s}_k$ are again nests on a sphere of radius ρ_q, but the nests define a rectangular parallelepiped obtain by rotating the original one about the third axis through an angle ϕ, see Figure 1(b).

The explanation for these different outcomes lies in the fact that the particle itself cannot be localised, since measurements of its coordinates do not mutually commute. It is the choice of coordinate to be measured that leads to the observed value corresponding to the associated eigenvalue.

6. The n-particle WQO [12]

We now return to the n-particle WQO, with Hamiltonian given by (7). Its energy spectrum can be computed using the basis (17), with actions (20)-(21):

$$\hat{H}\,|\,p;\Theta\rangle = E_q\,|\,p;\Theta\rangle \quad \text{with} \quad E_q = \hbar\omega\left(\frac{3np}{3n-1} - q\right) \tag{33}$$

for $q = 0, 1, 2, \ldots, \min(3n, p)$ and determined by (18).

This is to be compared with the spectrum of the canonical solution in the Fock space with basis vectors (22):

$$\hat{H}\,|\,\Phi\rangle = \bar{E}_q\,|\,\Phi\rangle \quad \text{where} \quad \bar{E}_q = \hbar\omega\left(\frac{3}{2}n + q\right) \tag{34}$$

with $q = \sum_{\alpha=1}^{n}\sum_{i=1}^{3}\phi_{\alpha i}$.

These energy spectra, together with their multiplicities, can also be found from branching rules with respect to a $\mathfrak{gl}(1)$ subalgebra of the Lie superalgebra under consideration. For the WQO, this branching rule is described by (see e.g. [13])

$$\begin{aligned} \mathfrak{sl}(1|3n) &\longrightarrow \mathfrak{gl}(1) \oplus \mathfrak{sl}(3n) \\ W(p) &\longrightarrow \sum_{q=0}^{\min(p,3n)} V_{\mathfrak{gl}(1)}^{q+3n(p-q)} \otimes V_{\mathfrak{sl}(3n)}^{1^q} \end{aligned}$$

where the superscripts refer to the highest weights of the representation in partition notation. The $\mathfrak{gl}(1)$ generator is $(3n-1)/(\hbar\omega)$ times the Hamiltonian. The branching rule also implies that the multiplicity of E_q is equal to the dimension of the $\mathfrak{sl}(3n)$ irrep with partition labels (1^q), thus $\text{mult}(E_q) = \binom{3n}{q}$.

For the canonical solution, the Fock space described by the basis vectors (22) is the infinite-dimensional irrep of $\mathfrak{osp}(1|6n)$ which decomposes into the sum of the two infinite-dimensional irreducible metaplectic or oscillator representations of $\mathfrak{sp}(6n)$ [14, 15], and then into finite-dimensional irreducible representations of $\mathfrak{gl}(1) \oplus \mathfrak{sl}(3n)$, according to the following branching rules:

$$\begin{aligned} &\mathfrak{osp}(1|6n) \longrightarrow \mathfrak{sp}(6n) \longrightarrow \mathfrak{gl}(1) \oplus \mathfrak{sl}(3n) \\ &\quad \text{Fock} \longrightarrow V^{\varepsilon+}_{\mathfrak{sp}(6n)} \oplus V^{\varepsilon-}_{\mathfrak{sp}(6n)} \longrightarrow \sum_{q=0}^{\infty} V^{\frac{3}{2}n+q}_{\mathfrak{gl}(1)} \otimes V^{q}_{\mathfrak{sl}(3n)}. \end{aligned}$$

So now the energy multiplicity is equal to the dimension of the $\mathfrak{sl}(3n)$ irrep with partition labels (q), i.e. $\mathrm{mult}(\bar{E}_q) = \binom{3n-1+q}{q}$.

Next, we turn our attention to the angular momentum of the n-particle WQO. Similarly to the one particle case, one obtains

$$\hat{\mathbf{M}}^2_{\alpha}|p;\Theta\rangle = \delta_{\alpha}\, 2\, |p;\Theta\rangle, \tag{35}$$

where

$$\delta_{\alpha} = \begin{cases} 0 & \text{if } \theta_{\alpha 1} = \theta_{\alpha 2} = \theta_{\alpha 3}; \\ 1 & \text{otherwise.} \end{cases}$$

In other words, the WQO behaves like a collection of spin zero and spin one particles.

Although the stationary states $|p;\Theta\rangle$ are eigenvectors of $\hat{\mathbf{M}}^2_{\alpha}$, they are not eigenstates of either $\hat{M}_3$ or $\hat{\mathbf{M}}^2$. This makes it more difficult to compute the possible values of the total angular momentum of the system. One classical way to determine these values, is to compute the decomposition of the representation space according to a branching rule with respect to the $\mathfrak{so}(3)$ subalgebra generated by the total angular momentum components. For the WQO, this branching rule is described by

$$\begin{aligned} &\mathfrak{sl}(1|3n) \to \mathfrak{gl}(1) \oplus \mathfrak{sl}(3n) \to \mathfrak{gl}(1) \oplus \mathfrak{sl}(3) \oplus \mathfrak{sl}(n) \\ &\to \mathfrak{gl}(1) \oplus \mathfrak{so}(3) \oplus \mathfrak{sl}(n) \to \mathfrak{gl}(1) \oplus \mathfrak{so}(3) \end{aligned}$$

and the decomposition of the irrep $W(p)$ can be computed for each n (although no closed formula can be given for all n). The total angular momentum ranges from 0 to n (with multiplicities).

Observe that for the canonical case, one should examine the branching rule

$$\mathfrak{osp}(1|6n) \to \mathfrak{sp}(6n) \to \mathfrak{gl}(1) \oplus \mathfrak{sl}(3n) \to \mathfrak{gl}(1) \oplus \mathfrak{sl}(3) \oplus \mathfrak{sl}(n)$$
$$\to \mathfrak{gl}(1) \oplus \mathfrak{so}(3) \oplus \mathfrak{sl}(n) \to \mathfrak{gl}(1) \oplus \mathfrak{so}(3)$$

for the Fock space spanned by (22). In this case, the total angular momentum can take infinitely many values.

Finally, we shall make some observations on the possible oscillator configurations for the n-particle WQO. As before, it is convenient to use dimensionless operators:

$$\hat{r}_{\alpha k}(t) = \sqrt{\frac{(3n-1)m\omega}{\hbar}}\hat{R}_{\alpha k}(t) = A^+_{\alpha k}\, e^{-i\omega t} + A^-_{\alpha k}\, e^{i\omega t}. \tag{36}$$

Then

$$\hat{r}^2_{\alpha k} = \{A^+_{\alpha k}, A^-_{\alpha k}\} \tag{37}$$

and the basis vectors $|p;\Theta\rangle$ are directly eigenvectors of these coordinate operators:

$$\hat{r}^2_{\alpha k}\,|p;\Theta\rangle = r^2_{\alpha k}\,|p;\Theta\rangle, \qquad \text{with } r^2_{\alpha k} = p - q + \theta_{\alpha k}. \tag{38}$$

Since the squares of the coordinate operators all commute, $[\hat{r}^2_{\alpha i}, \hat{r}^2_{\beta j}] = 0$ for all α, β, i, j, the conclusions about particle coordinates are similar as in the $n = 1$ case: as far as coordinates is concerned, one simply gets a superposition of n single particle systems. In other words, if the system is in the state $|\,p;\Theta\rangle$, measurements of the coordinates $r_{\alpha k}$ of the αth particle can yield only

$$r_{\alpha k} = \pm\sqrt{p - q + \theta_{\alpha k}} \quad \text{for } k = 1, 2, 3.$$

This corresponds to measuring the particle α in one of 8 nests on a sphere with radius $\rho_\alpha = \sqrt{3p - 3q + q_\alpha}$, where $q_\alpha = \theta_{\alpha 1} + \theta_{\alpha 2} + \theta_{\alpha 3}$.

As in the single particle case, however, one has to be careful about conclusions related to the position of particles, since the actual position operators do not commute (only their squares commute), and so the particles cannot be absolutely localised. The non-commutativity of the underlying geometry is also reflected by the action of the following commutator upon a basis state (for $(\alpha i) < (\beta j)$):

$$\begin{aligned}
&[\hat{r}_{\alpha i}(t), \hat{r}_{\beta j}(t)]\,|\,p; \dots, \theta_{\alpha i}, \dots, \theta_{\beta j}, \dots\rangle \\
&= (-1)^{\psi_{\beta j}-\psi_{\alpha i}} \Bigg(2e^{i2\omega t}\theta_{\alpha i}\theta_{\beta j}\sqrt{(p-q+1)(p-q+2)} \\
&\quad +(\theta_{\alpha i}-\theta_{\beta j})^2(2p-2q+1) \\
&\quad +2e^{-i2\omega t}\overline{\theta}_{\alpha i}\overline{\theta}_{\beta j}\sqrt{(p-q-1)(p-q)} \Bigg)\,|\,p; \dots, \overline{\theta}_{\alpha i}, \dots, \overline{\theta}_{\beta j}, \dots\rangle,
\end{aligned}$$

where $\overline{\theta}_{\gamma k} = 1 - \theta_{\gamma k}$. The right hand side of this expression is nonzero for all $p \geq q+2$.

In these circumstances, we must expect some difficulties over the interpretation of the measurement of the distance between two particles α and β. Let us, in line with classical notions of distance, define $\hat{d}^2_{\alpha\beta}(t)$ as the square distance operator for particles α and β, where

$$\hat{d}^2_{\alpha\beta}(t) = \sum_{i=1}^{3} \left(\hat{r}_{\alpha i}(t) - \hat{r}_{\beta i}(t)\right)^2 . \tag{39}$$

In particular, consider the case $n = 2$, with $\alpha = 1$ and $\beta = 2$. A computation of the eigenvalues of $\hat{d}^2_{12}(t)$ in a typical irrep $W(p)$ leads to the following list:

$$\begin{array}{l|l|l}
(6p)_1 & (6p-12)_4 & (6p-22)_3 \\
(6p-4)_3 & (6p-14)_9 & (6p-24)_3 \\
(6p-6)_3 & (6p-16)_9 & (6p-26)_3 \\
(6p-8)_3 & (6p-18)_4 & (6p-30)_1 \\
(6p-10)_9 & (6p-20)_9 &
\end{array} \tag{40}$$

The multiplicity of each eigenvalue is given here as a subscript.

Suppose that the system is in the state $|p;\dot{0}\rangle \equiv |p;0,0,0,0,0,0\rangle$. The nest coordinates for the two particles are $r_{1i} = \pm\sqrt{p}$ and $r_{2i} = \pm\sqrt{p}$ for $i = 1,2,3$. Thus, the square distances between these nests are determined by

$$\sum_{i=1}^{3} \left(r_{1i} - r_{2i}\right)^2 \in \{0, 4p, 8p, 12p\}. \tag{41}$$

So the spectrum of eigenvalues of $\hat{d}^2_{12}(t)$ given in (40) does not contain for general p the values of the squares of the distances between nests as given in (41). This is only an apparent contradiction. After all, the states $|p;\Theta\rangle$ are not eigenstates of $\hat{d}^2_{12}(t)$, and $\hat{d}^2_{12}(t)$ does not commute with $\hat{r}^2_{1i}(t)$ and $\hat{r}^2_{2j}(t)$.

There is, however, a way in which even these considerations about square distances between particles are consistent with the nest interpretation. For this purpose, let us return to the general case (arbitrary $n \geq 2$) with particles α and β. As usual in quantum mechanics, the expectation value of the square distance operator is given by:

$$\overline{d^2}_{\alpha\beta}(t) = \langle p;\Theta|\ \hat{d}^2_{\alpha\beta}(t)\ |p;\Theta\rangle, \tag{42}$$

and it yields the average value of the square distance of the two particles. A careful computation, expressing $\hat{d}^2_{\alpha\beta}(t)$ in terms of the operators $A^{\pm}_{\alpha k}$ and using their action, leads to the following result:

$$\overline{d^2}_{\alpha\beta}(t) = 6p - 6q + q_\alpha + q_\beta = \rho_\alpha^2 + \rho_\beta^2, \tag{43}$$

where $q_\alpha = \theta_{\alpha 1} + \theta_{\alpha 2} + \theta_{\alpha 3}$ and $\rho_\alpha = \sqrt{3p - 3q + q_\alpha}$ (similarly for q_β and ρ_β). Note that ρ_α and ρ_β are the radii of the spheres on which particles α resp. β are located.

As a consistency check, we will compare this to the classical average, i.e. the value $d^2_{\alpha\beta}$ being the average of the square of the distance between the nests available to particles α and β. In this case, another careful computation (including relations satisfied by occupancy probabilities) leads indeed to the same value:

$$d^2_{\alpha\beta} = 6p - 6q + q_\alpha + q_\beta = \rho_\alpha^2 + \rho_\beta^2. \tag{44}$$

Quite generally, the expectation values of all operators related to localisation of particles are in agreement with the computation of a classical average using the notion of nests as being the actual positions of the particles.

7. Conclusions

We have considered a major class of non-canonical solutions for the oscillator Hamiltonian of n particles in three-dimensional space, i.e. the n-particle WQO in 3D. Our class of solutions is formulated in the framework of the Lie superalgebra $\mathfrak{sl}(1|3n)$.

We have constructed Fock type representations $W(p)$ as state spaces for the WQO. These representations are finite-dimensional. The energy of the system is quantized, with equally spaced finite spectrum. The angular momentum is also quantized, with single particle angular momenta 0 or 1. But also the coordinates and momenta are quantized, having a finite spectrum and leading to a finite number of "nests". The coordinate operators do not commute. There is a sense in which WQSs provide a natural framework for considering non-commutative coordinates, since the non-commutativity comes automatically (and does not need to be introduced ad hoc as in many other approaches).

Acknowledgments

NIS was supported by a Marie Curie Individual Fellowship of the European Community Programme 'Improving the Human Research Potential and the Socio-Economic Knowledge Base' under contract number HPMF-CT-2002-01571.

References

1. E.P. Wigner, *Phys. Rev.* **77**, 711 (1950).
2. T.D. Palev, *J. Math. Phys.* **23**, 1778 (1982).
3. P. Ehrenfest, *Z. Phys.* **4**, 455 (1927).
4. S. Okubo, *Phys. Rev.* **D22**, 919 (1980).
5. P. Blasiak, A. Horzela and E. Kapuscik, *J. Opt.* **B5**, S245 (2003).
6. T.D. Palev and N.I. Stoilova, *J. Math. Phys.* **38**, 2506 (1997).
7. H.S. Green, *Phys. Rev.* **90**, 270 (1953).
8. A.Ch. Ganchev and T.D. Palev, *J. Math. Phys.* **21**, 797 (1980).
9. M.D. Gould and R.B. Zhang, *J. Math. Phys.* **31**, 2552 (1990).
10. T.D. Palev, *J. Math. Phys.* **21**, 1283 (1980).
11. R.C. King, T.D. Palev, N.I. Stoilova and J. Van der Jeugt, *J. Phys. A: Math. Gen.* **36**, 4337 (2003).
12. R.C. King, T.D. Palev, N.I. Stoilova and J. Van der Jeugt, *J. Phys. A: Math. Gen.* **36**, 11999 (2003).
13. P.H. Dondi and P.D. Jarvis, *J. Phys. A: Math. Gen.* **14**, 547 (1981).
14. M. Moshinsky and C. Quesne, *J. Math. Phys.* **12**, 1772 (1971).
15. R.C. King and B.G. Wybourne, *J. Math. Phys.* **41**, 5002 (2000).

BUNDLE GERBES AND BROWNIAN MOTION

RÉMI LÉANDRE
Institut Elie Cartan. Université de Nancy I
54000. Vandoeuvre-les-Nancy, FRANCE
email: leandre@iecn.u-nancy.fr

We define a stochastic line bundle on the loop space by using gerbes theory and we define the stochastic parallel transport along a diffusion on the loop space

1. Introduction

It is classical in theoretical physics (see the surveys of [15, 16, 17] for instance) that random surfaces realize the dynamic of the loop space. The simplest case is the case of the cylinder, which realizes a process on the loop space of a manifold M. We are concerned in this work by diffusion processes on the loop space.

Infinite dimensional processes over infinite dimensional manifolds has a long story initiated by Kuo in 1972 [26], and continued by the Russian school (See [5, 6, 12, 4] have constructed diffusion processes over infinite dimensional Hilbert manifolds.

We are concerned in this work by the case of the loop space. [3, 32, 27] have constructed the Ornstein-Uhlenbeck process over the loop space, by using the theory of Dirichlet forms. Airault-Malliavin [1] and Brzezniak-Elworthy [9] have constructed the Brownian motion on the space of Hoelder loop. It is a 1+1 dimensional theory.

Felder-Gawedzki-Kupiainen [14] and Konno [24] have shown that the W.Z.N.W. model is related to a line bundle over a loop space, and that the two dimensional quantum theory is involved with a generalized parallel transport along a Riemann surface for the loop space of this line bundle.

Stochastic interpretation of the considerations of these authors were done in the following works:

-)Brzezniak-Léandre [8] have considered a diffusion processes over the C^1 loop space and have constructed the stochastic parallel transport associated to the Felder-Gawedzki-Kupiainen line bundle along this diffusion process. The theory is 1+1 dimensional and the loops are C^1 such that the line bundle over the loop space is surely defined.

-)Léandre [28] has considered a diffusion process over the space of C^k maps from a Riemann surface Σ into a compact manifold M. The random boundary of Σ realize the Hilbert space of the theory: it is constituted of the Hilbert space of sections of the line bundle on the loop space of Felder-Gawedzki-Kupiainen. The theory is 1+2 dimensional, but the the line bundle on the loop space is surely defined. Combinatorial formulas of [14] are surely true in this context, because they do not require the use of stochastic integrals.

-)Léandre [29] has considered a diffusion process over the Hoelder torus group. [29] defined a stochastic line bundle over a loop group and the holonomy on a loop in the loop group. The theory is 1+2 dimensional, but the line bundle and the holonomy of this line bundle are almost surely defined, because they require the introduction to 1 and 2 dimensional stochastic integrals (See [24] for the study of the geometry involved in the deterministic case).

-)Léandre [30] considers the case of the $1 + n$ punctured sphere. It realizes an element of $Hom(H^{\otimes n}, H)$ where H is the Hilbert space of a stochastic line bundle over a loop group, and where we consider a generalized stochastic parallel transport for this line bundle along the punctured sphere (See in the deterministic case, the works of [20, 21, 23]). The theory is 1+2 dimensional, but the generalized parallel transport is almost surely defined, because it requires the introduction of stochastic integral on the punctured sphere.

We are concerned in this paper by a 1+1 dimensional theory, but with a line bundle almost surely defined.

[8] have used the definition of the bundle over the C^1 loop space of Felder-Gawedzki-Kupiainen which was defined by the help of Deligne cohomology. Brylinski pointed out that it is possible to define a line bundle over the loop space, by using the theory of gerbes of Grothendieck [7]. The presentation of Brylinski uses tools of algebraic geometry. Mathematical physicists have a differential geometrical approach to bundle gerbes (We refer to the survey of Carey-Mickelsson-Murray [11] about that topics).

In this work, we will follow the presentation of Gawedzki and Reis [18] to construct a line bundle on the loop space by using bundle gerbes. We consider a diffusion process on the Hoelder loop space, instead of the C^1 loop space as in [8], and we consider the construction of the line bundle over the loop space by using bundle gerbes instead of Deligne cohomology as in [8]. This leads to the study of two dimensional stochastic integrals on the cylinder in this 1+1 dimensional theory.

The reader interested by the relation between analysis on loop space and mathematical physic can see the review of S. Albeverio [2] and the two reviews of Léandre [27, 31].

2. A brief review of Gawedzki-Reis theory

Let us do in this part a brief review of the theory of [18], chapter 6.

We consider a compact Riemannian manifold M.

Suppose that we have a submersion $\pi : Y \to M$ with local smooth sections $\sigma_i : O_i \to Y$ associated to a finite cover O_i of M. Let $Y^{[n]} = Y \times_M .. \times_M Y$ be denoting the set of $(y_1, .., y_n)$ in Y^n such that $\pi(y_1) = .. = \pi(y_n)$. We denote by $p_{n_1,..,n_k}$ the projection from $Y^{[n]}$ to $Y^{[k]}$ which associates to $(y_1, .., y_n)$ $(y_{n_1}, .., y_{n_k})$. Let ω be a 3-form on M.

Definition II.1: A bundle gerbe G over M with connection of curvature ω is a quadruple (Y, α, L, μ) where:

i) α is a 2-form on Y such that

$$d\alpha = \pi^*\omega$$

ii)L is a hermitian line bundle with connection ∇ over $Y^{[2]}$ with curvature:

$$R = p_2^*\alpha - p_1^*\alpha$$

iii)μ is an isomorphism of hermitian line bundle with connection over $Y^{[3]}$

$$\mu : p_{1,2}^*L \otimes p_{2,3}^*L \to p_{1,3}^*L$$

iv) μ is associative in the sense that

$$\mu \circ (\mu \otimes Id) = \mu \circ (Id \circ \otimes \mu)$$

is an isomorphism of line bundles $p_{1,2}^*L \otimes p_{2,3}^*L \otimes p_{3,4}^*L$ and $p_{1,4}^*L$ over $Y^{[4]}$.

The 2-form α is called the curving of the bundle gerbe. The isomorphism μ defines a bilinear product

$$L_{(y_1,y_2)} \otimes L_{(y_2,y_3)} \to L_{(y_1,y_3)}$$

Let us introduce the free loop space $L_1(M)$ of finite energy maps from the circle S^1 into M: $s \to \gamma(s)$.

We consider a triangulation (b, v) of the oriented circle in oriented segments: if $v \in b$, we denote by y_v an element of Y such that $\pi(y_v) = \gamma(v)$

and we put $y_i(b) = \sigma_{i_b}(\gamma((b)))$ if we suppose that $\gamma(b) \subseteq O_{i_b}$ and where σ_i is the chosen section over O_i of Y.

The system of $O(b, v, (.)_b, (.)_v)$ such that each $\gamma(b)$ is included in O_{i_b} and such each $\gamma(v)$ is an element of O_{i_v} defines an open cover of $L_1(M)$. Over this cover, the fiber of the line bundle ξ over γ ξ_γ is given by (See [18] formula (6.7))

$$\xi_\gamma = \otimes_{v \in b} L_{(y_v, y_i(b)(v))}$$

If v is the beginning of a segment of the subdivision, we take the dual line in (2) and we take the line if v is the end of b.

Transition functional are given in term of cylindrical functional in μ and of the parallel transport on the line bundle L over the path in $Y^{[2]}$

$$b \cap b' \to Y^{[2]} : s \to (\sigma_{i_b}(\gamma(s)), \sigma_{i'_{b'}}(\gamma(s)))$$

We call the system of transition functional which are got by a mixture of cylindrical functional and parallel transports of the type (2) by $\rho_{b,v,i_v,i_b,b',v',i'_{v'},i'_{b'}}(\gamma)$. This system of transition functionals defines a complex line bundle ξ over $L_1(M)$.

Let us consider a path of finite energy over the loop space. This gives a map from the cylinder $S^1 \times [0,1] = T$ into M: If $S = (s,t)$, $S \to \gamma(S)$ is this map. s is the internal time of the loop and t is the dynamical time. Let (c, b) be a triangulation of the cylinder in small rectangles $[s, s'] \times [t, t']$. We do the restriction that the image of each rectangle c belongs to O_{i_c} and that the image of each segment b in the triangulation belongs to O_{i_b}. We denote:

$$\gamma_{i_c} = \sigma_{i_c} \circ \gamma|_c$$

$$\gamma_{i_b} = \sigma_{i_b} \circ \gamma|_b$$

which takes their values in Y. We denote

$$\gamma_{i_{c,b}} = (\sigma_{i_c} \circ \gamma|_b, \sigma_{i_b} \circ \gamma|_b)$$

which takes its values in $Y^{[2]}$. $\gamma_{i_{c,b}}$ is a path in $Y^{[2]}$, and we denote by $H(\gamma_{i_{c,b}})$ the parallel transport in L over it. It is an element of the tensor product of the dual of L taking in the starting point of b (i.e. the line L with our sign convention) and of L at the end point.

We introduce the amplitude:

$$A(\gamma) = \exp[i \sum_c \int_c \gamma_{i_c}^* \alpha] \otimes_{b \subseteq c} H(\gamma_{i_{c,b}})$$

From the triangulation of T, we deduce a triangulation of the boundary loops, and therefore a trivialization of the bundle ξ over the loop induced by the restriction on the boundary of the map $\gamma : T \to M$. In this trivialization, the amplitude $A(\gamma)$ is nothing else than the parallel transport along the path in the loop space $t \to (s \to \gamma(s,t))$ of the bundle ξ endowed with a suitable connection, whose curvature is

$$\int_{S1} \omega(d\gamma(s), ., .)$$

3. A stochastic line bundle on the loop space

Let us imbed isometrically the compact manifold M into R^n. If x belongs to M, let $\Pi(x)$ be the orthogonal projection from R^n into $T_x(M)$. It can be extended into a map from R^n into the linear applications over R^n bounded with bounded derivatives of all orders. We introduce the Hilbert space $H = H^{1;2}(S1; R^n)$ of the set of the maps from the circle S^1 into R^n h such that:

$$\int_0^1 |h(s)|^2 ds + \int_0^1 |d/dsh(s)|^2 ds = ||h||^2 < \infty$$

We can write $h = (h^1, .., h^n)$. Moreover,

$$h^j(0) =< h, e^j >$$

where

$$e^j(s) = (0, .., 0, \lambda \exp[-s] + \mu \exp[s], 0, .., 0)$$

for some λ and some μ for $0 \leq s \leq 1$. Therefore the function is smooth on $]0,1[$ with half derivatives of all orders in 0 and 1: $e(0) = e(1)$ but $e'(0) \neq e'(1)$.

We have

$$h^j(s) =< h, e^j(. - s) >$$

We consider the Brownian motion with values in H:

$$B_t(s) = (B_t^1(s), .., B_t^n(s))$$

The processes $B^j_.(.)$ are independent and $t \to B_t^j(s)$ is a Brownian motion with R such that the right bracket

$$d < B_t^j(s), B_t^j(s') >= e(s - s')dt$$

We consider Airault-Malliavin equation [1, 9]

$$dx_t(s) = \Pi(x_t(s))d_tB_t(s)$$

issued from x. It is a Stratonovitch equation (See [22, 13]). When s is fixed, $t \to x_t(s)$ is a Brownian motion with values in M.

We see that $s \to x_t(s)$ is Hoelder by Gronwall lemma and Kolmogorov lemma (See [28]).

Definition III.1: μ_t is the image law over the continuous free loop space $L(M)$ of $s \to x_t(s)$.

We would like to define a stochastic line bundle ξ_t over $L(M)$ endowed with the probability measure μ_t. The main problem is to define the stochastic parallel transport along the path in $Y^{[2]}$

$$s \to (\sigma_{i_b}(x_t(s)), \sigma_{i'_{b'}}(x_t(s)))$$

for the bundle L. But L is a line bundle, and in local coordinate, this parallel transport is given by

$$\exp[i \int_b <A(\sigma_{i_b}(x_t(s), \sigma_{i'_{b'}}(x_t(s)),$$
$$D\sigma_{i_b}(x_t(s))d_s x_t(s), D\sigma_{i'_{b'}}(x_t(s)), d_s x_t(s) >]$$

where b is a segment on the circle and A a linear form which depends smoothly of its argument.

The problem is put by this term, and not by the cylindrical terms, in the transition functionals: it leads namely to stochastic integrals. Let us consider a subdivision $s_i, i = 1,..,N$ of the interval b. Let $s \to x_t^N(s)$ be the polygonal approximation of $s \to x_t(s)$ (in R^n) and let β be a one form on R^n. Let

$$A_t^N = \int_b < \beta(x_t^N(s)), d_s x_t^N(s) >$$

for a one form β.

We have:

Proposition III.2: when $N \to \infty$, A_t^N tends in L^2 to a random variable called $\int_b < \beta(x_t(s)), d_s x_t(s) >$.

Proof: We write:

$$B_t(s + \Delta s) = B_t(s) + \Delta_s B_t(s)$$

Let us denote by $\phi_t(s)$ the derivative of the flow associated to $x_t(s)$.

We put:

$$I_t1(s,\Delta s) = \phi_t(s)\int_0^t D\Pi(x_u(s))\phi_u^{-1}(s)d_u\Delta_s B_u(s)$$
$$= \phi_t(s)\int_0^t r_u(s)d_u\Delta_s B_u(s)$$

and

$$I_t^2(s,\Delta s) = \phi_t(s)\int_0^t \phi_u(s)^{-1} < D\Pi(x_u(s), I_u^1(s,\Delta s), d_u\Delta_s B_u(s) >$$

By using the rules of differentiation along a parameters [25, 33] of a stochastic differential equations, we have that:

$$x_t(s+\Delta s) = x_t(s) + I_t^1(s,\Delta s) + I_t^2(s,\Delta s) + O((\Delta s)^{3/2})$$

$I_t1(s,\Delta s)$ behaves in $\sqrt{\Delta s}$ and $I_t2(s,\Delta s)$ behaves in Δs. From the properties of the correlation function $e(s)$, we deduce that:

Property H(1): If s' do not belong to $]s, s+\Delta s[$

$$d < B_t(s+\Delta s) - B_.(s), B_.(s') >= dt(C(s,s')\Delta s + O(\Delta s)2)$$

where $C(s,s'$ is continuous.

Property H(2): If $]s, s+\Delta s[\cap]s', s'+\Delta s'[=\emptyset$, we have

$$d < B_.(s+\Delta s) - B_.(s), B_.(s'+\Delta s') - B_.(s') >=$$
$$dt(C(s,s')\Delta s\Delta s' + O(\Delta s + \Delta s')^3)$$

Let us show Property H(2). It results from the fact that

$$e(s+\Delta s - s' - \Delta s') - e(s+\Delta s - s') - e(s - s' - \Delta s') - e(s-s') = \\ e"(s-s')\Delta s\Delta s' + O(\Delta s + \Delta s')3 \quad (1)$$

Properties H(1) and H(2) will allow us to show the following key lemma:

Lemma III.3: Let f and f' two smooth function on R^n bounded with bounded derivatives of all orders. Let us suppose that $]s, s+\Delta s[\cap]s', s'+\Delta s'[=\emptyset$ Then

$$E[f(x_t(s))f(x_t(s'))I_t^1(s,\Delta s)I_t^1(s',\Delta s')] = C(s,s')\Delta s\Delta s' + O(\Delta s + \Delta s')^{5/2}$$

Proof: In $I_t1(s,\Delta s)$, we isolate $\phi_t(s)$, which is $B_.(s)$ measurable from the stochastic integral, we write in Itô meaning. We get by property H(1) an expression of the shape

$$\int_0^t r_u(s)\delta_u\Delta_s B_u(s) + \Delta s\int_0^t g_u(s)du$$

$g_u(s)$ is $B_.(s)$ measurable. We write the term assembled which are not stochastic integral as

$$C\int_0^t < \alpha_u, \delta_u B_u(s), \delta_u B_u(s') >$$

by the Clark-Ocone formula [34]. The vector α_u is F_u measurable (where F_t denotes the filtration spanned by $t \to B_t(.)$ and $B_.(s), B_.(s')$ measurable.

In order to get our estimate, we apply successively Itô formula and Clark-Ocone formula. We begin to apply Itô formula to the product

$$\int_0^t r_u(s)\delta\Delta_s B(s) \int_0^t r_u(s')\delta\Delta_{s'} B(s')$$

The boring term by property H(2) comes from the double stochastic integrals. This leads to estimate quantities as, say,

$$E[\int_0^t \alpha_u \delta_u B_u(s) \int_{0<v<u<t} r_u(s)\delta_v \Delta_s B_v(s) r_u(s')\delta_u \Delta_{s'} B_v(s')]$$

We apply another time Itô formula in (3) and we arrive to estimate the quantity by property H(1):

$$\Delta s' E[\int_0^u R_v(s)\delta_v \Delta_s B_v(s)\alpha_u(s) r_u(s')]$$

But $\alpha_u(s)r_u(s')$ is $B_.(s)$ and $B_.(s')$ measurable. We apply the Ckark-Ocone formula to the product $\alpha_u(s)r_u(s')$, and we conclude by Itô formula. $\diamondsuit$

This lemma allows us to proof our theorem as in [29].

We can consider a more intrinsic approximation of the Stratonovitch integral $\int_b < \beta(x_t(s)), d_s x_t(s) >$. If $x_t(s_{i+1})$ is close from $x_t(s_i)$, we use the function

$$F_N(s, x_t(s_{i+1}), x_t(s_i)) = \exp_{x_t(s_i)}[\frac{s - s_i}{s_{i+1} - s_i}(x_t(s_{i+1}) - x_t(s_i))]$$

where exp is the Riemannian exponential conveniently extended over R^n and $x_t(s_{i+1}) - x_t(s_i)$ is the unique vector of the unique geodesic joining $x_t(s_i)$ to $x_t(s_{i+1})$. We extend the map $y \to y - x_t(s_i)$ conveniently. We put:

$$\tilde{x}_t^N(s) = F_N(s, x_t(s_{i+1}), x_t(s_i))$$

The technics of the Proof of Proposition III.2 (See [29] for details in o more complicated situation) works and allow to show:

Theorem III.4: When $N \to \infty$, $\int_b < \beta(\tilde{x}_t^N(s)), d_s\tilde{x}_t^N(s) >$ converges in L^2 to the Stratonovitch integral called $\int_b < \beta(x_t(s)), d_s x_t(s) >$.

Remark: We don't know if the Stratonovitch integral of Theorem III.4 is equal to the Stratonovitch integral of Proposition III.3.

We deduce that if γ belongs to the intersection of $O(b, v, (.)_b, (.)_v)$ and $O(b', v', (.)_{b'}, (.)_{v'})$, the transition functional $\rho_{b,v,i_b,i_v,b',v',i_{b'},i_{v'}}$ is almost surely defined for the measure μ_t. Moreover, since this stochastic transition functional is the limit in L^2 of the transition functional where we take the polygonal approximation γ^n of γ, we deduce that our stochastic transition functional check almost surely the rules of transition functional which allow to define a bundle.

Definition III.5: The Hilbert space $L^2(\xi_t)$ of L^2 section of the stochastic line bundle ξ_t over $L(M)$ endowed with the measure μ_t is given by the following data: a set of random variable from $L(M)$ from $O(b, v, (.)_b, (.)_v)$ into C called $\psi_{b,v,(.)_b,(.)_v}(\gamma)$ submitted to the following consistency relations checked almost surely:

$$\psi_{b,v,(.)_b,(.)_v} = \rho_{b,v,i_b,i_v,b',v',i_{b'},i_{v'}} \psi_{b',v',(.)_{b'},(.)_{v'}}$$

Since each transition functional is almost surely of modulus 1, we can define the norm of ψ as a random variable. The Hilbert space structure of $L^2(\xi_t)$ is defined by

$$E_{\mu_t}[|\psi|^2] = ||\psi||^2_{\xi_t}$$

4. Stochastic parallel transport along the cylinder

In order to define the amplitude $A(\gamma)$ associated to a path from $S^1 \times [t_1, t_2]$ into M, we should be able to define the integral over a rectangle $b \times l = c = [s_1, s_2] \times [t_1, t_2]$ for a two form. Let us namely consider a two form β and let us consider

$$A^N = \int_{b \times l} < \beta(x_t^N(s)), d_t x_t^N(s), d_s x_t^N(s) >$$

where we consider the polygonal approximation in time s of the process $s \to x_t(s)$ as in Proposition III.2.

We can consider the intrinsic polygonal approximation of $s \to x_t(s)$ $\tilde{x}_t^N(s)$ as in part III. We put:

$$F_N(s, x_y(s_i), x_t(s_{i+1})) = \exp_{x_t(s_i)}[\frac{s - s_i}{s_{i+1} - s_i}(x_t(s_{i+1} - x_t(s_i))] = \tilde{x}_t^N(s)$$

as in (3) if $x_t(s_i)$ and $x_t(s_{i+1})$ are close.

Theorem IV.1: When $N \to \infty$, the ordinary integral in time s and Stratonovitch integral in time t $\int_{b\times l} < \beta(\tilde{x}_t^N(s)), d_t\tilde{x}_t^N(s), d_s\tilde{x}_t^N(s) >$ tends in L^2 to a random variable called $\int_{b\times l} < \beta(x_t(s)), d_t x_t(s), d_s x_t(s) >$.

Let F_{t_1,t_2} be the σ- algebra spanned by $x_t(s), B_t(s)$ for $s \in S_1$ and $t \in [t_1, t_2]$. Theorem IV.2 allows us to define $\int_c \gamma_{i_c}^* \alpha$ where $\gamma(s,t) = x_t(s)$. We deduce that the amplitude $A(\gamma)$ of (2) is almost surely defined and realize a generalized stochastic parallel transport from ξ_{t_1} to ξ_{t_2} which is F_{t_1,t_2} measurable along the path $t \to \{s \to x_t(s)\}$.

Theorem IV.2: $A(\gamma)$ is an element of $L^2(\xi_{t_1}^*) \otimes L^2(F_{t_1,t_2}) \otimes L^2(\xi_{t_2})$.

References

1. H. Airault and P. Malliavin, *Integration on loop groups* (Publication Univ. Paris VI, 1990).
2. S. Albeverio, in: *Espaces de lacets*, eds. R. Léandre, S. Paycha and T. Wuerzbacher, Publication Univ Strasbourg, 5 (1996).
3. S. Albeverio, R. Léandre and M. Roeckner, *C.R.A.S. Série I*, **316**, 287-292 (1993).
4. M. Arnaudon and S. Paycha, *Com. Math. Phys.* **187**, 243 (1997).
5. Y.L. Belopolskaya and Y.E. Daletskii, *Stochastic equations and differential geometry* (Kluwer, 1990).
6. Y.L. Belopolskaya and Y.E. Gliklikh, *Inf. Dim. Anal. Quant. Probab. Rel. Topics.* **5**, 145 (2002).
7. J.L. Brylinski, *Loop spaces, characteristic classes and geometric quantization*, Prog. Math. vol. 107 (Birkhauser, 1993).
8. Z. Brzezniak and R. Léandre, *Potential Analysis*, **12**, 249 (2000).
9. Z. Brzezniak and K.D. Elworthy, *Meth. Funct. Anal. Topol.* **6**, 43 (2000).
10. R. Cairoli and J.B. Walsh, *Acta. Math.* **134**, 111 (1975).
11. A.L. Carey, J. Mickelsson and M. Murray, *Rev. Math. Phys.* **12** 65, (2000).
12. Y.L. Daletskii, in: *Espaces de lacets*, eds. R. Léandre, S. Paycha and T.Wuerzbacher, Publication Univ. Strasbourg, 45 (1996).
13. K.D. Elworthy, *Stochastic differential equations on manifolds*, L.M.S. Lectures Notes Series 20 (Cambridge Univ. Press, 1982).
14. G. Felder, K. Gawedzki and A. Kupiainen, *Com. Math. Phys.* **117**, 127 (1988).
15. K. Gawedzki, in: *Séminaire Bourbaki*, Astérisque, **177-178**, 95 (1989).
16. K. Gawedzki, in: *Quantum fields and strings: a course for mathematicians*, Vol. II, ed. P. Deligne, A.M.S. 727 (2000).
17. K. Gawedzki, in: *Conformal Field Theory: New Non-Perturbative Methods in String and Field theory*, eds. Y. Nutku , C. Saclioglu and A. Turgut (2000).
18. K. Gawedzki and N. Reis, *Rev. Math. Phys.* **14**, 1281 (2002).
19. M. Gradinaru, F. Russo and P. Vallois, "General covariation, local time and Stratonovitch Itô's formula for fractional Brownian motion", Preprint.
20. Y.Z. Huang, *Two dimensional field theory and vertex operator algebra*, Prog.

Math. Vol. 148 (Birkhauser, 1997).
21. Y.Z. Huang and Y. Lepowsky, in: *The Gelfand Mathematical Seminar. 1990-1992*, Prog. Math. Vol. 145 (Birkhauser, 1993).
22. N. Ikeda and S. Watanabe, *Stochastic Processes and Diffusion Processes* (North Holland, 1981).
23. T. Kimura, J. Stasheff and A.A. Voronov, *Com. Math. Phys.* **171**, 1 (1995).
24. H. Konno, in: *Symplectic Geometry and Quantization*, eds. Y. Maeda, H. Omori and A. Weinstein, Cont. Maths. **179**, 136 (1994).
25. H. Kunita, *Stochastic Flows and Stochastic Differential Equations* (Cambridge University Press, 1990).
26. H.H. Kuo, *Trans. Amer. Math. Soc.* **156**, 439 (1972).
27. R. Léandre, *Rev. Math. Phys.* **12**, 91 (2000).
28. R. Léandre, "Brownian surfaces with boundary and Deligne cohomology", in press: *Rep. Math. Phys.* (2003).
29. R. Léandre, *J. Math. Phys.* **44**, 5530 (2003)
30. R. Léandre, "Markov property and operads", Preprint.
31. R. Léandre, *Math. Notes*, **72**, 212 (2002).
32. R. Léandre and S.S. Roan, *J. Geom. Phys.* **16** 1995.
33. P.A. Meyer, in: *Séminaire de Probabilités XV*, Lect. Notes. Math. Vol. 850 (1981) p. 100.
34. D. Nualart, *Malliavin Calculus and Related Topics* (Springer, 1997).
35. M. Zaehle, *Probab. The. Rel. Fields.* **111**, 334 (1998).

ON A SPECTRAL TRIPLE RELATED TO DISCRETE QUANTUM MECHANICS

H.-D. DOEBNER, R. MATTHES

Department of Physics, Metallurgy and Material Science,
Technical University Clausthal,
Leibnizstr. 6, 38678 Clausthal-Zellerfeld, Germany
e-mails: asi@pt.tu-clausthal.de , ptrm@pt.tu-clausthal.de

The difference operator on a compactified one-dimensional lattice $\mathbb{Z}/N\mathbb{Z}$ gives rise to a finite spectral triple that is even if N is even. We describe the first order differential calculus related to the spectral triple. The bimodule of the calculus is two-dimensional, and the corresponding partial derivatives are again difference operators. We also compute the Connes distance of lattice points, which coincides up to a constant factor with the usual (flat) distance.

1. Introduction

Noncommutative geometry has been a major theme in mathematics and theoretical physics in recent years. The term has been coined by A. Connes in the beginning of the eighties of the last century, [4] and [5]. It stands for an attempt to generalize notions and methods of classical differential geometry and topology to a noncommutative (nonclassical) situation, in the spirit of Gelfand and Naimark.

The approach of A. Connes is deeply rooted both in functional analysis (operator algebras) and in differential topology and global analysis. Its main goal is to characterize noncommutative spaces by topological invariants, in the spirit of algebraic topology. This has lead to a new cohomology theory (cyclic cohomology), being related to K-theory and index theory of elliptic operators (on manifolds) in a very fruitful way. In fact, the theory was mainly motivated by situations with a classical geometrical background (foliations), where tools of classical point set topology may give no information.

A central notion in Connes' theory is that of a spectral triple. Such a triple consists of an associative $*$-algebra A, a Hilbert space $\mathcal{H}$, together with a $*$-representation π of A in the bounded operators $B(\mathcal{H})$, and an (unbounded) selfadjoint operator D on $\mathcal{H}$, fulfilling a certain set of axioms. If there is in addition a grading operator γ being in a certain sense compatible with the other data, one speaks of an even spectral triple. Further, in order to have some kind of integration, one needs summability conditions, which are encoded in spectral properties of the operator D. The main motivating example for such a triple comes from classical even-dimensional Riemannian manifolds with spin structure. Here, the Hilbert space is the space of L^2-sections of the spinor bundle, the algebra is the algebra of smooth functions on the manifold, acting on spinors by pointwise multiplication, and D is the usual Dirac operator. It turns out that one can derive the standard calculus of differential forms, the Riemannian distance, and the dimension of the manifold from these data. This suggests to define Riemannian spin manifolds in terms of spectral triples. To this end, one has to introduce more structure, namely a real structure, a first order condition for the operator D, some homological conditions ensuring orientability and Poincaré duality. The statement that all these data are equivalent to the data defining a Riemannian spin manifold is due to Connes [7] and is discussed rather detailed in [13]. Examples of spectral triples have been discussed in great variety, in particular related to the construction of unified field theoretical models, and also related to quantum groups. Some examples related to foliations have been analyzed in [15].

We are particularly interested in differential calculi related to spectral triples, because in good cases they yield partial derivatives, and the latter are natural candidates for momentum operators in quantum mechanics over a noncommutative space. The idea to use partial derivatives (derivations, first order differential operators) as momentum operators is more or less common to many quantization procedures, in particular to Borel quantization. This idea has also been followed in [16] (and later papers by Wess and his collaborators) and [8]. As was shown in [11] and [12], Borel quantization can be used as a guiding principle for quantization on discrete spaces. Our intention is to relate these ideas and to explore the usefulness of the notion of spectral triple for quantization procedures.

In the present note, we will consider a simple example of a spectral triple arising from difference operators on a compactified lattice $\mathbb{Z}/N\mathbb{Z}$, to be thought of as a discretization of the circle S^1. Our main result is the description of the first order differential calculus related to the spectral triple. The bimodule of the calculus is two-dimensional and yields two partial derivatives, corresponding to difference operators in the two possible directions of $\mathbb{Z}/N\mathbb{Z}$.

We also compute the Connes distance of lattice points, which coincides with the (flat) distance on the compactified lattice.

The structure of the paper is as follows: In Section 1, we recall the notion of a spectral triple. Section 2 is devoted to the example of a spectral triple mentioned above. In an appendix we recall some properties of the universal differential calculus on a finite set.

2. Spectral triples

We recall here the definition of a spectral triple and some notions (dimension, metric, differential calculus) related to such a triple [6],[13]:

Definition 1. *A spectral triple $(A, \mathcal{H}, D)$ consists of an associative $*$-algebra A, a Hilbert space $\mathcal{H}$ and a selfadjoint operator D on $\mathcal{H}$, such that*

(i) A acts on $\mathcal{H}$ by a $*$-representation $\pi \longrightarrow B(\mathcal{H})$ ($B(\mathcal{H})$ – algebra of bounded operators on $\mathcal{H}$),
(ii) the commutators $[D, \pi(a)]$, $a \in A$, are bounded and
(iii) the operator D has discrete spectrum with finite multiplicity.

A spectral triple has dimension n, if the eigenvalues μ_k (counted with multiplicity) of $|D|$ fulfil $\lim_{k\to\infty} \frac{\mu_k}{k^{1/n}} = C \neq 0$. It is said to be finite, if the algebra and the Hilbert space are finite dimensional [14]. A spectral triple is said to be even, if there is a selfadjoint operator $\gamma \in B(\mathcal{H})$ such that

(i) $\gamma^2 = 1$,
(ii) $\gamma\pi(a) = \pi(a)\gamma, \quad \forall a \in A$,
(iii) $D\gamma = -\gamma D$.

We will not refer to real structures and other additional data mentioned in the introduction that are usually included in the definition of a spectral triple, and also not to more general notions of dimension.

The representation $\pi : A \longrightarrow B(\mathcal{H})$ can be extended to a representation $\pi^* : \Omega(A) \longrightarrow B(\mathcal{H})$ of the universal differential calculus $\Omega(A)$ by

$$\pi^n(\sum_k a_0^k da_1^k \cdots da_n^k) = \sum_k \pi(a_0^k)[D, \pi(a_k^1)] \cdots [D, \pi(a_n^k)]. \tag{1}$$

If $J_0 := \oplus_n \mathrm{Ker}\, \pi^n$, then $J := J_0 + dJ_0$ is a differential ideal, and one arrives at the differential calculus $\Omega_D(A)$,

$$\Omega_D^n(A) := \Omega^n(A)/J. \tag{2}$$

Note that, if π is faithful, there are isomorphisms

$$\Omega_D^1(A) \simeq \pi^1(\Omega^1(A)) \tag{3}$$

and

$$\Omega_D^2(A) \simeq \pi^2(\Omega^2(A))/\pi^2(dJ_0^1). \tag{4}$$

The following formula defines a metric (a distance function) on the set $S(A)$ of states, i.e. of positive linear functionals, of A:

$$d(\varphi,\psi) = \sup_{a\in A, \|[D,\pi(a)]\|\leq 1} |\varphi(a)-\psi(a)|. \tag{5}$$

In the case of the classical spectral triple $(C^\infty(M), L^2(S), D_{\mathrm{Dirac}})$, the differential calculus defined above is isomorphic with the calculus of differential forms. The set of pure states can be identified with M, and the metric (5) coincides with the Riemannian distance. The operator D of a spectral triple is also called generalized Dirac operator.

3. Spectral triples from difference operators

The idea to use difference operators defined on lattices or more general discrete sets as generalized Dirac operators is not new. In [2] and [1], distances on a one-dimensional lattice $\mathbb{Z}$ were computed from a formula of type (5), using the difference operator

$$D_1\psi(k) = \frac{1}{2i}(\psi(k+1)-\psi(k-1)), \quad k\in\mathbb{Z}. \tag{6}$$

The result of the computation is

$$d(0,2n-1) = 2n, \quad d(0,2n) = 2\sqrt{n(n+1)}, \tag{7}$$

which does not agree with the equidistant metric on the lattice. In [10], on a finite set, with and without periodic boundary conditions, generalized Dirac operators reproducing the equidistant metric were given.

If one considers the operator D_1 of (6) as an operator in $\mathcal{H} = l^2(\mathbb{Z})$, this is not a generalized Dirac operator of a spectral triple in the above sense, because its spectrum is the interval $[-1,1]$, as follows from general results about Toeplitz operators, see the first two pages of [3]. This somehow reflects the noncompactness of $\mathbb{Z}$.

Thus it seems to be reasonable to "compactify" $\mathbb{Z}$ by going to $\mathbb{Z}/N\mathbb{Z}$ (which is simply the finite set $\{1,\ldots,N\}$ and should be considered as a discretized version of S^1) and "imposing periodic boundary conditions". To be precise, let us consider the following data: $\mathcal{H} = \mathbb{C}^N$, $A = \mathbb{C}^N$ (to be considered as the algebra of complex valued functions on $\{1,\ldots,N\}$), D_{1p} the linear operator on $\mathcal{H}$ given in the standard basis by the matrix

$$D_{1p} = \frac{1}{2i}\begin{pmatrix} 0 & 1 & 0 & 0 \ldots & 0 & -1 \\ -1 & 0 & 1 & 0 \ldots & 0 & 0 \\ 0 & -1 & 0 & 1 \ldots & 0 & 0 \\ 0 & 0 & -1 & 0 \ldots & 0 & 0 \\ \vdots & \vdots & \vdots & \vdots \ldots & \vdots & \vdots \\ 0 & 0 & 0 & 0 \ldots & 0 & 1 \\ 1 & 0 & 0 & 0 \ldots & -1 & 0 \end{pmatrix}. \tag{8}$$

The periodic boundary conditions are encoded by the ± 1 in the right upper and left lower corners. The algebra A acts on $\mathcal{H}$ via

$$\pi(f) = \begin{pmatrix} f(1) & 0 & \ldots & 0 \\ 0 & f(2) & \ldots & 0 \\ \vdots & \vdots & \vdots & \vdots \\ 0 & 0 & \ldots & f(N) \end{pmatrix} \tag{9}$$

Obviously $(A, \mathcal{H}, D_{1p})$ is a finite spectral triple. If N is even, there are two natural subspaces $\mathcal{H}^{\pm} \subseteq \mathcal{H}$ that are mapped onto each other by D_{1p}: $\mathcal{H}^{+} = \{\psi \in \mathcal{H} | \psi(k) = 0 \text{ for } k \text{ even}\}$, $\mathcal{H}^{-} = \{\psi \in \mathcal{H} | \psi(k) = 0 \text{ for } k \text{ odd}\}$. The corresponding grading operator γ has in the basis used for the matrices (8) and (9) 1 and -1 alternating on the main diagonal. It is easy to check that this gives an even spectral triple.

Proposition 1. *The eigenvalues of the matrix D_{1p} are $\pm \sin 2\pi \frac{k}{N}$, $k = 1,\ldots,N$.*

Proof. The matrix $2iD_{1p}$ is of the form $U - U^*$, with

$$U = \begin{pmatrix} 0\,1\,0\,0 \ldots 0 \\ 0\,0\,1\,0 \ldots 0 \\ \vdots\,\vdots\,\vdots\,\vdots\,\vdots\,\vdots \\ 0\,0\,0\,0 \ldots 1 \\ 1\,0\,0\,0 \ldots 0 \end{pmatrix}. \tag{10}$$

U is unitary, and its spectrum consists of the N-th roots of unity, as follows easily from solving the eigenvalue equation $U\psi = \lambda\psi$. Then, since U commutes with U^*, the eigenvalues of $2iD_{1p}$ are all possible differences of eigenvalues of U and U^* that are purely imaginary. This gives the above values. □

Note that for large N the values $\pm\sin 2\pi\frac{k}{N}$, $k = 1, \ldots, N$ tend to fill the interval $[-1, 1]$. In this sense, the spectrum of D_{1p} tends to the spectrum of the operator D_1.

For the distance function (5) as well as for the differential calculus, the commutators $[D_{1p}, \pi(f)]$ are important. An easy calculation gives for this commutator

$$\frac{1}{2i}\begin{pmatrix} 0 & f_2-f_1 & 0 & 0 & \ldots & 0 & 0 & f_1-f_N \\ f_2-f_1 & 0 & f_3-f_2 & 0 & \ldots & 0 & 0 & 0 \\ 0 & f_3-f_2 & 0 & f_4-f_3 & \ldots & 0 & 0 & 0 \\ \vdots & \vdots & \vdots & \vdots & \ldots & \vdots & \vdots & \vdots \\ 0 & 0 & 0 & 0 & \ldots & f_{N-1}-f_{N-2} & 0 & f_N-f_{N-1} \\ f_1-f_N & 0 & 0 & 0 & \ldots & 0 & f_N-f_{N-1} & 0 \end{pmatrix} \tag{11}$$

with $f_k := f(k)$.

Proposition 2. *The Connes distance on $\mathbb{Z}/N\mathbb{Z}$ is given by*

$$d(j, j+k) = \begin{cases} 2k & k \le \frac{N}{2} \\ 2(N-k) & \frac{N}{2} < k \le N, \end{cases} \tag{12}$$

for N even, and by

$$d(j, j+k) = \begin{cases} 2k & k \le \frac{N+1}{2} \\ 2(N-k) & \frac{N+1}{2} < k \le N, \end{cases} \tag{13}$$

for N odd.

Proof. It follows easily from (11) that

$$\|[D_{1p}, \pi(f)]\psi\|^2 = \frac{1}{4}(|f_2-f_1|^2|\psi_2|^2 + |f_3-f_2|^2|\psi_3|^2 + \ldots + |f_1-f_N|^2|\psi_1|^2). \tag{14}$$

Therefore,

$$\begin{aligned}\|[D_{1p}, \pi(f)]\|^2 &= \sup_{\|\psi\|=1} \|[D_p, \pi(f)]\psi\|^2 \\ &= \sup_{\|\psi\|=1} \frac{1}{4}(|f_2-f_1|^2|\psi_2|^2 + |f_3-f_2|^2|\psi_3|^2 + \ldots + |f_1-f_N|^2|\psi_1|^2) \\ &\ge \frac{1}{4}\max(|f_2-f_1|^2, |f_3-f_2|^2, \ldots, |f_1-f_N|^2).\end{aligned}$$

Thus, it follows from $|||[D_{1p}, \pi(f)]|||^2 \le 1$ that this maximum is ≤ 4. On the other hand, if the latter is true, then

$$\sup_{\|\psi\|=1} \frac{1}{4}(|f_2-f_1|^2|\psi_2|^2+|f_3-f_2|^2|\psi_3|^2+\ldots+|f_1-f_N|^2|\psi_1|^2) \le \frac{1}{4}4\|\psi\|^2 = 1. \tag{15}$$

Hence, we have $|||[D_{1p}, \pi(f)]|||^2 \le 1$ iff $\max(|f_2 - f_1|^2, |f_3 - f_2|^2, \ldots, |f_1 - f_N|^2) \le 4$. Thus, in order to determine the distance $d(j, j+k)$, our task is to place N points into the plane such that the distance of two consecutive points (in particular of the last and the first) is less than 2, trying to maximize the distance between points number j and $j+k$. It is evident from elementary geometry that this leads to the desired result. □

Thus, the metric d is like a discretized version of a flat metric on S^1. This is a bit in contrast with the results about the Connes metric on $\mathbb{Z}$ computed with D_1 mentioned above. The same result as in Proposition 2 has been obtained in [10] using a different Dirac operator (being a permutation and not a difference operator).

In order to describe the first order differential calculus, we have to determine the first order component of the differential ideal J, which coincides with $\operatorname{Ker}\pi^1$ (because $\operatorname{Ker}\pi = 0$). This kernel consists of all finite sums $\sum g_\alpha dh_\alpha$ such that $\sum \pi(g_\alpha)[D_{1p}, \pi(h_\alpha)] = 0$. By a straightforward computation, using (11), we obtain for one summand $\pi(g)[D_{1p}, \pi(h)]$

$$\frac{1}{2i}\begin{pmatrix} 0 & g_1(h_2-h_1) & 0 & \ldots & 0 & g_1(h_1-h_N) \\ g_2(h_2-h_1) & 0 & g_2(h_3-h_2) & \ldots & 0 & 0 \\ 0 & g_3(h_3-f_2) & 0 & \ldots & 0 & 0 \\ \vdots & \vdots & \vdots & \ldots & \vdots & \vdots \\ 0 & 0 & 0 & \ldots & 0 & g_{N-1}(h_N-h_{N-1}) \\ g_N(h_1-h_N) & 0 & 0 & \ldots & g_N(h_N-h_{N-1}) & 0 \end{pmatrix} \tag{16}$$

with $g_k := g(k)$, $h_k := h(k)$. It follows that the kernel of π^1 can be characterized as the space of finite sums $\sum_\alpha g_\alpha dh_\alpha$ fulfilling

$$\sum_\alpha g_\alpha(j)(h_\alpha(j+1)-h_\alpha(j)) = 0, \quad \sum_\alpha g_\alpha(j+1)(h_\alpha(j+1)-h_\alpha(j)) = 0, \tag{17}$$

for $j = 1, \ldots, N$ (with the identification $N+1 = 1$). Comparing with the $\mathbb{C}^N$-bimodule structure of $\Omega^1(\mathbb{C}^N)$ given by (A.2) and the formula for the universal differential (A.5), we can state

Proposition 3. *The kernel of π^1 coincides with the space of finite sums*

$\sum_\alpha g_\alpha dh_\alpha$ *satisfying*

$$(\sum_\alpha g_\alpha dh_\alpha)(j,j+1)=0, \quad (\sum_\alpha dh_\alpha g_\alpha)(j,j+1)=0 \tag{18}$$

for $j=1,\ldots,N$ *(identifying* $N+1$ *with* 1*).*

This can be rewritten using the basis $e_i de_j$ of $\Omega^1(\mathbb{C}^N)$. One finds that

$$\operatorname{Ker}\pi^1=\{\sum_{i\neq j} a_{ij}e_i de_j \mid a_{k,k+1}=0=a_{k,k-1},\ k=1,\ldots,N\}. \tag{19}$$

We can conclude the following

Proposition 4. *The bimodule* $\Omega^1_D(\mathbb{C}^N)=\Omega^1(\mathbb{C}^N)/Ker\pi^1$ *of the first order differential calculus belonging to the spectral triple can be identified with* $\{\sum_{|i-j|=1} a_{ij}e_i de_j \mid a_{ij}\in\mathbb{C}\}$*. The bimodule structure is given by*

$$f\sum_{|i-j|=1} a_{ij}e_i de_j=\sum_{|i-j|=1} f(i)a_{ij}e_i de_j, \tag{20}$$

$$(\sum_{|i-j|=1} a_{ij}e_i de_j)f=\sum_{|i-j|=1} a_{ij}f(j)e_i de_j, \tag{21}$$

and the differential is given by

$$d^1 f=\sum_{|i-j|=1}(f(j)-f(i))e_i de_j. \tag{22}$$

Note that the bimodule is still noncommutative. The differential d^1 contains only differences of values at nearest neighbours and thus remembers the "one-dimensionality" of $\mathbb{Z}/N\mathbb{Z}$ (in contrast to the universal differential). With other words, the difference operator reappears in the differential calculus of the spectral triple. Clearly, this fact is a consequence of the structure of $[D_{1p},\pi(f)]$, and Proposition 4 could also be derived by matrix calculations, using equation (3).

Now, one proves by straightforward computations

Proposition 5. *The bimodule* $\Omega^1_D(\mathbb{C}^N)$ *is two-dimensional. A basis is given by the two elements*

$$\alpha_+=\sum_{i=1}^N e_i de_{i+1}, \quad \alpha_-=\sum_{i=1}^N e_{i+1}de_i. \tag{23}$$

The bimodule structure is given by

$$f\alpha_+=\alpha_+ f^-, \quad f\alpha_-=\alpha_- f^+, \tag{24}$$

with $f^{\pm}(i) = f(i \pm 1)$. *(As above, here* $N+1$ *is always identified with 1, and 0 with* N.*)*

Note that there do not exist elements f of the algebra such that $\alpha_+ = d^1 f$ or $\alpha_- = d^1 f$.

We can now introduce partial derivatives by

$$d^1 f = \partial_+ f \; \alpha_+ + \partial_- f \; \alpha_- \tag{25}$$

and obtain

$$\partial_\pm f = f^\pm - f, \quad \text{i.e.} \quad \partial_\pm f(i) = f(i \pm 1) - f(i). \tag{26}$$

Thus the differential calculus yields two difference operators corresponding to the two possible directions.

Acknowledgments

We like to thank G. Rudolph, M. Schmidt and V.K. Dobrev for helpful discussions. This work was partly supported by the Clausthal-Leipzig-Sofia collaboration of the Alexander-von-Humboldt-Stiftung.

Appendix A. Some properties of the universal differential calculus on finite sets

Let A be a unital algebra. The universal differential calculus $\Omega(A)$ can be defined as follows: Let $A^{\otimes^k} := A \otimes \cdots \otimes A$ (k factors), and let $m_{j,j+1} : A^{\otimes^{k+1}} \to A^{\otimes^k}$ be the multiplication of the j-th and $j+1$-th factor. Then

$$\Omega^k(A) := \bigcap_{j=1}^{k} \operatorname{Ker} m_{j,j+1}. \tag{A.1}$$

Product and differential in $\Omega(A) := \oplus_k \Omega^k(A)$ are given by

$$(a_0 \otimes \cdots \otimes a_k)(b_0 \otimes \cdots \otimes b_l) = a_0 \otimes \cdots \otimes a_k b_0 \otimes \cdots \otimes b_l,$$

$$\begin{aligned} d(a_0 \otimes \cdots \otimes a_k) = {} & 1 \otimes a_0 \otimes \cdots \otimes a_k \\ & + \sum_{j=1}^{k} (-1)^j a_0 \otimes \cdots \otimes a_{j-1} \otimes 1 \otimes a_j \otimes \cdots \otimes a_k \\ & + (-1)^{k+1} a_0 \otimes \cdots \otimes a_k \otimes 1. \end{aligned}$$

Let now $F(X)$ be the algebra of complex valued functions on a finite set X. Then $\Omega^k(F(X))$ is the space of the functions on X^{k+1} that vanish if two neighbouring arguments coincide. Product and differential are given by

$$(fg)(x_0, \ldots, x_{k+l}) = f(x_0, \ldots, x_k)g(x_k, \ldots, x_{k+l}), \tag{A.2}$$

$$df(x_0, \ldots, x_{k+1}) = f(x_1, \ldots, x_{k+1}) + \sum_{j=1}^{k} f(x_0, \ldots, x_{j-1}, x_{j+1}, \ldots, x_{k+1}) + (-1)^{k+1} f(x_0, \ldots, x_k). \tag{A.3}$$

For $X = \{1, \ldots, N\}$ we have $F(X) = \mathbb{C}^N$. The elements e_j of the canonical basis of $\mathbb{C}^N$ are generators of the algebra, fulfilling

$$e_i e_j = \delta_{ij} e_i, \qquad \sum_{i=1}^{N} e_i = 1. \tag{A.4}$$

Then the elements $e_i de_j$, $i \neq j$ are a basis of the vector space (over $\mathbb{C}$) $\Omega^1(\mathbb{C}^N)$. The universal differential of $f \in \mathbb{C}^N$ is, according to (A.3), given by

$$df(j, k) = f(k) - f(l) \tag{A.5}$$

and can be written in terms of the basis $e_i de_j$ as follows:

$$df = \sum_{i \neq j} (f(j) - f(i)) e_i de_j. \tag{A.6}$$

We refer to [9] for more information about differential calculus on finite sets.

References

1. E. Atzmon, "Distances on a one-dimensional lattice from noncommutative geometry", *Lett. Math. Phys.* **37**, 341-348 (1996).
2. G. Bimonte, F. Lizzi and G. Sparano, "Distances on a lattice from noncommutative geometry", *Phys. Lett.* **B341**, 139-146 (1994).
3. A. Böttcher, "Toeplitz operators with piecewise continuous symbols - a neverending story?", *Jber. d. Dt. Math.-Verein.* **97**, 115-129 (1995).
4. A. Connes, "C*-algebres et geometrie differentielle", *C. R. Acad. Sci. Paris serie A*, **290**, 599-604 (1980).
5. A. Connes, "Noncommutative differential geometry", *Publ. Math. IHES*, **62**, 41-144 (1986).
6. A. Connes, *Noncommutative Geometry*, (Academic Press, Inc., San Diego, New York etc., 1994).

7. A. Connes, "Gravity coupled with matter and the foundation of non commutative geometry", *Commun. Math. Phys.* **182**, 155-176 (1996).
8. A. Dimakis and F. Müller-Hoissen, "Quantum mechanics on a lattice and q-deformations", *Phys. Lett.* **B295**, 242-248 (1992).
9. A. Dimakis and F. Müller-Hoissen, "Discrete differential calculus: Graphs, topologies and gauge theory", *J. Math. Phys.* **35**, 6703-6735 (1994).
10. A. Dimakis and F. Müller-Hoissen, "Connes' distance function on one-dimensional lattices", *Int. J. Theor. Phys.* **37**, 907-913 (1998).
11. V.K. Dobrev, H.-D. Doebner and R. Twarock, "Quantization of kinematics and dynamics on S^1 with difference operators and a related q-deformation of the Witt algebra", *J. Phys. A: Math. Gen.* **30**, 6841-6859 (1997).
12. V.K. Dobrev, H.-D. Doebner and R. Twarock, Quantum mechanics with difference operators. *Rep. Math. Phys.* **50**, 409-431 (2002).
13. J.M. Gracia-Bondía, J.C. Várilly and H. Figueroa, *Elements of noncommutative geometry*, (Birkhäuser, Basel, 2000).
14. T. Krajewski, "Classification of finite spectral triples", *J. Geom. Phys.* **28**, 1-30 (1998).
15. R. Matthes, O. Richter and G. Rudolph, "Spectral triples and differential calculi related to the Kronecker foliation", *J. Geom. Phys.* **46**, 48-73 (2003).
16. J. Schwenk and J. Wess, "A q-deformed quantum mechanical toy model", *Phys. Lett.* **B291**, 273-277 (1992).

PATH INTEGRAL APPROACH TO NONCOMMUTATIVE QUANTUM MECHANICS

BRANKO DRAGOVICH

Institute of Physics, P.O.Box 57, 11001 Belgrade
Serbia and Montenegro
E-mail: dragovich@phy.bg.ac.yu

ZORAN RAKIĆ

Faculty of Mathematics, University of Belgrade
Studentski trg 16, P.O.Box 550, 11001 Belgrade
Serbia and Montenegro
E-mail: zrakic@matf.bg.ac.yu

We consider Feynman's path integral approach to quantum mechanics with a noncommutativity in position and momentum sectors of the phase space. We show that a quantum-mechanical system with this kind of noncommutativity is equivalent to the another one with usual commutative coordinates and momenta. We found connection between quadratic classical Hamiltonians, as well as Lagrangians, in their commutative and noncommutative regimes. The general procedure to compute Feynman's path integral on this noncommutative phase space with quadratic Lagrangians (Hamiltonians) is presented. Using this approach, a particle in a constant field, ordinary and inverted harmonic oscillators are elaborated in detail.

1. Introduction

Quantum theories with noncommuting position and momentum coordinates have been investigated intensively during the recent years. Most of the research has been devoted to noncommutative (NC) field theory (for a review, see e.g. [1]). Noncommutative quantum mechanics (NCQM) has been also investigated, because it can be regarded as the corresponding one-particle nonrelativistic sector of NC quantum field theory and with relevance to concrete quantum systems.

In a very general NCQM one has that not only $[\hat{x_a}, \hat{p_b}] \neq 0$, but also $[\hat{x_a}, \hat{x_b}] \neq 0$ and $[\hat{p_a}, \hat{p_b}] \neq 0$. We consider here D-dimensional NCQM which is based on the following algebra:

$$[\hat{x_a}, \hat{p_b}] = i\,\hbar\,(\delta_{ab} - \frac{1}{4}\,\theta_{ac}\,\sigma_{cb}), \quad [\hat{x_a}, \hat{x_b}] = i\,\hbar\,\theta_{ab}, \quad [\hat{p_a}, \hat{p_b}] = i\,\hbar\,\sigma_{ab}, \quad (1)$$

where $\Theta = (\theta_{ab})$ and $\Sigma = (\sigma_{ab})$ are the antisymmetric matrices with constant elements.

We investigate the Feynman path integral [2] approach to NCQM,

$$\mathcal{K}(x'', t''; x', t') = \int_{x'}^{x''} \exp\left(\frac{i}{\hbar}\int_{t'}^{t''} L(\dot{q}, q, t)\, dt\right) \mathcal{D}q\,, \qquad (2)$$

where $\mathcal{K}(x'', t''; x', t')$ is the kernel of the unitary evolution operator $U(t)$ and $x'' = q(t'')$, $x' = q(t')$ are end points. In ordinary quantum mechanics (OQM) Feynman path integral for quadratic Lagrangians can be evaluated analytically and the exact form [3] is

$$\mathcal{K}(x'', t''; x', t') = \frac{1}{(ih)^{\frac{D}{2}}}\sqrt{\det\left(-\frac{\partial^2 \bar{S}}{\partial x''_a \partial x'_b}\right)} \exp\left(\frac{2\pi i}{h}\,\bar{S}(x'', t''; x', t')\right), \qquad (3)$$

where $\bar{S}(x'', t''; x', t')$ is the action for the classical trajectory.

In this article we search the form of an effective quadratic Lagrangian which corresponds to a system with noncommutative phase space coordinates (1). This is necessary to know before to employ Feynman's path integral method in NCQM. To this end, let us note that algebra (1) of operators $\hat{x_a}$, $\hat{p_b}$ can be replaced by the equivalent one

$$[\hat{q_a}, \hat{k_b}] = i\,\hbar\,\delta_{ab}, \quad [\hat{q_a}, \hat{q_b}] = 0, \quad [\hat{k_a}, \hat{k_b}] = 0\,, \qquad (4)$$

where linear transformations

$$\hat{x_a} = \hat{q_a} - \frac{\theta_{ab}\,\hat{k_b}}{2}, \quad \hat{p_a} = \hat{k_a} + \frac{\sigma_{ab}\,\hat{q_b}}{2} \qquad (5)$$

are used and summation over repeated indices is assumed. According to (1), (4) and (5), NCQM related to the quantum phase space ($\hat{p}$, $\hat{x}$) can be regarded as an OQM on the standard phase space ($\hat{k}$, $\hat{q}$). Similar aspects of noncommutativity with $\sigma_{ab} = 0$ are considered in [4] and [5] .

2. Dynamics on noncommutative phase space and path integral

Let the most general quadratic Lagrangian for a D-dimensional system be

$$L(\dot{x}, x, t) = \frac{1}{2}\,\left(\dot{x}^T\,\alpha\,\dot{x} + \dot{x}^T\,\beta\,x + x^T\,\beta^T\,\dot{x} + x^T\,\gamma\,x\right) + \delta^T\,\dot{x} + \eta^T\,x + \phi, \quad (6)$$

where coefficients of the $D \times D$ matrices $\alpha = ((1+\delta_{ab})\,\alpha_{ab}(t))$, $\beta = (\beta_{ab}(t))$, $\gamma = ((1+\delta_{ab})\,\gamma_{ab}(t))$, D-dimensional vectors $\delta = (\delta_a(t))$, $\eta = (\eta_a(t))$ and a scalar $\phi = \phi(t)$ are some analytic functions of the time t. Matrices α and γ are symmetric, α is nonsingular ($\det \alpha \neq 0$) and index ^{T} denotes transpose map.

The Lagrangian (6) can be presented in the more compact form:

$$L(X,t) = \frac{1}{2}\, X^T\, M\, X + N^T\, X + \phi, \tag{7}$$

where $2D \times 2D$ matrix M and $2D$-dimensional vectors X, N are defined as

$$M = \begin{pmatrix} \alpha & \beta \\ \beta^T & \gamma \end{pmatrix}, \quad X^T = (\dot{x}^T\,,\, x^T)\,, \quad N^T = (\delta^T\,,\, \eta^T). \tag{8}$$

Using the equations $p_a = \frac{\partial L}{\partial \dot{x}_a}$, one can express $\dot{x}$ as $\dot{x} = \alpha^{-1}\,(p - \beta\, x - \delta)$. Since $\dot{x}$ is linear in p and x, the corresponding classical Hamiltonian $H(p,x,t) = p^T\,\dot{x} - L(\dot{x},x,t)$ becomes also quadratic, i.e.

$$H(p,x,t) = \frac{1}{2}\,\left(p^T\, A\, p + p^T\, B\, x + x^T\, B^T\, p + x^T\, C\, x\right) + D^T\, p + E^T\, x + F, \tag{9}$$

where:

$$\begin{aligned} A &= \alpha^{-1}, & B &= -\,\alpha^{-1}\,\beta, & C &= \beta^T\,\alpha^{-1}\,\beta - \gamma, \\ D &= -\,\alpha^{-1}\,\delta, & E &= \beta^T\,\alpha^{-1}\,\delta - \eta, & F &= \frac{1}{2}\,\delta^T\,\alpha^{-1}\delta\; - \phi\,. \end{aligned} \tag{10}$$

Due to the symmetry of matrices α and γ one has that matrices $A = ((1+\delta_{ab})\,A_{ab}(t))$ and $C = ((1+\delta_{ab})\,C_{ab}(t))$ are also symmetric ($A^T = A$, $C^T = C$). The nonsingular ($\det \alpha \neq 0$) Lagrangian $L(\dot{x},x,t)$ implies nonsingular ($\det A \neq 0$) Hamiltonian $H(p,x,t)$. Note that the inverse map, i.e. $H \to L$, is given by the same relations (10).

The Hamiltonian (9) can be also presented in the compact form

$$H(\Pi,t) = \frac{1}{2}\,\Pi^T\,\mathcal{M}\,\Pi + \mathcal{N}^T\,\Pi + F, \tag{11}$$

where matrix $\mathcal{M}$ and vectors Π, $\mathcal{N}$ are

$$\mathcal{M} = \begin{pmatrix} A & B \\ B^T & C \end{pmatrix}, \quad \Pi^T = (p^T\,,\, x^T)\,, \quad \mathcal{N}^T = (D^T\,,\, E^T). \tag{12}$$

One can easily show that

$$\mathcal{M} = \sum_{i=1}^{3} \Upsilon_i^T(M)\, M\, \Upsilon_i(M), \tag{13}$$

where

$$\Upsilon_1(M) = \begin{pmatrix} \alpha^{-1} & 0 \\ 0 & -I \end{pmatrix}, \ \Upsilon_2(M) = \begin{pmatrix} 0 & \alpha^{-1}\beta \\ 0 & 0 \end{pmatrix}, \ \Upsilon_3(M) = \begin{pmatrix} 0 & 0 \\ 0 & i\sqrt{2}\, I \end{pmatrix}.$$

One has also $\mathcal{N} = Y(M)\, N$, where

$$Y(M) = \begin{pmatrix} -\alpha^{-1} & 0 \\ \beta^T \alpha^{-1} & -I \end{pmatrix} = -\Upsilon_1(M) + \Upsilon_2^T(M) + i\sqrt{2}\Upsilon_3(M) \tag{14}$$

and $F = N^T\, Z(M)\, N - \phi$, where

$$Z(M) = \begin{pmatrix} \frac{1}{2}\,\alpha^{-1} & 0 \\ 0 & 0 \end{pmatrix} = \frac{1}{2}\,\Upsilon_1(M) - \frac{i}{2\sqrt{2}}\,\Upsilon_3(M)\,. \tag{15}$$

Eqs. (5) can be rewritten in the compact form as

$$\hat{\Pi} = \Xi\, \hat{K}, \quad \Xi = \begin{pmatrix} I & \frac{1}{2}\,\Sigma \\ -\frac{1}{2}\,\Theta & I \end{pmatrix}, \quad \hat{K} = \begin{pmatrix} \hat{k} \\ \hat{q} \end{pmatrix}. \tag{16}$$

Since Hamiltonians depend on canonical variables, the transformation (16) lead to the transformation of Hamiltonians (9) and (11). To this end, let us quantize the Hamiltonian (9) and it easily becomes $H(\hat{p},\hat{x},t) = \frac{1}{2}\,\left(\hat{p}^T A\, \hat{p} + \hat{p}^T B\, \hat{x} + \hat{x}^T B^T \hat{p} + \hat{x}^T C\, \hat{x}\right) + D^T \hat{p} + E^T \hat{x} + F$ because (9) is already written in the Weyl symmetric form.

Performing linear transformations (5) in the above Hamiltonian we again obtain quadratic quantum Hamiltonian

$$\begin{aligned} H_{\theta\sigma}(\hat{k},\hat{q},t) = \frac{1}{2}\, & \left(\hat{k}^T A_{\theta\sigma}\, \hat{k} + \hat{k}^T B_{\theta\sigma}\, \hat{q} + \hat{q}^T B_{\theta\sigma}^T\, \hat{k} + \hat{q}^T C_{\theta\sigma}\, \hat{q}\right) \\ & + D_{\theta\sigma}^T\, \hat{k} + E_{\theta\sigma}^T\, \hat{q} + F_{\theta\sigma}\,, \end{aligned} \tag{17}$$

where

$$\begin{aligned} A_{\theta\sigma} &= A - \frac{1}{2}\, B\,\Theta + \frac{1}{2}\,\Theta\, B^T - \frac{1}{4}\,\Theta\, C\,\Theta, & D_{\theta\sigma} &= D + \frac{1}{2}\,\Theta\, E, \\ B_{\theta\sigma} &= B + \frac{1}{2}\,\Theta\, C + \frac{1}{2}\, A\,\Sigma + \frac{1}{4}\,\Theta\, B^T\,\Sigma, & E_{\theta\sigma} &= E - \frac{1}{2}\,\Sigma\, D, \\ C_{\theta\sigma} &= C - \frac{1}{2}\,\Sigma\, B + \frac{1}{2}\, B^T\,\Sigma - \frac{1}{4}\,\Sigma\, A\,\Sigma, & F_{\theta\sigma} &= F. \end{aligned} \tag{18}$$

Note that for the nonsingular Hamiltonian $H(\hat{p},\hat{x},t)$ and for sufficiently small θ_{ab} the Hamiltonian $H_{\theta\sigma}(\hat{k},\hat{q},t)$ is also nonsingular. Classical analogue of (17) maintains the same form $H_{\theta\sigma}(k,q,t) = \frac{1}{2}\,\left(k^T A_{\theta\sigma}\, k + k^T B_{\theta\sigma}\, q + q^T B_{\theta\sigma}^T\, k + q^T C_{\theta\sigma}\, q\right) + D_{\theta\sigma}^T\, k + E_{\theta\sigma}^T\, q + F_{\theta\sigma}\,.$

In the more compact form, Hamiltonian (17) is

$$\hat{H}_{\theta\sigma}(\hat{K},t) = \frac{1}{2}\,\hat{K}^T\,\mathcal{M}_{\theta\sigma}\,\hat{K} + \mathcal{N}_{\theta\sigma}^T\,\hat{K} + F_{\theta\sigma}, \tag{19}$$

where $2D \times 2D$ matrix $\mathcal{M}_{\theta\sigma}$ and $2D$-dimensional vectors $\hat{K}$, $\mathcal{N}_{\theta\sigma}$ are

$$\mathcal{M}_{\theta\sigma} = \begin{pmatrix} A_{\theta\sigma} & B_{\theta\sigma} \\ B_{\theta\sigma}^T & C_{\theta\sigma} \end{pmatrix}, \quad \hat{K}^T = (\hat{k}^T, \hat{q}^T), \quad \mathcal{N}_{\theta\sigma}^T = (D_{\theta\sigma}^T, E_{\theta\sigma}^T). \quad (20)$$

From (11), (16) and (19) one can find connections between $\mathcal{M}_{\theta\sigma}$, $\mathcal{N}_{\theta\sigma}$, $F_{\theta\sigma}$ and $\mathcal{M}$, $\mathcal{N}$, F, which are given by the following relations:

$$\mathcal{M}_{\theta\sigma} = \Xi^T \mathcal{M}\, \Xi, \qquad \mathcal{N}_{\theta\sigma} = \Xi^T \mathcal{N}, \qquad F_{\theta\sigma} = F. \quad (21)$$

To compute a path integral one can start from its Hamiltonian formulation on the phase space. However, such path integral on a phase space can be reduced to the Lagrangian path integral on configuration space whenever Hamiltonian is a quadratic polynomial with respect to momentum p (see, e.g. [5]). Hence, we need the corresponding classical Lagrangians related to the Hamiltonians (17) and (19). Using equations $\dot{q}_a = \frac{\partial H_{\theta\sigma}}{\partial p_a}$ which give $k = A_{\theta\sigma}^{-1}(\dot{q} - B_{\theta\sigma}\, q - D_{\theta\sigma})$ we can pass from Hamiltonian (17) to the corresponding Lagrangian by relation $L_{\theta\sigma}(\dot{q}, q, t) = k^T \dot{q} - H_{\theta\sigma}(k, q, t)$. Note that coordinates q_a and x_a coincide when $\theta = \sigma = 0$. Performing necessary computations we obtain

$$\begin{aligned} L_{\theta\sigma}(\dot{q}, q, t) = \frac{1}{2}\, &\left(\dot{q}^T \alpha_{\theta\sigma}\, \dot{q} + \dot{q}^T \beta_{\theta\sigma}\, q + q^T \beta_{\theta\sigma}^T\, \dot{q} + q^T \gamma_{\theta\sigma}\, q\right) \\ &+ \delta_{\theta\sigma}^T\, \dot{q} + \eta_{\theta\sigma}^T\, q + \phi_{\theta\sigma}, \end{aligned} \quad (22)$$

or in the compact form:

$$L_{\theta\sigma}(Q, t) = \frac{1}{2}\, Q^T M_{\theta\sigma}\, Q + N_{\theta\sigma}^T\, Q + \phi_{\theta\sigma}, \quad (23)$$

where

$$M_{\theta\sigma} = \begin{pmatrix} \alpha_{\theta\sigma} & \beta_{\theta\sigma} \\ \beta_{\theta\sigma}^T & \gamma_{\theta\sigma} \end{pmatrix}, \quad Q^T = (\dot{q}^T, q^T), \quad N^T = (\delta_{\theta\sigma}^T, \eta_{\theta\sigma}^T). \quad (24)$$

Then the connection between $M_{\theta\sigma}, N_{\theta\sigma}, \phi_{\theta\sigma}$ and M, N, ϕ are given by the following relations:

$$M_{\theta\sigma} = \sum_{i,j=1}^{3} \Xi_{ij}^T\, M\, \Xi_{ij}, \qquad \Xi_{ij} = \Upsilon_i(M)\, \Xi\, \Upsilon_j(\mathcal{M}_{\theta\sigma}), \quad (25)$$

and

$$N_{\theta\sigma} = Y(\mathcal{M}_{\theta\sigma})\, \Xi^T\, Y(M)\, N, \quad \phi_{\theta\sigma} = \mathcal{N}_{\theta\sigma}^T\, Z(\mathcal{M}_{\theta\sigma})\, \mathcal{N}_{\theta\sigma} - F. \quad (26)$$

In more detail, the connection between coefficients of the Lagrangians $L_{\theta\sigma}$ and L is given by the relations:

$$\alpha_{\theta\sigma} = \left[\alpha^{-1} - \frac{1}{2}\,(\Theta\,\beta^T\,\alpha^{-1} - \alpha^{-1}\,\beta\,\Theta) - \frac{1}{4}\,\Theta\,(\beta^T\,\alpha^{-1}\,\beta - \gamma)\,\Theta\right]^{-1},$$

$$\beta_{\theta\sigma} = \alpha_{\theta\sigma}\left(\alpha^{-1}\,\beta - \frac{1}{2}\,(\alpha^{-1}\,\Sigma - \Theta\,\gamma + \Theta\,\beta^T\,\alpha^{-1}\,\beta) + \frac{1}{4}\,\Theta\,\beta^T\,\alpha^{-1}\,\Sigma\right),$$

$$\gamma_{\theta\sigma} = \gamma + \beta^T_{\theta\sigma}\,\alpha^{-1}_{\theta\sigma}\,\beta_{\theta\sigma} - \beta^T\,\alpha^{-1}\,\beta + \frac{1}{4}\,\Sigma\,\alpha^{-1}\,\Sigma$$

$$-\frac{1}{2}\,(\Sigma\,\alpha^{-1}\,\beta - \beta^T\,\alpha^{-1}\,\Sigma),$$

$$\delta_{\theta\sigma} = \alpha_{\theta\sigma}\left(\alpha^{-1}\,\delta + \frac{1}{2}\,(\Theta\,\eta - \Theta\,\beta^T\,\alpha^{-1}\,\delta)\right),$$

$$\eta_{\theta\sigma} = \eta + \beta^T_{\theta\sigma}\,\alpha^{-1}_{\theta\sigma}\,\delta_{\theta\sigma} - \beta^T\,\alpha^{-1}\,\delta - \frac{1}{2}\,\Sigma\,\alpha^{-1}\,\delta,$$

$$\phi_{\theta\sigma} = \phi + \frac{1}{2}\,\delta^T_{\theta\sigma}\,\alpha^{-1}_{\theta\sigma}\,\delta_{\theta\sigma} - \frac{1}{2}\,\delta^T\,\alpha^{-1}\,\delta. \tag{27}$$

Note that $\alpha_{\theta\sigma}$, $\delta_{\theta\sigma}$ and $\phi_{\theta\sigma}$ do not depend on σ.

If we know Lagrangian (6) and algebra (1) we can obtain the corresponding effective Lagrangian (22) suitable for the path integral in NCQM. Exploiting the Euler-Lagrange equations $\frac{\partial L_{\theta\sigma}}{\partial q_a} - \frac{d}{dt}\frac{\partial L_{\theta\sigma}}{\partial \dot{q}_a} = 0$ one can obtain classical trajectory $q_a = q_a(t)$ connecting end points $x' = q(t')$ and $x'' = q(t'')$, and the corresponding action $\bar{S}_{\theta\sigma}(x'', t''; x', t') = \int_{t'}^{t''} L_{\theta\sigma}(\dot{q}, q, t)\,dt$. Path integral in NCQM is a direct analogue of (3) and its exact expression in the form of quadratic actions $\bar{S}_{\theta\sigma}(x'', t''; x', t')$ is

$$\mathcal{K}_{\theta\sigma}(x'', t''; x', t') = \frac{1}{(ih)^{\frac{D}{2}}}\sqrt{\det\left(-\frac{\partial^2 \bar{S}_{\theta\sigma}}{\partial x''_a\,\partial x'_b}\right)}$$

$$\times \exp\left(\frac{2\,\pi\,i}{h}\,\bar{S}_{\theta\sigma}(x'', t''; x', t')\right). \tag{28}$$

2.1. *A particle in a constant field and on noncommutative phase space*

The starting Lagrangian is

$$L(\dot{x}, x) = \frac{m}{2}\,(\dot{x}_1^2 + \dot{x}_2^2) - \eta_1\,x_1 - \eta_2\,x_2. \tag{29}$$

Using (27), one can easily find the Lagrangian $L_{\theta\sigma}(\dot{q}, q, t)$:

$$L_{\theta\sigma} = \frac{m}{2}\,(\dot{q}_1^2 + \dot{q}_2^2) + \frac{\sigma}{2}\,(q_1\,\dot{q}_2 - q_2\,\dot{q}_1) + \frac{m\,\theta}{2}\,(\eta_1\,\dot{q}_2 + \eta_2\,\dot{q}_1)$$

$$-\left(1 - \frac{\theta\,\sigma}{4}\right)(\eta_1\,q_1 + \eta_2\,q_2) + \frac{m\,\theta^2}{8}\,({\eta_1}^2 + {\eta_2}^2). \tag{30}$$

The corresponding equations of motion are $\ddot{q}_1 - \xi\,\dot{q}_2 + \eta_1\,\phi = 0$, $\ddot{q}_2 + \xi\,\dot{q}_1 + \eta_2\,\phi = 0$, where $\xi = \frac{\sigma}{m}$, $\phi = \frac{1}{m}\left(1 - \frac{\theta\,\sigma}{4}\right)$. One can transform the above system of equations to $q_1^{(3)} + \xi^2\,\dot{q}_1 + \eta_2\,\xi\,\phi = 0$, $q_2^{(3)} + \xi^2\,\dot{q}_2 - \eta_1\,\xi\,\phi = 0$, Their solutions are: $q_1(t) = C_1 + C_2\,\cos[\,\xi\,t\,] + C_3\,\sin[\,\xi\,t\,] - \frac{\phi\,\eta_2}{\xi}\,t$, $q_2(t) = D_1 + D_2\,\cos[\,\xi\,t\,] + D_3\,\sin[\,\xi\,t\,] + \frac{\phi\,\eta_1}{\xi}\,t$, where C_1, C_2, C_3, D_1, D_2 and D_3 are constants. Since the functions q_1 and q_2 have to satisfy coupled differential equations one has $D_2 = C_3$, $D_3 = -\,C_2$. Employing the conditions $q_1(0) = x_1'$, $q_1(T) = x_1''$, $q_2(0) = x_2'$, $q_2(T) = x_2''$ one obtains:

$$C_1 = \frac{(x_1'+x_1'')\,\xi + \phi\,\eta_2\,T}{2\,\xi} - \frac{\left((x_2'-x_2'')\,\xi + \phi\,\eta_1\,T\right)}{2\,\xi}\,\cot\left[\frac{\xi\,T}{2}\right],$$

$$C_3 = \frac{(x_2'-x_2'')\,\xi + \phi\,\eta_1\,T}{2\,\xi} - \frac{\left((x_1'-x_1'')\,\xi - \phi\,\eta_2\,T\right)}{2\,\xi}\,\cot\left[\frac{\xi\,T}{2}\right],$$

$$D_1 = \frac{(x_2'+x_2'')\,\xi - \phi\,\eta_1\,T}{2\,\xi} - \frac{\left((x_1''-x_1')\,\xi + \phi\,\eta_2\,T\right)}{2\,\xi}\,\cot\left[\frac{\xi\,T}{2}\right],$$

$$D_3 = \frac{(x_1''-x_1')\,\xi + \phi\,\eta_2\,T}{2\,\xi} - \frac{\left((x_2'-x_2'')\,\xi + \phi\,\eta_1\,T\right)}{2\,\xi}\,\cot\left[\frac{\xi\,T}{2}\right].$$

The Lagrangian for classical trajectory is $L_{\theta\sigma}(\dot{q}, q) = -\frac{m\,\xi\,\phi\,t}{2}\Big((C_3\,\eta_1 + D_3\,\eta_2)\,\cos[\,\xi\,t\,] + (D_3\,\eta_1 - C_3\,\eta_2)\,\sin[\,\xi\,t\,]\Big) - \frac{m}{2}\Big((C_3\,D_1 - C_1\,D_3)\,\xi^2 + (\theta\,\xi + 3\,\phi)(C_3\,\eta_2 - D_3\,\eta_1)\,\cos[\,\xi\,t\,] + (C_1\,C_3 + D_1\,D_3)\,\xi^2 + (\theta\,\xi + 3\,\phi)(C_3\,\eta_1 + D_3\,\eta_2)\,\sin[\,\xi\,t\,]\Big) - \frac{m\,\phi}{2}\,(C_1\,\eta_1 + D_1\,\eta_2) + \frac{m\,(\theta\,\xi + 2\,\phi)^2\,(\eta_1^2 + \eta_2^2)}{8\,\xi^2}$. Using this Lagrangian we compute the classical action

$$\begin{aligned}\bar{S}_{\theta\sigma}(x'', T; x', 0) &= \int_0^T L_{\theta\sigma}(\dot{q}, q)\,dt = \frac{m\,\xi}{2}\left(x_2''\,x_1' - x_1''\,x_2'\right) + \frac{m\,\xi\cot\left[\frac{\xi T}{2}\right]}{4} \\ &\times\Big((x_1'' - x_1')^2 + (x_2'' - x_2')^2\Big) + \frac{2\,\lambda + m\,\theta\,\xi - \lambda\,\xi\,T\cot\left[\frac{\xi T}{2}\right]}{2\,\xi} \\ &\times\Big((x_2'' - x_2')\,\eta_1 - (x_1'' - x_1')\,\eta_2\Big) - \frac{\lambda T}{2}\Big((x_1'' + x_1')\,\eta_1 + (x_2'' + x_2')\,\eta_2\Big) \\ &+ \frac{T}{8\,m\,\xi^2}\left(2\,\xi\,\lambda^2\,T\cot\left[\frac{\xi\,T}{2}\right] + \left(m^2\,\theta^2\xi^2 - 4\,\lambda^2\right)\right)(\eta_1^2 + \eta_2^2)\,. \qquad (31)\end{aligned}$$

Finally, we have

$$\mathcal{K}_{\theta\sigma}(x'', T; x', 0) = \frac{m\,|\xi\,|}{2\,i\,h\left|\sin\left[\frac{\xi T}{2}\right]\right|}\,\exp\left(\frac{2\pi i}{h}\,\bar{S}_{\theta\sigma}(x'', T; x', 0)\right). \qquad (32)$$

2.2. *Ordinary and inverted harmonic oscillator on noncommutative plane*

The Lagrangian in the question is

$$L(\dot{x},x) = \frac{m}{2}\left(\dot{x}_1^2+\dot{x}_2^2\right) - \varepsilon\frac{m\,\omega^2}{2}\left(x_1^2+x_2^2\right), \tag{33}$$

where $\varepsilon = 1$ and $\varepsilon = -1$ are for ordinary and inverted harmonic oscillators, respectively. Using formulas (27), one can easily find the corresponding noncommutative Lagrangian

$$\begin{aligned} L_{\theta\sigma}(\dot{q},q) = {} & \frac{m}{2\,\kappa}\left(\dot{q}_1^2+\dot{q}_2^2\right) + \frac{\sigma+\varepsilon\,m^2\,\omega^2\,\theta}{2\,\kappa}\left(\dot{q}_2\,q_1-\dot{q}_1\,q_2\right) \\ & - \frac{\varepsilon\,m\,\omega^2}{2\,\kappa}\,\lambda^2\left({q_1}^2+{q_2}^2\right), \end{aligned} \tag{34}$$

where $\kappa = 1+\frac{\varepsilon\,m^2\,\omega^2\,\theta^2}{4}$ and $\lambda = 1-\frac{\theta\,\sigma}{4}$.

From (34), we obtain the Euler-Lagrange equations, $\ddot{q}_1 - \chi\,\dot{q}_2 + \mu\,q_1 = 0, \quad \ddot{q}_2 + \chi\,\dot{q}_1 + \mu\,q_2 = 0$, where $\chi = \frac{\sigma}{m} + \varepsilon\,m\,\omega^2\,\theta\,, \quad \mu = \varepsilon\,\omega^2\,\lambda^2$. Let us note that these equations form a coupled system of second order differential equations, which is more complicated than in the commutative case ($\theta = \sigma = 0$). One can transform this system to $q_1^{(4)} + \left(\chi^2+2\,\mu\right)q_1^{(2)} + \mu^2\,q_1 = 0, \quad q_2^{(4)} + \left(\chi^2+2\,\mu\right)q_2^{(2)} + \mu^2\,q_2 = 0$. The solution of these equations for $\left|\theta\,\sigma\right| < 4$, $\omega > 0\,, \chi \neq 0$ and $\varepsilon > 0$ has the form $\left(\nu = \frac{\chi^2+4\,\mu}{4}\right)$:

$$\begin{aligned} q_1(t) &= C_1\cos[\tfrac{\chi}{2}\,t]\cos[\sqrt{\varepsilon\,|\nu|}\,t] + C_2\cos[\tfrac{\chi}{2}\,t]\Big|\sqrt{\varepsilon}\,\sin[\sqrt{\varepsilon\,|\nu|}\,t]\Big| \\ &+ C_3\sin[\tfrac{\chi}{2}\,t]\cos[\sqrt{\varepsilon\,|\nu|}\,t] + C_4\sin[\tfrac{\chi}{2}\,t]\Big|\sqrt{\varepsilon}\,\sin[\sqrt{\varepsilon\,|\nu|}\,t]\Big|\,, \\ q_2(t) &= D_1\cos[\tfrac{\chi}{2}\,t]\cos[\sqrt{\varepsilon\,|\nu|}\,t] + D_2\cos[\tfrac{\chi}{2}\,t]\Big|\sqrt{\varepsilon}\,\sin[\sqrt{\varepsilon\,|\nu|}\,t]\Big| \\ &+ D_3\sin[\tfrac{\chi}{2}\,t]\cos[\sqrt{\varepsilon\,|\nu|}\,t] + D_4\sin[\tfrac{\chi}{2}\,t]\Big|\sqrt{\varepsilon}\,\sin[\sqrt{\varepsilon\,|\nu|}\,t]\Big|\,. \end{aligned} \tag{35}$$

Let us note that in the case of inverted harmonic oscillator ($\varepsilon = -1$) the trigonometric functions in $\sqrt{\varepsilon\,|\nu|}\,t$ from (35) become the corresponding hyperbolic functions in $\sqrt{|\,\nu\,|}\,t$. After imposing connections between q_1 and q_2 given by coupled differential equations, we obtain the following relations between constants C and D: $D_1 = C_2, \;\; D_2 = -C_1, \;\; D_3 = C_4, \;\; D_4 = -C_3$. The unknown constants C_1, C_2, C_3 and C_4 one can find from end conditions $q_1(0) = x_1', \quad q_1(T) = x_1'', \quad q_2(0) = x_2', \quad q_2(T) = x_2''$. Then one obtains the solutions

$$q_1 = \frac{1}{\sin[\sqrt{\varepsilon\,|\nu|}\;T]}\Big(\Big(x_1''\,\cos\Big[\frac{\chi}{2}\,(t-T)\Big] + x_2''\,\sin\Big[\frac{\chi}{2}\,(t-T)\Big]\Big)\,\sin[\sqrt{\varepsilon\,|\nu|}\;t]$$

$$-\Big(x_1'\,\cos\Big[\frac{\chi\,t}{2}\Big] + x_2'\,\sin\Big[\frac{\chi\,t}{2}\Big]\Big)\,\sin\big[\sqrt{\varepsilon\,|\nu|}\,(t-T)\big]\Big), \qquad (36)$$

$$q_2 = \frac{1}{\sin[\sqrt{\varepsilon\,|\nu|}\;T]}\Big(\Big(x_2''\,\cos\Big[\frac{\chi}{2}\,(t-T)\Big] - x_1''\,\sin\Big[\frac{\chi}{2}\,(t-T)\Big]\Big)\,\sin[\sqrt{\varepsilon\,|\nu|}\;t]$$

$$+\Big(-x_2'\,\cos\Big[\frac{\chi\,t}{2}\Big] + x_1'\,\sin\Big[\frac{\chi\,t}{2}\Big]\Big)\,\sin\big[\sqrt{\varepsilon\,|\nu|}\,(t-T)\big]\Big). \qquad (37)$$

Inserting the above expressions and their time derivatives into (34) we find

$$L_{\theta\sigma}(\dot q, q) = \frac{m\,\varepsilon\,\nu}{2\,\kappa\,\sin^2\big[\sqrt{\varepsilon\,|\nu|}\;T\big]}\Big((x_1''^2 + x_2''^2)\,\cos\big[2\sqrt{\varepsilon\,|\nu|}\;t\big] + ({x_1'}^2 + {x_2'}^2)$$

$$\times\,\cos\big[2\,\sqrt{\varepsilon\,|\nu|}\,(t-T)\big] - 2\cos\big[\sqrt{\varepsilon\,|\nu|}\,(2\,t-T)\big]\Big((x_1'\,x_1'' + x_2'\,x_2'')$$

$$\times\,\cos\Big[\frac{\chi T}{2}\Big] + (x_1''\,x_2' - x_1'\,x_2'')\,\sin\Big[\frac{\chi T}{2}\Big]\Big)\Big). \qquad (38)$$

Using (38), we finally compute the corresponding classical action

$$\bar S_{\theta\sigma}(x'', T; x', 0) = \int_0^T L_{\theta\sigma}(\dot q, q)\,dt \qquad (39)$$

$$= \frac{m\,\sqrt{\varepsilon\,|\nu|}}{2\,\kappa\,\sin\big[\sqrt{\varepsilon\,|\nu|}\;T\big]}\Big(-2\,\Big((x_1'\,x_1'' + x_2'\,x_2'')\cos\Big[\frac{\chi T}{2}\Big]$$

$$+\,({x_1'}^2 + x_1''^2 + {x_2'}^2 + x_2''^2)\cos\big[\sqrt{\varepsilon\,|\nu|}\;T\big] + 2\,(-x_1''\,x_2' + x_1'\,x_2'')\sin\Big[\frac{\chi T}{2}\Big]\Big).$$

If we take into account expression (39) then we finally have

$$\mathcal{K}_{\theta\sigma}(x'', T; x', 0) = \frac{1}{i\,h}\frac{m\,\big|\sqrt{\varepsilon\,|\nu|}\,\big|}{\kappa\,\big|\sin\big[\sqrt{\varepsilon\,|\nu|}\,T\big]\big|}\,\exp\Big(\frac{2\pi i}{h}\,\bar S_{\theta\sigma}(x'', T; x', 0)\Big). \quad (40)$$

3. Concluding remarks

Let us mention that taking $\sigma = 0$, $\theta = 0$ in the above formulas we recover expressions for the Lagrangian $L(\dot q, q)$, action $\bar S(x'', T; x', 0)$ and probability amplitude $\mathcal{K}(x'', T; x', 0)$ of the ordinary commutative case.

Note that a similar path integral approach with $\sigma = 0$ has been considered in the context of the Aharonov-Bohm effect, the Casimir effect, a quantum system in a rotating frame, and the Hall effect (for references on these and some other related subjects, see [4, 5]). Our investigation includes all systems with quadratic Lagrangians (Hamiltonians).

On the basis of the expressions presented in this article, there are many possibilities to discuss noncommutative quantum-mechanical systems with respect to various values of noncommutativity parameters θ and σ. For example, in the case $\theta\sigma = 4$ one has a critical point in the above our particular two-dimensional models (see also [6]).

Acknowledgments

B.D. would like to thank the organizers of the *Fifth International Workshop* "Lie Theory and its Applications in Physics" for invitation and hospitality, as well as for a stimulating and pleasant atmosphere. A part of this paper was done during stay of B.D. in the Steklov Mathematical Institute, Moscow. The work on this article was partially supported by the Serbian Ministry of Science, Technologies and Development under contracts No 1426 and No 1646. The work of B.D. was also supported in part by RFFI grant 02-01-01084.

References

1. M.R. Douglas and N.A. Nekrasov, *Rev. Mod. Phys.* **73**, 977 (2001); hep-th/0106048.
2. R.P. Feynman and A.R. Hibbs, *Quantum Mechanics and Path Integrals*, (McGraw-Hill Book Company, New York, 1965).
3. C. Grosche and F. Steiner, *Handbook of Feynman Path Integrals* (Springer-Verlag, Berlin, 1998).
4. B. Dragovich and Z. Rakić, "Lagrangian Aspects of Quantum Dynamics on a Noncommutative Space", hep-th/0302167.
5. B. Dragovich and Z. Rakić, "Path Integrals in Noncommutative Quantum Mechanics", to appear in the *Theoretical and Mathematical Physics*; hep-th/0309204 .
6. S. Bellucci, A. Nersessian and C. Sochichiu, *Phys. Lett.* **B522** 345 (2001); hep-th/0106138.

VI. Quantum Groups and Related Objects

MATRIX CHAIN MODELS AND THEIR Q-DEFORMATIONS

HANS PLESNER JAKOBSEN
Mathematics Institute
Universitetsparken 5
DK-2100 Copenhagen Ø, Denmark,
jakobsen@math.ku.dk

The Lie Algebras for the open, respectively closed, Matrix Chain Models offer an interesting laboratory for the study of various non-trivial mixtures of Heisenberg, Kac-Moody, and Virasoro algebras as well as $sl(\infty)$. The quantum deformed analogues of these algebras offer a similar opportunity.

1. Introduction

In a number of papers, [19], [20], [14], [15], and [16], Rajeev and Lee, and subsequently Lee together with the present author, have studied various aspects of the open string model, also known as the Open Matrix Chain model, both from the point of view of physics and mathematics. We refer to these for additional motivation, applications to physics, and background results.

Here, we first give some additional results relating to orderings, to approximately finite, and to the non-split nature of various algebras. Moreover, a remarkable scaling property is revealed. After that we discuss various possible quantizations, and in special cases investigate relations to quantized Heisenberg, Virasoro, and Kac-Moody algebras. In this special case, for the Virasoro algebra, there is a close relation to a recent preprint by J.T. Hartwig, D. Larson, and S. Silvestrov ([11]), the details of which still remain to be investigated.

Much work has been done on the cohomology of conformal algebras ([1], [3], [4], [5], [7], [8], [9]), see in particular the remarks on p. 78–79 in "Vertex Algebras for Beginners" by V. Kac ([17]) where non-split extensions of the Virasoro Algebra by conformal modules are classified. Our work must evidently be related to this although our structures a priori are simpler. The analogies for the deformed algebras remain to be worked out.

2. The open string algebra

Consider the set $\{0, 1, \ldots, (\Lambda - 1)\}$, and let $\mathcal{W}_\Lambda$ denote the set of all finite words (of arbitrary length) made up of the integers in this set, together with the empty word $\emptyset$. For $K \in \mathcal{W}_\Lambda$, let s^K be the function which is 1 on the symbol K and 0 on all different symbols.

Set $|K| = r$ for a word $K = k_1 k_2 \ldots k_r \in \mathcal{W}$, $|\emptyset| = 0$, and define the *momentum* p of K by

$$\sum_{i=1}^{|K|=r} k_i.$$

If $K = k_1, \ldots, k_r$ then in a natural way $s^K \equiv e_{k_1} \otimes \cdots \otimes e_{k_r}$. We define:

$$\sigma^I_J s^K = \left(\sum_{K_1 K_2 K_3 = K} \delta^{K_2}_J s^{K_1 I K_3} \right). \tag{1}$$

Definition 2.1. The **open string algebra** Ξ_o is the Lie algebra generated by these operators under the usual operator bracket. The above representation is **the defining representation**, denoted by $\mathcal{T}_o$.

2.1. *Scaling*

That the Open String Algebra is very big may also be ascertained from the following fact which shows that it contains proper injective images of itself: Choose Λ *incommensurable* words $W_0, W_1, \ldots W_{\Lambda-1} \in \mathcal{W}$. Consider the map

$$\mathcal{W} \ni K = k_1 k_2 \ldots k_n \overset{\mathcal{I}}{\mapsto} W_1 W_2 \ldots W_n \in \mathcal{W}.$$

This map is clearly injective, and also proper unless the words $W_0, W_1, \ldots W_{\Lambda-1}$ are just a permutation of the letters in the alphabet. We suppress the words from $\mathcal{I}$. One simple example would be $\forall I = 0 \ldots \Lambda - 1$: $W_i = ii$ (doubling). It then follows from the defining commutators of the algebra (see also (8)) that

Proposition 2.2. *The map, also denoted $\mathcal{I}$, from Ξ_o to itself given by*

$$\mathcal{I} : \sigma^I_J \mapsto \sigma^{\mathcal{I}(I)}_{\mathcal{I}(J)}$$

is a Lie algebra monomorphism. It is proper unless $\mathcal{I}$ is a permutation of the letters in the alphabet.

2.2. *Orderings, Weyl decomposition, involutions, Verma-like modules, unitarity*

We say about two sequences A, B that $A < B$ if either $|A| < |B|$ or, when $|A| = |B|$, if, at the first index from the left where they are different, A has the biggest index.

$X^A_B > X^C_D \Leftrightarrow AD < BC.$

$$X^A_B \text{ is } \textbf{positive} \text{ if and only if } A < B.$$

$$s^A > s^B \Leftrightarrow A < B.$$

Then X^A_B is positive iff $\forall K : X^A_B(s^K) > s^K$. We refer to this as the *standard ordering.*

More generally, one may consider a function $\mathcal{O}$,

$$\mathcal{S}_\Lambda \ni K \mapsto \mathcal{O}(K) \in \mathbb{N}_0$$

(or, $\mathcal{O}$ may even take values in some $\mathbb{N}^r$ for some r, where the target space is ordered by some total ordering.)

Assume that $\mathcal{O}$ satisfies

$$\begin{gathered} \mathcal{O}(A) > \mathcal{O}(B) \Rightarrow \\ \mathcal{O}(XA) > \mathcal{O}(XB) \text{ and } \mathcal{O}(AY) > \mathcal{O}(BY) \quad \forall X, Y \end{gathered} \tag{2}$$

as well as

$$\begin{gathered} \mathcal{O}(A) = \mathcal{O}(B) \Rightarrow \\ \mathcal{O}(XA) = \mathcal{O}(XB) \text{ and } \mathcal{O}(AY) = \mathcal{O}(BY) \quad \forall X, Y. \end{gathered} \tag{3}$$

We may then define an ordering by $\mathcal{O}$:

$$X^A_B \text{ is } \mathcal{O}\textbf{-positive} \text{ if and only if } \mathcal{O}(A) < \mathcal{O}(B).$$

Notice that ordinary lexicographical ordering does not fit into this scheme. For a thorough discussion of length lexicographic as well as other orderings, see ([12]).

We now define

$$\begin{aligned} \Xi^+_{\mathcal{O}} &= \text{Span}\{X^A_B \mid X^A_B \text{ is positive}\}, \\ \Xi^0_{\mathcal{O}} &= \text{Span}\{X^A_B \mid \mathcal{O}(A) = \mathcal{O}(B)\}, \\ \Xi^-_{\mathcal{O}} &= \text{Span}\{X^A_B \mid X^A_B \text{ is negative}\}. \end{aligned}$$

It follows from the assumptions on the ordering that Each of these space is a Lie subalgebra and $[\Xi^0_{\mathcal{O}}, \Xi^{\pm}_{\mathcal{O}}] \subseteq \Xi^{\pm}_{\mathcal{O}}$. Moreover, $[\Xi^0_{\mathcal{O}}, \Xi^0_{\mathcal{O}}] \subseteq \Xi^0_{\mathcal{O}}$. Moreover, we have a Weyl decomposition:

$$\Xi_o = \Xi^-_{\mathcal{O}} \oplus \Xi^0_{\mathcal{O}} \oplus \Xi^+_{\mathcal{O}}. \tag{4}$$

We have the following converse:

Proposition 2.3. *Suppose a Weyl decomposition into three subalgebras as in (4) and with is induced from an ordering $\mathcal{O}$ of $\mathcal{W}$ as above and such that $[\Xi^0_{\mathcal{O}}, \Xi^{\pm}_{\mathcal{O}}] \subseteq \Xi^{\pm}_{\mathcal{O}}$. Then $\mathcal{O}$ satisfies (2) and (3).*

Proof. *This follows from e.g. (8) below. Indeed, given K, L and one of the situations in (2) or (3). Then one may choose $A = I = J$ prudently in (8) below to cover this situation.*

Remark 2.4. For the standard ordering, we actually get $[\Xi^0_{\mathcal{O}}, \Xi^0_{\mathcal{O}}] = 0$ since here, $\Xi^0_{\mathcal{O}}$ acts diagonally in the defining representation. On the other hand, e.g. $\Xi^0_{\mathcal{O}}$ is not diagonalizable in the adjoint representation as is easy to see by specific computations. The general decomposition (4) includes cases of parabolic subalgebras rather than just Borel subalgebras.

3. The limits

In the cases above with (W_p some Lie algebra)

$$0 \to sl(\infty) \to W_p \to \left(\mathbb{C}[t, t^{-1}] \otimes gl(\infty) + \mathbb{C} \cdot c\right) \to 0$$

Furthermore, to any element $w \in W_p$ it is possible to find an increasing sequence of matrices $X_n(w) \in gl(n); n = 1, 2, \ldots, r, \ldots$ with only finitely many non-zero entries in each column and row and such that $w - \lim_{n\to\infty} X_n(w) = \overline{X(w)}$ commutes with $sl(\infty)$ and such that

$$(\left[\overline{X(w_p)}, \overline{X(w_p')}\right] - \overline{X([w_p, w_p'])}) \in sl(\infty).$$

This means that some completed algebra $\overline{W_p}$ is actually split over $\overline{sl(\infty)}$. Specifically, we assume that each $\overline{X(w)}$ stays within a finite distance from the diagonal. Any such algebra can be handled by the methods of ([14]). Recall that we call a unitary highest weight representation of first $sl(\infty)$ and then W_p, approximately finite if it originates in a unitary highest weight representation of some $gl(N)$. Then,

Proposition 3.1. *Any (irreducible) unitary highest weight representation of an algebra W_p in this class is the tensor product of an approximately finite unitary representation with a unitary representation arising as a representation of* $\widehat{\mathrm{L}}'(gl(\infty)) = \left(\mathbb{C}[t,t^{-1}] \otimes gl(\infty) + \mathbb{C} \cdot c\right)$.

We have the following commutation relations in $\mathfrak{g}_{0,\infty}$:

$$\begin{aligned}
[\sigma_a, \sigma_b] &= (a-b)\sigma_{a+b} \text{ if } (\operatorname{sgn} ab) = 1 \qquad (5)\\
[\sigma_a, T_b] &= -b \cdot T_{a+b} \text{ if } a+b \neq 0\\
[T_a, T_b] &= 0 \text{ if } a+b \neq 0\\
[\sigma_a, \sigma_b] &= (a-b)\sigma_{a+b} - (\operatorname{sgn} a)\min\{a^2, b^2\}T_{a+b}\\
&\qquad \text{if } (\operatorname{sgn} ab) = -1 \text{ and } a+b \neq 0,\\
[\sigma_{-a}, \sigma_a] &= -2a\sigma_0 + a^2 T_0 - \frac{(a-1)a(2a-1)}{6}F_0\\
&\quad - \left(\frac{1}{12}(a+1)a(a-1)Z_0\right) \text{ for } a \geq 0\\
[T_{-a}, \sigma_a] &= [\sigma_{-a}, T_a] = -aT_0 + \frac{a(a-1)}{2}F_0 \text{ for } a \geq 0\\
[T_{-a}, T_a] &= -aF_0 \text{ for } a \geq 0\\
[\sigma_a, F_0] &= [T_a, F_0] = 0. \qquad (6)
\end{aligned}$$

Observe that if for $a < 0 : \tilde{\sigma}_a := \sigma_a - aT_a$ and for $a \geq 0 : \tilde{\sigma}_a := \sigma_a$ then

$$\begin{aligned}
[\tilde{\sigma}_a, \tilde{\sigma}_b] &= (a-b)\tilde{\sigma}_{a+b} \text{ for } a+b \neq 0, \text{ and}\\
[\tilde{\sigma}_{-a}, T_a] &= -aT_0 - \frac{a(a+1)}{2}F_0 \text{ for } a \geq 0,\\
[\tilde{\sigma}_{-a}, \tilde{\sigma}_a] &= -2a\tilde{\sigma}_0 + \left(\frac{1}{6}(a+1)a(a-1)F_0\right)\\
&\quad - \left(\frac{1}{12}(a+1)a(a-1)Z_0\right) \text{ for } a \geq 0.
\end{aligned}$$

Due to the term with F_0 in the last formula above, the Virasoro is indeed sitting in a non-split position. Also observe that the term with T_0 plays the role of a central extension. We study the extensions further below where, indeed, the most general extension is classified for the q analogue. The classical result follows easily from this.

4. Quantization of the open string algebra

Our point of departure is the defining representation and three observations

$$\sigma_J^I = \sum_{E,F\in\mathcal{W}} f_{EJF}^{EIF}, \tag{7}$$

$$(\sigma_J^I)_{B,A} = \sum_{E,F\in\mathcal{W}} \delta_{EIF}^{B}\delta_{EJF}^{A} = \sum_{B=B_1B_2B_3\in\mathcal{W}} \delta_I^{B_2}\delta_{B_1JB_3}^{A},$$

$$\begin{aligned} \sigma_J^I\sigma_L^K = & \sum_{\substack{X,Y\in\mathcal{W}\\ X,Y\neq\emptyset}} \delta_{XJY}^{K}\sigma_L^{XIY} + \sum_{\substack{X,Y\in\mathcal{W}\\ Y\neq\emptyset}} \delta_{JY}^{XK}\sigma_{XL}^{IY} \\ & + \sum_{\substack{X,Y\in\mathcal{W}\\ X\neq\emptyset}} \delta_{XJ}^{KY}\sigma_{LY}^{XI} + \sum_{X,Y\in\mathcal{W}} \delta_J^{XKY}\sigma_{XLY}^{I}. \end{aligned} \tag{8}$$

Observe that

$$\sum_{\substack{X,Y\in\mathcal{W}\\ X,Y\gg\emptyset}} \delta_{JY}^{XK}\sigma_{XL}^{IY} = \sum_{\substack{W\in\mathcal{W}\\ W\gg\emptyset}} \sigma_{JWL}^{IWK}, \text{ and}$$

$$\sum_{\substack{X,Y\in\mathcal{W}\\ X,Y\gg\emptyset}} \delta_{XJ}^{KY}\sigma_{LY}^{XI} = \sum_{\substack{W\in\mathcal{W}\\ W\gg\emptyset}} \sigma_{LWJ}^{KWI}.$$

These formulas explain why the commutator $[\sigma_J^I\sigma_L^K, \sigma_L^K\sigma_J^I]$ is a finite sum.

We discuss briefly three possible quantizations of this algebra. All take their point of departure at (7). In the first case, which is the simplest, we simply set

$$\overrightarrow{\sigma}_J^I(q^a) = \sum_{E,F\in\mathcal{W}} q^{a\phi(EJF)} f_{EJF}^{EIF}. \tag{9}$$

Here, and throughout, ϕ denotes a function from $\mathcal{W}$ to $\mathbb{Z}$ that has the additivity property: $\forall A,B\in\mathcal{W}: \phi(AB)=\phi(A)+\phi(B)$. Simple computations then yield that the analogue of (8) becomes

$$\begin{aligned} & q^{-a(\phi(K)-\phi(L))}\overrightarrow{\sigma}_J^I(q^a)\overrightarrow{\sigma}_L^K(q^b) = \\ & \Big(\sum_{\substack{X,Y\in\mathcal{W}\\ X,Y\neq\emptyset}} \delta_{XJY}^{K}\overrightarrow{\sigma}_L^{XIY}(q^{a+b}) + \sum_{\substack{X,Y\in\mathcal{W}\\ Y\neq\emptyset}} \delta_{JY}^{XK}\overrightarrow{\sigma}_{XL}^{IY}(q^{a+b}) \\ & + \sum_{\substack{X,Y\in\mathcal{W}\\ X\neq\emptyset}} \delta_{XJ}^{KY}\overrightarrow{\sigma}_{LY}^{XI}(q^{a+b}) + \sum_{X,Y\in\mathcal{W}} \delta_J^{XKY}\overrightarrow{\sigma}_{XLY}^{I})(q^{a+b})\Big). \end{aligned} \tag{10}$$

Thus,

$$[\overrightarrow{\sigma}^I_J(q^a), \overrightarrow{\sigma}^K_L(q^b)]_{q,1} = \tag{11}$$

$$q^{-a(\phi(K)-\phi(L))}\overrightarrow{\sigma}^I_J(q^a)\overrightarrow{\sigma}^K_L(q^b) - q^{-b(\phi(I)-\phi(J))}\overrightarrow{\sigma}^K_L(q^b)\overrightarrow{\sigma}^I_J(q^a)$$

gives a finite sum of operators $\overrightarrow{\sigma}^X_Y(q^a)$. Also notice that the set of operators $\overrightarrow{\sigma}^X_Y(q^a)$ for which $a = \phi(X) - \phi(Y)$ form a subalgebra.

The second quantization is on the face of it a little more dubious. Here, we namely set

$$\overleftarrow{\sigma}^I_J(q^a) = \sum_{E,F\in\mathcal{W}} q^{a\phi(E)} f^{EIF}_{EJF}. \tag{12}$$

and get the following analogue of (8):

$$\overleftarrow{\sigma}^I_J(q^a)\overleftarrow{\sigma}^K_L(q^b) =$$

$$\sum_{\substack{X,Y\in\mathcal{W}\\ X,Y\neq\emptyset}} q^{a\phi(X)}\delta^K_{XJY}\overleftarrow{\sigma}^{XIY}_L(q^{a+b}) + \sum_{\substack{X,Y\in\mathcal{W}\\ Y\neq\emptyset}} q^{b\phi(X)}\delta^{XK}_{JY}\overleftarrow{\sigma}^{IY}_{XL}(q^{a+b})$$

$$+ \sum_{\substack{X,Y\in\mathcal{W}\\ X\neq\emptyset}} q^{a\phi(X)}\delta^{KY}_{XJ}\overleftarrow{\sigma}^{XI}_{LY}(q^{a+b}) + \sum_{X,Y\in\mathcal{W}} \delta^{XKY}_J q^{b\phi(X)}\overleftarrow{\sigma}^I_{XLY})(q^{a+b}).$$

Now we have:

$$\sum_{\substack{X,Y\in\mathcal{W}\\ X,Y\gg\emptyset}} q^{b(\phi(X))}\delta^{XK}_{JY}\overleftarrow{\sigma}^{IY}_{XL}(q^{a+b}) = \sum_{\substack{W\in\mathcal{W}\\ W\gg\emptyset}} q^{b(\phi(J))}q^{b(\phi(W))}\overleftarrow{\sigma}^{IWK}_{JWL},$$

and

$$\sum_{\substack{X,Y\in\mathcal{W}\\ X,Y\gg\emptyset}} q^{a(\phi(X))}\delta^{KY}_{XJ}\overleftarrow{\sigma}^{XI}_{LY}(q^{a+b}) = \sum_{\substack{W\in\mathcal{W}\\ W\gg\emptyset}} q^{a(\phi(K))}q^{a(\phi(W))}\overleftarrow{\sigma}^{KWI}_{LWJ}.$$

This means that if we want to define a "quantized Lie bracket" $[\cdot,\cdot]_{2,q}$ in analogy to (10), we have got to multiply the second term in (13) by $q^{-b\phi(J))}$ and the third term in (13) by $q^{-a\phi(K))}$. After that we subtract the analogous expression with the interchanges $a \leftrightarrow b, I \leftrightarrow K$, and $J \leftrightarrow L$.

Finally, there is a third quantization, where we set $\{z\}_q = \frac{1-q^z}{1-q}$ and define

$$\sigma^I_J(q^a) = \sum_{E,F\in\mathcal{W}} \{\phi(EJF)\}_{q^a} f^{EIF}_{EJF}. \tag{13}$$

In some sense, this is the one which seems most in spirit with quantum combinatorics, but which also is farthest from the usual quantizations. Still, it is well behaved in many ways. We study it below in a simple case.

Here we get the following analogue of (8):

$$\overleftrightarrow{\sigma}^I_J(q^a)\overleftrightarrow{\sigma}^K_L(q^b)$$

$$\sum_{\substack{X,Y\in\mathcal{W}\\X,Y\neq\emptyset}} \delta^K_{XJY}(q^{a\phi(X+J+Y)}\overleftrightarrow{\sigma}^{XIY}_L(q^a) + q^{b\phi(L)}\overleftrightarrow{\sigma}^{XIY}_L(q^b))$$

$$+\sum_{\substack{X,Y\in\mathcal{W}\\X,Y\neq\emptyset}} \delta^K_{XJY}(1 - q^{a\phi(X+J+Y)} - q^{b\phi(L)}\overleftrightarrow{\sigma}^{XIY}_L(q^0)) +$$

$$+\sum_{\substack{X,Y\in\mathcal{W}\\Y\neq\emptyset}} q^{b\phi(X)}\delta^{XK}_{JY}\overleftrightarrow{\sigma}^{IY}_{XL}(q^{a+b}) + \sum_{\substack{X,Y\in\mathcal{W}\\X\neq\emptyset}} q^{a\phi(X)}\delta^{KY}_{XJ}\overleftrightarrow{\sigma}^{XI}_{LY}(q^{a+b})$$

$$+\sum_{X,Y\in\mathcal{W}} \delta^{XKY}_J q^{b\phi(X)}\overleftrightarrow{\sigma}^I_{XLY})(q^{a+b})$$

+ terms with equal the corresponding terms for $\overleftrightarrow{\sigma}^K_L(q^b)\overleftrightarrow{\sigma}^I_J(q^a)$.

4.1. *Left and right operators.*

Recall that in the classical model we have the right and left operators in the space spanned by the operators σ^X_Y. To save time, we give their quantum analogues, where we use the quantization (9). The classical limit is obvious.

$$\ell^I_J(q^a) = \sum_{F\in\mathcal{W}} q^{a\phi(JF)} f^{IF}_{JF} \tag{14}$$

$$r^I_J(q^a) = \sum_{E\in\mathcal{W}} q^{a\phi(EJ)} f^{EI}_{EJ}$$

$$f^I_J(q^a) = q^{a\phi(J)} f^I_J.$$

These operators, as do the f^I_J's actually span ideals. We also have the equations

$$\sum_{i=0}^{\Lambda-1} \sigma^{Ii}_{Ji}(q^a) = \sigma^I_J(q^a) - r^I_J(q^a), \tag{15}$$

$$\sum_{i=0}^{\Lambda-1} \sigma^{iI}_{iJ}(q^a) = \sigma^I_J(q^a) - \ell^I_J(q^a),$$

$$\sum_{i=0}^{\Lambda-1} \ell^{Ii}_{Ji}(q^a) = \ell^I_J(q^a) - f^I_J(q^a), \text{ and}$$

$$\sum_{i=0}^{\Lambda-1} r^{iI}_{iJ}(q^a) = r^I_J(q^a) - f^I_J(q^a)$$

If one uses the second mentioned quantization, the operators ℓ^I_J in the second equation of (15) become replaced by their classical limits and, subsequently, so does the third equation. The rest remain formally unchanged.

Returning to (14) we have

$$\ell^I_J(q^a)\ell^K_L(q^b) = \sum_{W\in\mathcal{W}} \delta^{JW}_K \ell^{IW}_L(q^{a+b})q^{a(\phi(K)-\phi(L))} + \sum_{\emptyset\neq W\in\mathcal{W}} \delta^J_{KW}\ell^I_{LW}(q^{a+b})q^{a(\phi(K)-\phi(L))}.$$

These sums are clearly finite. Again the operators $\ell^I_J(q^a)$ where $\phi(I)-\phi(J)\equiv a$ form a subalgebra.

To cut down on the complexity we may, as in the classical case, restrict ourselves to the algebras of finite momentum. Also, we will just look at the left operators. Roughly speaking, *The momentum-P algebra* $\mathcal{B}_{\leq P}$ is the algebra spanned by all f^I_J, l^I_J, r^I_J, and σ^I_J such that

$$p(I) = p(J) \leq P \text{ or } p(I) = p(J) \leq P.$$

It contains the algebra $\mathcal{A}_P$ defined to be the operators of exact momentum P.

The general structure is as follows:

$$0 \to \mathcal{A}_p \to \mathcal{B}_{\leq p} \to \mathcal{B}_{\leq p-1} \to 0.$$

Let us finish by a look at the case of the operators ℓ^I_J at momentum $P=1$: Specifically, set $I=0^i10^t$ and $J=0^j1$. Assume that $i=j$. We have

$$\ell^I_J(q^a)\ell^J_I(q^b) - \ell^J_I(q^b)\ell^I_J(q^a) = \ell^I_I(q^{a+b})q^{-r} - \ell^J_J(q^{a+b})q^r.$$

If we assume that $f^{0^i10^r}_{0^i10^r} \equiv q^{0^i10^r} f^1_1$ then it follows from (15) that *modulo momentum* ≥ 2,

$$\begin{aligned}\ell^{0^i10^r}_{0^i10^r} &= \ell^{0^i10^{r-1}}_{0^i10^{r-1}} - f^1_1 q^{i+r} \\ &= \ell^{0^i1}_{0^i1} - f^1_1 q^{i+1}\frac{1-q^r}{1-q}.\end{aligned}$$

This equation should be compared to the "classical" ($q=1$) situation, where one sees a Kac-Moody algebra at this place, the r corresponding to the r in $\otimes t^r$ - in standard notation.

5. q-Virasoro

We have

$$[a]_q = (1 - q^{2a}) \tag{16}$$
$$[V_a, V_b]_q = ([b]_q - [a]_q)V_{a+b} \tag{17}$$
$$[V_a, T_b]_q = [b]_q T_{a+b} \tag{18}$$
$$[T_a, T_b]_q = 0 \tag{19}$$
$$(1 + q^{2c}) \cdot \Big[[V_a, V_b]_q, V_c\Big]_q + \text{ cyclic} = 0 \tag{20}$$
$$(1 + q^{2c}) \Big[[V_a, T_b]_q, V_c\Big]_q + \text{ cyclic} = 0 \tag{21}$$
$$(1 + q^{2c}) \Big[[T_a, T_b]_q, V_c\Big]_q + \text{ cyclic} = 0. \tag{22}$$

We now introduce a new bracket $[\cdot,\cdot]_q^{\tilde{}}$ corresponding to a central extension. We assume that the new bracket satisfies the same q-Jacobi identities as does the old. We assume that

$$[V_a, V_b]_q^{\tilde{}} = [V_a, V_b]_q + g_b \delta_{a+b,0}$$
$$[V_a, T_b]_q^{\tilde{}} = [V_a, T_b]_q + h_b \delta_{a+b,0}$$
$$[T_a, T_b]_q^{\tilde{}} = [T_a, T_b]_q + k_b \delta_{a+b,0}.$$

Working from the bottom, it follows from the analogue of (22) that

$$(1 + q^{2b})[a]_q k_b - (1 + q^{2a})[b]_q k_a = 0, \text{ hence}$$

Proposition 5.1.

$$k_a = \frac{[a]_q}{(1 + q^{2a})} F_0$$

for some constant (central element) F_0.

Remark 5.2. The computation of k_a rests on equation (22). But the equation actually just says $0 +$ cyclic $= 0$. Thus, the factor $(1 + q^{-2c})$ might be replaced by any nonzero function of c, and e.g. the following extension given by $\tilde{k}_a$ might be considered:

$$\tilde{k}_a = (q^q - q^{-a})\tilde{F}_0.$$

Now consider the analogue of (21) for the new bracket:

$$(1 + q^{2c})[b]_q(-h_{a+b}) + (1 + q^{2b})([a]_q - [c]_q)(h_b)$$
$$+(1 + q^{2a})[b]_q(h_{b+c}) = 0.$$

It follows that $h_0 = 0$. By making sign changes and observing that $(1 + q^{2a+2c})([-a]_q - [-c]_q) = (1 + q^{-2a-2c})([c]_q - [a]_q)$ it follows that

$$(1 + q^{-2c})[a + c]_q(-h_c) + (1 + q^{-2a-2c})([c]_q - [a]_q)(h_{a+c})$$
$$+(1 + q^{-2a})[a + c]_q(h_a) = 0.$$

We set $h_a = \frac{[a]_q}{(1+q^{-2a})} H_a$ and obtain, for $a + c \neq 0$,

$$([a]_q - [c]_q)H_{a+c} - [a]_q H_a = -[c]_q H_c.$$

It follows that

$$\begin{aligned} H_{a+1} &= \frac{1}{(q^2 - q^{2a})}([a]_q H_a - [1]_q H_1) \qquad (a \neq 1) \\ &= \frac{q^{-2(r+1)}[a]_q}{[a - (r + 1)]_q} H_{a-r} + \frac{[-(r + 1)]_q}{[a - (r + 1)]_q} H_1 \qquad (a \neq r + 1). \end{aligned}$$

It follows that

$$H_a = \alpha + \beta[-a - 1]_q$$

for arbitrary constants α, β, and hence

Proposition 5.3.

$$h_a = (F_1 + [-a - 1]_q F_2)\frac{[a]_q}{(1 + q^{-2a})}$$

for central elements F_1, F_2.

Finally, the analogue of (20) becomes

$$(1 + q^{2c})([b]_q - [a]_q)g_c + (1 + q^{2a})([c]_q - [b]_q)g_a$$
$$+(1 + q^{2b})([a]_q - [c]_q)g_b = 0.$$

It is easy to see that $g_c = \alpha$ is a solution for any constant α, but we assume that $g_{-c} = -g_c$ so $\alpha = 0$. Equally easy it may be seen that $\frac{[c]_q}{1+q^{2c}}$ is a solution. In our quest for the most general solution we will from now on assume that $g_1 = 0$. We set $G_c = (1 + q^{2c})g_c$. Then

$$([b]_q - [a]_q)G_c + ([c]_q - [b]_q)G_a + ([a]_q - [c]_q)G_b = 0. \qquad (23)$$

It follows easily that

$$G_{a+1} = q^{-2}\frac{[a + 2]_q}{[a - 1]_q} G_a$$

and then, that

$$\forall b : G_b = q^{-2b}[b + 1]_q[b]_q[b - 1]_q\gamma$$

for some constant γ. Thus, (see also ([11])

Proposition 5.4.

$$g_b = \frac{q^{-2b}[b+1]_q[b]_q[b-1]_q}{1+q^{2b}} F_4.$$

6. General setup

Let $\mathcal{P}$ be a (finite-dimensional) algebra with basis p_i and suppose given a representation by automorphisms $O_a, a \in \mathbb{Z}$ of $\mathbb{Z}$. This could e.g. be the space of (tails of) functions $p : \mathbb{N} \mapsto \mathbb{Z}[q]$ and with the automorphism $(O_a p)(i) = q^{ra}p(i+a)$ if p is homogeneous of degree r. Define

$$[p_i \otimes e_a, p_j \otimes e_b] = (p_i \cdot O_a(p_j) - p_j \cdot O_b(p_i)) \otimes e_{a+b}. \tag{24}$$

Proposition 6.1. *The formula (24) defines the structure of a Lie algebra on* $\text{Span}\{p \otimes e_a \mid p \in \mathcal{P}, a \in \mathbb{Z}\}$.

The above formulas can be obtained by first letting considering the span of the (tails of the) sequences $(1)_{n\in\mathbb{N}}$, $(n)_, n \in \mathbb{N}, (q^{rn})_{n\in\mathbb{N}}$. Specifically, $T_a^r \equiv q^{ri}E_{i,i+a}, V_b \equiv (1-q^{-2j})E_{j,j+b}$. For later reference, we also introduce $W_c = i \cdot E_{k,k+c}$. Throughout, $E_{i,j}$ denotes the usual matrix units. The equations then results from insisting that $O_a V_b(j) = q^{2a}V_b(j+a)$ and by considering the subalgebra generated by the functions T_a^r, W_c and V_b. The claim is, indeed, that they are closed under the Lie bracket. The remaining Lie brackets are:

$$\begin{aligned}
[T_a^r, T_b^s] &= 0,\\
[T_a^r, W_b] &= aT_{a+b}^r,\\
[W_a, W_b] &= (b-a)W_{a+b},\\
[T_a^r, V_b] &= (1-q^{2a})T_{a+b}^r, \text{ and}\\
[V_a, W_b] &= (1-q^{2a})W_{a+b} + aV_{a+b}.
\end{aligned}$$

Set $T_a = T_a^0$. The commutators may then also be written:

$$\begin{aligned}
q^{2a}V_aV_b - q^{2b}V_bV_a &= ([b]_q - [a]_q)V_{a+b},\\
V_aT_b - q^{2b}W_bV_a &= [b]_qW_{a+b} + aV_{a+b},\\
V_aT_b - q^{2b}T_bV_a &= [b]_qT_{a+b},
\end{aligned}$$

while the remaining relations involving only the elements T_a and W_b are the usual, "classical" ones. Observe that when we cut down the algebra of functions to the one generated by the above mentioned elements, since the

Lie bracket between homogeneous elements may be of lower degree - and inhomogeneous, it is not true that $O_a [X_b, Y_c] = [O_a(X_b), O_a(Y_c)]$. But there will still be relations such as the quantized Jacobi relations (20 -22) above. In some sense we are of course promoting the defining representation to a higher level - in line with $W_b W_{a+c} - W_c W_{a+b} = (c-b)W_{a+b}$ in the defining representation of the Witt (Virasoro) algebra. We get the following:

$$\left[V_a, [W_b, W_c]_q\right]_q + \text{ cyclic} = (c-b)[b]_q[c]_q W_{a+b+c}.$$

Likewise, there is a simple relation for the commutators with two W's and one V:

$$q^{2a}\left[W_a, [V_b, V_c]_q\right]_q + \left[V_b, [V_c, W_a]_q\right]_q + \left[V_c, [W_a, B_B]_q\right]_q = (b[a]_q[c]_q - c[a]_q[b]_q)V_{a+b+c}.$$

7. q-Sugawara

It is well known that in the $q=1$ situation, the T_x operators may be used to define the Virasoro algebra. See e.g. [18, Exercise 14.8]:

$$\underline{a>0:}\quad \sigma_a = \sum_{n=0}^{\infty} T_{-n}T_{n+a} + \frac{1}{2}\sum_{n=1}^{a-1} T_n T_{a-n}$$

$$\underline{a<0:}\quad \sigma_a = \sum_{n=0}^{\infty} T_{-n+a}T_n + \frac{1}{2}\sum_{n=1}^{|a|-1} T_{n+a}T_{-n}$$

$$\sigma_0 = \sum_{n=0}^{\infty} T_{-n}T_n - \frac{1}{2}T_0^2$$

Moreover,

$$[\sigma_a, \sigma_b] = (a-b)\sigma_{a+b} + \delta_{a+b,0}\left(\frac{1}{12}(a-1)a(a+1)Z_0\right).$$

Moving to the q generic situation, one is lad to new considerations. First of all, we set

$$q^{-a}T_aT_b - q^{-b}T_bT_a = \delta_{a+b,0}\,[a]'_q\,F_0,$$

where $[a]'_q$ is some q integer which is odd as a function of a. Immediately one is led to the consistency requirement

$$\forall a \in \mathbb{Z} : F_0 T_a = q^{-2a}T_a F_0.$$

Thus we see that the q analogue of a central extension may be an extension by an element satisfying covariance rather than invariance.

The q analogue of the Sugawara operators then becomes

$$\underline{a > 0:} \quad \sigma_a(q) = \sum_{n=0}^{\infty} T_{-n} q^{n+a} T_{n+a} + \frac{1}{2} \sum_{n=1}^{a-1} T_n q^{a-n} T_{a-n},$$

$$\underline{a < 0:} \quad \sigma_a(q) = \sum_{n=0}^{\infty} T_{-n+a} q^{n} T_{n} + \frac{1}{2} \sum_{n=1}^{a-1} T_{n+a} q^{-n} T_{-n},$$

$$\sigma_0(q) = \sum_{n=0}^{\infty} T_{-n} q^{n} T_{n} - \frac{1}{2} T_0^2.$$

Acknowledgments

We thank L. Hellström for furnishing us with the interesting reference ([12]) and S. Silvestrov for making the preprint ([11]) available to us. Finally, we thank the Mittag-Leffler Institute where parts of this work was carried out.

References

1. B. Bakalov, V.G. Kac and A.A. Voronov, math.QA/9803022.
2. V. Chari and A. Pressley, *Compositio Math.* **67**, 315 (1988).
3. B.L. Feigin, *Uspekhi Mat. Nauk.* **35**, no2(212), 225–226 (1980).
4. B.L. Feigin, *Selecta Math. Soviet,* **7** no. 1, 49 (1988).
5. B.L. Feigin and D.B. Fuchs, *(Russian) Funkts. Anal. i Priložen.* **14** no. 3, 45 (1980).
6. D, Friedan, Q.Z. Qiu and S. Shenker, in: *Vertex operators in mathematics and physics* (Berkeley, Calif., 1983), Vol. 419, (Springer, New York, 1985); Phys. Rev. Lett. **52**, 1575 (1984); Commun. Math. Phys. **107**, 535 (1986).
7. D.B. Fuks, *Cohomology of infinite-dimensional Lie algebras,* Contemporary Soviet Mathematics (Consultants Bureau, New York, 1986).
8. I.M. Gelfand and D.B. Fuchs, *Funkts. Anal. i Priložen.* **2**, 92 (1968).
9. I.M. Gelfand and D.B. Fuchs, *(Russian), Izv. Akad. Nauk SSSR Ser. Mat.* **34**, 322 (1970).
10. M. Hausner and J.T. Schwartz, *Lie Groups; Lie Algebras,* (Gordon and Breach, New York, 1968).
11. J.T. Hartwig, D. Larson and S. Silvestrov, "Deformations of Lie algebras using σ-derivations", Preprint 2003:32, Lund Institute of Technology (2003).
12. L. Hellström, http://urn.kb.se/resolve?urn=urn:nbn:se:umu:diva-92 (2004-03-14) (2003)
13. J.E. Humphreys, *Introduction to Lie Algebras and Representation Theory,* 2nd ed. (Springer-Verlag, New York, 1978).

14. H.P. Jakobsen and C.-W. H. Lee, *J. Math. Phys.* **42**, 3817 (2001).
15. H.P. Jakobsen and C.-W. H. Lee, in: *Theoretical high energy physics; MRST 2000*, AIP Conference Proceedings 541, 130 (2000)
16. H.P. Jakobsen and C.-W. H. Lee, in: *Vertex Operators and Kac-Moody algebras*, Proceedings, Madras 2002, Contemp. Math., A.M.S.
17. V.G. Kac, *Vertex Algebras for Beginners*, 2nd ed., University Lecture Series, Vol 10, (A.M.S., R.I., 1998).
18. V.G. Kac, *Infinite Dimensional Lie Algebras*, 2nd ed. (Cambridge University Press, Cambridge, 1985).
19. C.-W. H. Lee and S.G. Rajeev, *Nucl. Phys.* **B529**, 656 (1998).
20. C.-W. H. Lee and S.G. Rajeev, *Int. J. Mod. Phys.* **A14**, 4395 (1999).

EXOTIC BIALGEBRAS : NON-DEFORMATION QUANTUM GROUPS

D. ARNAUDON

Laboratoire d'Annecy-le-Vieux de Physique Théorique LAPTH,
UMR 5108 du CNRS associée à l'Université de Savoie,
BP 110, F-74941 Annecy-le-Vieux Cedex, France,
daniel.arnaudon@lapp.in2p3.fr

A. CHAKRABARTI

Centre de Physique Théorique, École Polytechnique,
F-91128 Palaiseau Cedex, France,
chakra@cpht.polytechnique.fr

V.K. DOBREV

School of Informatics, University of Northumbria,
Newcastle-upon-Tyne NE1 8ST, UK,
vladimir.dobrev@unn.ac.uk,
Permanent address:
Institute of Nuclear Research and Nuclear Energy,
Bulgarian Academy of Sciences,
72 Tsarigradsko Chaussee, 1784 Sofia, Bulgaria,
dobrev@inrne.bas.bg

S.G. MIHOV

Institute of Nuclear Research and Nuclear Energy,
Bulgarian Academy of Sciences,
72 Tsarigradsko Chaussee, 1784 Sofia, Bulgaria,
smikhov@inrne.bas.bg

In the classification of solution of the Yang–Baxter equation, there are solutions that are not deformations of the trivial solution (essentially the identity). We consider the algebras defined by these solutions, and the corresponding dual algebras. We then study the representations of the latter. We are also interested in the Baxterisation of these R-matrices and in the corresponding quantum planes.

1. Introduction: Five rank-4 R-matrices which are not deformations of the identity

We are interested in the study of algebraic structures coming from R-matrices (solutions of the Yang–Baxter equation) that are not deformations of classical ones (i.e., the identity up to signs). Those matrices were obtained by Hlavatý [1] and are also in the classification of Hietarinta [2]. There are five such R-matrices that are invertible. Most of this text is based on the articles [3, 4, 5].

The first cases of interest are

$$R_{H2,3} = \begin{pmatrix} 1 & x_1 & x_2 & x_3 \\ & 1 & 0 & x_2 \\ & & 1 & x_1 \\ & & & 1 \end{pmatrix} \tag{1}$$

denoted by

- *Exotic 1* (E1) if $x_1 = -x_2 = -h, h_3 \neq -h^2$ [3],
- *Exotic 2* (E2) if $x_1 \neq -x_2$ [3].

Note that if $x_1 = -x_2 = -h, h_3 = -h^2$, then this is $R_{H1,3}$ with $g = -h$ in the notation of Hietarinta, i.e., Jordanian deformation with two parameters.)

The case denoted by *Exotic 3* (E3) is related to the R-matrix

$$R_{S0,2} = \begin{pmatrix} 1 & 0 & 0 & 1 \\ & -1 & 0 & 0 \\ & & -1 & 0 \\ & & & 1 \end{pmatrix} \tag{2}$$

and the algebraic structure was studied in [3].

The cases denoted by "S03" and "S14" come from the R-matrices

$$R_{S0,3} \equiv \begin{pmatrix} 1 & 0 & 0 & 1 \\ 0 & 1 & 1 & 0 \\ 0 & 1 & -1 & 0 \\ -1 & 0 & 0 & 1 \end{pmatrix}, \qquad R_{S1,4} \equiv \begin{pmatrix} 0 & 0 & 0 & q \\ 0 & 0 & 1 & 0 \\ 0 & 1 & 0 & 0 \\ q & 0 & 0 & 0 \end{pmatrix}. \tag{3}$$

The structures corresponding to S03 and S14 were studied in [4].

2. Algebra and co-algebra structures

The algebra relations are obtained using

$$R_{12}T_1T_2 = T_2T_1R_{12} \tag{4}$$

$$\text{with} \qquad T = \begin{pmatrix} a & b \\ c & d \end{pmatrix} \qquad T_1 = T \otimes 1 \qquad T_2 = 1 \otimes T \tag{5}$$

The coalgebra structure is as usual given by $\Delta(T) = T \dot{\otimes} T$
For $R = \mathbb{1}_2 \otimes \mathbb{1}_2$, the relations $RTT = TTR$ are just the commutativity of a, b, c, d ($ab = ba$, $ac = ca$, ...), i.e., T is a matrix of commuting objects. For R of "quantum" type (with two parameters q, p), the relations are

$$ab = qba \qquad ac = pca \qquad bd = pdb \tag{6}$$
$$cd = qdc \qquad qbc = pcb \qquad ad - da = (q - p^{-1})bc \tag{7}$$

which are deformations of simple commutativity relations. The relations we will obtain in the exotic cases are not such deformations.
For the following we use the generators: $\tilde{a} = \frac{1}{2}(a + d)$, $\quad \tilde{d} = \frac{1}{2}(a - d)$.

3. The E1 Case

3.1. *Algebra and coalgebra relations*

The algebra relations (4) in the E1 case are:

$$\tilde{a}c = c\tilde{a} = \tilde{d}c = c\tilde{d} = \tilde{a}\tilde{d} = \tilde{d}\tilde{a} = 0 \ , \quad cb = bc \ , \quad c^2 = 0 \ , \tag{8}$$
$$\tilde{d}b = b\tilde{d} + 2h\tilde{d}^2 + hbc \qquad\qquad \tilde{a}b = b\tilde{a}, \tag{9}$$

and a basis of the enveloping algebra is

$$b^n\tilde{a}^k \ , \quad b^n\tilde{d}^\ell \ , \quad b^n c \ , \qquad n, k \in \mathbb{Z}_+ \ , \quad \ell \in \mathbb{N} \ . \tag{10}$$

h_3 is not explicitly in the relations, but some relations exist uniquely because $h_3 \neq -h_1^2$. This algebra has ideals $I = \mathcal{A}_1 b\tilde{d} \oplus \mathcal{A}_1\tilde{d}^2 \oplus \mathcal{A}_1 bc$, $\; I_2 = \mathcal{A}_1\tilde{d}^2 \oplus \mathcal{A}_1 bc$ and $I_1 = \mathcal{A}_1 bc$ such that $I_1 \subset I_2 \subset I \subset \mathcal{A}_1$. The dual algebra, or, more precisely, the *algebra in duality*, is defined via non-degenerate pairing $\langle \cdot, \cdot \rangle$, consistent with the co-algebraic structure, i.e. such that,

$$\langle\, u \,,\, ab \,\rangle \;=\; \langle\, \Delta(u) \,,\, a \otimes b \,\rangle \ , \quad \langle\, uv \,,\, a \,\rangle \;=\; \langle\, u \otimes v \,,\, \Delta(a) \,\rangle \tag{11a}$$
$$\langle 1_{\mathcal{U}}, a \rangle \;=\; \varepsilon_{\mathcal{A}}(a) \ , \quad \langle u, 1_{\mathcal{A}} \rangle \;=\; \varepsilon_{\mathcal{U}}(u) \tag{11b}$$

The algebraic relations are then:

$$[\tilde{D}, C] = -2C \ , \quad [B, C] = \tilde{D} \ , \quad [B, C]_+ = \tilde{D}^2 \tag{12}$$
$$[\tilde{D}, B] = 2B\tilde{D}^2 \ , \quad [\tilde{D}, B]_+ = 0 \ , \quad \tilde{D}^3 = \tilde{D} \ , \quad C^2 = 0 \ , \tag{13}$$
$$[\tilde{A}, Z] = 0 \ , \quad Z = B, C, \tilde{D} \ , \tag{14}$$
$$EZ = ZE = 0 \ , \quad Z = \tilde{A}, B, C, \tilde{D} \tag{15}$$

with co-algebraic structure:

$$\Delta(\tilde{A}) = \tilde{A} \otimes 1_{\mathcal{U}} + 1_{\mathcal{U}} \otimes \tilde{A}, \quad \Delta(B) = B \otimes 1_{\mathcal{U}} + 1_{\mathcal{U}} \otimes B,$$
$$\Delta(C) = C \otimes E + E \otimes C, \quad \Delta(\tilde{D}) = \tilde{D} \otimes E + E \otimes \tilde{D}, \quad \Delta(E) = E \otimes E$$
$$\varepsilon_{\mathcal{U}}(Z) = 0 \ , \quad Z = \tilde{A}, B, C, \tilde{D} \ , \quad \varepsilon_{\mathcal{U}}(E) = 1$$

The extra operator E is defined by: $\langle E, 1_A \rangle = 1$ with all other pairings being zero. Strictly speaking the above algebra is in duality with the factor-algebra since it has zero pairings with the ideals. The full dual is infinitely generated and is under investigation.

Let us make a comparison with FRT duality [7]. We use the relation $\langle L^{\pm} \ , \ T \rangle = R^{\pm}$, where: $R^{+} \equiv PRP = R(-h) \ , \quad R^{-} \equiv R^{-1}$. The comparison leads to

$$L^{\pm}_{11} = L^{\pm}_{22} = e^{-hB} \ , \quad L^{\pm}_{12} = ((h_{\pm} + h^2)B + h\tilde{A})e^{-hB} \tag{16}$$

where $h_+ = h_3$ and $h_- = -h_3 - 2h^2$. We see that $\tilde{A}$ and B are taken into account in this formalism, but that it says nothing about the generators $C, \tilde{D}$.

3.2. *R-matrix minimal polynomials and quantum planes*

In order to address the question of the quantum planes corresponding to the exotic bialgebras we have to know the minimal identity relations which the R-matrices fulfil. As we know the R-matrices producing deformations of the $GL(2)$ and $GL(1|1)$ fulfil second order relations. However, in the cases at hand we have higher order relations. We have:

$$(\hat{R} - \mathrm{id})^2(\hat{R} + \mathrm{id}) = 0, \quad h_1 = -h_2 = h, h_3 \neq -h^2 \tag{17a}$$
$$(\hat{R} - \mathrm{id})^3(\hat{R} + \mathrm{id}) = 0, \quad h_1 + h_2 \neq 0 \tag{17b}$$
$$(\hat{R} - \mathrm{id})(\hat{R} + \mathrm{id}) = 0, \quad h_1 = -h_2 = h, h_3 = -h^2 \tag{17c}$$

where $\hat{R} \equiv PR$, $R = R_{H2,3}$, id is the 4×4 unit matrix. The first case is E1, the second case is E2, the third case is the Jordanian subcase which produces $GL_{h,h}(2)$.

To derive the corresponding quantum planes we shall apply the formalism of [6]. The commutation relations between the coordinates z^i and differentials ζ^i , $(i = 1, 2)$, are given as follows:

$$z^i z^j = \mathcal{P}_{ijk\ell}\, z^k z^\ell \ , \quad \zeta^i \zeta^j = -\mathcal{Q}_{ijk\ell}\, \zeta^k \zeta^\ell \ , \quad z^i \zeta^j = \mathcal{Q}_{ijk\ell}\, \zeta^k z^\ell \tag{18}$$

where the operators $\mathcal{P}$, $\mathcal{Q}$ are functions of $\hat{R}$ and must satisfy:

$$(\mathcal{P} - \mathrm{id})\,(\mathcal{Q} + \mathrm{id}) = 0 \ . \tag{19}$$

Thus, there are different choices: four for E1, six for E2 (and just one for $GL_{h,h}(2)$). Choosing $\mathcal{P} - \mathrm{id} = (\hat{R} - \mathrm{id})^a$ with $a = 2, 3, 1$, respectively, and $\mathcal{Q} = \hat{R}$ in all cases. and denoting $(x, y) = (z^1, z^2)$ we obtain

$$xy - yx = hy^2 \ , \qquad h_1 = -h_2 = h \tag{20}$$

for E1 (and the Jordanian), or

$$xy - yx = \frac{1}{2}(h_1 - h_2)y^2 \ , \qquad h_1 \neq -h_2 \tag{21}$$

for E2. We note that the quantum planes corresponding to the three cases are not essentially different.

Denoting $(\xi, \eta) = (\zeta^1, \zeta^2)$ we obtain

$$\xi^2 + \frac{h_1 - h_2}{2}\, \xi\eta = 0 \ , \quad \eta^2 = 0 \ , \quad \xi\eta = -\eta\xi \tag{22}$$

for E2, while for E1 $\xi^2 + h\,\xi\eta = 0$, which is valid also for the Jordanian subcase.

Finally, for the coordinates-differentials relations we obtain

$$x\xi = \xi x + h_1 \xi y + h_2 \eta x + h_3 \eta y \ , \quad x\eta = \eta x + h_1 \eta y \ , \tag{23}$$

$$y\xi = \xi y + h_2 \eta y \ , \qquad y\eta = \eta y \tag{24}$$

4. The S03 case

4.1. *Algebraic relations*

The algebra relations in the S03 case are:

$$\tilde{b}^2 = \tilde{c}^2 = 0 \ , \qquad \tilde{a}\tilde{d} = \tilde{d}\tilde{a} = 0 \ , \qquad \tilde{a}\tilde{b} = 0 \ , \tag{25}$$

$$\tilde{b}\tilde{d} = 0 \ , \qquad \tilde{d}\tilde{c} = 0 \ , \qquad \tilde{c}\tilde{a} = 0 \ . \tag{26}$$

where: $\tilde{b} = \frac{1}{2}(b + c)$, $\tilde{c} = \frac{1}{2}(b - c)$.

There is no PBW basis in this case. Indeed, the ordering is cyclic:

$$\tilde{a} > \tilde{c} > \tilde{d} > \tilde{b} > \tilde{a} \tag{27}$$

Thus, the basis consists of building blocks like $\tilde{a}^k\, \tilde{c}\, \tilde{d}^\ell\, \tilde{b}$ and cyclic. Explicitly the basis can be described by the following monomials:

$$\tilde{a}^{k_1}\, \tilde{c}\tilde{d}^{\ell_1}\, \tilde{b} \cdots \tilde{a}^{k_n}\, \tilde{c}\tilde{d}^{\ell_n}\, \tilde{b}\tilde{a}^{k_{n+1}} \ , \qquad \tilde{d}^{\ell_1}\, \tilde{b}\tilde{a}^{k_1}\, \tilde{c} \cdots \tilde{d}^{\ell_n}\, \tilde{b}\tilde{a}^{k_n} \ ,$$

$$\tilde{a}^{k_1}\, \tilde{c}\tilde{d}^{\ell_1}\, \tilde{b} \cdots \tilde{a}^{k_n}\, \tilde{c}\tilde{d}^{\ell_n} \ , \qquad \tilde{d}^{\ell_1}\, \tilde{b}\tilde{a}^{k_1}\, \tilde{c} \cdots \tilde{d}^{\ell_n}\, \tilde{b}\tilde{a}^{k_n}\, \tilde{c}\tilde{d}^{\ell_{n+1}} \ ,$$

where in all cases $n, k_i, \ell_i \in \mathbb{Z}_+$.
The algebra in duality is given by

$$
\begin{aligned}
&[\tilde{A}, Z] = 0 \,, \quad Z = \tilde{B}, \tilde{C} \,, \quad \tilde{A}\tilde{D} = \tilde{D}\tilde{A} = \tilde{D}^3 = \tilde{B}^2\tilde{D} = \tilde{D}\tilde{B}^2 = \tilde{D} \,, \\
&[\tilde{B}, \tilde{C}] = -2\tilde{D} \,, \quad \tilde{D}\tilde{B} = -\tilde{B}\tilde{D} = \tilde{C}\tilde{D}^2 = \tilde{D}^2\tilde{C} \,, \quad \{\tilde{C}, \tilde{D}\} = 0 \,, \\
&\tilde{B}^2 + \tilde{C}^2 = 0 \,, \quad \tilde{B}^3 = \tilde{B} \,, \quad \tilde{C}^3 = -\tilde{C} \,, \quad \tilde{B}^2\tilde{A} = \tilde{A} \,,
\end{aligned}
\tag{29}
$$

with coproduct:

$$\Delta_{\mathcal{U}}(\tilde{A}) = \tilde{A} \otimes 1_{\mathcal{U}} + 1_{\mathcal{U}} \otimes \tilde{A} \tag{30}$$
$$\Delta_{\mathcal{U}}(\tilde{B}) = \tilde{B} \otimes 1_{\mathcal{U}} + (1_{\mathcal{U}} - \tilde{B}^2) \otimes \tilde{B} \tag{31}$$
$$\Delta_{\mathcal{U}}(\tilde{C}) = \tilde{C} \otimes (1_{\mathcal{U}} - \tilde{B}^2) + 1_{\mathcal{U}} \otimes \tilde{C} \tag{32}$$
$$\Delta_{\mathcal{U}}(\tilde{D}) = \tilde{D} \otimes (1_{\mathcal{U}} - \tilde{B}^2) + (1_{\mathcal{U}} - \tilde{B}^2) \otimes \tilde{D} \tag{33}$$
$$\varepsilon_{\mathcal{U}}(Z) = 0 \,, \qquad Z = \tilde{A}, \tilde{B}, \tilde{C}, \tilde{D} \,. \tag{34}$$

$\tilde{A}$, $\tilde{B}^2 = -\tilde{C}^2$ and $\tilde{D}^2$ are Casimir operators.
The bialgebra $s03$ is not a Hopf algebra (since there is no antipode).

The algebra generated by the generator $\tilde{A}$ is a sub-bialgebra of $s03$. The algebra $s03'$ generated by $\tilde{B}, \tilde{C}, \tilde{D}$ is a nine-dimensional sub-bialgebra of $s03$ with PBW basis:

$$1_{\mathcal{U}}, \ \tilde{B}, \ \tilde{C}, \ \tilde{D}, \ \tilde{B}\tilde{C}, \ \tilde{B}\tilde{D}, \ \tilde{D}\tilde{C}, \ \tilde{B}^2, \ \tilde{D}^2 \tag{35}$$

The algebra $s03$ is not the direct sum of the two subalgebras described above since both subalgebras have nontrivial action on each other, e.g., $\tilde{B}^2\tilde{A} = \tilde{A}$, $\tilde{A}\tilde{D} = \tilde{D}$. The algebra $s03$ is a nine-dimensional associative algebra over the central algebra generated by $\tilde{A}$.

Let us again make a comparison with FRT duality.
$L^{\pm}$ are matrices of operators $L^{\pm}_{ij}$ $(i,j = 1,2)$ satisfying the relations

$$R^+ L_1^+ L_2^+ = L_2^+ L_1^+ R^+ \tag{36}$$
$$R^+ L_1^- L_2^- = L_2^- L_1^- R^+ \tag{37}$$
$$R^+ L_1^+ L_2^- = L_2^- L_1^+ R^+ \tag{38}$$

with $L_1 \equiv L \otimes 1$, $L_2 \equiv 1 \otimes L$. Explicitly, these RLL relations read

$$
\begin{aligned}
&(L^{\pm}_{11})^2 = (L^{\pm}_{22})^2 && [L^{\pm}_{11}, L^{\pm}_{22}] = 0 \\
&(L^{\pm}_{12})^2 = -(L^{\pm}_{21})^2 && [L^{\pm}_{12}, L^{\pm}_{21}]_+ = 0 \\
&L^{\pm}_{11} L^{\pm}_{12} = L^{\pm}_{22} L^{\pm}_{21} && L^{\pm}_{11} L^{\pm}_{21} = L^{\pm}_{22} L^{\pm}_{12} \\
&L^{\pm}_{12} L^{\pm}_{11} = -L^{\pm}_{21} L^{\pm}_{22} && L^{\pm}_{12} L^{\pm}_{22} = -L^{\pm}_{21} L^{\pm}_{11}
\end{aligned}
\tag{39}
$$

and for the RL^+L^- ones

$$L^+_{ij}L^-_{kl} - L^-_{ij}L^+_{kl} + \theta_i L^+_{\bar{i}j}L^-_{\bar{k}l} + \theta_j L^-_{i\bar{j}}L^+_{k\bar{l}} = 0 \tag{40}$$

with $\bar{n} \equiv 3-n$, $\theta_1 = 1$, $\theta_2 = -1$. In the case of $s03$, the FRT relations in the dual algebra are richer than the relations given using only (11). They will indeed lead to more irreducible finite dimensional representations. A convenient basis is given by

$$\begin{aligned} \tilde{L}^\pm_{11} &= L^\pm_{11} + L^\pm_{22} & \tilde{L}^\pm_{22} &= L^\pm_{11} - L^\pm_{22} \\ \tilde{L}^\pm_{12} &= L^\pm_{12} + L^\pm_{21} & \tilde{L}^\pm_{21} &= L^\pm_{12} - L^\pm_{21} \end{aligned} \tag{41}$$

In this basis, the relations (39) read

$$\begin{aligned} \tilde{L}^\pm_{11}\,\tilde{L}^\pm_{22} &= 0 & \tilde{L}^\pm_{22}\,\tilde{L}^\pm_{11} &= 0 \\ (\tilde{L}^\pm_{12})^2 &= 0 & (\tilde{L}^\pm_{21})^2 &= 0 \\ \tilde{L}^\pm_{11}\,\tilde{L}^\pm_{21} &= 0 & \tilde{L}^\pm_{12}\,\tilde{L}^\pm_{11} &= 0 \\ \tilde{L}^\pm_{21}\,\tilde{L}^\pm_{22} &= 0 & \tilde{L}^\pm_{22}\,\tilde{L}^\pm_{12} &= 0 \end{aligned} \tag{42}$$

whereas the relations (40) become

$$\begin{aligned} [\tilde{L}^+_{11}, \tilde{L}^-_{11}] &= 0 & \tilde{L}^-_{21}\,\tilde{L}^+_{11} &= \tilde{L}^+_{21}\,\tilde{L}^-_{11} \\ \tilde{L}^-_{11}\,\tilde{L}^+_{12} &= \tilde{L}^+_{11}\,\tilde{L}^-_{12} & \tilde{L}^-_{21}\,\tilde{L}^+_{12} &= \tilde{L}^+_{21}\,\tilde{L}^-_{12} \\ \tilde{L}^-_{11}\,\tilde{L}^+_{21} &= \tilde{L}^+_{21}\,\tilde{L}^-_{22} & \tilde{L}^-_{21}\,\tilde{L}^+_{21} &= -\tilde{L}^+_{11}\,\tilde{L}^-_{22} \\ \tilde{L}^-_{11}\,\tilde{L}^+_{22} &= \tilde{L}^+_{21}\,\tilde{L}^-_{21} & \tilde{L}^-_{21}\,\tilde{L}^+_{22} &= -\tilde{L}^+_{11}\,\tilde{L}^-_{21} \\ \\ \tilde{L}^-_{12}\,\tilde{L}^+_{11} &= -\tilde{L}^+_{22}\,\tilde{L}^-_{12} & \tilde{L}^-_{22}\,\tilde{L}^+_{11} &= \tilde{L}^+_{12}\,\tilde{L}^-_{12} \\ \tilde{L}^-_{12}\,\tilde{L}^+_{12} &= -\tilde{L}^+_{22}\,\tilde{L}^-_{11} & \tilde{L}^-_{22}\,\tilde{L}^+_{12} &= \tilde{L}^+_{12}\,\tilde{L}^-_{11} \\ \tilde{L}^-_{12}\,\tilde{L}^+_{21} &= \tilde{L}^+_{12}\,\tilde{L}^-_{21} & \tilde{L}^-_{22}\,\tilde{L}^+_{21} &= \tilde{L}^+_{22}\,\tilde{L}^-_{21} \\ \tilde{L}^-_{12}\,\tilde{L}^+_{22} &= \tilde{L}^+_{12}\,\tilde{L}^-_{22} & [\tilde{L}^+_{22}, \tilde{L}^-_{22}] &= 0 \end{aligned} \tag{43}$$

Denote for $n \geq 1$:

$$F_n(k_i; l_i) \equiv \prod_{i=1}^{n} \tilde{L}^{+k_i}_{11}\,\tilde{L}^+_{12}\tilde{L}^{+l_i}_{22}\,\tilde{L}^+_{21} \tag{44}$$

$$G_n(l_i; k_i) \equiv \prod_{i=1}^{n} \tilde{L}^{+l_i}_{22}\,\tilde{L}^+_{21}\tilde{L}^{+k_i}_{11}\,\tilde{L}^+_{12} \tag{45}$$

and for $n = 0$

$$F_0(k_i; l_i) \equiv 1; \quad G_0(l_i; k_i) \equiv 1 \tag{46}$$

The basis elements of the algebra generated by the $\tilde{L}^+$'s are (following [4])

$$\begin{aligned} &F_n(k_i;l_i)\tilde{L}_{11}^{+k_{n+1}}; \quad F_{n-1}(k_i;l_i)\tilde{L}_{11}^{+k_n}\tilde{L}_{12}^{+}\tilde{L}_{22}^{+l_n};\\ &G_n(l_i;k_i)\tilde{L}_{22}^{+l_{n+1}}; \quad G_{n-1}(l_i;k_i)\tilde{L}_{22}^{+l_n}\tilde{L}_{21}^{+}\tilde{L}_{11}^{+k_n} \end{aligned} \tag{47}$$

Defining also $K_n = \sum_{i=1}^n k_i$, $\quad L_n = \sum_{i=1}^n l_i$ the actions of generators $\tilde{L}^-$ on the basis elements are, e.g.,

$$\begin{aligned} \tilde{L}_{11}^- F_n(k_i;l_i) &= F_{n-1}(k_1+1,k_i;l_i)\tilde{L}_{11}^{+k_n}\tilde{L}_{12}^{+}\tilde{L}_{22}^{+l_n}\tilde{L}_{21}^-\\ \tilde{L}_{12}^- F_n(k_i;l_i) &= (-1)^{K_n+L_n+1}G_{n-1}(k_1+1,k_i;l_i)\tilde{L}_{22}^{+k_n}\tilde{L}_{21}^{+}\tilde{L}_{11}^{+l_n}\tilde{l}_{22}^-\\ \tilde{L}_{21}^- F_n(k_i;l_i) &= G_n(0,..,l_{n-1};k_i)\tilde{L}_{22}^{+l_n}\tilde{L}_{21}^-\\ \tilde{L}_{22}^- F_n(k_i,;l_i) &= (-1)^{K_n+L_n}F_n(0,..,l_{n-1};k_i)\tilde{L}_{11}^{+l_n}\tilde{L}_{22}^- \end{aligned} \tag{48}$$

$$\begin{aligned} \tilde{L}_{11}^- G_n(l_i;k_i) &= (-1)^{K_n+L_n}G_n(0,..,k_{n-1};l_i)\tilde{L}_{22}^{+k_n}\tilde{L}_{11}^-\\ \tilde{L}_{12}^- G_n(l_i;k_i) &= F_n(0,..,k_{n-1};l_i)\tilde{L}_{11}^{+k_n}\tilde{l}_{12}^- \end{aligned} \tag{49}$$

$$\begin{aligned} \tilde{L}_{21}^- G_n(l_i;k_i) &= (-1)^{K_n+L_n+1}F_{n-1}(l_1+1,l_i;k_i)\tilde{L}_{11}^{+l_n}\tilde{L}_{12}^{+}\tilde{L}_{22}^{+k_n}\tilde{L}_{11}^-\\ \tilde{L}_{22}^- G_n(l_i;k_i) &= G_{n-1}(l_1+1,l_i;k_i)\tilde{L}_{22}^{+l_n}\tilde{L}_{21}^{+}\tilde{L}_{11}^{+k_n}\tilde{L}_{12}^- \end{aligned} \tag{50}$$

These equations allow one to order the $\tilde{L}^-$ with respect to the $\tilde{L}^+$. For the $\tilde{L}^-$ among themselves, there exists a basis similar to (47).

4.2. *Finite dimensional irreducible representations*

Let us first consider the algebra generated by A, B, C, D and the relations (29). Since $\tilde{D}^3 = \tilde{D}$, there exists a weight vector v_0 such that: $\tilde{D}\,v_0 = \lambda\,v_0$ where $\lambda^3 = \lambda$. The finite dimensional irreps are then

- one-dimensional trivial
- two-dimensional with Casimir values $\mu, 1, 1$ for $\tilde{A}, \tilde{B}^2, \tilde{D}^2$, respectively, $\mu \in \mathbb{C}$.
- one-dimensional with Casimir values $\mu, 1, 0$ for $\tilde{A}, \tilde{B}^2, \tilde{D}^2$, respectively, $\mu \in \mathbb{C}$.

A two-dimensional representation for the algebra generated by the L-operators is provided by the R-matrix itself, setting $\pi(L^+) = R_{21}$, $\pi(L^-) = R^{-1}$ (see [7, 8])

$$\begin{aligned} \pi(L_{11}^\pm) &= \begin{pmatrix} 1 & 0 \\ 0 & \mp 1 \end{pmatrix} \qquad & \pi(L_{12}^\pm) &= \begin{pmatrix} 0 & \pm 1 \\ 1 & 0 \end{pmatrix}\\ \pi(L_{21}^\pm) &= \begin{pmatrix} 0 & 1 \\ \mp 1 & 0 \end{pmatrix} \qquad & \pi(L_{22}^\pm) &= \begin{pmatrix} \pm 1 & 0 \\ 0 & 1 \end{pmatrix} \end{aligned} \tag{51}$$

Note that this does not exhaust the set of two-dimensional representations.

Let N_1 and N_2 be two non negative integers. Here is an example (in the $\tilde{L}$ basis) of a finite dimensional irreducible representation of arbitrary dimension $N_1 + N_2$.

$$\pi(\tilde{L}_{11}) = \mathrm{diag}(\rho_1, \cdots, \rho_{N_1}, \underbrace{0, \cdots, 0}_{N_2}) \qquad \rho_i \neq \rho_j \quad \text{for} \quad i \neq j \tag{52}$$

$$\pi(\tilde{L}_{22}) = \mathrm{diag}(\underbrace{0, \cdots, 0}_{N_1}, \lambda_1, \cdots, \lambda_{N_2}) \qquad \lambda_i \neq \lambda_j \quad \text{for} \quad i \neq j \tag{53}$$

$$\left(\pi(\tilde{L}_{12})\right)_{ij} \neq 0 \quad \text{iff} \quad i \in \{1, \cdots, N_1\}, \quad j \in \{N_1 + 1, \cdots, N_1 + N_2\} \tag{54}$$

$$\left(\pi(\tilde{L}_{21})\right)_{ij} \neq 0 \quad \text{iff} \quad i \in \{N_1 + 1, \cdots, N_1 + N_2\}, \quad j \in \{1, \cdots, N_1\} \tag{55}$$

4.3. *Baxterisation*

We introduce the following Ansatz (choosing a convenient normalisation):

$$\hat{R}(x) = I + c(x)\hat{R} \tag{56}$$

and we try to find $c(x)$ such that $\hat{R}(x)$ would satisfy the parametrised Yang–Baxter equation:

$$\hat{R}_{(12)}(x)\hat{R}_{(23)}(xy)\hat{R}_{(12)}(y) = \hat{R}_{(23)}(y)\hat{R}_{(12)}(xy)\hat{R}_{(23)}(x) \tag{57}$$

The result is:

$$\hat{R}(x) = (\sqrt{2}x)^{-1}\hat{R} + (\sqrt{2}x)\hat{R}^{-1} = \frac{1}{\sqrt{2x}} \begin{pmatrix} x+1 & 0 & 0 & 1-x \\ 0 & x+1 & x-1 & 0 \\ 0 & 1-x & x+1 & 0 \\ x-1 & 0 & 0 & x+1 \end{pmatrix} \tag{58}$$

4.4. *Spectral decomposition and noncommutative planes*

The minimal polynomial identity satisfied by $\mathcal{R}$ is

$$\hat{R}^2 - 2\hat{R} + 2I = 0 \tag{59}$$

and we have the spectral decomposition:

$$\hat{R} = (1-i)P_{(+)} + (1+i)P_{(-)} = (1+i)I - 2iP_{(+)} \tag{60}$$

$$\text{where} \qquad P_{(\pm)} \equiv \frac{1}{2}(I \pm i(\hat{R} - I)) \tag{61}$$

are projectors resolving the identity: $P_{(i)}P_{(j)} = \delta_{ij}P_{(i)}, \quad i,j = \pm,\ P_{(+)} + P_{(-)} = I$. The quantum plane relations in the case of S03 are then

$$x_1{}^2 = x_1x_2, \quad x_2{}^2 = -x_2x_1\ , \quad \xi_1{}^2 = -\xi_1\xi_2, \quad \xi_2{}^2 = \xi_2\xi_1 \tag{62}$$

$$x_1\xi_1 = (\nu-1)\xi_1x_1 + \nu\xi_1x_2\ , \quad x_1\xi_2 = (\nu-1)\xi_1x_2 + \nu\xi_1x_1 \tag{63}$$

$$x_2\xi_1 = (\nu-1)\xi_2x_1 - \nu\xi_2x_2\ , \quad x_2\xi_2 = (\nu-1)\xi_2x_2 - \nu\xi_2x_1 \tag{64}$$

where ν is arbitrary real parameter.

5. The S14 case

The relations in the S14 case are:

$$\tilde{b}\tilde{c} + \tilde{c}\tilde{b} = 0 \qquad \tilde{a}\tilde{d} + \tilde{d}\tilde{a} = 0$$
$$\tilde{a}\tilde{b} = \tilde{b}\tilde{a} = \tilde{a}\tilde{c} = \tilde{c}\tilde{a} = \tilde{b}\tilde{d} = \tilde{d}\tilde{b} = \tilde{c}\tilde{d} = \tilde{d}\tilde{c} = 0$$

and the dual algebra is:

$$\tilde{C} = \tilde{D}\tilde{B} = -\tilde{B}\tilde{D}\ , \quad [\tilde{A},\tilde{D}] = 0\ , \quad EZ = ZE = 0 \tag{65a}$$

$$\tilde{A}\tilde{B} = \tilde{B}\tilde{A} = \tilde{D}^2\tilde{B} = \tilde{B}^3 = \tilde{B}\ , \quad Z = \tilde{A},\tilde{B},\tilde{D}\ . \tag{65b}$$

The dual coalgebra is given by (with $K \equiv (-1)^{\tilde{A}}$)

$$\Delta_{\mathcal{U}}(\tilde{A}) = \tilde{A}\otimes 1_{\mathcal{U}} + 1_{\mathcal{U}}\otimes\tilde{A}\ , \quad \Delta_{\mathcal{U}}(\tilde{B}) = \tilde{B}\otimes E + E\otimes\tilde{B} \tag{66a}$$

$$\Delta_{\mathcal{U}}(\tilde{D}) = \tilde{D}\otimes K + 1_{\mathcal{U}}\otimes\tilde{D}\ , \qquad \Delta(E) = E\otimes E \tag{66b}$$

$$\varepsilon_{\mathcal{U}}(Z) = 0 \ \text{ for } \ Z = \tilde{A},\tilde{B},\tilde{D}\ ; \qquad \varepsilon_{\mathcal{U}}(E) = 1 \tag{66c}$$

The irreducible representations of $s14$ follow the classification

- one-dimensional with Casimir values $\mu, 0, \lambda^2$ for $\tilde{A}, \tilde{B}^2, \tilde{D}^2$, respectively, $\mu, \lambda \in \mathbb{C}$.
- two two-dimensional with all Casimirs $\tilde{A}, \tilde{B}^2, \tilde{D}^2$ having the value 1.

Acknowledgments

This work was supported in part by the CNRS-BAS France/Bulgaria agreement number 12561. It was also inspired by the TMR Network EUCLID: "Integrable models and applications: from strings to condensed matter", contract number HPRN-CT-2002-00325.

References

1. L. Hlavatý, *J. Phys.* **A20**, 1661 (1987); *J. Phys.* **A25**, L63 (1992).
2. J. Hietarinta, *J. Math. Phys.* **34**, 1725 (1993).
3. D. Arnaudon, A. Chakrabarti, V.K. Dobrev and S.G. Mihov, *J. Phys.* **A34**, 4065 (2001); `math.QA/0101160`.
4. D. Arnaudon, A. Chakrabarti, V.K. Dobrev and S.G. Mihov, *J. Math. Phys.* **43**, 6238 (2002); `math.QA/0206053`.
5. D. Arnaudon, A. Chakrabarti, V.K. Dobrev and S.G. Mihov, *Int. J. Mod. Phys.* **A18**, 4201 (2003); `math.QA/0209321`.
6. J. Wess and B. Zumino, *Covariant differential calculus on the quantum hyperplane*, *Nucl. Phys. (Proc. Suppl.)* **18**, 302 (1990).
7. L.D. Faddeev, N.Yu. Reshetikhin and L.A. Takhtajan, "Quantization of Lie groups and Lie algebras", *Alg. Anal.* **1**, 178-206 (1989) (in Russian) and in: *Algebraic Analysis*, Vol. 1 (Academic Press, 1988) pp. 129-139.
8. A. Chakrabarti, "A nested sequence of projectors and corresponding braid matrices $\hat{R}(\theta)$ (1) odd dimensions", `math.QA/0401207`.

THE Q–BOSON–FERMION REALIZATION OF THE QUANTUM SUPERALGEBRA $U_Q(\mathrm{GL}(M/N))$

Č. BURDíK
Department of Mathematics and Doppler Institute, FNSPE, Czech Technical University,
Trojanova 13, CZ-120 00 Prague 2, Czech Republic
e-mail: burdik@kmalpha.fjfi.cvut.cz

O. NAVRÁTIL
Department of Mathematics, FTS, Czech Technical University,
Na Florenci 25, CZ-110 00 Prague 1, Czech Republic
e-mail: navratil@fd.cvut.cz

We show that our construction of realizations for Lie algebras and quantum algebras can be generalized to quantum superalgebras, too. We study an example of quantum superalgebra $U_q(\mathrm{gl}(m/n))$ and give the boson–fermion realization with respect to $m-1$ pairs of q–boson operators, n pairs of fermions and with respect to $U_q(\mathrm{gl}(m-1/n))$.

1. Introduction

Quantum groups or q–deformed Lie algebras imply some specific deformations of classical Lie algebras. From a mathematical point of view, it is a noncommutative associative Hopf algebra. The structure and representation theory of quantum groups have been developed extensively staring from works by Jimbo [1] and Drinfeld, [2] and many authors.

In the course of studying quantum algebras, a lot of attention has been paid to the case of quantum superalgebras [3], [4]. These algebras provide solutions to the Yang–Baxter equations and there may serve as a source of new exactly solvable models in statistical mechanics [5]. The applications in field theory are connected with the WZW models [6] and we can find some application of quantum superalgebra $U_q(\mathrm{gl}(m/n))$ for construction of the Alexander–Conway polynomial in the knot theory [7].

The Weyl and Clifford algebras also admit quantum deformation [8] with q–analogues of the Bose, and respectively, Fermi oscillator operators as generators [8, 9, 10]. These quantized algebras have been used to construct oscillator realizations of the quantum algebras that correspond to all classical Lie algebras [8]. These realizations are of the Jordan–Schwinger type [11]. The realizations of this type for quantum superalgebras were constructed in [12]. In the literature there exist Dyson [13] and Holstein–Primakoff [14] realizations which were first written for the algebra sl(2).

The first "quantum" version of Holstein–Primakoff was worked out for $U_q(\mathrm{sl}(2))$ [15] and then for $U_q((\mathrm{sl}(3))$ [16]. These realizations found immediate applications [17–22].

In our paper [23] we formulated the method starting from the Verma modules for obtaining boson realizations and in [24] we obtained explicitly a braid class of realizations which generalized the results from [25, 26].

Later the idea of boson–fermion realizations was extended to the Lie superalgebra, and the Dyson type boson–fermion realizations were explicitly given in [27], generalizing the results to sl(2/1) ([28], [29]).

In our papers [30, 31, 32] we studied the Dyson realizations of the series algebras $U_q(\mathrm{sl}(2))$, $U_q(\mathrm{gl}(n))$, $U_q(B_n)$, $U_q(C_n)$ and $U_q(D_n)$. There is some special case [31] for which the realization of the subalgebra $U_q(\mathrm{gl}(n-1))$ in the recurrence is trivial. Such special realizations of the quantum algebra $U_q(\mathrm{sl}(n))$ of Dyson type were studied in [33].

The aim of the present paper is to show that it is possible to generalize our method [23] to deriving the boson–fermion realization, too. This will be exemplified by the quantum superalgebra $U_q(\mathrm{gl}(m/n))$. This superalgebra can be applied to physical problems such as strongly correlated electron systems [34, 35, 36]. We explicitly see the recurrence with respect to $U_q(\mathrm{gl}(m-1/n))$ and consequently we will show that again it is a generalization of the result from [37].

2. Preliminaries

In this article, we will use the definition of the quantum superalgebra $U_q(\mathrm{gl}(m/n))$ which can be found in [37].

Let q be an independent variable, $\mathcal{A} = C[q, q^{-1}]$ and $\mathcal{C}(q)$ be a division field of $\mathcal{A}$. The superalgebra $U_q(\mathrm{gl}(m/n))$ is the associative superalgebra over $\mathcal{C}(q)$ generated by even generators K_i, K_i^{-1}, $i = 1, 2, \ldots, m+n$, E_k, F_k, $k = 1, \ldots, m+n-1$, $k \neq m$, and odd generators E_m, F_m, which satisfy the following relations:

$$K_iK_j = K_jK_i\,, \qquad K_iK_j^{-1} = K_j^{-1}K_i\,,$$
$$K_iK_i^{-1} = 1\,, \qquad K_i^{-1}K_j^{-1} = K_j^{-1}K_i^{-1}$$

$$
\begin{aligned}
&K_iE_i = qE_iK_i\,, && K_i^{-1}E_i = q^{-1}E_iK_i^{-1}\,,\\
&K_iF_i = q^{-1}F_iK_i\,, && K_i^{-1}F_i = qF_iK_i^{-1}\\
&K_iE_{i-1} = q^{-1}E_{i-1}K_i\,, && K_i^{-1}E_{i-1} = qE_{i-1}K_i^{-1}\,,\\
&K_iF_{i-1} = qF_{i-1}K_i\,, && K_i^{-1}F_{i-1} = q^{-1}F_{i-1}K_i^{-1}\\
&\big[E_i, E_j\big] = 0\,, \qquad \big[F_i, F_j\big] = 0 \qquad \text{for } \ j \neq i,\, i \pm 1\\
&\big[E_i, F_j\big] = 0 \qquad \text{for } \ i \neq j\\
&\big[E_i, F_i\big] = \frac{K_iK_{i+1}^{-1} - K_i^{-1}K_{i+1}}{q - q^{-1}} \qquad \text{for } \ i \neq m\\
&E_mF_m + F_mE_m = \frac{K_mK_{m+1} - K_m^{-1}K_{m+1}^{-1}}{q - q^{-1}}\\
&E_i^2E_{i\pm1} - \big(q + q^{-1}\big)E_iE_{i\pm1}E_i + E_{i\pm1}E_i^2 = 0 \qquad \text{for } \ i \neq m\\
&F_i^2F_{i\pm1} - \big(q + q^{-1}\big)F_iF_{i\pm1}F_i + F_{i\pm1}F_i^2 = 0 \qquad \text{for } \ i \neq m\\
&E_m^2 = F_m^2 = 0\,,\\
&\big\{E_{m-1}E_m - qE_mE_{m-1}, E_mE_{m+1} - qE_{m+1}E_m\big\} = 0\,,\\
&\big\{F_{m-1}F_m - qF_mF_{m-1}, F_mF_{m+1} - qF_{m+1}F_m\big\} = 0\,,
\end{aligned}
\tag{1}
$$

where $[X, Y] = XY - YX$ is the commutator and $\{X, Y\} = XY + YX$ is the anticommutator of two elements.

The Hopf structure of this superalgebra is defined by the following operations:

1. *Coproduct* Δ

$$
\begin{aligned}
&\Delta(1) = 1 \otimes 1 && \Delta\big(K_r^{\pm1}\big) = K_r^{\pm1} \otimes K_r^{\pm1} &\\
&\Delta\big(E_r\big) = E_r \otimes 1 + K_rK_{r+1}^{-1} \otimes E_r && \Delta\big(F_r\big) = F_r \otimes K_r^{-1}K_{r+1} + 1 \otimes F_r & r < m\\
&\Delta\big(E_m\big) = E_m \otimes 1 + K_mK_{m+1} \otimes E_m && \Delta\big(F_m\big) = F_m \otimes K_m^{-1}K_{m+1}^{-1} + 1 \otimes F_m &\\
&\Delta\big(E_s\big) = E_s \otimes 1 + K_s^{-1}K_{s+1} \otimes E_s && \Delta\big(F_s\big) = F_s \otimes K_sK_{s+1}^{-1} + 1 \otimes F_s & s > m
\end{aligned}
$$

2. *Counit* ε

$$
\begin{aligned}
&\varepsilon(1) = \varepsilon\big(K_r\big) = \varepsilon\big(K_r^{-1}\big) = 1\\
&\varepsilon\big(E_r\big) = \varepsilon\big(F_r\big) = 0
\end{aligned}
$$

3. *Antipode* S

$$
\begin{aligned}
&S(1) = 1 && S\big(K_r\big) = K_r^{-1} && S\big(K_r^{-1}\big) = K_r\\
&S\big(E_r\big) = -K_r^{-1}K_{r+1}E_r && S\big(F_r\big) = -F_rK_rK_{r+1}^{-1} && r < m\\
&S\big(E_m\big) = -K_m^{-1}K_{m+1}^{-1}E_m && S\big(F_m\big) = -F_mK_mK_{m+1} &&\\
&S\big(E_s\big) = -K_sK_{s+1}^{-1}E_s && S\big(F_s\big) = -F_sK_s^{-1}K_{s+1} && s > m
\end{aligned}
$$

However we do not use these operations for construction of the realization.

The method of construction used is the same as in the case of the Lie algebras [23] or quantum algebra [32] and is based on using the induced representation. The difference from quantum algebra is that together with q–deformed boson operators [9], [10] we also use fermion operators.

The algebra $\mathcal{B}$ of the q–deformed boson operators is the associative algebra over the field $\mathcal{C}(q)$ generated by the elements of a^+, $a^- = a$, q^x and q^{-x}, satisfying the commutation relations

$$\begin{aligned} &q^x q^{-x} = q^{-x} q^x = 1\,, &&q^x a^+ q^{-x} = q a^+\,, &&q^x a q^{-x} = q^{-1} a, \\ &a a^+ - q^{-1} a^+ a = q^x\,, &&a a^+ - q a^+ a = q^{-x}\,, && \end{aligned} \tag{2}$$

The algebra $\mathcal{B}$ has faithful representation on vector space with the basic elements $|n\rangle$, where $n = 0, 1, \ldots,$ of the form

$$q^x|n\rangle = q^n|n\rangle\,, \quad a^+|n\rangle = |n+1\rangle\,, \quad a|n\rangle = [n]\,|n-1\rangle\,, \tag{3}$$

where $[n] = \dfrac{q^n - q^{-n}}{q - q^{-1}}$.

Because of odd generators E_m and F_m we construct realization by means of the algebra $\mathcal{B}$ for even elements, and by fermion algebra $\mathcal{F}$ with elements b^+ and b for odd ones. These fermion elements commute with the elements of $\mathcal{B}$ and together fulfil the relations

$$bb = b^+ b^+ = 0\,, \quad bb^+ + b^+ b = 1\,. \tag{4}$$

The algebra $\mathcal{F}$ has faithful representation on vector space with the basis $|0\rangle$ and $|1\rangle$ of the form

$$b|M\rangle = \frac{1-(-1)^M}{2}\,|M-1\rangle\,, \qquad b^+|M\rangle = \frac{1+(-1)^M}{2}\,|M+1\rangle\,. \tag{5}$$

We use superalgebra $\mathcal{H}(r,s)$ for realization superalgebra. We mean the superalgebra of r copies of the algebras $\mathcal{B}$ of q–deformed bosons and s copies of the algebras $\mathcal{F}$ of the fermions. We suppose that the elements from $\mathcal{B}_i$ commute with the elements $\mathcal{F}_k$ for all i, k, and for elements $x_i \in \mathcal{B}_i$ and $y_i \in \mathcal{F}_i$ the relations $[x_i, x_k] = \{y_i, y_k\} = 0$ for $i \neq k$ hold.

As in the case of the Lie algebras or quantum groups our realizations contain elements of quantum sub–superalgebra of $\mathcal{A}_0 \subset U_q(\mathrm{gl}(m/n))$ namely, quantum superalgebra $U_q(\mathrm{gl}(m-1/n))$. The element x of this subalgebra commutes with the elements from $\mathcal{B}_i$, and for the fermion elements $b^\pm$ the relation

$$x b^\pm = (-1)^{\deg x} b^\pm x\,, \tag{6}$$

holds. This superalgebra is denoted by $\mathcal{W}(r,s,\mathcal{A}_0)$.

In general, we define

Definition. Realization of the quantum superalgebra $\mathcal{A}$ is called the homomorphism ρ of $\mathcal{A}$ to the superalgebra $\mathcal{W}(r,s,\mathcal{A}_0)$, where $\mathcal{A}_0$ is the sub–superalgebra of $\mathcal{A}$.

3. Construction of the realization of $U_q(\mathrm{gl}(m/n))$

First, for construction of the realization we find the induced representation of $U_q(\mathrm{gl}(m/n))$. As subalgebra $\mathcal{A}_0$ of $U_q(\mathrm{gl}(m/n))$ we choose a quantum superalgebra generated by E_k, $k>1$, F_r, $r=1,\ldots,m+n-1$, K_i and K_i^{-1}, $i=1,\ldots,m+n$. Let φ be a representation of $\mathcal{A}_0$ on vector space V. Let λ be the left regular representation on $U_q(\mathrm{gl}(m/n))\otimes V$, i.e. for x, $y\in U_q(\mathrm{gl}(m/n))$ and $v\in V$ the representation λ is defined by

$$\lambda(x)\big(y\otimes v\big)=xy\otimes v\,. \tag{7}$$

Let $\mathcal{I}$ be subspace of $U_q(\mathrm{gl}(m/n))\otimes V$ generated by the relations

$$xy\otimes v=x\otimes\varphi(y)v\,,$$

for all $x\in U_q(\mathrm{gl}(m/n))$, $y\in\mathcal{A}_0$ and $v\in V$. It is easy to see that the subspace $\mathcal{I}$ is λ–invariant. Therefore, (7) gives the representation on the factor–space $W=\big[U_q(\mathrm{gl}(m/n))\otimes V\big]/\mathcal{I}$.

To find the induced representation of $U_q(\mathrm{gl}(m/n))$ explicitly we define the elements

$$\begin{aligned}
&X_1=E_1\\
&X_k=X_{k-1}E_k-q^{-1}E_kX_{k-1}\qquad \text{for } k=2,\ldots,m\\
&X_k=X_{k-1}E_k-qE_kX_{k-1}\qquad \text{for } k=m+1,\ldots,m+n-1\,.
\end{aligned}\tag{8}$$

It follows from (8) that X_k for $k<m$ are even and for $k\geq m$ are odd elements. We can derive from generating relations (1) that the elements X_k fulfil the relations

$$\begin{aligned}
&X_rX_s=q^{-1}X_sX_r\,, &&\text{for } s<r \text{ and } s<m\,;\\
&X_k^2=0\,, &&\text{for } r\geq m\,;\\
&X_rX_s=-q^{-1}X_sX_r\,, &&\text{for } m\leq r<s\,.
\end{aligned}\tag{9}$$

Let $|N_1,N_2,\ldots,N_{m+m-1}\rangle=|N\rangle=X_1^{N_1}X_2^{N_2},\ldots,X_{m+n-1}^{N_{m+n-1}}$. Due to the Poincaré–Birkhoff–Witt theorem the space W of the induced representation is generated by the elements $|N\rangle\otimes v$, where $N_k=0,1,2,\ldots$ for $k<m$, $N_k=0,1$ for $k\geq m$ and $v\in V$.

To obtain the explicit form of the induced representation, we give some relations. They can be proved by mathematical induction from relations (1).

Lemma 1. For any $n = 0, 1, 2, \dots$ the following formulae hold:

$$E_r X_{r-1}^N = q^N X_{r-1}^N E_r - q[N] X_{r-1}^{N-1} X_r \qquad \text{for } r \le m$$
$$E_r X_r^N = q^{-N} X_r^N E_r \qquad \text{for } r < m$$
$$E_m X_m^N = (-q)^N X_m^N E_m$$
$$E_m X_r^N = (-1)^N X_r^N E_m \qquad \text{for } r > m$$
$$E_r X_{r-1}^N = q^{-N} X_{r-1}^n E_r - \frac{1-(-1)^N}{2} q^{-N} X_{r-1}^{N-1} X_r \qquad \text{for } r > m$$
$$E_r X_r^N = q^N X_r^N E_r \qquad \text{for } r > m$$
$$K_1 X_r^N = q^N X_r^N K_1$$
$$K_r X_{r-1}^N = q^{-N} X_{r-1}^N K_r \qquad \text{for } r > 1$$
$$F_r X_r^N = X_r^N F_r - q^{-1}[N] X_{r-1} X_r^{N-1} K_r K_{r+1}^{-1} \qquad \text{for } r < m$$
$$F_m X_m^N = (-1)^N X_m^N F_m + \frac{1-(-1)^N}{2} q^{-N} X_{m-1} X_m^{N-1} K_m K_{m+1}$$
$$F_r X_r^N = X_r^N F_r - \frac{1-(-1)^N}{2} q^{-N+2} X_{r-1} X_r^{N-1} K_r^{-1} K_{r+1} \qquad \text{for } r > m$$
$$F_1 X_1^N = X_1^N F_1 - \frac{[N]}{q-q^{-1}} X_1^{N-1}\left(q^{N-1} K_1 K_2^{-1} - q^{-N+1} K_1^{-1} K_2\right)$$
$$F_1 X_r^N = X_r^N F_1 + q^{-N+1}[N] X_r^{N-1} Y_r K_1^{-1} K_2 \qquad \text{for } 1 < r < m$$
$$F_1 X_s^N = X_s^N F_1 + \frac{1-(-1)^N}{2} X_s^{N-1} Y_s K_1^{-1} K_2 \qquad \text{for } s > m$$

where

$$Y_2 = E_2$$
$$Y_k = Y_{k-1} E_k - q^{-1} E_k Y_{k-1} \qquad \text{for } k = 3, \dots, m$$
$$Y_k = Y_{k-1} E_k - q E_k Y_{k-1} \qquad \text{for } k = m+1, \dots, m+n-1\,.$$

We omit the details of the calculations and write the result for the action of the induced representation on the basis elements $|N\rangle \otimes v$.

Theorem 1. Let $1 < r < m$, $s > m$, $t > 1$ and $S_k = \sum_{i=k}^{m+n-1} N_i$. Then the formulae

$$E_1 |N\rangle \otimes v = |N + 1_1\rangle \otimes v$$
$$E_r |N\rangle \otimes v = -q[N_{r-1}] |N - 1_{r-1} + 1_r\rangle \otimes v + q^{N_{r-1}-N_r} |N\rangle \otimes \varphi(E_r) v$$

$$
\begin{aligned}
E_m|N\rangle \otimes v &= -q[N_{m-1}]|N-1_{m-1}+1_m\rangle \otimes v + (-1)^{S_m} q^{N_{m-1}+N_m}|N\rangle \otimes \varphi(E_m)v \\
E_s|N\rangle \otimes v &= -\frac{1-(-1)^{N_{s-1}}}{2}\, q^{-N_{s-1}}|N-1_{s-1}+1_s\rangle \otimes v + q^{N_s-N_{s-1}}|N\rangle \otimes \varphi(E_s)v \\
K_1|N\rangle \otimes v &= q^{S_1}|N\rangle \otimes \varphi K_1 v \\
K_t|N\rangle \otimes v &= q^{-N_{t-1}} \otimes \varphi K_t v \\
F_r|N\rangle \otimes v &= -q^{-1}[N_r]|N+1_{r-1}-1_r\rangle \otimes (K_r K_{r+1}^{-1})v + |N\rangle \otimes \varphi(F_r)v \\
F_m|N\rangle \otimes v &= \frac{1-(-1)^{N_m}}{2} q^{-N_m}|N+1_{m-1}-1_m\rangle \otimes \varphi(K_m K_{m+1})v + \\
&\quad +(-1)^{S_m}|N\rangle \otimes \varphi(F_m)v \\
F_s|N\rangle \otimes v &= -\frac{1-(-1)^{N_s}}{2} q^{-N_s+2}|N+1_{s-1}-1_s\rangle \otimes \varphi(K_s^{-1}K_{s+1})v + |N\rangle \otimes \varphi(F_s)v \\
F_1|N\rangle \otimes v &= -\frac{[N_1]q^{S_1-1}}{q-q^{-1}}|N-1_1\rangle \otimes \varphi(K_1 K_2^{-1})v + \\
&\quad + \frac{[N_1]q^{-S_1+1}}{q-q^{-1}}|N_1-1\rangle \otimes \varphi(K_1^{-1}K_2)v + \\
&\quad + \sum_{r=2}^{m-1} q^{-S_r+1}[N_r]|N-1_r\rangle \otimes \varphi(Y_r K_1^{-1}K_2)v + \\
&\quad + \sum_{s=m}^{m+n-1} \frac{1-(-1)^{N_s}}{2}(-q)^{-S_s+1}|N-1_s\rangle \otimes \varphi(Y_s K_1^{-1}K_2)v + |N\rangle \otimes \varphi(F_1)v
\end{aligned}
$$

give the induced representation of the quantum superalgebra $U_q(\mathrm{gl}(m/n))$. We use the notation $|N \pm 1_r\rangle = |N_1, \ldots, N_r \pm 1, \ldots, N_{m+n-1}\rangle$.

To obtain the realization of quantum superalgebra $U_q(\mathrm{gl}(m/n))$ we choose the representation φ, for which $\varphi(F_1)v = 0$ and $\varphi(K_1)v = q^\lambda v$ and rewrite the induced representation given in Theorem 1 by means of the elements from $\mathcal{W}(m-1, n, \mathcal{A}_0)$. It follows from (3) and (5) that we substitute

$$
\begin{aligned}
&q^{\pm N_k} \to q^{\pm x_k} && \text{for } k < m \\
&q^{\pm N_k} \to b_k b_k^+ + q^{\pm 1} b_k^+ b_k \equiv q^{\pm x_k} && \text{for } k \geq m \\
&|N+1_k\rangle \to a_k^+ && \text{for } k < m \\
&[N_k]\,|N-1_k\rangle \to a_k && \text{for } k < m \\
&(-1)^{N_m+\cdots+N_{k-1}}|N+1_k\rangle \to b_k^+ && \text{for } k \geq m \\
&(-1)^{N_m+\cdots+N_{k-1}} \frac{1-(-1)^{N_k}}{2}|N-1_k\rangle \to b_k && \text{for } k \geq m
\end{aligned}
\qquad (10)
$$

$$
\begin{array}{ll}
\varphi(F_1)v \to 0 & \\
\varphi(K_1^{\pm 1})v \to q^{\pm\lambda} & \\
\varphi(K_r^{\pm 1})v \to k_r^{\pm 1} & \text{for } k \geq 2 \\
\varphi(E_k)v \to e_k & \text{for } k \neq m \\
\varphi(F_k)v \to f_k & \text{for } k \neq m \\
(-1)^{S_m}\varphi(E_m)v \to e_m & \\
(-1)^{S_m}\varphi(F_m)v \to f_m & \\
\varphi(Y_k)v \to y_k & \text{for } k < m \\
(-1)^{S_m}\varphi(Y_k)v \to y_k & \text{for } k \geq m
\end{array}
$$

The factors $(-1)^{N_m+\cdots+N_{k-1}}$ reflect the fact that the corresponding elements are fermions.

By substitutions (10) we obtain the realization of the quantum superalgebra $U_q(\mathrm{gl}(m/n))$, which is given in the following theorem.

Theorem 2. Let be $r = 2, \ldots, m-1$ and $s = m+1, \ldots, m+n-1$. The mapping $\rho : U_q(\mathrm{gl}(m/n)) \to \mathcal{W}\Big(m-1, n, U_q\big(\mathrm{gl}(m-1/n)\big)\Big)$ defined by the formulae

$$
\begin{aligned}
\rho(E_1) &= a_1^+ \\
\rho(E_r) &= -qa_{r-1}a_r^+ + q^{x_{r-1}-x_r}e_r \\
\rho(E_m) &= -qa_{m-1}b_r^+ + q^{x_{m-1}+x_m}e_m \\
\rho(E_s) &= q^{-1}b_{s-1}b_s^+ + q^{x_s-x_{s-1}}e_s \\
\rho(K_1) &= q^{\lambda+x_1+x_2+\cdots+x_{m+n-1}} \\
\rho(K_r) &= q^{-x_{r-1}}k_r \\
\rho(K_m) &= q^{-x_{m-1}}k_m \\
\rho(K_s) &= q^{-x_{s-1}}k_s \\
\rho(F_r) &= -q^{-1}a_{r-1}^+a_rk_rk_{r+1}^{-1} + f_r \\
\rho(F_m) &= q^{-1}a_{m-1}^+b_mk_kk_{m+1} + f_m \\
\rho(F_s) &= -qb_{s-1}^+b_sk_s^{-1}k_{s+1} + f_s \\
\rho(F_1) &= -\frac{a_1}{q-q^{-1}}\Big(q^{\lambda+x_1+\cdots+x_{m+n-1}-1}k_2^{-1} - q^{-\lambda-x_1-\cdots-x_{m+n-1}+1}k_2\Big) + \\
&\quad + \sum_{r=2}^{m-1} a_r q^{-\lambda-x_r-\cdots-x_{m+n-1}+1}y_rk_2 + \sum_{r=m}^{m+n-1} b_r q^{-\lambda-x_{r+1}-\cdots-x_{m+n-1}}y_sk_2
\end{aligned}
$$

is the realization of the quantum superalgebra $U_q\big(\mathrm{gl}(m/n)\big)$.

In the formulae we used the aberrations $q^{\pm x_k} = b_kb_k^+ + q^{\pm 1}b_k^+b_k$ for $k \geq m$ and $q^{x_r+x_s} = q^{x_r}q^{x_s}$ for simplicity.

Proof: Since the representations of $\mathcal{B}$ and $\mathcal{F}$ given in (3) and (5) are faithful, the representation $\mathcal{W}\Big(m-1, n, U_q\big(\mathrm{gl}(m-1/n)\big)\Big)$ is faithful on the vector space W generated by $|N\rangle \otimes x$, where $x \in U_g\big(\mathrm{gl}(m-1/n)\big)$. Theorem follows from the fact that the representation $U_q\big(\mathrm{gl}(m/n)\big)$ on W, which we obtain by means of the inverse formulae to (10), is the representation given in Theorem 1.

4. Conclusion

In this paper we gave the method of construction of the q–boson–fermion realization of quantum superalgebras and applied it to the quantum superalgebra $U_q\big(\mathrm{gl}(m/n)\big)$. One of the advantages of this method, in comparison with [37], is that we automatically obtain a realization and we do not need to verify the generating relation.

The other advantage we see in the fact that our realization is expressed by means of polynomials of q–deformed bosons and fermions. On the other hand, we can easily obtain the Dyson realization of quantum superalgebra. For this purpose, it is sufficient to choose a realization of the generators of the algebra $\mathcal{B}$ in the form

$$a^+ = A^+\,, \quad a = \frac{[N+1]}{N+1}\,A\,, \quad q^x = q^N\,, \tag{11}$$

where $\big[A, A^+\big] = 1$ and $N = A^+A$. It is easy to verify that the realization of $U_q\big(\mathrm{gl}(m/n)\big)$ from Theorem 2 with realization (11) of the algebra $\mathcal{B}$ and with a trivial realization of the subalgebra $U_q\big(\mathrm{gl}(m-1/n)\big)$ leads, after homomorphism of $U_q\big(\mathrm{gl}(m/n)\big)$, to the realization given in [37]. In this case, the realization is of course expressed by means of a series in the operators A^+ and A. Therefore, we prefer our form of realizations.

Finally, our realizations contain, in contrast with those in [37], quantum sub–superalgebras. Various forms of realizations of this sub–superalgebra give various realizations of the quantum superalgebra. For instance, by the presented method we can find realization of $U_q\big(\mathrm{gl}(m-1/n)\big)$ on $\mathcal{W}\Big(m-2, n, U_q\big(\mathrm{gl}(m-2,n)\big)\Big)$ and use this realization in our formulae. In this case we obtain the realization of $U_q\big(\mathrm{gl}(m/n)\big)$ on $\mathcal{W}\Big(2m-3, n, U_q\big(\mathrm{gl}(m-2/n)\big)\Big)$. On the other hand, we can construct the realization of $U_q\big(\mathrm{gl}(m-1/n)\big)$ on $\mathcal{W}\Big(n-1, m-1, U_q\big(\mathrm{gl}(m-1,n-1)\big)\Big)$. Using this realization in our formulae we obtain realization $U_q\big(\mathrm{gl}(m/n)\big)$ on $\mathcal{W}\Big(m+n-1, m+n-1, U_q\big(\mathrm{gl}(m-$

$1, n-1))\big)$. The possibility of using similar recurrence is, in our opinion, one of the most important advantages of presented construction.

Acknowledgments

Partial support from grant 201/01/0130 of the Czech Grant Agency is gratefully acknowledged.

References

1. M. Jimbo, *Lett. Math. Phys.* **10**, 63 (1985); **11**, 247 (1986).
2. V. Drinfeld, *Proc. Intern. Congress of Mathematicians*, Berkeley, 798 (1986).
3. P. Kulish and N. Reshetikhin, *Lett. Math. Phys.* **18**, 143 (1989).
4. M. Chaichian and P. Kulish, *Phys. Lett.* **234B**, 72 (1990).
5. S. Sauler, *Nucl. Phys.* **B336**, 363 (1990).
6. M. Bershadsky and H. Ooguri, *Phys. Lett.* **229B**, 374 (1989).
7. L. Kauffman and H. Saleur, "Free Fermions and the Alexander-Conway Polynomial", preprint EFI 90-42.
8. T. Hayashi, *Commun. Math. Phys.* **128**, 129 (1990).
9. A.J. Macfarlane, *J. Phys. A: Math. Gen.* **22**, 4581 (1989).
10. L.C. Biedenharn, *J. Phys. A: Math. Gen.* **22**, L873 (1989).
11. J. Schwinger, in: *Quantum Theory of Angular Momentum*, (Acad. Press, New York–London, 1965).
12. R. Floreanini, V.P. Spiridonov and L. Vinet, *Commun. Math. Phys.* **137**, 149 (1991).
13. F.J. Dyson, *Phys. Rev.* **102**, 1217 (1956).
14. T. Holstein and H. Primakoff, *Phys. Rev.* **58**, 1098 (1949).
15. M. Chaichian, D. Ellinas and P.P. Kulish, *Phys. Rev. Lett.* **65**, 980 (1990).
16. J. da-Providencia, *J. Phys. A: Math. Gen.* **26**, 5845 (1993).
17. C. Quesne, *Phys. Lett.* **A153**, 303 (1991).
18. R. Chakrabarti and R. Jagannathan, *J. Phys. A: Math. Gen.* **24**, L711 (1991).
19. J. Katriel and A.I. Solomon, *J. Phys. A: Math. Gen.* **24**, 2093 (1991).
20. Z.R. Yu, *J. Phys. A: Math. Gen.* **24**, L1321 (1991).
21. A. Kundu and M. Basu Mallik, *Phys. Lett.* **A156**, 175 (1991).
22. F. Pan. Own, *Phys. Lett.* **8**, 56 (1991).
23. Č. Burdík, *J. Phys. A: Math. Gen.* **18** 3102, (1985).
24. Č. Burdík, *Czech. J. Phys.* **B36**, 1235 (1986); *J. Phys. A: Math. Gen.* **19**, 2465 (1986); *J. Phys. A: Math. Gen.* **21**, 289 (1988).
25. M. Havlíček and W. Lassner, *Rep. Math. Phys.* **8**, 391 (1975); *Int. J. Theor. Phys.* **15**, 867 (1976).
26. P. Exner and M. Havlíček, *Ann. Inst. H. Poincare, Sect. A (N.S.)* **23**, 335 (1975).
27. T.D. Palev, *J. Phys. A: Math. Gen.* **30**, 8273 (1997), hep–th/9607222.
28. A. Angelucci and R. Link, *Phys. Rev.* **B46**, 3809 (1992).
29. N.I. Karchev, *Teor. Mat. Fiz.* **92**, 988 (1992).

30. Č. Burdík and O. Navrátil, *J. Phys. A: Math. Gen.* **23**, L1205 (1990).
31. Č. Burdík, L. Černý and O. Navrátil, *J. Phys. A: Math. Gen.* **26**, L83 (1993).
32. Č. Burdík and O. Navrátil, *Czech. J. Phys.* **B47**, 1301 (1998); *J. Phys. A: Math. Gen.* **32**, 6141 (1999); *Int. J. Mod. Phys. Lett.* **14**, 4491 (1999).
33. T.D. Palev, *J. Phys. A: Math. Gen.* **31**, 5145 (1998).
34. A.J. Bracken, M.D. Gould and J.R. Links, *Phys. Rev. Lett.* **74**, 2768 (1994); cond-mat/9410026.
35. M.D. Gould, K.E. Hibberd and J.R. Links, *Phys. Lett.* **A212**, 156 (1996); cond-mat/9506119.
36. A. Kümper and K. Sakai, *J. Phys. A: Math. Gen.* **34**, 8015 (2001); cond-mat/0105416.
37. T.D. Palev, *Mod. Phys. Lett.* **A14**, 299 (1999).

IRREDUCIBLE REPRESENTATIONS OF LIFTINGS OF QUANTUM PLANES

N. ANDRUSKIEWITSCH
Facultad de Matemática, Astronomía y Física,
Universidad Nacional de Córdoba
CIEM - CONICET
(5000) Ciudad Universitaria, Córdoba, Argentina
andrus@mate.uncor.edu

MARGARET BEATTIE
Department of Mathematics and Computer Science
Mount Allison University,
Sackville, NB, Canada E4L 1E6
mbeattie@mta.ca

In this note, the irreducible representations of a lifting of a quantum plane are determined.

1. Introduction

In [5, 1], the structure of the coradicals of the duals of liftings of some quantum linear spaces was studied, and several examples were explicitly constructed. In this note, we describe the irreducible representations for any lifting of a quantum plane. If A is the lifting of a quantum linear plane, then the approach in [5, 1] was via the coradical of A^*. The coalgebra $A^* = C$ was written as a direct sum of certain sub-coalgebras $C(\eta)$ with $\eta \in \hat{\Gamma}$ and the coradicals of the $C(\eta)$ were determined. Here we write the lifting A as a product of algebras $A(\xi)$, $\xi \in \hat{\Gamma}$, and determine irreducible representations of each of six possible types of algebra $A(\xi)$ that can arise.

Notation

Let k be an algebraically closed field of characteristic 0. If G is a finite group then ${}_G^G\mathcal{YD} = {}_{k[G]}^{k[G]}\mathcal{YD}$ will denote the category of Yetter-Drinfel'd modules over the group algebra $k[G]$.

Throughout, we let Γ be a finite abelian group and let $\hat{\Gamma}$ denote the group of characters of Γ. For $V \in_{\Gamma}^{\Gamma} \mathcal{YD}$, $g \in \Gamma, \chi \in \hat{\Gamma}$, we write V_g^{χ} for the set of $v \in V$ with the action of Γ on v given by $h \rightarrow v = \chi(h)v$ and the coaction by $\delta(v) = g \otimes v$. Since Γ is an abelian group, $V = \oplus_{g\in\Gamma,\chi\in\hat{\Gamma}} V_g^{\chi}$ [2, Section 2].

The coradical of a coalgebra C is denoted C_0.

If v is a complex number and a is a non-negative integer, then we set as usual

$$(0)_v = 0, \qquad (a)_v = 1 + v + \dots v^{a-1} = 1 + v(a-1)_v = (a-1)_v + v^{a-1}.$$

2. Preliminaries

Let A be a finite-dimensional pointed Hopf algebra; let V be the infinitesimal braiding of A [2]. Recall that $V \in_{G(A)}^{G(A)} \mathcal{YD}$ is the Yetter-Drinfeld module of right coinvariant elements in the $kG(A)$-Hopf module A_1/A_0.

Let Λ be a subgroup of $G(A)$ central in A. For $\xi \in \widehat{\Lambda}$, let

$$e_{\xi} = |\Lambda|^{-1} \sum_{g\in\Lambda} \xi^{-1}(g) g$$

denote the minimal idempotent corresponding to ξ. If also $\eta \in \widehat{\Lambda}$ then

$$e_{\eta} \leftharpoonup \xi = \langle \xi, e_{\eta,(1)} \rangle e_{\eta,(2)} = e_{\xi^{-1}\eta}. \tag{1}$$

Clearly, Ae_{ξ} is a two-sided ideal of A and $A = \oplus_{\xi\in\widehat{\Lambda}} Ae_{\xi}$. Furthermore, let $J_{\xi} = \oplus_{\xi\neq\zeta\in\widehat{\Lambda}} Ae_{\zeta}$ and let $A(\xi) := A/J_{\xi}$; we have an isomorphism of algebras $A \simeq \prod_{\xi\in\widehat{\Lambda}} A(\xi)$. Hence, we have an isomorphism of coalgebras $C := A^* \simeq \oplus_{\xi\in\widehat{\Lambda}} C(\xi)$ where $C(\xi) := A(\xi)^* = J_{\xi}^{\perp}$. Clearly, $C(\epsilon)$ is a Hopf subalgebra of C; it is dual to the quotient Hopf algebra $A/J_{\epsilon} = A/Ak[\Lambda]^+$. It turns out that a useful approach to the description of the coalgebra structure of C is via the coalgebras $C(\xi)$ for a well-chosen subgroup Λ. Indeed, $C_0 := \oplus_{\xi\in\widehat{\Lambda}} C(\xi)_0$ by [9, 9.0.1].

Assume that $\xi \in \widehat{\Lambda}$ is the restriction of $\widetilde{\xi} \in G(A)$. Then $\widetilde{\xi}C(\zeta) = C(\xi\zeta)$ for any $\zeta \in \widehat{\Lambda}$. Indeed, $\widetilde{\xi}C(\zeta) \subset (Ae_{\eta})^{\perp}$ for any $\eta \in \widehat{\Lambda}$, $\eta \neq \xi\zeta$, by (1); thus $\widetilde{\xi}C(\zeta) = J_{\xi\zeta}^{\perp}$.

Let us consider the following hypotheses.

(a) The group $\Gamma := G(A)$ is abelian. Then there exists a basis $v_1, \dots v_{\theta}$ of V with $v_i \in V_{g_i}^{\chi_i}$ for all i. Let $r_i > 1$ be the order of $\chi_i(g_i)$.

(b) V is a quantum linear space; that is, $\chi_i(g_j)\chi_j(g_i) = 1$ for $i \neq j$.

Furthermore, assuming (a) and (b), we choose the subgroup Λ of Γ as follows:

(c) $\Lambda = \{g \in \Gamma : \chi_i(g) = 1 \text{ for } 1 \leq i \leq \theta\}$.

Note that by the definition in (c), Λ is central in A, and if $\chi_j^{r_j} = \epsilon$ or if $\chi_i\chi_j = \epsilon$ for all $i \neq j$, then

$$g_j^{r_j} \in \Lambda \text{ and } \chi_j \in \widehat{\Gamma/\Lambda}. \tag{2}$$

Also since the sequence $1 \to \widehat{\Gamma/\Lambda} \to \widehat{\Gamma} \to \widehat{\Lambda} \to 1$ is exact, we can always choose a preimage $\widetilde{\xi}$ in $\widehat{\Gamma}$ of $\xi \in \widehat{\Lambda}$; then $C(\xi) = \widetilde{\xi}C(\epsilon)$.

Proposition 2.1. *Let A be a finite-dimensional pointed Hopf algebra; let V be the infinitesimal braiding of A. Assume (a) and (b) above, i.e. A is a lifting of a quantum linear space with abelian group of grouplikes. Then A is generated by the grouplike elements $h \in \Gamma$ and by $(1, g_i)$-primitives $x_i, 1 \leq i \leq \theta$ with defining relations*

$$\begin{aligned} hx_i &= \chi_i(h)x_ih; \\ x_i^{r_i} &= \alpha_{ii}(g_i^{r_i} - 1); \\ x_ix_j &= \chi_j(g_i)x_jx_i + \alpha_{ij}(g_ig_j - 1), \end{aligned}$$

for all $h \in \Gamma$, $1 \leq i \leq \theta$. We have dim $A = |\Gamma|r_1 \ldots r_\theta$.

We may assume $\alpha_{ii} \in \{0, 1\}$ and then we must have that

$$\begin{aligned} \alpha_{ii} &= 0 \text{ if } g_i^{r_i} = 1 \text{ or } \chi_i^{r_i} \neq \epsilon; \\ \alpha_{ij} &= 0 \text{ if } g_ig_j = 1 \text{ or } \chi_i\chi_j \neq \epsilon. \end{aligned}$$

Note that $\alpha_{ji} = -\chi_j(g_i)^{-1}\alpha_{ij} = -\chi_i(g_j)\alpha_{ij}$. Thus the lifting A is described by the *lifting matrix* $\mathcal{A} = (\alpha_{ij})$ with 0's or 1's on the diagonal and with $\alpha_{ji} = -\chi_i(g_j)\alpha_{ij}$ for $i \neq j$.

PROOF. By the same argument as in [4], A is generated by group-like and skew-primitive elements. See [3] for the rest of the proof. These liftings were independently constructed in [6] by iterated Ore extensions. □

In the rest of the paper we assume that A is a lifting of a quantum linear space with notation as in Proposition 2.1.

Since Γ is finite abelian, there exist elements $h_1, \dots, h_t$ in Γ, and non-negative integers $a_1, \dots, a_t$, $b_1, \dots, b_t$ with $a_u = \operatorname{ord} h_u$, $\Gamma = \langle h_1 \rangle \oplus \cdots \oplus \langle h_t \rangle$ and $\Lambda = \langle h_1^{b_1} \rangle \oplus \cdots \oplus \langle h_t^{b_t} \rangle$.

Let $\xi \in \widehat{\Lambda}$; the algebra $A(\xi)$ is then generated by x_i, $1 \le i \le \theta$, h_u, $1 \le u \le t$, and relations

$$h_u h_\ell = h_\ell h_u, \quad h_u^{b_u} = \xi(h_u^{b_u}), \quad h_u x_i = \chi_i(h_u) x_i h_u,$$
$$x_i^{r_i} = \alpha_{ii}(g_i^{r_i} - 1), \quad x_i x_j = \chi_j(g_i) x_j x_i + \alpha_{ij}(g_i g_j - 1),$$

$1 \le u, \ell \le t$, $1 \le i \le \theta$. Let V be any finite-dimensional $A(\xi)$-module; we denote the action of the elements x_i, h_u on V by the same letters. Thus V is a Γ-module, since $A(\xi)$ is a quotient of A; and

$$V = \oplus_{\eta \in F(\xi)} V^\eta$$

where

$$F(\xi) = \{\eta \in \widehat{\Gamma} : \eta|_\Lambda = \xi\} = \widetilde{\xi}\,\widehat{\Gamma/\Lambda}; \tag{3}$$

see above. We have $\dim A(\xi) = \dfrac{|\Gamma| r_1 \dots r_\theta}{|\Lambda|}$.

The case when the rank $\theta = 1$ is known [8, 5]. We shall investigate the case $\theta = 2$ in the next Section. Consider the condition

(d) The rank $\theta > 2$ and $\operatorname{ord} r_i > 2$, $1 \le i \le \theta$.

We say that $i, j \in I := \{1, \dots, \theta\}$ are *linked* [4] if $\alpha_{ij} \neq 0$. By (d), if i is linked to j and k then $j = k$ [4]. Thus I is a disjoint union of the set of vertices which are linked, which has even cardinal, and the rest. Roughly speaking, the representation theory of $A(\xi)$ looks like the "tensor product" of representation theories of similar algebras with rank $\theta = 1$ or 2.

3. The rank 2 case

In this section we assume that $\theta = 2$ and write $x = x_1$, $y = x_2$, $r = r_1$, $s = r_2$. The lifting matrix of A has the form $\mathcal{A} = \begin{bmatrix} \alpha_{11} & \nu \\ -\chi_2(g_1)\nu & \alpha_{22} \end{bmatrix}$.

Let $\xi \in \widehat{\Lambda}$; we shall determine the irreducible representations of the algebra $A(\xi)$ generated by x, y, h_u, $1 \leq u \leq t$, and relations

$$h_u h_\ell = h_\ell h_u, \quad h_u^{b_u} = \xi(h_u^{b_u}); \tag{4}$$

$$h_u x = \chi_1(h_u) x h_u; \tag{5}$$

$$h_u y = \chi_2(h_u) y h_u; \tag{6}$$

$$x^r = \alpha_{11}(g_1^r - 1); \text{ we denote } \alpha_{11}(\xi(g_1^r) - 1) = \alpha; \tag{7}$$

$$y^s = \alpha_{22}(g_2^s - 1); \text{ we denote } \alpha_{22}(\xi(g_2^s) - 1) = \beta; \tag{8}$$

$$xy = \chi_2(g_1) yx + \nu(g_1 g_2 - 1). \tag{9}$$

We have $\dim A(\xi) = \dfrac{|\Gamma| rs}{|\Lambda|}$.

Let $q = \chi_1(g_1)$. If $\chi_1\chi_2 = \epsilon$, then $q = \chi_2(g_1)^{-1} = \chi_1(g_2) = \chi_2(g_2)^{-1}$ (hence $r = s$), and relation (9) becomes $yx = q\,(xy - \nu(g_1g_2 - 1))$.

We distinguish six cases; up to change of variables, these six cases cover all possibilities for $A(\xi)$.

(I) $\alpha = \beta = 0$, $\nu = 0$.

(II) $\alpha = 1$, $\beta = 0$, $\nu = 0$. Necessarily, $\chi_1^r = \epsilon$ and $g_1^r \neq 1$.

(III) $\alpha = \beta = 1$, $\nu = 0$. Necessarily, $\chi_1^r = \chi_2^s = \epsilon$, $g_1^r \neq 1$ and $g_2^s \neq 1$.

(IV) $\alpha = \beta = 0$, $\nu = 1$. Necessarily, $\chi_1\chi_2 = \epsilon$ and $g_1g_2 \neq 1$.

(V) $\alpha = 1$, $\beta = 0$, $\nu = 1$. Necessarily, $\chi_1^r = \epsilon$ and $g_1^r \neq 1$; $\chi_1\chi_2 = \epsilon$ and $g_1g_2 \neq 1$.

(VI) $\alpha = \beta = 1$, $\nu \neq 0$. Necessarily, $\chi_1^r = \chi_2^s = \epsilon$, $g_1^r \neq 1$, $g_2^s \neq 1$; $\chi_1\chi_2 = \epsilon$ and $g_1g_2 \neq 1$.

Now we examine each of the 6 cases listed above.

Case (I) Here the lifting is trivial. In this case we have the following well-known theorem.

Theorem 3.1. *The irreducible representations of $A(\xi)$ have dimension one and are parametrized by $\widehat{\Gamma/\Lambda}$. Indeed, $A(\xi) \simeq \mathcal{B}(V)\#k[\Gamma]/(\mathcal{B}(V)\#k[\Gamma])e_\xi$.*

Case (II) Here $\alpha = 1$ and $\beta = \nu = 0$. Thus $\chi_1^r = \epsilon$ and since $q = \chi_1(g_1)$ is a primitive $r-$th root of unity, r is the order of χ_1.

It is now convenient to consider the algebra $B(\xi)$ presented by generators x, h_u, $1 \leq u \leq t$, and relations

$$h_u h_\ell = h_\ell h_u, \qquad h_u^{b_u} = \xi(h_u^{b_u}),$$
$$h_u x = \chi_1(h_u) x h_u, \quad x^r = 1.$$

The representation theory of $B(\xi)$ is well-known; we include it for completeness. (For example, see [10], [8].)

Lemma 3.2. *Let $\eta \in F(\xi)$. Let $W(\eta)$ be a vector space with a basis f_i, $0 \leq i \leq r-1$, with subscripts $i \in \mathbb{Z}/r$. There is a representation of $B(\xi)$ on $W(\eta)$ defined by the following rules:*

$$x.f_i = f_{i+1}, \quad h_u.f_i = (\chi_1^i \eta)(h_u) f_i, \quad 1 \leq u \leq t$$

where $i \in \mathbb{Z}/r$ means that $x.f_{r-1} = f_0$. Furthermore, $W(\eta)$ is irreducible; all irreducibles are of this kind; $W(\eta) \simeq W(\eta')$ if and only if $\eta = \eta' \chi_1^m$ for some m; and $B(\xi)$ is semisimple.

PROOF. It is straightforward to verify that $W(\eta)$ is a representation of $B(\xi)$. We see that $W(\eta)$ is irreducible because the f_i's belong to different isotypical components for Γ.

Let V be an irreducible representation of $A(\xi)$. Since $V = \oplus_{\eta \in F(\xi)} V^\eta$, we can choose $v \in V^\eta - 0$ for some $\eta \in \widehat{\Gamma}$. Let t be the order of x in V; clearly $x^i.v \in V^{\chi_1^i \eta} - 0$, $0 \leq i \leq t$; and $x^t.v = v$. Then $r|t|r$, thus $t = r$ and $V \simeq W(\eta)$. If $\phi : W(\eta) \to W(\eta')$ is an isomorphism of $B(\xi)$-modules, then $\phi(f_0) \in k f'_m$; thus $\eta = \eta' \chi_1^m$. Conversely, assume that $\eta = \eta' \chi_1^m$ and define $\phi : W(\eta) \to W(\eta')$ by $\phi(f_i) = f'_{m+i}$; ϕ is an isomorphism of $B(\xi)$-modules. Finally, $\dim B(\xi) \leq |\Gamma/\Lambda| r$ by the defining relations, but the dimension of the quotient of $B(\xi)$ by its Jacobson radical is $\geq |\Gamma/\Lambda| r$ by what we have already proved; thus $B(\xi)$ is semisimple. □

Theorem 3.3. *Let $A(\xi)$ be such that $\alpha = 1$, $\beta = \nu = 0$. Let $\pi : A(\xi) \to B(\xi)$ be the algebra map sending y to 0. Then any irreducible representation of $A(\xi)$ factorizes through π. In particular, all irreducible representations of $A(\xi)$ are described by Lemma 3.2.*

PROOF. Let V be an irreducible representation of $A(\xi)$. Since $0 \neq \ker y$ is stable under the action of Γ and x, we see that y acts as 0 on V. □

Case (III) Here $\alpha = \beta = 1$ and $\nu = 0$. Let $w := \chi_2(g_1)$. Since $\alpha = \beta = 1$, as noted in previous cases, we have that χ_1 has order r and χ_2 has order s. Thus $w^r = w^s = 1$.

Theorem 3.4. *$A(\xi)$ is semisimple.*

PROOF. There exists a unique algebra automorphism Y of $B(\xi)$ such that $Y(h_u) = \chi_2(h_u)^{-1}h_u$, $1 \leq u \leq t$, and $Y(x) = w^{-1}x$; clearly Y^s =id. It is well-known that the smash product $B(\xi)\#kY$ is semisimple, see [7]. But $B(\xi)\#kY$ is isomorphic as an algebra to $A(\xi)$, say by dimension counting. □

The explicit description of all simple $A(\xi)$-modules can be obtained by means of Clifford theory, see [7].

Case (IV) In the next three cases, we have $\nu \neq 0$ so that $\chi_1\chi_2 = \epsilon$ and so $r = s$. For all of the remaining cases, we will want to define a set of scalars c_i by the following recursive definition.

Assume that $\chi_1\chi_2 = \epsilon$. Fix $c = c_0$ and for $\eta \in F(\xi)$ and $i > 0$, define

$$c_i = q\left(c_{i-1} - \nu\eta\chi_1^{i-1}(g_1g_2 - 1)\right) = q\left(c_{i-1} + \nu - \nu q^{2(i-1)}\eta(g_1g_2)\right). \quad (10)$$

The second equality follows from the fact that $q = \chi_1(g_1) = \chi_1(g_2)$. A simple induction shows that for $i > 0$, we have

$$c_i = q^i c + q(i)_q\nu\left(1 - q^{i-1}\eta(g_1g_2)\right). \quad (11)$$

Thus if q is a primitive r-th root of unity, if $i \equiv k \mod r$, then $c_i = c_k$.

In this case, $\alpha = \beta = 0$ and $\nu = 1$. Here, the representation theory is similar to that of a Frobenius-Lusztig kernel of type $\mathfrak{sl}(2)$.

Theorem 3.5. *Let $\eta \in F(\xi)$ as defined in (3). Let $(c_i)_{i\geq 0}$ be scalars defined recursively by (10) with $c_0 = 0$.*

Let N be the least positive integer such that $c_N = 0$. Note that since $(r)_q = 0$, then $N \leq r$. Let $L(\eta)$ be a vector space with a basis $(v_i)_{0\leq i\leq N-1}$; set $v_{-1} = v_N = 0$ in $L(\eta)$. Then there exists a representation of $A(\xi)$ on $L(\eta)$ given by

$$h_u.v_i = \eta\chi_1^i(h_u)v_i, \quad 1 \leq u \leq t, \qquad y.v_i = c_iv_{i-1}, \qquad x.v_i = v_{i+1}. \quad (12)$$

Furthermore, $L(\eta)$ is irreducible. Also any irreducible $A(\xi)$-module is isomorphic to $L(\eta)$ for some η; and $L(\eta)$ is isomorphic to $L(\eta')$ only when $\eta = \eta'$.

PROOF. The verification that (12) defines a representation of $A(\xi)$ is straightforward. The fact that $y.v_0 = 0 = x.v_{N-1}$ ensures that relations (7) and (8) hold while (10) guarantees that relation (9) is respected. We leave the reader to check the details and to check that $A(\xi)$ is irreducible.

Let $\rho : A(\xi) \to \operatorname{End} V$ be an irreducible representation. Since $\ker y \neq 0$ and is Γ-stable, there exists $v \in \ker y - 0$, $v \in V^{\eta}$ for some $\eta \in F(\xi)$. Set $v_0 = v$, $v_i = x^i.v$, $i > 0$; $v_i \in V^{\eta\chi_1^i}$ for all i. It follows from the fact that relation (9) must be respected and from a simple induction that for $i > 0$, we have $y.v_i = d_i v_{i-1}$, where the d_i's satisfy the recursive relation (10).

Now, $v_0, v_1 \dots v_{m-1}$ generate a submodule of V where $v_m = x^m.v_0 = 0$ and $m \leq r$. If $m > N$, then $v_{m-1}, \dots, v_N$ is a submodule of V. Thus $m = N$ and since $d_i = c_i$, $i \geq 0$, then V coincides with the submodule generated by the v_i's, $i \geq 0$ which is isomorphic to $L(\eta)$.

Finally, $L(\eta)$ is presented as $A(\xi)$-module by generator v_0 with relations $h_u(v_0) = \eta(h_u)v_0$, $1 \leq u \leq t$, $y.v_0 = 0$, $x^N.v_0 = 0$. Thus, $L(\eta) \simeq L(\eta')$ implies $\eta = \eta'$. □

Case (V) In this case, we have $\alpha = \nu = 1$ and $\beta = 0$.

Theorem 3.6. *Let $\eta \in F(\xi)$. Let $W(\eta)$ be the $B(\xi)$-module defined in Lemma 3.2 with basis $f_0, \dots, f_{r-1}$ with subscripts taken modulo r. Set $c_i = 0$ for $i = 0$ and define scalars $(c_i)_{0<i\leq r}$ recursively by (10). Define an operator y on $W(\eta)$ by $y.f_i = c_i f_{i-1}$. Then this defines a representation of $A(\xi)$ and we denote this $A(\xi)$-module by $L(\eta)$. Then $L(\eta)$ is irreducible; all irreducibles are of this kind; If $L(\eta) \simeq L(\eta')$ then $\eta = \eta'\chi_1^m$ for some m.*

PROOF. As usual, the verification that $L(\eta)$ is an irreducible representation is straightforward. Recall from (11) that it makes sense to compute subscripts modulo r.

Next, let V be an irreducible $A(\xi)$-module. Then there exists $\eta \in F(\xi)$ and $v \neq 0$ such that $v \in \ker y \cap V^{\eta}$. Set $f_0 = v$, $f_i = x^i v$. Arguing as in Lemma 3.2 we see that the f_i's span a $B(\xi)$-submodule U isomorphic to $W(\xi)$. Relation (9) implies the description of the action of y on U by $y.f_i = \alpha_i f_{i-1}$, where the α_i's are defined by (10). Hence $V = U \simeq L(\xi)$. □

Case (VI) In this case, $\alpha = \beta = 1$ and $\nu \neq 0$.

Theorem 3.7. *Let $\eta \in F(\xi)$. Let $W(\eta)$ be the $B(\xi)$-module defined in Lemma 3.2. Set $c_0 = c \neq 0$ and define a family of scalars $(c_i)_{0\leq i\leq r-1}$ inductively by (10). Define an operator y on $W(\eta)$ by $y.f_i = c_i f_{i-1}$.*

(i). This defines a representation of $A(\xi)$ if and only if c is a solution of the equation

$$c_0 c_1 \dots c_{r-1} = 1. \tag{13}$$

If this is the case, we denote this $A(\xi)$-module by $L(\eta, c)$. Furthermore, $L(\eta, c)$ is irreducible.

(ii). If $L(\eta, c) \simeq L(\eta', c')$ if and only if $\eta = \eta' \chi_1^m$ for some m, and $c = c'_m$.

(iii). Assume that the equation (13), a polynomial in c, has simple roots. Then all irreducibles are of this kind and $A(\xi)$ is semisimple.

PROOF. (i). We have to check the relations (6), (8) and (9). Here (6) is clear and (8) is equivalent to (13). We evaluate both sides of (9) on f_i; if $0 \leq i < r - 1$ the equality follows from the defining condition (10); otherwise it follows from (11). The irreducibility is clear.

(ii). Left to the reader.

(iii). In general, the dimension of the semisimple quotient of $A(\xi)$ corresponding to all the representations of the type $L(\eta, c)$ is $\frac{|\Gamma| \#\{\text{solutions of (13)}\} r^2}{|\Lambda| r}$, and this equals $\dim A(\xi)$ if and only if the equation (13), a polynomial in c, has simple roots. □

4. Conclusions

Let A be a lifting of a quantum plane. Then $A \simeq \prod_{\xi \in \widehat{\Lambda}} A(\xi)$; hence the irreducible representations of A are the union of the irreducible representations of $A(\xi)$, $\xi \in \widehat{\Lambda}$. We have determined the last ones in Section 3, up to a finite number of exceptions in Case (VI). As a consequence, we can also determine the coradical of the dual Hopf algebra $C = A^*$.

Acknowledgments

The first author thanks Sonia Natale for conversations on Case (III). He also thanks Vlado Dobrev, Toshko Popov and all the local organizers of the V. International Workshop LIE THEORY AND ITS APPLICATIONS IN PHYSICS for the beautiful organization.

The work of N. A. was partially supported by CONICET, Agencia Córdoba Ciencia, ANPCyT and Secyt (UNC).

The work of M. B. was partially supported by NSERC Grant 9137.

References

1. N. Andruskiewitsch and M. Beattie, "The coradical of the dual of a lifting of a quantum plane", preprint (2003), to appear in: *Hopf algebras*, Proceedings of an International Conference Held at DePaul University (M. Dekker, 2004).
2. N. Andruskiewitsch and H.-J. Schneider, "Pointed Hopf algebras", in: *Recent developments in Hopf algebra theory*, eds. S. Montgomery and H.-J. Schneider, MSRI Publications 43, (Cambridge Univ. Press, 2002) pp. 1-68.
3. N. Andruskiewitsch and H.-J. Schneider, "Lifting of quantum linear spaces and pointed Hopf algebras of order p^3", *J. Algebra*, **209**, 659 (1998).
4. N. Andruskiewitsch and H.-J. Schneider, "Finite quantum groups over abelian groups of prime exponent", *Ann. Sci. Ec. Norm. Super.* **35**, 1 (2002).
5. M. Beattie, "Duals of pointed Hopf algebras", *J. Algebra*, **262**, 54 (2003).
6. M. Beattie, S. Dăscălescu and L. Grünenfelder, "Constructing pointed Hopf algebras by Ore extensions", *J. Algebra*, **225**, 743 (2000).
7. S. Montgomery and S. Witherspoon, "Irreducible representations of crossed products", *J. Pure Appl. Algebra*, **129**, 315 (1998).
8. D.E. Radford, "On the coradical of a finite-dimensional Hopf algebra", *Proc. Amer. Math. Soc.* **53**, 8 (1975).
9. M.E. Sweedler, *Hopf algebras* (W.A. Benjamin, New York, 1969).
10. Shilin Yang, "Representations of simple pointed Hopf algebras", *J. Algebra and its Applications*, to appear.

List of Participants

1. Nicolas Andruskiewitsch (Cordoba)
2. Shigeru Aoki (Tokyo)
3. Daniel Arnaudon (Annecy)
4. Bojko Bakalov (Berkeley)
5. Nadezda Bazunova (Tallinn)
6. Alexander Belavin (Chernogolovka)
7. Cestmir Burdik (Prague)
8. Rutwig Campoamor-Stursberg (Mulhouse)
9. Stanimir Dimitrov (Sofia Univ.)
10. Vladimir Dobrev (INRNE)
11. Heinz-Dietrich Doebner (Clausthal)
12. Branko Dragovich (Belgrade)
13. Kurusch Ebrahimi-Fard (IHES)
14. Alexander Elashvili (Tbilisi)
15. Alexander Ganchev (INRNE)
16. Lachezar Georgiev (INRNE)
17. Gerald Goldin (Rutgers)
18. Cesar Gomez (Madrid)
19. Harald Grosse (Vienna)
20. Ludmil Hadjiivanov (INRNE)
21. Gerard Helminck (Twente)
22. Emil Horozov (IMI, Sofia)
23. Faheem Hussain (ICTP)
24. Hans Jakobsen (Copenhagen)
25. Victor Kac (MIT)
26. Ivan Kostov (Saclay)
27. Remi Leandre (Nancy)
28. Andrei Marshakov (Moscow)
29. Rainer Matthes (Clausthal)
30. Stefan Mihov (INRNE)
31. Shiraz Minwalla (Harvard)
32. Antonia Miteva (Sofia Univ.)
33. Dimitar Mladenov (Dubna)
34. Ivailo Mladenov (IBP, BAS)
35. Els Moens (Gent)
36. Vladimir Molotkov (INRNE)

37. Nikolay Nikolov (INRNE)
38. Masatoshi Noumi (Kobe)
39. Emil Nissimov (INRNE)
40. Svetlana Pacheva (INRNE)
41. Tchavdar Palev (INRNE)
42. Valentina Petkova (INRNE)
43. George Pogosyan (Yerevan)
44. Todor Popov (INRNE)
45. Eric Ragoucy (Annecy)
46. Emeri Sokatchev (Annecy)
47. Matthias Staudacher (Potsdam)
48. Michail Stoilov (INRNE)
49. Orlin Stoytchev (Amer. Univ. Blagoevgrad)
50. Ivan Todorov (INRNE)
51. Tuong Ton-That (Iowa)
52. Joris Van der Jeugt (Gent)
53. Vasil Velev (Sofia Univ.)
54. Stanislaw Woronowicz (Warsaw)
55. Boyko Zlatev (INRNE)

---

56. Daniela Dimitrova (INRNE)
57. Virginia Doseva (INRNE)
58. Preslav Konstantinov (INRNE)
59. Elena Puncheva (INRNE)